Proteostasis

Nidhi Puranik
Editor

Proteostasis

Investigating Molecular Dynamics in Neurodegenerative Disorders

Springer

Editor
Nidhi Puranik
Department of Life Science
Yeungnam University
Gyeongsan, Gyeongsangbuk-do
Korea (Republic of)

ISBN 978-981-96-6201-2 ISBN 978-981-96-6202-9 (eBook)
https://doi.org/10.1007/978-981-96-6202-9

This Springer imprint is published by the registered company Springer Nature Singapore Pte Ltd.
The registered company address is: 152 Beach Road, #21-01/04 Gateway East, Singapore 189721, Singapore

Dedicated to Almighty, who gave us the strength to complete this work.

Preamble

Worldwide, neurodegenerative illnesses impact millions of individuals. According to estimates, there were 55 million cases of dementia globally in 2019; by 2050, that number is expected to rise to 139 million cases due to neurodegenerative illnesses. The two most prevalent neurodegenerative illnesses are Parkinson's and Alzheimer's diseases. Neurodegenerative disorders arise from the gradual loss of function and eventual death of nerve cells inside the brain or peripheral nervous system. While there is presently no cure for neurodegenerative diseases, some of their physical or psychological symptoms may be lessened by various therapies, and the disease's progression cannot be slowed down.

The proteostasis network is particularly important in nondividing, long-lived cells, such as neurons, as its failure is implicated with the development of neurodegenerative diseases, such as Alzheimer's disease, Parkinson's disease, and amyotrophic lateral sclerosis. The etiology of several neurological disorders may be related to general processes controlling proteostasis, as many disorders have common risk factors such as oxidative stress, age, environmental stress, and protein malfunction, all of which change cellular proteostasis. Proteostasis, also known as protein homeostasis, is the vital balance of protein levels in cells, regulated by a vast network of biochemical pathways that govern various stages of protein quality control, from synthesis to destruction. Every protein in a cell undergoes constant turnover, which promotes growth, differentiation, and aging. Owing to the numerous linkages and interactions among the proteostasis pathways, several kinds of protein damage can result from stress exposure, which can change cellular homeostasis and upset the network as a whole by adding more cellular stress. Additionally, during protein synthesis, proteins can misfold or change, producing toxic or inert proteins that can overwhelm the systems for protein breakdown. The UPS and the autophagy-lysosomal system are the two main catabolic pathways that maintain a balanced proteome and stop the synthesis of defective proteins.

The book will discuss about epidemiology and management of neurodegenerative diseases in the initial chapters. In further chapters, it will discuss the role of proteostasis in neurodegenerative disorders, and the role of ubiquitin protease systems, chaperons, and co-chaperons in protein folding and misfolding. Further chapter will discuss how the impairment in functioning of these molecules leads to neuroinflammation and neurodegenerative disorders.

Acknowledgments

All contributors to the respective chapters of this book are thankfully acknowledged by the editor. Their honest efforts, hard work, and analytical approach have been appreciated and recognized. Dr. Nidhi Puranik is thankful to Yeungnam University, Gyeongsan, South Korea, for providing infrastructure facilities and all relevant assistance.

studies have detected abnormal expression of HDAC [illegible] mal activity of HDACs leads to the disruption of [illegible] ance of a variety of proteins in cells, which are [illegible] tumors (Table 18.5).

[illegible]

[illegible]

References

[illegible]

Boyers EM, Yan G, Mukherjee [illegible] Hrozek G, Man H [illegible]

Mantdoe [illegible] acetyltransferase [illegible]

Cai LY, Chen SJ, Xiao SH, Sun QJ, Ding CH, Zheng [illegible] B, Luo C, Zhang X, Xie WF (2021) Targeting p300/CBP [illegible] progression through epigenetic regulation [illegible]

Cao K, Wang G, Li W, Zhang L, Wang R, Huang Y, Du L, [illegible] Robson AB, Wang Y, Jia Y (2015) Histone deacetylase [illegible] cell death and promote autophagic immunity [illegible]

Contents

1 **Neurodegenerative Disorder: An Overview** 1
Monika Verma, Sangeeta Shukla, Sadhana Shrivastava, and Nidhi Puranik

2 **Alzheimer's: Epidemiology, Pathophysiology, Diagnosis, and Treatments** 39
Ayesha Rahman, Arpita Jaiswal, Pankaj Keshari, and Dhiraj Kumar Singh

3 **Parkinson's Disease: Epidemiology, Pathophysiology, Diagnosis, and Treatment** 73
Anushri Sharma, Agina Nair, and Dharini Dedhia

4 **A Proteasome System and Impairment in Ubiquitin-Proteasome System in Neurodegeneration and Neuroprotection** 93
Shiv Kumar Yadav, Manish Kumar Maurya, Chhavi Tyagi, Riddhi Rai, Binu Bhat, and Mukul Das

5 **Protein Misfolding and Cell Death in Neurodegenerative Diseases: Consequences and Strategies** 109
Nidhi Saxena and Sagarkumar Joshi

6 **Proteostasis Disturbances and Inflammation in Neurodegenerative Diseases** 123
Shivani Malviya, Rupali Yadav, Neha Arya, Rajesh Singh, Priyanka V. Kashyap, and Ashok Kumar

7 **Proteostasis: Investigating Molecular Dynamics in Neurodegenerative Disorders** 139
Dhiraj Kumar Singh, Ayesha Rahman, Rashmi Gupta, Amit Kumar Verma, Qulsum Akhter Mir, and Riyaz Ahmad Mir

8 **Biomarkers for Proteostasis Disruption: An Overview** 163
Deena Prakash and Divya Bisht

9 **Insights into Proteostasis System in Ageing and Neurodegenerative Disorders** 185
Yashaswee Mishra, Rahul Kumar, and Ritesh Kumar Netam

10 **The Role of Chaperones and Co-Chaperones in Synaptic Proteostasis and Neurodegenerative Disease** 231
Neha Sharma, Ragini Singh, Sharmistha Banerjee, and Shuchi Kaushik

11 **Proteasomal Degradation and Autophagy in Neurodegenerative Diseases** 243
Neha Sharma, Sharmistha Banerjee, and Shuchi Kaushik

12 **The Insight into the Role of Protein Misfolding and Tau Oligomers in Alzheimer's Disease** 253
Pooja Pandey, Chandan Kumar, and Rekha Khandia

13 **Therapeutic Strategies Targeting Amyloid-β in Alzheimer's Disease** 277
Neha Tyagi, Shilpi Pal, and Divya Bisht

14 **A Comprehensive Review of the Ubiquitin-Proteasome System in Synapse Remodeling and Neurodegenerative Diseases** 297
Meenakshi Kumari

15 **Endoplasmic Reticulum Stress and Protein Misfolding in Neurodegenerative Diseases** 329
Suman Samaddar, Moqbel Ali Moqbel Redhwan, and Mohan Muttanahally Eraiah

16 **Ameliorative Effects of Heat Shock Proteins in Therapeutic Implications of Neurodegenerative Diseases** 353
Divya Bisht and Deena Prakash

17 **Degradation of Misfolded Proteins in Neurodegenerative Diseases: Therapeutic Targets and Strategies** 371
Aradhana Rajak

18 **Future Perspectives and Challenges in the Field of Proteostasis and Neurodegenerative Disorders** 407
Suruchi Chaudhary, Vikas Shrivastava, Manish Kumar, Ekansh Agrawal, Niharika Singh, and Rajesh S. Tomar

Editor and Contributors

About the Editor

Nidhi Puranik is a Research Assistant Professor in the Department of Life Science at Yeungnam University in Gyeongsan, South Korea. She did her Ph.D. at Defence Research & Development Establishment-DRDO (Research center) and her degree was awarded from Bharathiar University, Coimbatore, Tamil Nadu, India. She has more than 10 years of research and teaching experience in the field of disease therapeutics and diagnostics. Her current research focuses on neuroscience, particularly the identification of protein targets and treatment of neurodevelopmental disorders (NDDs). She has been awarded the prestigious Graduate Aptitude Test in Engineering (GATE), Council of Scientific & Industrial Research-National Eligibility Test (CSIR-NET), Indian Council of Agricultural Research National Eligibility Test (ICAR-NET), Defence Research and Development Organisation (DRDO) Junior research fellowship, Madhya Pradesh Council of Science and Technology (MPCST) Young Scientist award, MPCST Young Scientist training fellowship (National), and Young Investigator award (International). She has worked in various prestigious institutes of India—ICMR-National Institute for Research in Environmental Health (ICMR-NIREH), Bhopal, Indian Institute of Soybean Research (ICAR-IISR), Indore, Jiwaji University, Gwalior, and Barkatullah University, Bhopal. She has published over 40 research articles with high citation scores, edited three books with IOP publishing, one book for Springer and CRC Press and has 1 Indian patent. The relevance and importance of her research work to human clinician medicine in treating and diagnosing diseases cannot be underestimated.

Contributors

Ekansh Agrawal Amity Institute of Biotechnology, Amity University, Gwalior, Madhya Pradesh, India

Neha Arya Department of Translational Medicine, All India Institute of Medical Sciences, Bhopal, India

Sharmistha Banerjee Chhattisgarh Swami Vivekanand Technical University, Bhilai, Chhattisgarh, India

Binu Bhat Shriram Institute for Industrial Research, Delhi, India

Divya Bisht Biochemistry Discipline, School of Sciences, Indira Gandhi National Open University, New Delhi, India

Suruchi Chaudhary Amity Institute of Biotechnology, Amity University, Gwalior, Madhya Pradesh, India

Mukul Das Shriram Institute for Industrial Research, Delhi, India

Dharini Dedhia Department of Molecular Medicine and Stem Cell Technologies, Jain University, Bangalore, India

Mohan Muttanahally Eraiah Department of Medicine, BGS Global Institute of Medical Sciences, Bengaluru, Karnataka, India

Rashmi Gupta Department of Biochemistry, All India Institute of Medical Sciences, New Delhi, India

Arpita Jaiswal Department of NMR, All India Institute of Medical Sciences, New Delhi, India

Sagarkumar Joshi Department of Microbiology and Biotechnology, Gujarat University, Ahmedabad, Gujarat, India
Department of Microbiology, Darshan Institute of Science, Darshan University, Rajkot, Gujarat, India

Priyanka V. Kashyap Department of Neurology, All India Institute of Medical Sciences, Bhopal, India

Shuchi Kaushik State Forensic Science Laboratory, Madhya Pradesh, India

Pankaj Keshari School of Life Sciences, Jawaharlal Nehru University, New Delhi, India

Rekha Khandia Department of Biochemistry and Genetics, Barkatullah University, Bhopal, India

Ashok Kumar Department of Biochemistry, All India Institute of Medical Sciences, Bhopal, India

Chandan Kumar Department of Biochemistry and Genetics, Barkatullah University, Bhopal, India

Manish Kumar Amity Institute of Biotechnology, Amity University, Gwalior, Madhya Pradesh, India

Rahul Kumar Department of Zoology, Sheodeni Sao College (Magadh University), Kaler, India

Meenakshi Kumari Department of Botany, Career Point University, Kota, Rajasthan, India

Shivani Malviya Department of Translational Medicine, All India Institute of Medical Sciences, Bhopal, India

Manish Kumar Maurya Shriram Institute for Industrial Research, Delhi, India

Qulsum Akhter Mir Department of Biochemistry, All India Institute of Medical Sciences, New Delhi, India

Riyaz Ahmad Mir Department of Biochemistry, All India Institute of Medical Sciences, New Delhi, India

Yashaswee Mishra Department of Physiology, All India Institute of Medical Sciences, New Delhi, India

Agina Nair Department of Biotechnology, K.J Somaiya College of Science & Commerce, University of Mumbai, Mumbai, India

Ritesh Kumar Netam Department of Physiology, All India Institute of Medical Sciences, New Delhi, India

Shilpi Pal Hi-Tech Dental College and Hospital, Utkal University, Bhubaneswar, Odisha, India

Pooja Pandey Department of Biochemistry and Genetics, Barkatullah University, Bhopal, India

Deena Prakash Biochemistry Discipline, School of Sciences, Indira Gandhi National Open University, New Delhi, India

Nidhi Puranik Department of Life Science, Yeungnam University, Gyeongsan, Gyeongsangbuk-do, Korea (Republic of)

Ayesha Rahman Department of Biochemistry, All India Institute of Medical Sciences, New Delhi, India

Riddhi Rai Shriram Institute for Industrial Research, Delhi, India

Aradhana Rajak Bharathiar University, Coimbatore, Tamil Nadu, India

Moqbel Ali Moqbel Redhwan Department of Pharmacology, KLE College of Pharmacy, Bengaluru, Karnataka, India
KLE Academy of Higher Education and Research, Belagavi, Karnataka, India

Suman Samaddar Research Institute, BGS Global Institute of Medical Sciences, BGS Health & Education City, Bengaluru, Karnataka, India

Nidhi Saxena Department of Microbiology, Atmiya University, Rajkot, Gujarat, India

Anushri Sharma Department of Biochemistry and Genetics, Barkatullah University, Bhopal, India

Neha Sharma Amity Institute of Biotechnology, Amity University, Gwalior, Madhya Pradesh, India

Sadhana Shrivastava USIC, Jiwaji University, Gwalior, Madhya Pradesh, India

Vikas Shrivastava Amity Institute of Biotechnology, Amity University, Gwalior, Madhya Pradesh, India

Sangeeta Shukla Chaudhary Charan Singh University, Meerut, Uttar Pradesh, India
Trace-Element-Institute for UNESCO, Lyon, France

Dhiraj Kumar Singh Department of Biochemistry, All India Institute of Medical Sciences, New Delhi, India

Niharika Singh Amity Institute of Biotechnology, Amity University, Gwalior, Madhya Pradesh, India

Ragini Singh Amity Institute of Biotechnology, Amity University, Gwalior, Madhya Pradesh, India

Rajesh Singh Department of Human Genetics, Banaras Hindu University, Varanasi, India

Rajesh S. Tomar Amity Institute of Biotechnology, Amity University, Gwalior, Madhya Pradesh, India

Chhavi Tyagi Shriram Institute for Industrial Research, Delhi, India

Neha Tyagi Department of Pharmaceutics, School of Pharmaceutical Education & Research, Jamia Hamdard, New Delhi, India

Amit Kumar Verma Department of Biosciences, Jamia Millia Islamia, New Delhi, India

Monika Verma Department of Biotechnology, GDRCST, Rungta Institute (R1), Bhillai, Chhattisgarh, India

Rupali Yadav Department of Biochemistry, All India Institute of Medical Sciences, Bhopal, India

Shiv Kumar Yadav Shriram Institute for Industrial Research, Delhi, India

Neurodegenerative Disorder: An Overview

1

Monika Verma, Sangeeta Shukla, Sadhana Shrivastava, and Nidhi Puranik

1.1 Introduction

Neurodegenerative disorders (NDDs) are a broad range of diseases characterized by progressive degeneration and loss of neurons in the central nervous system (CNS) [1]. These disorders tend to affect specific brain regions, resulting in distinct clinical presentations. Despite their differences, most NDDs share common pathophysiology, such as protein misfolding [2], mitochondrial dysfunction [3], and inflammation [4, 5], which ultimately results in neuronal death [6]. This chapter provides an in-depth exploration of key NDDs, including Parkinson's disease (PD), Alzheimer's disease (AD), multiple sclerosis (MS), Huntington's disease (HD), amyotrophic lateral sclerosis (ALS), prion, and others. Further, this chapter will discuss the current therapeutic strategies and the future of NDD research.

Based on many classifications, NDDs can be widely comprehended. These may include dementia, PD, AD, and others based on the main clinical characteristics. These are classified as spinocerebellar degenerations, extrapyramidal diseases, or frontotemporal degenerations based on the anatomic distribution of neurodegeneration. NDDs can also be classified as tauopathies, α-synucleinopathies, amyloidoses, and TDP-43 proteinopathies based on the protein abnormalities seen in these conditions [7].

M. Verma (✉)
Department of Biotechnology, GDRCST, Rungta Institute (R1), Bhillai, Chhattisgarh, India

S. Shukla
Chaudhary Charan Singh University, Meerut, Uttar Pradesh, India

Trace-element-Institute for UNESCO, Lyon, France

S. Shrivastava
USIC, Jiwaji University, Gwalior, Madhya Pradesh, India

N. Puranik
Department of Life Sciences, Yeungnam University, Gyeongsan, Gyeongsangbuk-do, Korea (Republic of)

N. Puranik (ed.), *Proteostasis*, https://doi.org/10.1007/978-981-96-6202-9_1

Over the last 10 years, the prevalence of NDDs has significantly increased. Among these is path etiology, which is produced in reaction to various CNS injuries to restore homeostasis and protect and restore the damaged tissues. A key mechanism in the development of several NDDs is a protracted inflammatory response or the transition from acute to chronic neuroinflammation. Primary immune cells, the blood-brain barrier (BBB), and signaling molecules interact to orchestrate the CNS inflammatory response, resulting in a central and peripheral reaction [4].

In some cases, it was observed that an individual has multiple neurodegenerations [8]. Neurodegenerative protein anomalies may also exist prior to a disease's clinical presentation or manifestation. The most reliable method for diagnosing NDDs is an autopsy. The creation and validation of biomarkers for NDDs have received a lot of attention in recent decades [1]. The variety and breadth of biomarkers for disorders of the CNS have kept growing, including those derived from biofluids like blood or cerebrospinal fluid (CSF), nucleic acids, tissues, and imaging [9]. Still, there is a need for further evolution and further discovery of biomarkers and suitable techniques for early diagnosis. Figure 1.1 summarizes the major causes and management of NDDs.

Key Features of Neurodegenerative Disorders

- *Memory loss*: The inability to recall recently taught material is one of the first and most noticeable signs.
- *Cognitive decline*: As the illness worsens, people's ability to reason, think, and make decisions is compromised.
- *Behavioral and personality changes*: Mood swings, depression, apathy, and social withdrawal are common.

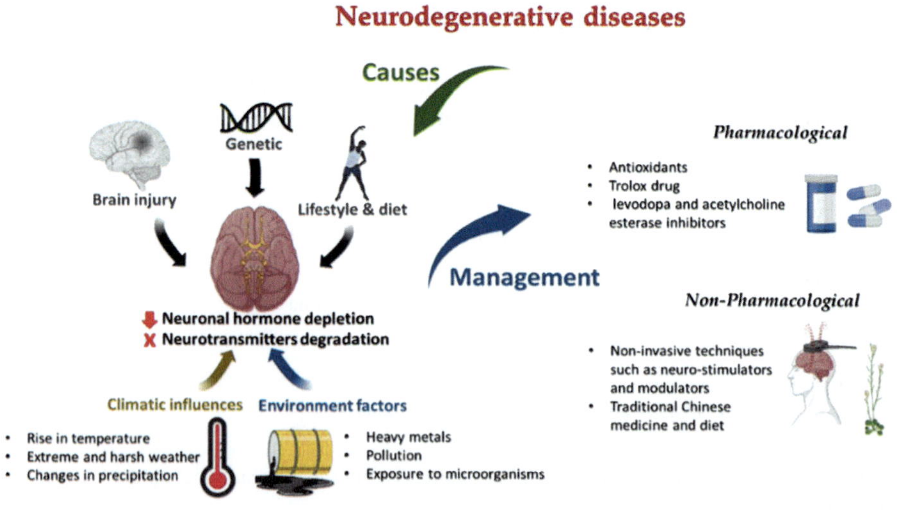

Fig. 1.1 Schematic representation of causes and management of neurodegenerative diseases [10]

- *Disorientation*: Patients often become confused about time and place.
- *Difficulty with Tasks*: Simple tasks like managing finances or preparing meals can become challenging.

Further, we have discussed each NDD individually, highlighting their key characteristics.

1.2 Alzheimer's Disease

AD is the most prevalent NDD and the primary cause of dementia worldwide [11]. It accounts for 60–80% of dementia cases globally, especially those over the age of 65 [12]. AD is a progressive and irreversible neuro condition that impairs the functions of memory, cognition, behavior, and ultimately, the ability to perform basic functions [12]. AD was first identified by Dr. Alios Alzheimer in the year 1906, this disease has since been a major focus of neuroscience research, given its widespread prevalence and profound impact on individuals, families, and healthcare systems [13].

1.2.1 Stages of Alzheimer's [14, 15]

Mild (Early Stage): Memory loss, trouble with simple planning and even organizing.
Moderate (Middle Stage): Increased loss of memory, disorientation, trouble identifying family members, significant sleep disturbances, and roaming.
Severe (Late Stage): Loss of ability to respond to the environment, communicate, and control movement.

1.2.2 Risk Factors

Several factors contribute to the development of AD, with both genetic and environmental components playing important roles:

Age: age considerably raises the risk, especially after age 65.
Genetics: genetic factors, apolipoprotein E ε4 allele, angiotensin I converting enzymes, α2-macroglobulin, or apolipoprotein E receptors [16] and family history.
Lifestyle: Cardiovascular factors, like high blood pressure and diabetes, can amplify the risk [17].
Head Trauma: A history of severe head injury is a greater risk [18].

1.2.3 Pathophysiology

AD is characterized by two major pathological hallmarks: the accumulation of beta-amyloid (Aβ) plaques and neurofibrillary tangles (NFTs) composed of hyperphosphorylated tau proteins. These protein aggregates disrupt normal cellular function, leading to synaptic loss, neuroinflammation, and ultimately neuronal death [19, 20]. In addition to these protein-related neuropathies, other factors such as neuroinflammation, oxidative stress, and vascular damage play roles in the pathogenesis of AD (Fig. 1.2). These processes lead to extensive atrophy in the regions of the brain, particularly the hippocampus and cerebral cortex, which are critical for memory and higher cognitive functions [21, 22].

- *Amyloid Plaques*

Amyloid-beta Plaques:
Aβ peptides are derived from the cleavage of amyloid precursor protein (APP) by beta-secretase and gamma-secretase. Normally, these peptides are cleared from the brain. However, in the case of AD, the clearance mechanism fails, which leads to the accumulation of Aβ and the formation of extracellular plaques. These plaques interfere with the communication system between neurons and thereby triggering further inflammatory responses hence damaging the brain tissues [24, 25].

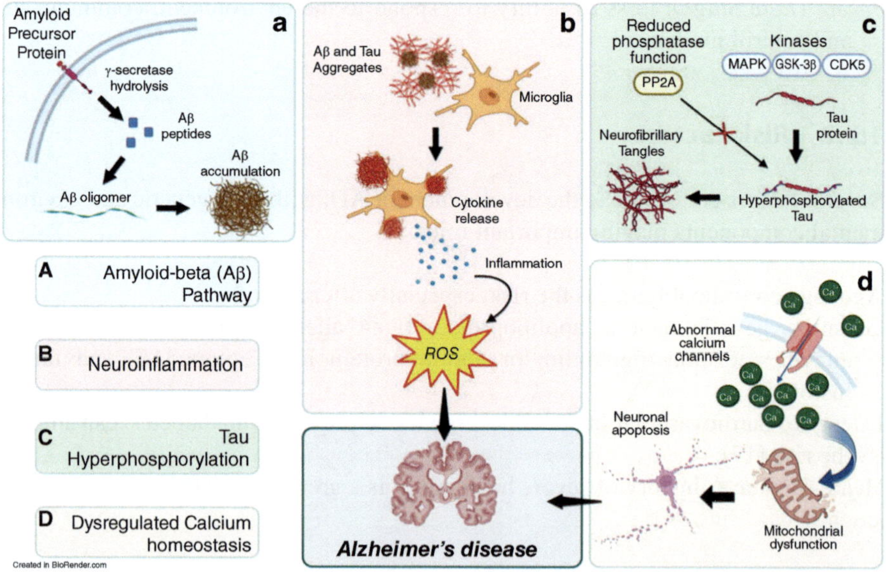

Fig. 1.2 Pathogenesis of AD [23]

Plaque formation:
These amyloid-beta peptides aggregate to form extracellular plaques, which are toxic to neurons. The plaques interfere with cell-to-cell communication and trigger immune responses that lead to inflammation and further damage [24, 26].

- *Neurofibrillary Tangles*

Tau protein hyperphosphorylation:
Neuronal microtubules are stabilized by the microtubule-associated protein tau. When tau is hyperphosphorylated abnormally in AD, it separates from microtubules and forms insoluble tangles inside the structure of neurons. The accumulation of these tangles disrupts the transport of nutrients and signals within neurons, contributing to cell death [27, 28].

Tangle Formation:
The abnormal tau proteins aggregate into insoluble twisted fibers known as NFTs inside neurons. These tangles disrupt the microtubule network, impairing the transport of essential nutrients and molecules within the cell, ultimately leading to cell death [29, 30].

- *Neuroinflammation*

The buildups of the amyloid plaques and tau tangles lead to the activation of microglia, the brain's resident immune cells. While microglia initially attempt to clear the abnormal proteins, chronic activation results in the release of pro-inflammatory cytokines and chemokines, which further contribute to neuronal damage. Astrocytes, another type of glial cell, become activated in response to damage and contribute to neuroinflammation. However, their prolonged activation can exacerbate neurodegeneration. AD-related astrocytes may exhibit a range of degenerative symptoms throughout AD due to the functional diversity, morphological heterogeneity, and physiological variability of astrocytes in the CNS. The development of abnormal phenotypes by astrocytes may accelerate the course of AD [31]. In vitro and in vivo animal models of AD, as well as in the brains of AD patients, alterations in astrocyte function have been noted [32].

- *Synaptic Dysfunction*

The toxic effects of amyloid plaques and tau tangles lead to a progressive loss of synapses, which are the connections between neurons. This synaptic loss is one of the earliest events in Alzheimer's and correlates strongly with the severity of the cognitive decline [33].

- *Oxidative Stress*

AD is significantly influenced by oxidative stress. Reactive oxygen species (ROS) overproduction damages neurons' lipids, proteins, and DNA. This exacerbates the effects of tau tangles and amyloid plaques by causing brain cell degeneration [34].

- *Neurotransmitters*

One of the earliest neurotransmitter systems affected in Alzheimer's is the cholinergic system. Acetylcholine, a neurotransmitter involved in learning and memory, is reduced due to the degeneration of cholinergic neurons in the basal forebrain. This contributes to cognitive deficits, and many treatments aim to increase acetylcholine levels [35]. Glutamate is a crucial excitatory neurotransmitter that is involved in synaptic transmission, brain activity, and plasticity. However, it is now recognized that changes in the glutamatergic signaling pathway play a key role in the pathophysiology of Alzheimer's disease [36].

- *Blood-Brain Barrier Dysfunction*

The brain is shielded from dangerous compounds by the BBB, a selective barrier. In AD, the integrity of the BBB is compromised, allowing the entry of toxic substances, inflammatory cells, and proteins that contribute to neurodegeneration [37].

1.3 Parkinson's Disease (PD)

PD is a degenerative, long-term neurological condition that mostly affects movement. It happens when dopamine-producing neurons in the substantia nigra, a part of the brain that is essential for controlling movement, degenerate [38]. PD is progressive and gradually worsens over time. The speed of progression varies widely between individuals. With treatment, many people can maintain a good quality of life for many years [39]. It is a complex condition with both motor and non-motor symptoms, affecting movement and cognitive functions.

1.3.1 Key Symptoms

1. *Motor Symptoms*:

   Tremor: Usually starting in a hand or fingers, often called "resting tremor" [40].

   Bradykinesia: Slowness of motion, which makes regular chores challenging and time-consuming [41].

   Muscle Rigidity: Muscle stiffness resulting in a decreased range of motion [42].

   Postural instability: Increases the risk of falls by impairing balance and coordination [43].

2. *Non-Motor Symptoms*:
   Cognitive changes: Include trouble focusing, memory issues, and perhaps dementia in later stages [44].
   Mood Disorders: Anxiety, depression, and apathy are common [45].
   Sleep Disorders: Insomnia, REM sleep behavior disorder, and excessive daytime sleepiness [46].
   Autonomic Dysfunction: Problems with digestion, bladder control, and blood pressure regulation [47].

1.3.2 Risk Factors

Although the precise etiology of Parkinson's disease is still unknown, a number of established factors may have a role in its onset:

Genetics: Certain genetic mutations (α-synuclein, LRRK-2, and GBA) have been linked to the disease, although they are more common in people with a family history of Parkinson's [48].
Environmental Factors: Exposure to pesticides, heavy metals, or industrial chemicals may increase risk [49].
Aging: Risk increases with age, particularly after the age of 60 [50].

1.3.3 Pathophysiology

The pathophysiology of PD primarily involves the degeneration of dopamine-producing neurons in the substantia nigra pars compacta, a region of the brain that is part of the basal ganglia. The basal ganglia are crucial for regulating movement, and the loss of dopamine in this area leads to the characteristic motor symptoms of Parkinson's [43]. Figure 1.3 provides the major factors that lead to the loss of neuron and degeneration in PD.

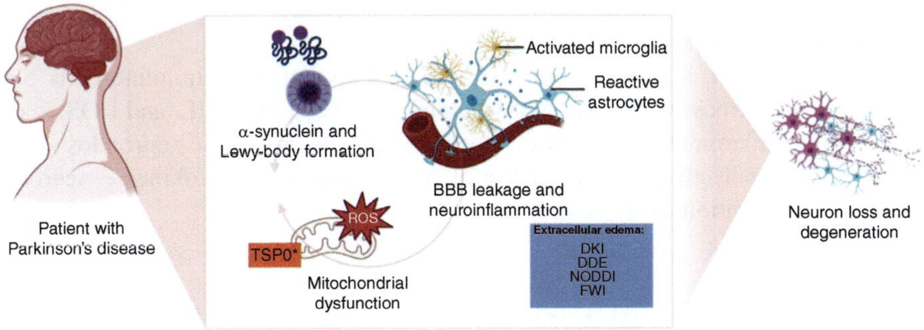

Fig. 1.3 Schematic representation of major pathophysiology of PD [51]

Degeneration of Dopaminergic Neurons:
Dopamine, a neurotransmitter essential for sending signals that coordinate voluntary, smooth movement, is produced by the substantia nigra, a structure in the midbrain [38]. Neurons in the substantia nigra begin to degenerate and die, leading to a reduction in dopamine levels [42]. Typically, an 80% loss of dopaminergic neurons occurs before the motor symptoms become noticeable [50].

Impact on Basal Ganglia Circuitry:
A group of subcortical nuclei called the basal ganglia are involved in a number of processes, such as emotions, executive behaviors and functions, motor learning, and motor control. They regulate movement through connections with the cortex and thalamus. Dopamine must be produced at the input nuclei for the basal ganglia, which regulate movement, to function properly. Numerous movement problems are linked to dopamine deficiency [52]. Dopamine depletion affects two main pathways within the basal ganglia: the direct and indirect pathways. The disruption of these pathways leads to bradykinesia, rigidity, and tremor [53].

Formation of Lewy Bodies:
Lewy bodies, which are aberrant clusters of the protein alpha-synuclein inside neurons, are a hallmark pathogenic characteristic of PD. These aggregates disrupt cellular functions and contribute to neuron death [54].

Non-Dopaminergic Involvement:
Loss of serotonergic, cholinergic, and noradrenergic neurons contributes to non-motor symptoms like mood disturbances, cognitive decline, and autonomic dysfunction [55].

Mitochondrial Dysfunction and Oxidative Stress:
Mitochondrial dysfunction is a key factor in the pathogenesis of PD, causing energy deficits and increased susceptibility to oxidative stress. According to research on PD in people and several animal models, mitochondrial dysfunction may be a flaw that arises early in the disease's etiology and seems to be present in both sporadic and monogenic forms of the disease [56].

Neuroinflammation:
Neuroinflammation, a key component of PD, is regulated by immune cells like microglia and astrocytes, releasing pro-inflammatory cytokines, IL, and ROS, leading to a toxic inflammatory response, apoptosis, and dopaminergic neuron loss [57]. Activation of microglia and the release of pro-inflammatory cytokines exacerbate neuronal degeneration [43].

Genetic Factors:
Mutations in genes like *SNCA, LRRK2, PINK1, PARK7*, and *Parkin* are linked to familial and sporadic PD cases [58].

1.4 Huntington's Disease

A hereditary mutation in the HTT gene, which codes for the huntingtin protein, causes HD, a progressive NDD. Because it is an autosomal dominant disorder, the illness can be brought on by a single copy of the defective gene that is inherited from either parent [59, 60]. The condition typically manifests in mid-adulthood (30–50 years), although symptoms can appear earlier or later, depending on the number of CAG repeats in the mutated gene. The greater the number of repeats, the earlier the onset tends to be [61]. HD is a complex disorder rooted in a genetic mutation that leads to widespread neurodegeneration, particularly in motor and cognitive regions of the brain. The hallmark features of HD result from the interplay between toxic protein accumulation, disrupted cellular mechanisms, and the eventual death of neurons critical for motor and cognitive function. Despite extensive research, there is currently no cure, but ongoing studies aim to target these pathogenic mechanisms to slow or halt the progression of the disease [62]. Figure 1.4 gives an overview of major causes of pathogenesis of HD.

1.4.1 Key Symptoms

Motor symptoms: Uncontrolled movements, such as chorea (involuntary, jerky movements), rigidity, and dystonia [63, 64].

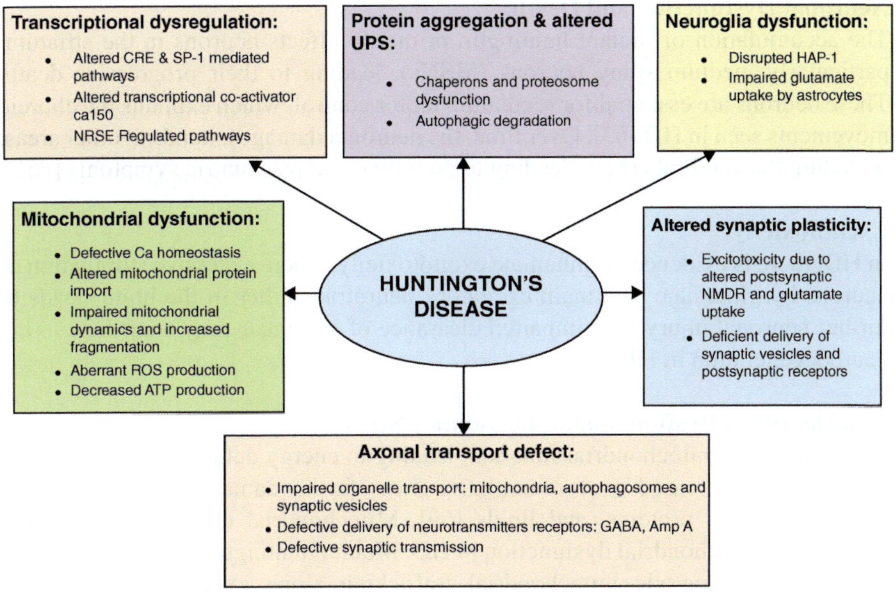

Fig. 1.4 Schematic of various pathways leading to the pathogenesis of HD [60]

Cognitive decline: Memory loss, impaired judgment, and difficulty with concentration and decision-making [65].

Psychiatric disturbances: Depression, anxiety, irritability, and sometimes psychosis [66].

The disease is progressive and eventually leads to severe disability and death, often 15–20 years after symptom onset, primarily due to complications like pneumonia, falls, or malnutrition [67].

1.4.2 Pathophysiology

Genetic Mutation:
The CAG trinucleotide repeats in the HTT gene on chromosome 4, resulting in HD. In healthy individuals, the CAG sequence is repeated fewer than 35 times. In people with HD, the repeats exceed 40, leading to the production of an abnormally long huntingtin protein [64]. This mutant huntingtin protein (mHTT) becomes toxic to neurons, particularly in the basal ganglia and the cortex (involved in cognition) [67].

Protein Aggregation:
The elongated mHTT protein undergoes misfolding and forms aggregates in neurons. These aggregates impair cellular functions, including protein degradation pathways, mitochondrial function, and neurotransmission [63].

Neuronal Dysfunction and Death:
The accumulation of mutant huntingtin primarily affects neurons in the striatum, particularly medium spiny neurons (MSNs), leading to their progressive death. These neurons are essential for regulating motor control, which explains the choreic movements seen in HD [63]. Over time, the neuronal damage extends to other areas, including the cerebral cortex, leading to cognitive and psychiatric symptoms [65].

Excitotoxicity:
In HD, there is evidence of glutamate excitotoxicity, where excessive stimulation of neurons by glutamate (the main excitatory neurotransmitter in the brain) leads to further neuronal injury. The impaired clearance of glutamate may contribute to the neurodegeneration in HD [68].

Mitochondrial Dysfunction and Oxidative Stress:
mHTT disrupts mitochondrial function, leading to energy deficits in neurons. This contributes to increased levels of oxidative stress, further damaging cellular components like DNA, proteins, and lipids [69]. Mitochondrial trafficking is another instance of mitochondrial dysfunction in HD. Mutant huntingtin inhibits both retrograde and anterograde mitochondrial trafficking along axons, which disrupts

mitochondrial maintenance and decreases mitochondrial deposition at locations with high energy demand, like synapses [62].

Impaired Cellular Mechanisms:
Age-dependent neurological symptoms and the buildup of misfolded proteins, which are frequently detected in the nucleus, are hallmarks of HD. The increased accumulation of the disease protein huntingtin in neuronal cells is also linked to the selective neuropathology in HD. While it is evident that the ubiquitin-proteasome system (UPS) plays a crucial role in eliminating mutant huntingtin, it is yet unknown if aging or HD has an impact on the neuronal UPS's ability to eliminate harmful and misfolded proteins. In this review, we concentrate on the connection between HD and aging and the UPS function [70]. Autophagy and proteasomal degradation are key cellular mechanisms for clearing misfolded proteins. In HD, these pathways are impaired, allowing the buildup of toxic protein aggregates [64, 71].

Inflammation:
There is growing evidence that HD involves a chronic, low-level inflammatory response in the brain, with activation of microglia, further contributing to neuronal damage [72]. These glial cells' reactive shape is a hallmark of neuroinflammation in the HD brain. To control neuronal activity and preserve the ideal environment for neuronal function, microglia and astrocytes are essential. According to earlier research, reactive astrocytes and activated microglia prolong a chronic inflammatory state by transcriptionally activating pro-inflammatory genes, which contributes to HD pathogenesis. Additionally, reactive astrocytes exhibit functional alterations in energy metabolism and glutamate and ion homeostasis. The neuronal loss seen as HD progresses may also be attributed to astrocyte and microglial alterations. The extent to which these neuroinflammatory alterations harm neurons and advance HD pathogenesis is a crucial question that remains unclear [73].

1.5 Motor Neuron Disease (MND)

MND refers to a group of progressive NDDs that specifically affect motor neurons, the nerve cells responsible for controlling voluntary muscles [74]. Amyotrophic lateral sclerosis (ALS) is the most well-known and prevalent type of MND, although there are other subtypes as well, including progressive bulbar palsy (PBP), progressive muscular atrophy (PMA), and primary lateral sclerosis (PLS) [75].

MND leads to the degeneration of both upper motor neurons (UMNs) [76] and lower motor neurons (LMNs) [77], causing progressive weakness, muscle wasting, and eventually, paralysis [78].

1.6 Amyotrophic Lateral Sclerosis (ALS)

ALS is a progressive NDD that primarily affects motor neurons, the nerve cells responsible for controlling voluntary muscles. It is characterized by the progressive degeneration of both UMNs in the brain and LMNs in the spinal cord and brainstem.

ALS generally manifests in adulthood, typically between the ages of 40 and 70. It is sporadic in about 90–95% of cases, but 5–10% of cases are familial ALS (fALS), linked to inherited genetic mutations. The disease progresses over time, leading to muscle weakness, paralysis, and ultimately, respiratory failure, which is the usual cause of death in patients within 2–5 years of diagnosis [79].

1.6.1 Key Symptoms

- Muscle weakness and atrophy in the limbs
- Fasciculations (muscle twitches)
- Difficulty with speech, swallowing, and breathing
- Loss of reflexes and spasticity (stiffness)
- Emotional changes, including pseudobulbar affect (uncontrollable laughing or crying)
- Some cognitive changes, particularly frontotemporal dementia in some cases

1.6.2 Pathophysiology (Fig. 1.5)

Motor Neuron Degeneration:

- ALS is defined by the progressive loss of both UMNs and LMNs.
- Upper motor neuron loss leads to spasticity, hyperreflexia, and weakness.
- Lower motor neuron degeneration results in muscle atrophy, fasciculations, and weakness [81].

Genetic Mutations:
In familial ALS, specific genetic mutations have been identified. Some of the most well-known genes involved in ALS include the following:

- *Superoxide dismutase 1 (SOD1)*: Mutations in SOD1 gene lead to oxidative stress and toxic gain-of-function mechanisms, causing neuronal damage [82].
- *C9ORF72*: Mutation in C9ORF72 is common in ALS and frontotemporal dementia, involving an abnormal expansion of a hexanucleotide repeat [83].
- *TDP-43 and FUS*: These mutations affect proteins involved in RNA processing, leading to abnormal aggregation in neurons [84].

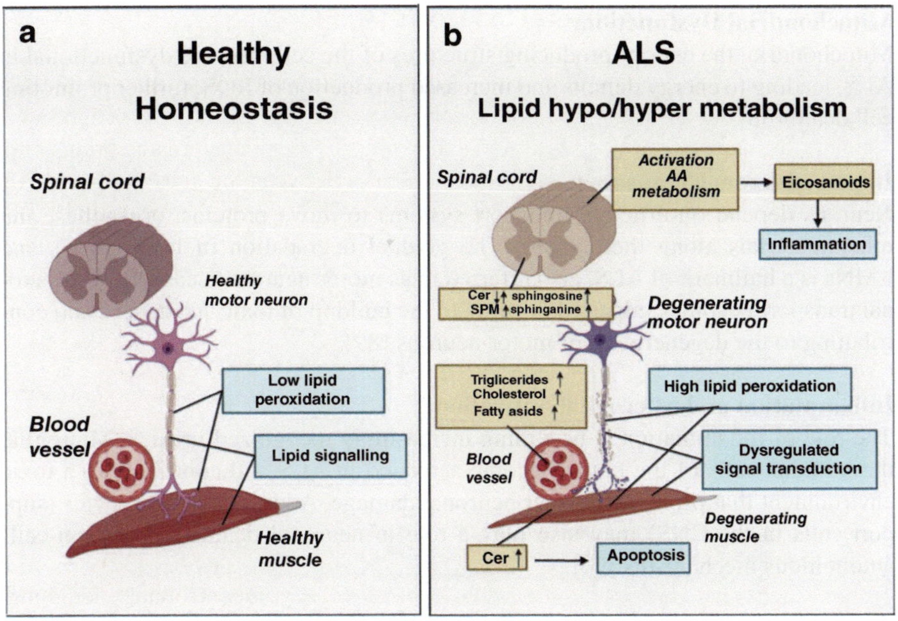

Fig. 1.5 Lipid homeostasis in healthy and ALS individual [80]

Protein Aggregation:

Aggregation of abnormal protein in motor neurons is also observed in ALS patients. Proteins like TDP-43, SOD1, and FUS can misfold and accumulate, disrupting cellular function. These aggregates are toxic to cells and interfere with cellular processes such as protein degradation and RNA processing [85].

Glutamate Excitotoxicity:

In ALS, there is impaired regulation of glutamate, the main excitatory neurotransmitter. Excessive glutamate leads to overstimulation of neurons (excitotoxicity), causing calcium overload, oxidative stress, and eventually, cell death. This is one of the mechanisms targeted by the drug Riluzole, which reduces glutamate activity and modestly slows disease progression [86].

Oxidative Stress:

Oxidative damage is thought to play a key role in ALS. Neurons are particularly vulnerable to damage from ROS. Mutations in SOD1, an antioxidant enzyme, lead to reduced ability to detoxify ROS, contributing to oxidative stress and neuronal death [87].

Mitochondrial Dysfunction:
Mitochondria, the energy-producing structures of the cell, become dysfunctional in ALS, leading to energy deficits and increased production of ROS, further promoting cell death [88].

Impaired Axonal Transport:
Neurons depend on efficient transport systems to move proteins, organelles, and other materials along their axons. The gradual degradation of both UMNs and LMNs is a hallmark of ALS, also referred to as motor neuron disease. In ALS, axonal transport becomes impaired, leading to the buildup of toxic substances and contributing to the degeneration of motor neurons [89].

Inflammation and Microglial Activation:
The role of inflammation is becoming increasingly recognized in ALS. Microglia, the immune cells of the brain, become activated in ALS and contribute to a toxic environment that promotes further neuronal damage. Additionally, astrocytes (support cells in the CNS) may also play a role in neuronal death through non-cell-autonomous mechanisms [90].

Role of RNA Dysregulation:
Proteins such as TDP-43 and FUS are involved in RNA metabolism, including transcription, splicing, and transport. In ALS, these proteins become mislocalized and form aggregates, disrupting normal RNA processing. This leads to impaired cellular function and motor neuron death [91].

ALS is a complex, multifactorial disease that involves motor neuron degeneration due to various mechanisms, including genetic mutations, protein aggregation, excitotoxicity, oxidative stress, and impaired cellular processes. The gradual loss of motor neurons leads to progressive muscle weakness, paralysis, and eventual death, typically due to respiratory failure. Although there is no cure, treatments aim to slow the disease progression and manage symptoms.

1.7 Prion Disease

Prion diseases are a unique class of progressive NDDs caused by misfolded prion proteins. Unlike other NDDs such as AD or PD, prion diseases are transmissible and can be inherited, sporadic, or acquired. The hallmark of prion diseases is the accumulation of an abnormal form of a normally benign protein, known as the prion protein (PrP), which leads to brain damage [92].

Creutzfeldt-Jakob disease (CJD) in humans, bovine spongiform encephalopathy (BSE), sometimes known as “mad cow disease” in cattle, and scrapie in sheep are examples of prion diseases that can affect both people and animals [93].

1.7.1 Types of Prion Diseases in Humans

Creutzfeldt-Jakob Disease (CJD):
The most common human prion disease. It occurs in several forms [94]:

Sporadic CJD (sCJD): occurring spontaneously without a known cause (85–90% of cases).
Familial CJD (fCJD): caused by inherited mutations in the PRNP gene.
Variant CJD (vCJD): acquired from consuming contaminated meat products from cattle infected with BSE (mad cow disease).
Iatrogenic CJD (iCJD): transmitted via medical procedures, such as contaminated surgical instruments or tissue grafts.
Gerstmann-Sträussler-Scheinker Syndrome (GSS): causes progressive ataxia, dementia, and motor deficits.
Fatal Familial Insomnia (FFI): a hereditary prion disease that predominantly affects the thalamus, leading to progressive insomnia, autonomic dysfunction, and ultimately death.

Kuru:
A prion disease historically found in the Fore people of Papua New Guinea spread through ritualistic cannibalism.

1.7.2 Pathophysiology

Prion is caused by the misfolding of a normal cellular prion protein (PrP^{C}) into a misfolded, disease-causing form (PrP^{Sc}), where "Sc" stands for "scrapie," one of the prion diseases in animals. The PrP^{Sc} form is resistant to normal cellular degradation processes and can induce other PrP^{C} proteins to adopt the same misfolded structure. This leads to a self-propagating cascade of protein misfolding [95] (Fig. 1.6).

Prion Protein Misfolding:
The prion protein is a normal protein found in many tissues but is most abundant in the brain. Under normal conditions, PrP^{C} has a specific three-dimensional (3D) structure dominated by alpha-helices. In prion diseases, PrP^{C} converts into an abnormally folded form (PrP^{Sc}) dominated by beta sheets. This misfolded protein has a propensity to aggregate, forming amyloid fibrils and plaques in the brain [97].

Propagation of Misfolded Prions:
The hallmark of prion diseases is the ability of the misfolded PrP^{Sc} protein to act as a template, causing the normally folded PrP^{C} to misfold and adopt the PrP^{Sc} form. This prion replication process leads to the accumulation of PrP^{Sc} in the brain, which is neurotoxic and leads to neuronal dysfunction and death [98].

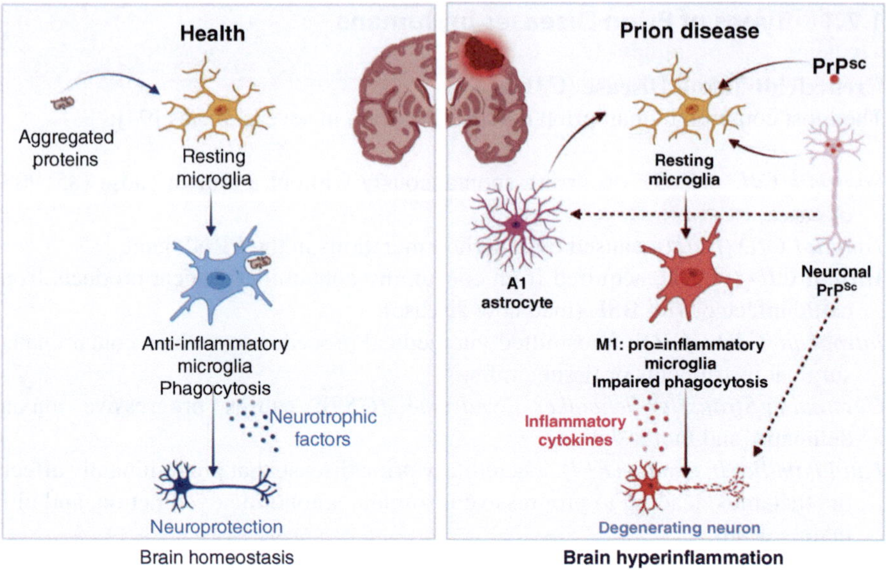

Fig. 1.6 Schematic representation of inflammatory response in a healthy brain and a prion disease patient's brain [96]

Neurodegeneration:
PrP^{Sc} aggregates disrupt normal brain function by forming amyloid plaques and spongiform changes, which give the brain a sponge-like appearance. The brain undergoes extensive neurodegeneration, particularly in regions involved in cognition, movement, and memory, leading to rapid neurological decline [99].

Neuronal loss, synaptic dysfunction, and gliosis (proliferation of glial cells) contribute to the overall brain damage.

Transmission:
Prion diseases are transmissible and can be spread through contaminated tissue (e.g., from eating infected meat, medical procedures, or genetic inheritance) [100]. Prions are highly resistant to common sterilization techniques and can survive in extreme conditions, making them difficult to eliminate in infected environments.

1.7.3 Key Symptoms

Prion diseases present with a range of neurological and psychiatric symptoms, which worsen over time. The clinical course is often rapid, leading to death within months to a few years. Key symptoms include the following:

Cognitive Decline:

Dementia is a prominent feature, particularly in CJD, with rapid progression over weeks to months. Memory loss, confusion, disorientation, and behavioral changes are common [100].

Motor Dysfunction:

Ataxia (loss of coordination), tremors, and muscle stiffness develop as the disease progresses.

Myoclonus (sudden, involuntary muscle jerks) is a hallmark feature of many prion diseases.

Psychiatric Symptoms:

Depression, anxiety, mood swings, and personality changes are often seen, particularly in vCJD.

Insomnia and Autonomic Dysfunction:

In Fatal Familial Insomnia (FFI), patients experience progressive, untreatable insomnia, autonomic dysfunction (abnormal heart rate, blood pressure), and eventually, coma.

Seizures and Visual Disturbances:

Seizures may occur in advanced stages, along with blurred or double vision in some cases.

At autopsy, the brain of a person with prion disease shows the following characteristics:

Spongiform changes: Vacuolization or formation of tiny holes in the brain tissue, leading to the characteristic sponge-like appearance.

Neuronal loss: Significant neuronal degeneration.

Astrogliosis: Proliferation of astrocytes as part of the brain's response to injury.

Amyloid plaques: In some forms of prion disease, such as vCJD, amyloid plaques containing PrP^{Sc} are present.

1.7.4 Diagnosis

Diagnosing prion diseases is challenging because symptoms often resemble other neurodegenerative conditions. The diagnosis is typically based on a combination of clinical signs, imaging studies, and laboratory tests. Definitive diagnosis usually requires postmortem examination of brain tissue, but several tools can aid in diagnosis during life [101, 102]:

1. *Magnetic Resonance Imaging (MRI)*:
   - MRI may show characteristic changes in the brain, such as hyperintensities in the basal ganglia or cerebral cortex in CJD.
2. *Cerebrospinal Fluid (CSF) Analysis*:
   - CSF tests, such as detecting 14-3-3 protein, tau protein, and real-time quaking-induced conversion (RT-QuIC), can help support a prion disease diagnosis.
3. *Electroencephalography (EEG)*:
   - In sCJD, an EEG may show characteristic periodic sharp wave complexes (PSWC).
4. *Genetic Testing*:
   - Inherited forms of prion disease (e.g., fCJD, GSS, FFI) can be diagnosed through genetic testing for mutations in the PRNP gene.

1.7.5 Treatment and Management of Prion Diseases

Prion illnesses cannot be cured or their progression slowed by any effective therapy. Currently, management focuses on symptomatic treatment and palliative care to maintain comfort and quality of life.

1. *Supportive Care*:
   - Symptomatic treatment includes medications to manage myoclonus, seizures, and psychiatric symptoms such as depression or anxiety.
   - Nutritional support and physical therapy may help manage some of the disease's physical complications.
2. *Experimental Therapies*:
   - Research is ongoing to find therapeutic agents that could halt prion replication or protect neurons from prion-related toxicity. Strategies include the following [103]:
     - *Anti-prion drugs*: Agents that stabilize the normal prion protein or prevent PrP^{C} from converting into PrP^{Sc}.
     - *Immunotherapy*: Attempts to stimulate the immune system to target and neutralize PrP^{Sc}.
     - *Gene silencing*: Techniques such as antisense oligonucleotides (ASOs) or RNA interference (RNAi) to reduce PrP production.

1.8 Spinal Muscular Atrophy (SMA)

The gradual degeneration of motor neurons in the brainstem and spinal cord is a hallmark of SMA, a hereditary NDD. This leads to muscle weakness and atrophy, affecting the muscles used for movement, breathing, and swallowing. SMA primarily impacts infants and children, but it can also occur in adults, with varying levels of severity depending on the type of SMA. The disease is caused by mutations in the survival motor neuron 1gene (SMN1), which is responsible for producing SMN protein, essential for the survival and proper functioning of motor neurons [104].

1.8.1 Types of Spinal Muscular Atrophy

SMA is classified into several types based on the age of onset and severity of symptoms. The most common classification includes five types [105]:

1. *SMA Type 0 (Prenatal SMA)*:
   - The rarest and most severe form.
   - Symptoms begin before birth, often resulting in fetal immobility, weak muscle tone, and respiratory failure at birth.
   - Infants with SMA Type 0 may not survive beyond a few weeks after birth.
2. *SMA Type 1 (Werdnig-Hoffmann Disease)*:
   - The most severe form of SMA that typically presents within the first 6 months of life.
   - Babies with SMA Type 1 experience severe muscle weakness, poor head control, difficulty breathing, and swallowing.
   - Without treatment, most infants do not survive past 2 years due to respiratory failure.
3. *SMA Type 2 (Intermediate SMA)*:
   - Symptoms appear between 6 and 18 months of age.
   - Children with SMA Type 2 can sit but cannot walk unaided.
   - Respiratory issues and scoliosis are common, and life expectancy varies widely depending on the severity and management of complications.
4. *SMA Type 3 (Kugelberg-Welander Disease)*:
   - Milder form of SMA with onset after 18 months (often in childhood or adolescence).
   - Patients can walk independently at some point but may lose the ability to walk as the disease progresses.
   - Life expectancy is usually near normal, but mobility becomes increasingly difficult over time.
5. *SMA Type 4 (Adult-onset SMA)*:
   - The mildest form of SMA, with symptoms typically appearing in adulthood.
   - Muscle weakness is less severe, and patients retain the ability to walk, but muscle function gradually deteriorates.

1.8.2 Pathophysiology

SMA results from mutations in the SMN1 gene, which provides instructions for producing the SMN protein. The SMN protein is essential for the survival and function of motor neurons in the spinal cord and brainstem, which control voluntary muscles. Without sufficient levels of SMN protein, motor neurons degenerate, leading to muscle atrophy and weakness [106].

1.8.3 Key Pathophysiological Features

1. *SMN1 Gene Mutation* [106]:
   - SMA is caused by homozygous deletions or mutations in the SMN1 gene located on chromosome 5q13. The gene is responsible for producing the majority of the SMN protein required for motor neuron function.
   - The SMN2 gene is a nearly identical copy of SMN1 but produces much lower levels of functional SMN protein due to a splicing error that excludes exon 7 in most transcripts. The number of SMN2 gene copies influences disease severity. More SMN2 copies tend to result in a milder form of SMA.
2. *Motor Neuron Degeneration* [107]:
   - The lack of SMN protein primarily affects alpha motor neurons in the anterior horn of the spinal cord. These neurons are in charge of coordinating movement by sending signals from the brain to the muscles.
   - Without enough SMN protein, motor neurons undergo progressive degeneration and apoptosis, leading to the loss of neural input to muscles.
   - As motor neurons die, muscles lose innervation, leading to muscle atrophy, weakness, and impaired motor function.
3. *Muscle Weakness and Atrophy*:
   - Motor neurons are essential for stimulating muscle activity. As neurons degenerate, muscles lose their stimulation, resulting in muscle wasting (atrophy).
   - Muscles used for movement (such as those in the arms and legs), as well as those involved in breathing and swallowing, become progressively weaker.
4. *Selective Vulnerability*:
   - While motor neurons are most affected, the selective vulnerability of these neurons is still not fully understood. Other cells in the body can survive with reduced SMN protein, but motor neurons are highly sensitive to its deficiency.

1.8.4 Clinical Features of Spinal Muscular Atrophy

SMA presents with a spectrum of symptoms depending on the type and severity of the disease. The primary symptoms include muscle weakness and atrophy, which affect movement, breathing, and swallowing [108].

1. *Muscle Weakness*:
   - Progressive muscle weakness is the hallmark feature of SMA. Weakness begins in the proximal muscles (those closest to the center of the body), such as the shoulders, hips, and trunk, and gradually spreads to other muscle groups.

2. *Hypotonia (Low Muscle Tone)*:
   - Babies with SMA often present with floppiness or poor muscle tone, known as hypotonia. They may have difficulty holding up their head or controlling their limbs.
3. *Delayed Motor Milestones*:
   - Children with SMA may show delayed developmental milestones, such as difficulty sitting up, standing, or walking.
4. *Breathing Difficulties*:
   - *Respiratory muscle weakness* is common, especially in SMA Type 1. Breathing problems often lead to respiratory infections, pneumonia, and respiratory failure, which is the primary cause of death in severe SMA.
5. *Feeding and Swallowing Difficulties*:
   - Weakness of the muscles involved in chewing and swallowing can lead to dysphagia, resulting in difficulties with feeding and an increased risk of aspiration pneumonia.
6. *Skeletal Deformities*:
   - Children with SMA are at risk for developing scoliosis (curvature of the spine) and joint contractures due to weak muscles and poor posture.
7. *Fasciculations and Tremors*:
   - In some cases, visible muscle twitches (fasciculations) or fine tremors of the tongue may be observed due to the denervation of the muscles.

1.8.5 Diagnosis of Spinal Muscular Atrophy

SMA can be diagnosed through a combination of clinical evaluation and genetic testing.

1. *Genetic Testing*:
   - Genetic testing is the gold standard for diagnosing SMA. It involves detecting deletions or mutations in the SMN1 gene. Testing for the number of SMN2 copies can help predict disease severity [109].
2. *Electromyography (EMG)*:
   - EMG can be used to assess the electrical activity of muscles and identify denervation due to motor neuron loss.
3. *Muscle Biopsy*:
   - In rare cases, a muscle biopsy may be performed to observe the characteristic changes in muscle fibers due to denervation, including muscle atrophy.
4. *Newborn Screening*:
   - Newborn screening programs for SMA have been put in place in several nations, enabling early identification and treatment before symptoms manifest.

1.8.6 Treatment and Management of Spinal Muscular Atrophy

Recent advancements in treatment have dramatically changed the prognosis for SMA, particularly with the development of gene therapies and drugs that increase SMN protein levels.

1. *Gene Therapy (Onasemnogene Abeparvovec—Zolgensma)*:

Using a viral vector, Zolgensma is a one-time gene treatment that introduces a functional copy of the SMN1 gene into motor neurons. Children with SMA younger than 2-years old can utilize it. The therapy has shown significant improvement in motor function and survival in infants treated early [110].

2. *SMN2* Splicing Modifiers (Nusinersen—Spinraza):

Spinraza is an antisense oligonucleotide treatment that modifies the splicing of the SMN2 gene to increase the generation of functional SMN protein, leading to the inclusion of exon 7 in the mRNA transcript. Spinraza is delivered via intrathecal injection and is approved for all types of SMA. It improves motor function and survival, particularly when started early [111].

3. *Risdiplam* (Evrysdi):

Evrysdi is an oral SMN2 splicing modifier that increases the production of SMN protein. It is the first oral treatment approved for all types of SMA, offering a more convenient administration route [112].

4. *Supportive Care*:
   - *Respiratory support*: Noninvasive ventilation (e.g., BiPAP) and mechanical ventilation may be necessary for patients with respiratory muscle weakness [113].
   - *Nutritional support*: Feeding tubes may be needed for those with swallowing difficulties to prevent malnutrition and aspiration pneumonia.
   - *Physical therapy and rehabilitation*: These interventions help maintain muscle strength and flexibility and prevent complications like contractures and scoliosis.

1.8.7 Future Directions and Research in SMA

1. *Early Diagnosis and Intervention*:
   - Newborn screening programs for SMA allow for early diagnosis, which is critical for the effectiveness of therapies like gene therapy and SMN2 splicing modifiers. Early treatment significantly improves outcomes [114].

2. *Combination Therapies*:
    - Ongoing research is exploring combination therapies that may enhance SMN.

1.9 Friedreich Ataxia (FRDA)

FRDA is a rare, inherited NDD that commonly affects the nervous system and the heart. It is characterized by progressive difficulty in walking, loss of coordination, muscle weakness, and various other neurological symptoms. FRDA also leads to cardiomyopathy (heart disease) and, in some cases, diabetes [115]. FRDA is a progressive disorder, and while its severity can vary, most patients experience increasing disability over time. The median life expectancy is 30–40 years, though some patients live into their 50s or 60s with proper care [116]. Cardiac complications are the leading cause of death, often from heart failure or arrhythmias.

FRDA is caused by mutations in the FXN gene that lead to reduced levels of frataxin, a protein critical for mitochondrial function. The disease typically begins in childhood or adolescence, and symptoms progressively worsen over time [117].

1.9.1 Etiology and Genetics of Friedreich Ataxia

FRDA is caused by mutations in the FXN gene, which is located on chromosome 9q13 and encodes the protein frataxin. Most cases (96%) are due to an expansion of GAA trinucleotide repeats in the first intron of the FXN gene. This mutation leads to reduced expression of frataxin, a protein involved in mitochondrial iron-sulfur cluster biogenesis, leading to mitochondrial dysfunction [117].

FRDA is an autosomal recessive disorder; therefore, disease cannot develop in people who only have one mutant copy of the FXN gene.

1.9.2 Pathophysiology

The reduction of frataxin levels in cells, especially in the CNS, heart, and pancreas is the primary feature of FRDA. Frataxin is crucial for the proper function of mitochondria, the energy-producing organelles of the cell [118]. Reduced frataxin leads to mitochondrial dysfunction due to the following:

1. *Iron Accumulation*

In the absence of frataxin, iron accumulates in the mitochondria, which leads to the production of ROS, causing oxidative stress and cellular damage [119].

2. *Impaired Energy Production*

Mitochondria with dysfunctional iron-sulfur clusters have reduced activity of key enzymes in the electron transport chain, leading to impaired ATP production (energy production). Cells, especially neurons and cardiac cells, become vulnerable to energy depletion and oxidative damage [118].

3. *Neuronal Degeneration*

The neurons most affected in FRDA are those in the dorsal root ganglia (sensory neurons), spinocerebellar tracts, and corticospinal tracts, leading to the loss of coordination and sensory deficits. As a result, patients experience progressive ataxia, loss of reflexes, and sensory impairment [120].

4. *Cardiac Involvement*

The heart muscle cells are also affected by mitochondrial dysfunction, leading to hypertrophic cardiomyopathy (thickening of the heart muscle). Cardiomyopathy is a leading cause of mortality in FRDA patients [121].

1.9.3 Clinical Features of Friedreich Ataxia

FRDA typically presents between the ages of 5 and 15, though adult-onset cases can occur. The severity and progression of symptoms vary based on the number of GAA repeats, with longer repeats associated with earlier onset and more severe disease.

1. *Ataxia*:
   - Gait ataxia is often the first symptom, manifesting as clumsiness, unsteady walking, and frequent falls. Over time, ataxia affects the arms and hands, leading to difficulty with fine motor tasks.
   - Dysarthria and dysphagia may also develop as the disease progresses.
2. *Loss of Reflexes*:
   - Loss of deep tendon reflexes is common, particularly in the lower limbs, due to degeneration of peripheral sensory nerves.
3. *Sensory Impairment*:
   - Patients lose proprioception (the sense of body position) and vibration sensation, which contributes to the loss of balance and coordination.
4. *Muscle Weakness*:
   - Progressive weakness and atrophy, particularly of the legs, develop over time, leading to difficulty standing or walking. Many patients become wheelchair-bound within 10–20 years of onset.
5. *Skeletal Abnormalities*:
   - Scoliosis and foot deformities are common in FRDA patients and may require surgical correction.

6. *Cardiomyopathy*:
   - Hypertrophic cardiomyopathy develops in most patients, leading to symptoms such as chest pain, shortness of breath, palpitations, and heart failure.
   - Arrhythmias and sudden cardiac death are significant causes of mortality in FRDA.
7. *Diabetes Mellitus*:
   - Up to 30% of patients may develop glucose intolerance or diabetes mellitus due to the involvement of pancreatic beta cells.

1.9.4 Diagnosis of Friedreich Ataxia

1. *Genetic Testing*:
   - The definitive diagnosis is made through genetic testing to identify the expanded GAA repeat in the FXN gene. Testing is highly sensitive and specific for FRDA.
2. *Neurological Examination*:
   - Clinical findings such as ataxia, loss of reflexes, and sensory deficits are important for initial suspicion of FRDA.
3. *Electrophysiological Studies*:
   - Nerve conduction studies may show reduced sensory nerve action potentials due to the degeneration of peripheral sensory neurons.
4. *MRI of the Brain and Spine*:
   - MRI may show atrophy of the spinal cord and cerebellum, although these findings are not specific to FRDA.
5. *Echocardiogram and ECG*:
   - These tests are used to assess the severity of hypertrophic cardiomyopathy and identify arrhythmias.
6. *Glucose Testing*:
   - Blood tests are performed to screen for diabetes in patients with FRDA.

1.9.5 Treatment and Management of Friedreich Ataxia

There is currently no cure for FRDA, and treatment is primarily symptomatic and supportive, focusing on managing complications and improving the quality of life. However, research into disease-modifying therapies is ongoing.

1. *Symptomatic Management*:
   - Physical therapy, speech therapy, and orthopedic interventions.
2. *Cardiac Management*:
   - Medications such as beta-blockers, ACE inhibitors, or antiarrhythmic drugs are used to manage cardiomyopathy and prevent heart failure and arrhythmias.
   - In advanced cases, cardiac transplantation may be considered.

3. *Diabetes Management*:
   - Patients with diabetes are managed with dietary modifications, oral hypoglycemic agents, or insulin therapy.
4. *Antioxidants and Mitochondrial Support*:
   - Given the role of oxidative stress in FRDA, antioxidant therapies such as idebenone, coenzyme Q10, and vitamin E have been explored, though their efficacy remains under investigation.
   - *Omaveloxolone*, a drug that activates pathways reducing oxidative stress, is being investigated for its potential in slowing disease progression.
5. *Gene Therapy and Disease-Modifying Therapies*:
   - Research is ongoing into gene therapy and frataxin replacement therapy to restore normal frataxin levels and prevent neurodegeneration. Early trials have shown promise in animal models, but human studies are still in progress.

1.9.6 Future Directions in Friedreich Ataxia Research

1. *Gene Therapy*:
   - Advances in CRISPR-Cas9 and other gene-editing technologies offer potential strategies for correcting the GAA repeat expansions in the FXN gene. Efforts to deliver functional frataxin or silence the abnormal gene are key areas of exploration [122].
2. *Frataxin Protein Replacement*:
   - Research is focused on finding ways to increase frataxin levels in affected tissues. Protein replacement therapy and small molecules that increase frataxin production or stability are being tested.
3. *Targeting Mitochondrial Dysfunction*:
   - Treatments that enhance mitochondrial performance and lower oxidative stress may delay the course of illness. Clinical trials of drugs like omaveloxolone and others are ongoing [123].

1.10 Current Therapeutic Strategies for Neurodegenerative Diseases

NDDs, such as AD, PD, ALS, and HD, have no known cures, but a variety of therapeutic strategies exist to manage symptoms, slow disease progression, and improve quality of life. While each disease has unique mechanisms, many treatments target common themes such as protein aggregation, oxidative stress, and inflammation.

1.10.1 Symptomatic Treatments

These therapies do not alter the disease course but help manage the symptoms to improve patient quality of life.

- *AD*

Cholinesterase inhibitors (e.g., Donepezil, Rivastigmine) increase acetylcholine levels, improving cognitive function in early and mid-stage AD. Memantine (NMDA receptor antagonist) helps in moderate-to-severe stages by reducing glutamate excitotoxicity [124].

- *PD*

Levodopa (converted to dopamine in the brain) and dopamine agonists (e.g., pramipexole, ropinirole) help manage motor symptoms by replacing or mimicking dopamine. MAO-B inhibitors (e.g., selegiline, rasagiline) prevent dopamine breakdown [124].

- *ALS*

Riluzole reduces glutamate excitotoxicity and modestly prolongs survival. Edaravone helps to reduce oxidative stress and slow disease progression in some patients [125].

- *HD*

Tetrabenazine and antipsychotics (e.g., risperidone) help manage chorea (involuntary movements). Antidepressants are often used to treat psychiatric symptoms like depression and anxiety [126].

1.10.2 Disease-Modifying Therapies

These therapies aim to slow or stop the progression of NDDs by targeting underlying mechanisms like protein misfolding, genetic mutations, and cellular dysfunctions.

- *AD*
  - *Anti-amyloid therapies*: The FDA has approved aducanumab and lecanemab, monoclonal antibodies targeting amyloid-beta plaques. These drugs aim to reduce amyloid accumulation, though their clinical benefit remains debated [127].
  - *Anti-tau therapies* are in development to target neurofibrillary tangles, another pathological hallmark of AD.
- *PD*
  - Research is focused on targeting alpha-synuclein, a misfolded protein that aggregates in Lewy bodies [128].
  - Glucocerebrosidase (GCase) therapies are being investigated to improve lysosomal function and protein degradation [129].

- *ALS*
  - Gene-targeting therapies such as tofersen target mutations in the SOD1 gene in familial ALS cases. It is an antisense oligonucleotide (ASO) designed to reduce mutant SOD1 protein levels [130].
  - Stem cell therapies (like NurOwn) are under investigation to replace or repair damaged motor neurons.
- *HD*
  - Gene silencing therapies like antisense oligonucleotides (ASOs) are being developed to lower levels of mutant huntingtin protein [131].
  - CRISPR-based approaches are being explored to directly target and correct the genetic mutations [132].

1.10.3 Neuroprotective Therapies

These treatments aim to protect neurons from degeneration, reducing cell death and slowing disease progression.

- *Antioxidants*: Medications like Edaravone (ALS) and dietary antioxidants like vitamin E or CoQ10 are believed to reduce oxidative stress, a common feature in neurodegeneration [133].
- *Anti-inflammatory approaches*: Chronic inflammation in the brain is a feature of many NDDs. Drugs targeting microglial activation and the inflammatory cascade (e.g., NLRP3 inhibitors) are in trials for conditions like Alzheimer's and ALS [4].
- *Neurotrophic factors*: Delivering growth factors such as brain-derived neurotrophic factor (BDNF) or glial cell line-derived neurotrophic factor (GDNF) is a potential approach to support neuron survival and regeneration in diseases like PD and ALS [134].

1.10.4 Cell-Based Therapies

- *Stem cell therapy*: Stem cells are being investigated for their ability to replace lost neurons, reduce inflammation, and promote repair in ALS, PD, and other NDDs [135].
  - In PD, dopaminergic neurons derived from stem cells are being explored as a way to replace damaged cells.
  - For ALS, NurOwn, which uses autologous mesenchymal stem cells, aims to secrete neurotrophic factors to protect motor neurons [136].

1.11 Future Directions in Neurodegenerative Disease Research

1.11.1 Gene Editing and Gene Therapy

- The CRISPR-Cas9 gene-editing technique has the potential to directly fix genetic mutations that cause diseases like HD and familial ALS [137].
- Gene therapy approaches using viral vectors to deliver functional copies of genes or silence mutant genes are being explored in several conditions. For example, AAV (adeno-associated viruses) are used to deliver therapeutic genes to target neurons [138].

1.11.2 Precision Medicine

- Research is moving toward more personalized approaches, tailoring treatments based on genetic profiles, biomarkers, and individual disease mechanisms. Biomarker discovery, particularly through advances in neuroimaging and fluid biomarkers (e.g., cerebrospinal fluid or blood markers), will enable earlier detection and more accurate tracking of disease progression [139].

1.11.3 Immunotherapy

- Immunotherapy is increasingly being explored, particularly in Alzheimer's and Parkinson's, where targeting abnormal proteins (amyloid, tau, and alpha-synuclein) through immune modulation or vaccination strategies may reduce pathological protein buildup. Ongoing trials are investigating monoclonal antibodies that target misfolded proteins [140].

1.11.4 Protein Misfolding and Autophagy Enhancement

- Autophagy is the process by which cells clear misfolded proteins. Enhancing autophagy is being studied as a potential strategy for diseases like ALS, HD, and PD, where abnormal protein aggregates contribute to neuronal death [141].
- Drugs like rapamycin and other autophagy enhancers are being investigated for their neuroprotective potential.

1.11.5 Regenerative Medicine

- Research in regenerative medicine is focused on promoting the repair and regeneration of damaged neurons through the use of growth factors, neuroprotective drugs, and cell therapies [142].

- Advanced approaches like 3D bioprinting or tissue engineering might allow the creation of neuron-replacement therapies or the development of models to study disease mechanisms in the lab [143].

1.12 Conclusion

The future of NDD research is moving toward a more integrated, multidisciplinary approach that combines genetic therapy, protein-targeting strategies, and regenerative medicine. While current therapies mostly manage symptoms or slow disease progression, future research holds promise for more targeted, disease-modifying treatments that could significantly alter the trajectory of these devastating conditions.

References

1. Sahoo RK, Gupta T, Kumar V, Rani S, Gupta U, Smily. Aetiology and pathophysiology of neurodegenerative disorders. In: Nanomedical drug delivery for neurodegenerative diseases. Academic Press; 2022. p. 1–16. https://doi.org/10.1016/B978-0-323-85544-0.00006-X.
2. Sweeney P, Park H, Baumann M, Dunlop J, Frydman J, Kopito R, et al. Protein misfolding in neurodegenerative diseases: implications and strategies. Transl Neurodegener. 2017;6:1–13. https://doi.org/10.1186/S40035-017-0077-5.
3. Klemmensen MM, Borrowman SH, Pearce C, Pyles B, Chandra B. Mitochondrial dysfunction in neurodegenerative disorders. Neurotherapeutics. 2023;21:e00292. https://doi.org/10.1016/J.NEUROT.2023.10.002.
4. Zhang W, Xiao D, Mao Q, Xia H. Role of neuroinflammation in neurodegeneration development. Signal Transduct Target Ther. 2023;8:1–32. https://doi.org/10.1038/s41392-023-01486-5.
5. Kwon HS, Koh SH. Neuroinflammation in neurodegenerative disorders: the roles of microglia and astrocytes. Transl Neurodegener. 2020;9:1–12. https://doi.org/10.1186/S40035-020-00221-2.
6. Chi H, Chang HY, Sang TK. Neuronal cell death mechanisms in major neurodegenerative diseases. Int J Mol Sci. 2018;19:3082. https://doi.org/10.3390/IJMS19103082.
7. Dugger BN, Dickson DW. Pathology of neurodegenerative diseases. Cold Spring Harb Perspect Biol. 2017;9:a028035. https://doi.org/10.1101/CSHPERSPECT.A028035.
8. Uchikado H, Lin WL, Delucia MW, Dickson DW. Alzheimer disease with amygdala lewy bodies: a distinct form of α-synucleinopathy. J Neuropathol Exp Neurol. 2006;65:685–97. https://doi.org/10.1097/01.JNEN.0000225908.90052.07.
9. Selvam S, Ayyavoo V. Biomarkers in neurodegenerative diseases: a broad overview. Explor Neuroprot Ther. 2024;4:119–47. https://doi.org/10.37349/ENT.2024.00075.
10. Ayeni EA, Aldossary AM, Ayejoto DA, Gbadegesin LA, Alshehri AA, Alfassam HA, et al. Neurodegenerative diseases: implications of environmental and climatic influences on neurotransmitters and neuronal hormones activities. Int J Environ Res Public Health. 2022;19:12495. https://doi.org/10.3390/IJERPH191912495.
11. Puranik N, Yadav D, Song M. Advancements in the application of nanomedicine in Alzheimer's disease: a therapeutic perspective. Int J Mol Sci. 2023;24:24. https://doi.org/10.3390/IJMS241814044.
12. 2024 Alzheimer's disease facts and figures. Alzheimers Dement. 2024;20:3708–821. https://doi.org/10.1002/ALZ.13809.

13. Hippius H, Neundörfer G. The discovery of Alzheimer's disease. Dialogues Clin Neurosci. 2003;5:101–8. https://doi.org/10.31887/DCNS.2003.5.1/HHIPPIUS.
14. Alzheimer's stages – early, middle, late Dementia symptoms | alz.org; n.d. https://www.alz.org/alzheimers-dementia/stages?gad_source=1&gclid=Cj0KCQiA2oW-BhC2ARIsADSIAWpk0GKERSB4BuHF5marvp1U48N-iPpRHlDTbBkNLAx1pe9AA4_XLAMaAgXvEALw_wcB. Accessed 28 Feb 2025.
15. Jack CR, Bennett DA, Blennow K, Carrillo MC, Dunn B, Haeberlein SB, et al. NIA-AA research framework: toward a biological definition of Alzheimer's disease. Alzheimers Dement. 2018;14:535. https://doi.org/10.1016/J.JALZ.2018.02.018.
16. Richard F, Amouyel P. Genetic susceptibility factors for Alzheimer's disease. Eur J Pharmacol. 2001;412:1–12. https://doi.org/10.1016/S0014-2999(00)00903-1.
17. Tini G, Scagliola R, Monacelli F, La Malfa G, Porto I, Brunelli C, et al. Alzheimer's disease and cardiovascular disease: a particular association. Cardiol Res Pract. 2020;2020:2617970. https://doi.org/10.1155/2020/2617970.
18. Gottlieb S. Head injury doubles the risk of Alzheimer's disease. BMJ. 2000;321:1100.
19. Zhang H, Wei W, Zhao M, Ma L, Jiang X, Pei H, et al. Interaction between Aβ and Tau in the pathogenesis of Alzheimer's disease. Int J Biol Sci. 2021;17:2181. https://doi.org/10.7150/IJBS.57078.
20. Kamatham PT, Shukla R, Khatri DK, Vora LK. Pathogenesis, diagnostics, and therapeutics for Alzheimer's disease: breaking the memory barrier. Ageing Res Rev. 2024;101:102481. https://doi.org/10.1016/J.ARR.2024.102481.
21. Misrani A, Tabassum S, Yang L. Mitochondrial dysfunction and oxidative stress in Alzheimer's disease. Front Aging Neurosci. 2021;13:617588. https://doi.org/10.3389/FNAGI.2021.617588.
22. Jurcău MC, Andronie-Cioara FL, Jurcău A, Marcu F, Ţiţ DM, Paşcalău N, et al. The link between oxidative stress, mitochondrial dysfunction and neuroinflammation in the pathophysiology of Alzheimer's disease: therapeutic implications and future perspectives. Antioxidants. 2022;11:2167. https://doi.org/10.3390/ANTIOX11112167.
23. Kazemeini S, Nadeem-Tariq A, Shih R, Rafanan J, Ghani N, Vida TA. From plaques to pathways in Alzheimer's disease: the mitochondrial-neurovascular-metabolic hypothesis. Int J Mol Sci. 2024;25:11720. https://doi.org/10.3390/IJMS252111720.
24. Zhang Y, Chen H, Li R, Sterling K, Song W. Amyloid β-based therapy for Alzheimer's disease: challenges, successes and future. Signal Transduct Target Ther. 2023;8:1–26. https://doi.org/10.1038/s41392-023-01484-7.
25. Gulisano W, Maugeri D, Baltrons MA, Fà M, Amato A, Palmeri A, et al. Role of amyloid-β and Tau proteins in Alzheimer's disease: confuting the amyloid cascade. J Alzheimers Dis. 2018;64:S611. https://doi.org/10.3233/JAD-179935.
26. Rahman MM, Lendel C. Extracellular protein components of amyloid plaques and their roles in Alzheimer's disease pathology. Mol Neurodegener. 2021;16:1–30. https://doi.org/10.1186/S13024-021-00465-0.
27. Mietelska-Porowska A, Wasik U, Goras M, Filipek A, Niewiadomska G. Tau protein modifications and interactions: their role in function and dysfunction. Int J Mol Sci. 2014;15:4671. https://doi.org/10.3390/IJMS15034671.
28. Gong C-X, Iqbal K. Hyperphosphorylation of microtubule-associated protein tau: a promising therapeutic target for Alzheimer disease. Curr Med Chem. 2008;15:2321. https://doi.org/10.2174/092986708785909111.
29. Medeiros R, Baglietto-Vargas D, Laferla FM. The role of tau in Alzheimer's disease and related disorders. CNS Neurosci Ther. 2010;17:514. https://doi.org/10.1111/J.1755-5949.2010.00177.X.
30. Chen Y, Yu Y. Tau and neuroinflammation in Alzheimer's disease: interplay mechanisms and clinical translation. J Neuroinflammation. 2023;20:1–21. https://doi.org/10.1186/S12974-023-02853-3.
31. Kim J, Yoo ID, Lim J, Moon JS. Pathological phenotypes of astrocytes in Alzheimer's disease. Exp Mol Med. 2024;56:95–9. https://doi.org/10.1038/s12276-023-01148-0.

32. González-Reyes RE, Nava-Mesa MO, Vargas-Sánchez K, Ariza-Salamanca D, Mora-Muñoz L. Involvement of astrocytes in Alzheimer's disease from a neuroinflammatory and oxidative stress perspective. Front Mol Neurosci. 2017;10:427. https://doi.org/10.3389/FNMOL.2017.00427.
33. Rajmohan R, Reddy PH. Amyloid beta and phosphorylated tau accumulations cause abnormalities at synapses of Alzheimer's disease neurons. J Alzheimers Dis. 2017;57:975. https://doi.org/10.3233/JAD-160612.
34. Tönnies E, Trushina E. Oxidative stress, synaptic dysfunction, and Alzheimer's disease. J Alzheimers Dis. 2017;57:1105. https://doi.org/10.3233/JAD-161088.
35. Ferreira-Vieira TH, Guimaraes IM, Silva FR, Ribeiro FM. Alzheimer's disease: targeting the cholinergic system. Curr Neuropharmacol. 2016;14:101. https://doi.org/10.2174/1570159X13666150716165726.
36. Puranik N, Song M. Glutamate: molecular mechanisms and signaling pathway in Alzheimer's disease, a potential therapeutic target. Molecules. 2024;29:5744. https://doi.org/10.3390/MOLECULES29235744.
37. Alkhalifa AE, Al-Ghraiybah NF, Odum J, Shunnarah JG, Austin N, Kaddoumi A. Blood–brain barrier breakdown in Alzheimer's disease: mechanisms and targeted strategies. Int J Mol Sci. 2023;24:16288. https://doi.org/10.3390/IJMS242216288.
38. Kalia LV, Lang AE. Parkinson's disease. Lancet. 2015;386:896–912. https://doi.org/10.1016/S0140-6736(14)61393-3.
39. Armstrong MJ, Okun MS. Diagnosis and treatment of Parkinson disease: a review. JAMA. 2020;323:548–60. https://doi.org/10.1001/JAMA.2019.22360.
40. Jankovic J. Parkinson's disease: clinical features and diagnosis. J Neurol Neurosurg Psychiatry. 2008;79:368–76. https://doi.org/10.1136/JNNP.2007.131045.
41. Postuma RB, Berg D, Stern M, Poewe W, Olanow CW, Oertel W, et al. MDS clinical diagnostic criteria for Parkinson's disease. Mov Disord. 2015;30:1591–601. https://doi.org/10.1002/MDS.26424.
42. Obeso JA, Stamelou M, Goetz CG, Poewe W, Lang AE, Weintraub D, et al. Past, present, and future of Parkinson's disease: a special essay on the 200th anniversary of the Shaking Palsy. Mov Disord. 2017;32:1264–310. https://doi.org/10.1002/MDS.27115.
43. Bloem BR, Okun MS, Klein C. Parkinson's disease. Lancet. 2021;397:2284–303. https://doi.org/10.1016/S0140-6736(21)00218-X.
44. Marras C, Chaudhuri KR. Nonmotor features of Parkinson's disease subtypes. Mov Disord. 2016;31:1095–102. https://doi.org/10.1002/MDS.26510.
45. Schapira AHV, Olanow CW, Greenamyre JT, Bezard E. Slowing of neurodegeneration in Parkinson's disease and Huntington's disease: future therapeutic perspectives. Lancet. 2014;384:545–55. https://doi.org/10.1016/S0140-6736(14)61010-2.
46. Armstrong MJ, Okun MS. Time for a new image of Parkinson disease. JAMA Neurol. 2020;77:1345–6. https://doi.org/10.1001/JAMANEUROL.2020.2412.
47. Dickson DW. Neuropathology of Parkinson disease. Parkinsonism Relat Disord. 2018;46(Suppl 1):S30–3. https://doi.org/10.1016/J.PARKRELDIS.2017.07.033.
48. Lees AJ, Hardy J, Revesz T. Parkinson's disease. Lancet. 2009;373:2055–66. https://doi.org/10.1016/S0140-6736(09)60492-X.
49. Pyatha S, Kim H, Lee D, Kim K. Association between heavy metal exposure and Parkinson's disease: a review of the mechanisms related to oxidative stress. Antioxidants. 2022;11:2467. https://doi.org/10.3390/ANTIOX11122467.
50. Poewe W, Seppi K, Tanner CM, Halliday GM, Brundin P, Volkmann J, et al. Parkinson disease. Nat Rev Dis Primers. 2017;3:1–21. https://doi.org/10.1038/NRDP.2017.13.
51. Pizarro-Galleguillos BM, Kunert L, Brüggemann N, Prasuhn J. Neuroinflammation and mitochondrial dysfunction in Parkinson's disease: connecting neuroimaging with pathophysiology. Antioxidants. 2023;12:1411. https://doi.org/10.3390/ANTIOX12071411/S1.
52. Lanciego JL, Luquin N, Obeso JA. Functional neuroanatomy of the Basal Ganglia. Cold Spring Harb Perspect Med. 2012;2:a009621. https://doi.org/10.1101/CSHPERSPECT.A009621.

53. Ramesh S, Arachchige ASPM. Depletion of dopamine in Parkinson's disease and relevant therapeutic options: a review of the literature. AIMS Neurosci. 2023;10:200. https://doi.org/10.3934/NEUROSCIENCE.2023017.
54. Calabresi P, Mechelli A, Natale G, Volpicelli-Daley L, Di Lazzaro G, Ghiglieri V. Alpha-synuclein in Parkinson's disease and other synucleinopathies: from overt neurodegeneration back to early synaptic dysfunction. Cell Death Dis. 2023;14:1–16. https://doi.org/10.1038/s41419-023-05672-9.
55. Goldman JG, Postuma R. Premotor and non-motor features of Parkinson's disease. Curr Opin Neurol. 2014;27:434. https://doi.org/10.1097/WCO.0000000000000112.
56. Subramaniam SR, Chesselet MF. Mitochondrial dysfunction and oxidative stress in Parkinson's disease. Prog Neurobiol. 2013:17. https://doi.org/10.1016/J.PNEUROBIO.2013.04.004.
57. Çınar E, Tel BC, Şahin G. Neuroinflammation in Parkinson's disease and its treatment opportunities. Balkan Med J. 2022;39:318–33. https://doi.org/10.4274/BALKANMEDJ.GALENOS.2022.2022-7-100.
58. Nuytemans K, Theuns J, Cruts M, Van Broeckhoven C. Genetic etiology of Parkinson disease associated with mutations in the SNCA, PARK2, PINK1, PARK7, and LRRK2 genes: a mutation update. Hum Mutat. 2010;31:763. https://doi.org/10.1002/HUMU.21277.
59. Tong H, Yang T, Xu S, Li X, Liu L, Zhou G, et al. Huntington's disease: complex pathogenesis and therapeutic strategies. Int J Mol Sci. 2024;25:3845. https://doi.org/10.3390/IJMS25073845/S1.
60. Irfan Z, Khanam S, Karmakar V, Firdous SM, El Khier BSIA, Khan I, et al. Pathogenesis of Huntington's disease: an emphasis on molecular pathways and prevention by natural remedies. Brain Sci. 2022;12:1389. https://doi.org/10.3390/BRAINSCI12101389.
61. Reiner A, Dragatsis I, Dietrich P. Genetics and neuropathology of Huntington's disease. Int Rev Neurobiol. 2011;98:325. https://doi.org/10.1016/B978-0-12-381328-2.00014-6.
62. Labbadia J, Morimoto RI. Huntington's disease: underlying molecular mechanisms and emerging concepts. Trends Biochem Sci. 2013;38:378. https://doi.org/10.1016/J.TIBS.2013.05.003.
63. Ross CA, Tabrizi SJ. Huntington's disease: from molecular pathogenesis to clinical treatment. Lancet Neurol. 2011;10:83–98. https://doi.org/10.1016/S1474-4422(10)70245-3.
64. Ross CA, Aylward EH, Wild EJ, Langbehn DR, Long JD, Warner JH, et al. Huntington disease: natural history, biomarkers and prospects for therapeutics. Nat Rev Neurol. 2014;10(4):204–16. https://doi.org/10.1038/nrneurol.2014.24.
65. Paulsen JS. Cognitive impairment in Huntington disease: diagnosis and treatment. Curr Neurol Neurosci Rep. 2011;11:474–83. https://doi.org/10.1007/S11910-011-0215-X.
66. Van Duijn E, Kingma EM, Van Der Mast RC. Psychopathology in verified Huntington's disease gene carriers. J Neuropsychiatry Clin Neurosci. 2007;19:441–8. https://doi.org/10.1176/JNP.2007.19.4.441.
67. Walker FO. Huntington's disease. Semin Neurol. 2007;27:143–50. https://doi.org/10.1055/S-2007-971176.
68. Estrada Sánchez AM, Mejía-Toiber J, Massieu L. Excitotoxic neuronal death and the pathogenesis of Huntington's disease. Arch Med Res. 2008;39:265–76. https://doi.org/10.1016/J.ARCMED.2007.11.011.
69. Joshi DC, Chavan MB, Gurow K, Gupta M, Dhaliwal JS, Ming LC. The role of mitochondrial dysfunction in Huntington's disease: implications for therapeutic targeting. Biomed Pharmacother. 2025;183:117827. https://doi.org/10.1016/J.BIOPHA.2025.117827.
70. Li XJ, Li S. Proteasomal dysfunction in aging and Huntington disease. Neurobiol Dis. 2011;43:4–8. https://doi.org/10.1016/J.NBD.2010.11.018.
71. Metcalf DJ, García-Arencibia M, Hochfeld WE, Rubinsztein DC. Autophagy and misfolded proteins in neurodegeneration. Exp Neurol. 2012;238:22. https://doi.org/10.1016/J.EXPNEUROL.2010.11.003.
72. Crotti A, Glass CK. The choreography of neuroinflammation in Huntington's disease. Trends Immunol. 2015;36:364–73. https://doi.org/10.1016/J.IT.2015.04.007.

73. Harvey RJ, Sassone J, Reid CA, Kwakowsky A, Palpagama TH, Waldvogel HJ, et al. The role of microglia and astrocytes in Huntington's disease. Front Mol Neurosci. 2019;12:258. https://doi.org/10.3389/FNMOL.2019.00258.
74. Rowland LP, Shneider NA. Amyotrophic lateral sclerosis. N Engl J Med. 2001;344:1688–700. https://doi.org/10.1056/NEJM200105313442207.
75. Chiò A, Calvo A, Bovio G, Canosa A, Bertuzzo D, Galmozzi F, et al. Amyotrophic lateral sclerosis outcome measures and the role of albumin and creatinine: a population-based study. JAMA Neurol. 2014;71:1134–42. https://doi.org/10.1001/JAMANEUROL.2014.1129.
76. Kiernan MC, Vucic S, Cheah BC, Turner MR, Eisen A, Hardiman O, et al. Amyotrophic lateral sclerosis. Lancet. 2011;377:942–55. https://doi.org/10.1016/S0140-6736(10)61156-7.
77. van Es MA, Hardiman O, Chio A, Al-Chalabi A, Pasterkamp RJ, Veldink JH, et al. Amyotrophic lateral sclerosis. Lancet. 2017;390:2084–98. https://doi.org/10.1016/S0140-6736(17)31287-4.
78. Hardiman O, Al-Chalabi A, Chio A, Corr EM, Logroscino G, Robberecht W, et al. Amyotrophic lateral sclerosis. Nat Rev Dis Primers. 2017;3:3. https://doi.org/10.1038/NRDP.2017.71.
79. Jankovska N, Matej R. Molecular pathology of ALS: what we currently know and what important information is still missing. Diagnostics. 2021;11:1365. https://doi.org/10.3390/DIAGNOSTICS11081365.
80. Alessenko AV, Gutner UA, Shupik MA. Involvement of lipids in the pathogenesis of amyotrophic lateral sclerosis. Life. 2023;13:510. https://doi.org/10.3390/LIFE13020510.
81. Van Es MA. Amyotrophic lateral sclerosis; clinical features, differential diagnosis and pathology. Int Rev Neurobiol. 2024;176:1–47. https://doi.org/10.1016/BS.IRN.2024.04.011.
82. Pansarasa O, Bordoni M, Diamanti L, Sproviero D, Gagliardi S, Cereda C. SOD1 in amyotrophic lateral sclerosis: "ambivalent" behavior connected to the disease. Int J Mol Sci. 2018;19:1345. https://doi.org/10.3390/IJMS19051345.
83. Renton AE, Majounie E, Waite A, Simón-Sánchez J, Rollinson S, Gibbs JR, et al. A hexanucleotide repeat expansion in C9ORF72 is the cause of chromosome 9p21-linked ALS-FTD. Neuron. 2011;72:257–68. https://doi.org/10.1016/J.NEURON.2011.09.010.
84. Arnold ES, Ling SC, Huelga SC, Lagier-Tourenne C, Polymenidou M, Ditsworth D, et al. ALS-linked TDP-43 mutations produce aberrant RNA splicing and adult-onset motor neuron disease without aggregation or loss of nuclear TDP-43. Proc Natl Acad Sci USA. 2013;110:E736–45. https://doi.org/10.1073/PNAS.1222809110/SUPPL_FILE/SM01.MOV.
85. Blokhuis AM, Groen EJN, Koppers M, Van Den Berg LH, Pasterkamp RJ. Protein aggregation in amyotrophic lateral sclerosis. Acta Neuropathol. 2013;125:777. https://doi.org/10.1007/S00401-013-1125-6.
86. Van Den Bosch L, Van Damme P, Bogaert E, Robberecht W. The role of excitotoxicity in the pathogenesis of amyotrophic lateral sclerosis. Biochim Biophys Acta (BBA) – Mol Basis Dis. 2006;1762:1068–82. https://doi.org/10.1016/J.BBADIS.2006.05.002.
87. Cunha-Oliveira T, Montezinho L, Mendes C, Firuzi O, Saso L, Oliveira PJ, et al. Oxidative stress in amyotrophic lateral sclerosis: pathophysiology and opportunities for pharmacological intervention. Oxidative Med Cell Longev. 2020;2020:5021694. https://doi.org/10.1155/2020/5021694.
88. Zhao J, Wang X, Huo Z, Chen Y, Liu J, Zhao Z, et al. The impact of mitochondrial dysfunction in amyotrophic lateral sclerosis. Cells. 2022;11:2049. https://doi.org/10.3390/CELLS11132049.
89. Chen X, Lv S, Liu J, Guan Y, Xu C, Ma X, et al. Exploring the role of axons in ALS from multiple perspectives. Cells. 2024;13:2076. https://doi.org/10.3390/CELLS13242076.
90. Gao C, Jiang J, Tan Y, Chen S. Microglia in neurodegenerative diseases: mechanism and potential therapeutic targets. Signal Transduct Target Ther. 2023;8:1–37. https://doi.org/10.1038/s41392-023-01588-0.
91. Suk TR, Rousseaux MWC. The role of TDP-43 mislocalization in amyotrophic lateral sclerosis. Mol Neurodegener. 2020;15:45. https://doi.org/10.1186/S13024-020-00397-1.

92. Poggiolini I, Saverioni D, Parchi P. Prion protein misfolding, strains, and neurotoxicity: an update from studies on mammalian prions. Int J Cell Biol. 2013;2013:910314. https://doi.org/10.1155/2013/910314.
93. Lee J, Kim SY, Hwang KJ, Ju YR, Woo HJ. Prion diseases as transmissible zoonotic diseases. Osong Public Health Res Perspect. 2013;4:57. https://doi.org/10.1016/J.PHRP.2012.12.008.
94. Burrell CJ, Howard CR, Murphy FA. Fenner and White's medical virology. Academic Press; 2017. p. 527–35. https://doi.org/10.1016/B978-0-12-375156-0.00038-2.
95. Baral PK, Yin J, Aguzzi A, James MNG. Transition of the prion protein from a structured cellular form (PrPC) to the infectious scrapie agent (PrPSc). Protein Sci. 2019;28:2055. https://doi.org/10.1002/PRO.3735.
96. Assis-de-Lemos G, Moura-do-Nascimento R, Amaral-do-Nascimento M, Miceli AC, Vieira TCRG. Interactions between cytokines and the pathogenesis of Prion diseases: insights and implications. Brain Sci. 2024;14:413. https://doi.org/10.3390/BRAINSCI14050413.
97. Moore RA, Taubner LM, Priola SA. Prion protein misfolding and disease. Curr Opin Struct Biol. 2009;19:14. https://doi.org/10.1016/J.SBI.2008.12.007.
98. Westergard L, Christensen HM, Harris DA. The cellular prion protein (PrPC): its physiological function and role in disease. Biochim Biophys Acta. 2007;1772:629. https://doi.org/10.1016/J.BBADIS.2007.02.011.
99. Soto C, Satani N. The intricate mechanisms of neurodegeneration in prion diseases. Trends Mol Med. 2011;17:14. https://doi.org/10.1016/J.MOLMED.2010.09.001.
100. Stephen CD, Vallabh S, Bian J, Appleby BS, Shetty S, Elkasaby M. Genetic aspects of human prion diseases. Front Neurol. 2022;13:1003056. https://doi.org/10.3389/FNEUR.2022.1003056.
101. Canas LS, Sudre CH, De Vita E, Nihat A, Mok TH, Slattery CF, et al. Prion disease diagnosis using subject-specific imaging biomarkers within a multi-kernel Gaussian process. Neuroimage Clin. 2019;24:102051. https://doi.org/10.1016/J.NICL.2019.102051.
102. Connor A, Wang H, Appleby BS, Rhoads DD. Clinical laboratory tests used to aid in diagnosis of human prion disease. J Clin Microbiol. 2019;57:e00769–19. https://doi.org/10.1128/JCM.00769-19.
103. Liu F, Lü W, Liu L. New implications for prion diseases therapy and prophylaxis. Front Mol Neurosci. 2024;17:1324702. https://doi.org/10.3389/FNMOL.2024.1324702.
104. Farrar MA, Kiernan MC. The genetics of spinal muscular atrophy: progress and challenges. Neurotherapeutics. 2015;12:290–302. https://doi.org/10.1007/S13311-014-0314-X.
105. Angilletta I, Ferrante R, Giansante R, Lombardi L, Babore A, Dell'Elice A, et al. Spinal muscular atrophy: an evolving scenario through new perspectives in diagnosis and advances in therapies. Int J Mol Sci. 2023;24:14873. https://doi.org/10.3390/IJMS241914873.
106. Keinath MC, Prior DE, Prior TW. Spinal muscular atrophy: mutations, testing, and clinical relevance. Appl Clin Genet. 2021;14:11. https://doi.org/10.2147/TACG.S239603.
107. Park GH, Maeno-Hikichi Y, Awano T, Landmesser LT, Monani UR. Reduced survival of motor neuron (SMN) protein in motor neuronal progenitors functions cell autonomously to cause spinal muscular atrophy in model mice expressing the human centromeric (SMN2) gene. J Neurosci. 2010;30:12005. https://doi.org/10.1523/JNEUROSCI.2208-10.2010.
108. Pierzchlewicz K, Kępa I, Podogrodzki J, Kotulska K. Spinal muscular atrophy: the use of functional motor scales in the era of disease-modifying treatment. Child Neurol Open. 2021;8:2329048X211008725. https://doi.org/10.1177/2329048X211008725.
109. Zhou Y, Jiang Y. Current advances in genetic testing for spinal muscular atrophy. Curr Genomics. 2023;24:273. https://doi.org/10.2174/0113892029273388231023072050.
110. Bitetti I, Lanzara V, Margiotta G, Varone A. Onasemnogene abeparvovec gene replacement therapy for the treatment of spinal muscular atrophy: a real-world observational study. Gene Ther. 2022;30:592. https://doi.org/10.1038/S41434-022-00341-6.
111. Chaytow H, Faller KME, Huang YT, Gillingwater TH. Spinal muscular atrophy: from approved therapies to future therapeutic targets for personalized medicine. Cell Rep Med. 2021;2:100346. https://doi.org/10.1016/J.XCRM.2021.100346.

112. Paik J. Risdiplam: a review in spinal muscular atrophy. CNS Drugs. 2022;36:401–10. https://doi.org/10.1007/S40263-022-00910-8.
113. Bach JR. Noninvasive respiratory management of patients with neuromuscular disease. Ann Rehabil Med. 2017;41:519. https://doi.org/10.5535/ARM.2017.41.4.519.
114. Shrestha B, Wang L, Brey EM, Uribe GR, Tang L. Smart nanoparticles for chemo-based combinational therapy. Pharmaceutics. 2021;13:853. https://doi.org/10.3390/PHARMACEUTICS13060853.
115. Parkinson MH, Boesch S, Nachbauer W, Mariotti C, Giunti P. Clinical features of Friedreich's ataxia: classical and atypical phenotypes. J Neurochem. 2013;126:103–17. https://doi.org/10.1111/JNC.12317.
116. Tsou AY, Paulsen EK, Lagedrost SJ, Perlman SL, Mathews KD, Wilmot GR, et al. Mortality in Friedreich ataxia. J Neurol Sci. 2011;307:46–9. https://doi.org/10.1016/J.JNS.2011.05.023.
117. Pandolfo M. Friedreich ataxia: an (almost) 30-year history after gene discovery. Neurol Genet. 2025;11:e200236. https://doi.org/10.1212/NXG.0000000000200236.
118. González-Cabo P, Palau F. Mitochondrial pathophysiology in Friedreich's ataxia. J Neurochem. 2013;126:53–64. https://doi.org/10.1111/JNC.12303.
119. Clark E, Johnson J, Dong YN, Mercado-Ayon E, Warren N, Zhai M, et al. Role of frataxin protein deficiency and metabolic dysfunction in Friedreich ataxia, an autosomal recessive mitochondrial disease. Neuronal Signal. 2018;2:NS20180060. https://doi.org/10.1042/NS20180060.
120. Cook A, Giunti P. Friedreich's ataxia: clinical features, pathogenesis and management. Br Med Bull. 2017;124:19. https://doi.org/10.1093/BMB/LDX034.
121. Payne RM, Wagner GR. Cardiomyopathy in Friedreich ataxia: clinical findings and research. J Child Neurol. 2012;27:1179. https://doi.org/10.1177/0883073812448535.
122. Mazzara PG, Muggeo S, Luoni M, Massimino L, Zaghi M, Valverde PTT, et al. Frataxin gene editing rescues Friedreich's ataxia pathology in dorsal root ganglia organoid-derived sensory neurons. Nat Comm. 2020;11:1–18. https://doi.org/10.1038/s41467-020-17954-3.
123. Naghipour S, Corben LA, Hulme AJ, Dottori M, Delatycki MB, Lees JG, et al. Omaveloxolone for the treatment of Friedreich ataxia: efficacy, safety, and future perspectives. Mov Disord. 2025;40:226–30. https://doi.org/10.1002/MDS.30070.
124. Puranik N, Song M. Therapeutic role of heterocyclic compounds in neurodegenerative diseases: insights from Alzheimer's and Parkinson's diseases. Neurol Int. 2025;17:26. https://doi.org/10.3390/NEUROLINT17020026.
125. Jaiswal MK. Riluzole and edaravone: a tale of two amyotrophic lateral sclerosis drugs. Med Res Rev. 2019;39:733–48. https://doi.org/10.1002/MED.21528.
126. Videnovic A. Treatment of Huntington disease. Curr Treat Options Neurol. 2013;15:424. https://doi.org/10.1007/S11940-013-0219-8.
127. Cummings J, Osse AML, Cammann D, Powell J, Chen J. Anti-amyloid monoclonal antibodies for the treatment of Alzheimer's disease. BioDrugs. 2023;38:5. https://doi.org/10.1007/S40259-023-00633-2.
128. O'Hara DM, Kalia SK, Kalia LV. Emerging disease-modifying strategies targeting α-synuclein for the treatment of Parkinson's disease. Br J Pharmacol. 2018;175:3080. https://doi.org/10.1111/BPH.14345.
129. Chen Y, Sam R, Sharma P, Chen L, Do J, Sidransky E. Glucocerebrosidase as a therapeutic target for Parkinson's disease. Expert Opin Ther Targets. 2020;24:287. https://doi.org/10.1080/14728222.2020.1733970.
130. Ludolph A, Wiesenfarth M. Tofersen and other antisense oligonucleotides in ALS. Ther Adv Neurol Disord. 2025;18:17562864251313916. https://doi.org/10.1177/17562864251313915.
131. Rook ME, Southwell AL. Antisense oligonucleotide therapy: from design to the Huntington disease clinic. BioDrugs. 2022;36:105. https://doi.org/10.1007/S40259-022-00519-9.
132. Alkanli SS, Alkanli N, Ay A, Albeniz I. CRISPR/Cas9 mediated therapeutic approach in Huntington's disease. Mol Neurobiol. 2022;60:1486. https://doi.org/10.1007/S12035-022-03150-5.

133. Ashok A, Andrabi SS, Mansoor S, Kuang Y, Kwon BK, Labhasetwar V. Antioxidant therapy in oxidative stress-induced neurodegenerative diseases: role of nanoparticle-based drug delivery systems in clinical translation. Antioxidants. 2022;11:408. https://doi.org/10.3390/ANTIOX11020408.
134. El Ouaamari Y, Van den Bos J, Willekens B, Cools N, Wens I. Neurotrophic factors as regenerative therapy for neurodegenerative diseases: current status, challenges and future perspectives. Int J Mol Sci. 2023;24:3866. https://doi.org/10.3390/IJMS24043866.
135. Rahimi Darehbagh R, Seyedoshohadaei SA, Ramezani R, Rezaei N. Stem cell therapies for neurological disorders: current progress, challenges, and future perspectives. Eur J Med Res. 2024;29:386. https://doi.org/10.1186/S40001-024-01987-1.
136. Frawley L, Taylor NT, Sivills O, McPhillamy E, To TD, Wu Y, et al. Stem cell therapy for the treatment of amyotrophic lateral sclerosis: comparison of the efficacy of mesenchymal stem cells, neural stem cells, and induced pluripotent stem cells. Biomedicines. 2024;13:35. https://doi.org/10.3390/BIOMEDICINES13010035.
137. Nojadeh JN, Eryılmaz NSB, Ergüder Bİ. CRISPR/Cas9 genome editing for neurodegenerative diseases. EXCLI J. 2023;22:567. https://doi.org/10.17179/EXCLI2023-6155.
138. Wang JH, Gessler DJ, Zhan W, Gallagher TL, Gao G. Adeno-associated virus as a delivery vector for gene therapy of human diseases. Signal Transduct Target Ther. 2024;9:1–33. https://doi.org/10.1038/s41392-024-01780-w.
139. Strafella C, Caputo V, Galota MR, Zampatti S, Marella G, Mauriello S, et al. Application of precision medicine in neurodegenerative diseases. Front Neurol. 2018;9:701. https://doi.org/10.3389/FNEUR.2018.00701.
140. Mortada I, Farah R, Nabha S, Ojcius DM, Fares Y, Almawi WY, et al. Immunotherapies for neurodegenerative diseases. Front Neurol. 2021;12:654739. https://doi.org/10.3389/FNEUR.2021.654739.
141. Guo F, Liu X, Cai H, Le W. Autophagy in neurodegenerative diseases: pathogenesis and therapy. Brain Pathol. 2017;28:3. https://doi.org/10.1111/BPA.12545.
142. Sivandzade F, Cucullo L. Regenerative stem cell therapy for neurodegenerative diseases: an overview. Int J Mol Sci. 2021;22:2153. https://doi.org/10.3390/IJMS22042153.
143. Islam A, Vakitbilir N, Almeida N, França R. Advances in 3D bioprinting for neuroregeneration: a literature review of methods, bioinks, and applications. Micro. 2024;4:490–508. https://doi.org/10.3390/MICRO4030031.

Alzheimer's: Epidemiology, Pathophysiology, Diagnosis, and Treatments

2

Ayesha Rahman, Arpita Jaiswal, Pankaj Keshari, and Dhiraj Kumar Singh

2.1 Introduction

Alzheimer's disease (AD) is a leading cause of dementia, characterized by a progressive decline in cognitive functions. Dementia manifests as memory impairment and challenges with reasoning, language, and problem-solving abilities. Pathologically, AD is marked by the accumulation of insoluble amyloid-beta (Aβ) protein in extracellular plaques and microtubule-associated protein tau in intracellular NFTs, leading to synaptic dysfunction and neuronal loss. Recent studies in biomarker development, including neuroimaging and fluid markers, have enhanced early diagnosis. Though there is no cure, treatments are available to manage symptoms and slow disease progression [1]. AD advances through several stages, from preclinical to severe dementia, with worsening cognitive and functional impairments. This gradual decline affects memory, comprehension, language, attention, reasoning, and judgment. Early symptoms often include short-term memory loss, followed by executive dysfunction, language issues, and neuropsychiatric symptoms as the disease progresses [1]. In the context these are the stages of Dementia:

A. Rahman (✉) · D. K. Singh
Department of Biochemistry, All India Institute of Medical Sciences, New Delhi, India
e-mail: ayesharahman@aiims.edu

A. Jaiswal
Department of NMR, All India Institute of Medical Sciences, New Delhi, India

P. Keshari
School of Life Sciences, Jawaharlal Nehru University, New Delhi, India

N. Puranik (ed.), *Proteostasis*, https://doi.org/10.1007/978-981-96-6202-9_2

AD Dementia: This stage involves significant cognitive impairment that affects daily life, with clear symptoms present.

Prodromal AD: At this stage, individuals experience mild cognitive impairment with noticeable but not debilitating symptoms, allowing them to still manage daily activities.

Preclinical AD: No symptoms are present at this stage, but biomarkers indicate the presence of Alzheimer's pathology.

2.2 The Global Landscape of Alzheimer's Disease

In 2022, Alzheimer's disease was the seventh leading cause of death in the United States. Globally, the statistics are equally alarming, with over ten million new cases of dementia diagnosed each year. This translates to a new case every 3.2 s, underscoring the urgent need for advancements in research, treatment, and support for those affected. These figures not only reflect the growing prevalence of dementia but also the immense emotional and financial burden it places on families and healthcare systems worldwide [2, 3]. Approximately 416.4 million individuals globally are estimated to be across the Alzheimer's disease continuum, including 32.3 million with AD dementia, 69.0 million with prodromal AD, and 315.2 million in the preclinical stage of AD [4] as described (Fig. 2.1) in the form of pie chart.

2.2.1 The Worldwide Impact of Alzheimer's: Prevalence and Emerging Cases

The global impact of AD is staggering, affecting an estimated 416.4 million individuals across various stages of the disease, from preclinical to full-blown dementia. Approximately 32.3 million people currently live with dementia due to AD, experiencing significant cognitive decline that impacts their daily lives. Another 69.0 million individuals are in the prodromal stage, where early

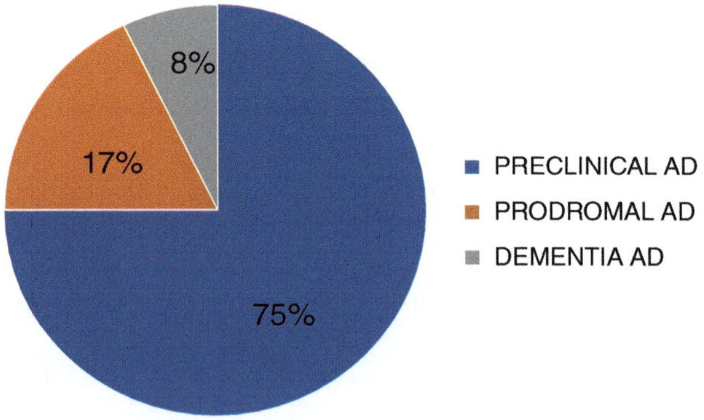

Fig. 2.1 Epidemiology of Alzheimer's

symptoms begin to manifest but do not yet meet the full criteria for dementia. The largest group, about 315.2 million people, are in the preclinical stage, showing no outward symptoms but having underlying pathological changes in the brain [4, 5]. In the United States alone, around 5.5 million individuals are affected, with global estimates reaching up to 24 million. As both developed and developing nations experience rapid aging, the incidence of Alzheimer's is projected to double every 20 years until 2040. This significant increase due to the aging population will pose a substantial and costly public health challenge in the future. In 2005, Alzheimer Disease International convened global experts to assess dementia prevalence and incidence across 14 WHO regions using recent data, finding that 24.2 million people had dementia, with 4.6 million new cases annually. The prevalence of Alzheimer's disease is notably higher in the United States compared to regions like Africa, Asia, and Europe [4].

In 2005, an estimated 24.2 million people worldwide were living with dementia, with Alzheimer's disease being the most prevalent form. Each year, approximately 4.6 million new cases of dementia are diagnosed. The prevalence of dementia varies by region, with North America and Western Europe showing the highest rates among people aged 60 and above, at 6.4% and 5.4%, respectively. Latin America follows with a prevalence rate of 4.9%, while China and its neighboring western-Pacific countries have a rate of 4.0% [6]. When considering the annual incidence rates of Alzheimer's disease, North America leads with 10.5 new cases per 1000 people, followed by Latin America at 9.2, Western Europe at 8.8, and China and its western-Pacific neighbors at 8.0. These rates increase significantly with age, particularly during the seventh and eighth decades of life [7]. Alzheimer's disease is the primary cause of dementia in Western countries, affecting about 5.5 million people in the United States alone. Globally, the number of people with Alzheimer's is estimated to be around 24 million. As populations in both developed and developing countries age rapidly, the incidence of Alzheimer's is expected to double every 20 years, reaching alarming levels by 2040. This trend poses a significant public health challenge, emphasizing the need for increased awareness, research, and resources to manage and mitigate the impact of this disease [8, 9].

2.2.2 Risk Factors

Age: The risk of developing Alzheimer's disease increases significantly after the age of 65. As people age, the likelihood of experiencing cognitive decline and dementia rises [10].

Genetics: Having a specific variant of the APOE gene, known as APOE ε4, can increase the risk of Alzheimer's. This genetic factor plays a crucial role in the development of the disease [11].

Sex: Women are more likely to develop Alzheimer's disease compared to men. This difference may be due to a combination of biological, genetic, and possibly lifestyle factors [12].

Education and Lifestyle: Individuals with lower levels of education and those who engage in unhealthy lifestyle habits, such as physical inactivity, smoking, and poor diet, are at a higher risk of developing Alzheimer's. Maintaining a healthy lifestyle and staying mentally active can help reduce this risk [13, 14].

2.2.3 Genetic Hypothesis

The genetic landscape of Alzheimer's disease reveals intriguing connections. Down syndrome, characterized by an extra chromosome 21, heightens the risk of AD due to increased production of amyloid precursor protein and amyloid-beta. Astonishingly, 40–80% of individuals with Down syndrome develop clinical AD by their fifth or sixth decade, with nearly all showing AD-related pathology [15].

Alzheimer's can also be inherited in an autosomal dominant manner through mutations in APP (chromosome 21), Presenilin1 (PSEN1, chromosome 14), and Presenilin2 (PSEN2, chromosome 1). APP mutations boost Aβ production, while PSEN1 and PSEN2 mutations disrupt Aβ processing, leading to early-onset AD in about 5–10% of cases, with PSEN1 being the most prevalent [16].

Apolipoprotein E (APOE) plays a significant role in Alzheimer's risk with its ε2, ε3, and ε4 alleles. The ε4 allele increases the risk of late-onset AD (LOAD) threefold for heterozygous carriers and 15-fold for homozygous carriers. This risk is even more pronounced for early-onset AD among ε4 carriers with a family history. Additionally, genetic variations in the SORT1 gene, essential for APP transport, have been linked to both familial and sporadic AD [15, 17].

2.2.4 Strategies to Prevent Alzheimer's Disease

To support brain health and mitigate cognitive decline, engage in regular aerobic exercises like walking or swimming, which enhance neuroplasticity. Cognitive stimulation through activities such as puzzles or learning new skills can help maintain synaptic connections. Adopting a diet rich in antioxidants and omega-3 fatty acids, like the Mediterranean or DASH diet, supports neuronal health. Prioritize sleep to facilitate memory consolidation and reduce amyloid-beta accumulation. Manage cardiovascular health by controlling blood pressure, glucose levels, and cholesterol to prevent vascular contributions to cognitive impairment. Maintain a healthy weight to reduce inflammation and oxidative stress. Avoid smoking and limit alcohol to prevent neurotoxicity. Social engagement and regular health check-ups are crucial for the early detection and management of cognitive decline. Stress management through mindfulness or relaxation techniques can reduce cortisol levels, which are detrimental to brain health (Fig. 2.2).

Fig. 2.2 Alzheimer's disease prevention

2.3 Alzheimer's Disease Pathophysiology

The pathophysiology of AD is characterized by a complex interplay of biological processes that lead to neurodegeneration. Initially, Alzheimer's disease required an autopsy to confirm by examining the brain after death and by observing symptoms during life [18]. However, this approach has been replaced by advancements that allow for earlier identification of the disease through specific biological markers and instrumental targets. This change is crucial for understanding the disease's progression and implementing timely intervention.

Pathophysiology of AD involves the accumulation of amyloid-beta plaques and tau neurofibrillary tangles, leading to the death of neurons and resulting in memory loss and cognitive decline. Various factors such as gene mutations, neuroinflammation, disruptions in the blood-brain barrier (BBB), mitochondrial dysfunction, oxidative stress, and imbalances in metal ions contribute to AD progression [19–21]. The disease progresses through several stages, starting with a long preclinical phase, followed by mild cognitive impairment (MCI), and ultimately reaching a stage where the individual cannot live independently. Despite all the research that has been done, we still do not completely understand how Alzheimer's disease works. This lack of understanding makes it challenging to develop effective treatments.

Key biomarkers for Alzheimer's disease include amyloid-beta 1-42, phosphorylated tau (p-tau), and total tau (t-tau) levels, along with indicators of neuroinflammation are essential for describing the pathophysiological processes and the complex cascade of events leading to Alzheimer's disease. These biomarkers provide a more accurate reflection of the underlying pathology compared to clinical symptoms alone [22]. The integration of these biomarkers into clinical practice could enhance early detection and intervention strategies.

2.3.1 Neuropathology of Alzheimer

The neuropathological diagnosis of AD is primarily based on the identification of specific intracellular and extracellular protein aggregates in the brain. The accumulation of NFTs and senile plaques contributes to neuronal death and cognitive decline [23] (Fig. 2.3).

Neurofibrillary tangles are a hallmark of Alzheimer's pathogenesis. These twisted fibres, found in areas like the entorhinal cortex, hippocampus, and neocortex, are indicative of disease progression and severity [24]. NFTs are primarily composed of hyperphosphorylated tau protein, forming thickened fibrils within neurons.

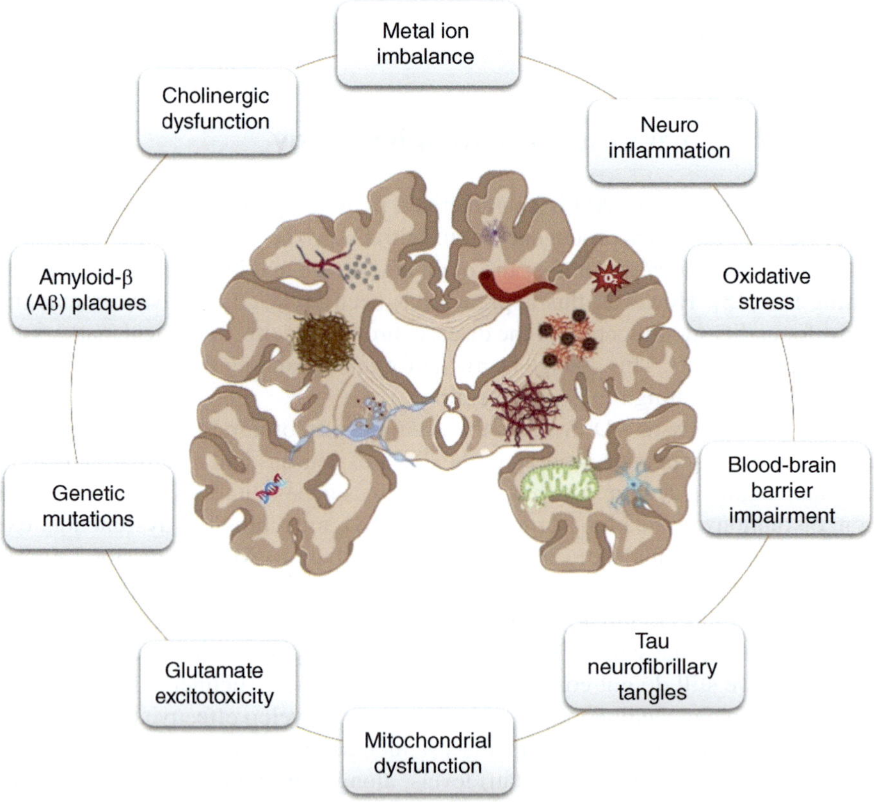

Fig. 2.3 Pathogenesis of Alzheimer's disease

They are considered a cardinal microscopic lesion within neurons in brain regions affected by Alzheimer's disease, which can be visualized using specialized staining techniques like silver impregnation or thioflavin S, and their identification is crucial in postmortem examinations of brain tissue [25]. The extent and distribution of these tangles correlate with the severity of dementia and the duration of Alzheimer's disease [26]. In a healthy brain, tau stabilizes microtubules, but in Alzheimer's disease, it becomes abnormally phosphorylated, leading to the formation of paired helical filaments that aggregate into tangles [27].

Senile plaques, also known as neuritic plaques, are one of the pathological features of Alzheimer's disease. These extracellular deposits predominantly consist of beta-amyloid (Aβ) peptides, which are generated through the proteolytic cleavage of amyloid precursor protein [28]. They accumulate in the brain, disrupting cell communication, triggering inflammation, and promoting neurodegeneration. Senile plaques are predominantly located in the cerebral cortex and hippocampus, areas crucial for memory and cognition [29]. Their presence is associated with synaptic loss and neuronal damage, contributing to cognitive decline [30]. While they are a diagnostic criterion for Alzheimer's, their presence in some elderly individuals without dementia raises questions about their exact role in neurodegeneration.

Vascular amyloid deposition refers to the accumulation of amyloid beta (Aβ) peptides in the walls of blood vessels, particularly in the brain, and is significantly associated with AD and cerebral amyloid angiopathy (CAA). These deposits primarily consist of Aβ peptides derived from the cleavage of APP, forming insoluble fibrils that accumulate in the leptomeningeal and cortical blood vessels, affecting small arteries and arterioles [31]. The pathological implications include vascular dysfunction, which can lead to reduced blood flow and ischemia; an increased risk of spontaneous haemorrhages, particularly lobar haemorrhages, due to vessel fragility; and cognitive decline, as this deposition contributes to the neurodegenerative processes in Alzheimer's disease [32]. Understanding vascular amyloid deposition is critical for elucidating the pathophysiology of AD and may guide therapeutic strategies aimed at mitigating its impact on vascular health and cognitive function.

Neuroinflammation is also a feature of AD, characterized by the activation of microglia and astrocytes. These cells release pro-inflammatory cytokines that causes neuronal damage. The presence of activated astrocytes in the early stages of the disease suggests that inflammatory processes may be involved in the progression of neurodegeneration.

Granulovacuolar degeneration (GVD) is a neuropathological feature that is difficult to identify using standard staining techniques. It is often visualized through specialized methods like immunohistochemistry, which highlights specific proteins associated with the granules [33]. GVD is considered a marker of neurodegenerative processes, and its presence is often correlated with cognitive decline [34]. The exact role of GVD in Alzheimer's disease pathogenesis is still being investigated, but it is believed to be related to impaired autophagy, oxidative stress, and protein homeostasis disruptions. These processes can lead to the accumulation of damaged cellular components [35], contributing to neuronal dysfunction and death, making it an important area of study and provide insights into disease progression and severity, aiding in the identification of therapeutic targets for neurodegeneration.

Hirano bodies are eosinophilic, rod-like inclusions, composed of actin and associated proteins [36], such as tropomyosin and vinculin, and found in the brains of individuals with AD and other neurodegenerative conditions, particularly in the CA1 region of the hippocampus [37]. The role of Hirano bodies in Alzheimer's disease remains unclear, as they are nonspecific lesions that can also be found in normal aging and other neurodegenerative diseases without a direct link to the severity of cognitive impairment. Current research is focused on elucidating their role and impact in neurodegenerative processes, making them an important area of study for understanding the pathology of these conditions.

Reactive astrocytes are specialized glial cells that, when exposed to inflammatory factors, go through a process known as reactive astrogliosis, which is also a characteristic sign of Alzheimer's disease at the neuropathological level [38]. Reactive astrocytes in Alzheimer's disease are linked to APP processing, leading to β-amyloid accumulation influence the progression of tau pathology. Understanding astrocyte reactivity could lead to new therapeutic strategies targeting both Aβ and tau pathologies in AD [39].

Research indicates that significant neuropathological changes can occur long before clinical symptoms manifest. This preclinical phase may last for years or even decades, during which individuals may exhibit no cognitive impairment despite the presence of pathological changes in the brain. The National Alzheimer's Coordinating Centre provides a standardized ABC scoring system for evaluating neuropathological changes, which is essential for accurate diagnosis [40].

Synaptic Dysfunction: Early stages of AD are characterized by synaptic dysfunction and loss, which can occur even in the absence of significant plaque or tangle formation. This synaptic loss is believed to contribute to the cognitive disabilities observed in patients and may precede the more overt pathological features [41, 42].

Vascular Changes: Emerging evidence suggests that vascular pathology, including alterations in the integrity of the blood-brain barrier and cerebral blood flow, might also play a role in the early stages of AD. These vascular changes can exacerbate neurodegenerative processes and may be linked to cognitive decline [43].

Macroscopic changes in AD denote observable modifications in brain structure throughout the progression of the disease, which can be detected via neuroimaging or postmortem examinations and are linked to clinical manifestations of cognitive decline and functional impairment. There is a notable reduction in brain tissue, especially within the cerebral cortex; as cortical tissue diminishes, the lateral and third ventricles of the brain experience enlargement, serving as a compensatory mechanism for the loss of adjacent brain tissue. The hippocampus, critical for memory formation, exhibits pronounced atrophy and shrinkage in AD, often

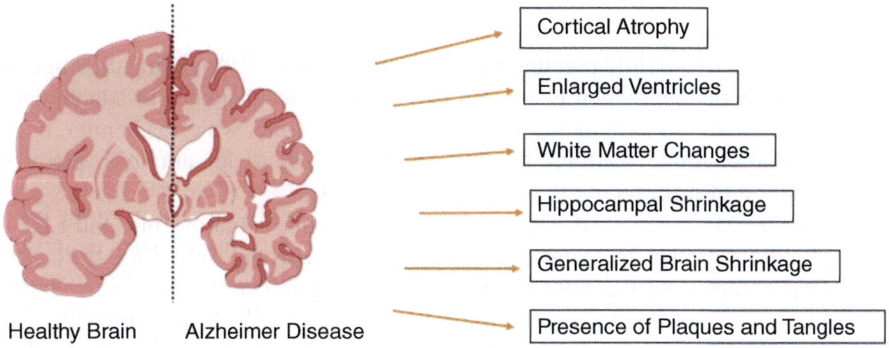

Fig. 2.4 Macroscopic changes in an Alzheimer's brain

representing one of the earliest indicators of the disorder and correlating with memory deficits [44]. Alterations in white matter may occur, including myelin loss and changes in the integrity of white matter tracts, potentially disrupting communication among various brain regions. In sum, there is a decline in both brain weight and volume [45, 46] (Fig. 2.4).

2.3.2 Familial Alzheimer's Disease (FAD)

FAD is primarily linked to mutations in the APP and presenilin genes (PSEN1 and PSEN2), which significantly influence disease progression through various mechanisms.

Previous research elucidates the molecular processes of FAD, emphasizing the effects of mutations associated with FAD on the processing of presenilin and the APP. Cryoelectron microscopy and molecular dynamics simulations revealed that Mutations in PSEN1 and PSEN2 alter γ-secretase activity, affecting APP cleavage and increasing the production of the toxic C99 fragment, stabilizing enzyme–substrate complexes, which is implicated in synaptic loss [47, 48]. In Research conducted in vivo on *Caenorhabditis elegans* has shown that the presence of FAD-mutant transgenes in conjunction with C99 and presenilin-1 expression is linked to synapse loss. These findings suggest that targeting processing pathways rather than solely Aβ may be a promising therapeutic strategy for AD.

Mutations in the amyloid β-peptide (Aβ) related to familial AD can lead to earlier onset and more aggressive forms of the disease. Specific mutations, such as E22G (Arctic) and D23N (Iowa), have been shown to accelerate the aggregation of Aβ compared to the wild type peptide [49]. While the exact mechanisms by which these mutations facilitate increased aggregation are still being studied, it is known that they alter the overall charge of the Aβ peptide, which could potentially influence its behavior in the brain.

2.3.3 Processing of Amyloid Precursor Protein

The processing of amyloid precursor protein (APP) involves sequential cleavage by α, β, and γ-secretases, which is key to the underlying mechanisms of Alzheimer's disease. APP can follow either a non-amyloidogenic route, which is benign, or an amyloidogenic route leading to the formation of toxic amyloid plaques. Dysregulation of this process is a hallmark of AD [50].

Several key interactors of APP, including neurexins, semaphorins, and LRP4, play significant roles in synapse formation and neuronal signaling [51]. The binding partners of APP can modulate its processing and subsequent neurodevelopmental processes, such as promoting neurite outgrowth, facilitating synaptogenesis [52], guiding axonal growth [53], and supporting neurogenesis The aberrant processing of APP results in the production of neurotoxic amyloid-beta peptides. However, underscoring the dual role of APP in both healthy brain function and disease.

Understanding the alterations in APP's normal functions in disease states is crucial for developing therapeutic strategies. It is essential to consider the interplay of genetic, environmental, and molecular factors that collectively influence its processing and the progression of the disease. Environmental factors, such as arsenic exposure, have been shown to enhance APP processing expression and promote β-secretase cleavage, triggering enhanced Aβ production and cognitive decline [54]. Genetic factors also play a significant role. For instance, the Purkinje cell protein 4 (PCP4), located on chromosome 21, has been found to enhance Aβ production by altering APP processing, thereby contributing to AD pathology [55].

2.3.4 Amyloid-Beta (Aβ) Toxicity

The neurotoxicity of beta-amyloid (Aβ) is linked to Alzheimer's disease pathology, primarily driven by the aggregation of Aβ peptides into toxic forms. Research indicates that Aβ causes neuronal degeneration and death, primarily through fibril formation, oxidative stress, and conformational changes in the peptide.

Mechanisms of Aβ Toxicity

Secondary Structure: The neurotoxic activity of Aβ is significantly influenced by its conformation. Aβ adopts a beta-sheet structure, which correlates with increased toxicity, while random coil conformations exhibit reduced toxicity [56, 57].

Oxidative Stress: Aβ increases levels of hydrogen peroxide and lipid peroxides, leading to oxidative damage in neurons. Antioxidants have shown protective effects against Aβ toxicity, highlighting oxidative stress role in its neurotoxic effects [58, 59].

Cellular Stress Responses: Aβ1–42 fibrils induce distinct unfolded protein responses (UPR) in neuronal cells, contributing to enhanced production of reactive oxygen species (ROS) and subsequent cell death [60].

Neurovascular Dysfunction: Aβ accumulation disrupts the blood-brain barrier integrity and impairs neurovascular coupling, resulting in reduced cerebral blood flow and neurovascular uncoupling [61].

Protofibril Toxicity: Aβ protofibrils are particularly toxic, activating the plasma contact system, which contributes to vascular inflammation and exacerbates AD pathology [62].

The findings suggest that targeting oxidative pathways and stabilizing Aβ conformation could be potential therapeutic strategies in mitigating Aβ toxicity and its associated neurodegeneration in Alzheimer's disease [63]. N-acetyl cysteine (NAC) has demonstrated potential in alleviating oxidative stress induced by amyloid-β oligomers [64]. Additionally, the chaperone protein HSP110 plays a role in modulating Aβ aggregation dynamics; its depletion has been shown to reduce aggregation and promote autophagic clearance [65], indicating a possible therapeutic approach for diminishing Aβ toxicity.

While the focus on Aβ toxicity is prevalent, some studies suggest that other factors, such as tau pathology, also play significant roles in AD progression, indicating a multifaceted approach is necessary for effective treatment strategies.

2.3.5 Tau Pathology

Amyloid precursor protein facilitates tau spread in Alzheimer's models. APP binds to tau fibrils and promotes their release and spreading between neurons [66]. Pathogenic APP mutations increase tau release, while the absence of APP reduces tau spread, underscoring its part in the disease's progression.

Tau pathology plays a crucial role in the progression of AD through mechanisms involving neuroinflammation, neuronal communication, and the spread of misfolded tau proteins, impacting cognitive decline. This condition is marked by the accumulation of hyperphosphorylated tau proteins, which form paired helical filaments (PHFs). These filaments are a defining feature of Alzheimer's disease [67, 68].

The accumulation of tau tangles is directly proportional to the severity of dementia, indicating a strong link between tau pathology and cognitive impairments [67]. In AD, tau proteins become hyperphosphorylated, which prevents them from stabilizing microtubules, resulting in the development of insoluble NFTs, leading to impaired axonal transport and neuronal degeneration [68, 69]. These tangles can propagate through neural circuits, affecting adjacent neurons and contributing to cognitive decline [70]. Misfolded tau can spread prion-like, causing further neurodegeneration and inflammation. Additionally, Tau pathology significantly impacts neuron-glia interactions, primarily through adenosine receptor 1 (A1R) upregulation, which activates astrocytes and contributes to synaptic dysfunction [71]. These activated astrocytes release inflammatory cytokines, worsening neuroinflammation, which can lead to cholinergic dysfunction and oxidative stress, further damaging neuronal health [72]. Tau pathology triggers a neuroinflammatory response, with microglia initially targeting early tau aggregates and astrocytes becoming increasingly involved as the disease progresses [73].

2.3.6 Sporadic Alzheimer's Disease

Sporadic Alzheimer disease is influenced by various modifiable risk factors, such as genetic, lifestyle, biological mechanisms, and social elements, which can significantly impact its prevalence and progression. Following are risk factors associated with sporadic Alzheimer disease, which are crucial for developing preventive strategies (Table 2.1).

While these factors provide insight into the risk landscape for sporadic AD, it is essential to consider that not all individuals with these risk factors will develop AD, indicating a complex interplay of genetics and environment.

Table 2.1 Risk factors associated with sporadic Alzheimer's disease

Exercise	Regular physical activity has been shown to increase the level of BDNF, which are essential for neuronal health.	[74]
Mental disorders	Individuals with mental disorders have a higher risk of developing cardiovascular diseases, which can worsen AD symptoms, underscoring the need for integrated care approaches.	[75]
Metabolic syndrome	Conditions like insulin resistance, dyslipidemia, and hypertension are linked to both metabolic syndrome and AD, indicating shared pathophysiological mechanisms.	[76]
Body mass index	High body mass index (BMI) and metabolic syndrome are associated with increased AD risk.	[77]
Hypertension	Elevated systolic blood pressure is a major risk factor for AD, contributing to vascular damage and cognitive impairment.	[78]
Dietary factors	Poor dietary habits, such as high carbohydrate intake and low antioxidant consumption (e.g., vitamin E), are linked to increased AD risk.	[79]
Amyloid-targeting therapies	New treatments like lecanemab show promise but may also pose risks, highlighting the importance of careful patient selection and monitoring.	[80]
Homocysteine levels	Elevated plasma homocysteine is a significant risk factor for vascular diseases and may also contribute to AD through vascular mechanisms.	[78]
Loneliness	Loneliness, particularly exacerbated by social isolation during the COVID-19 pandemic, is a modifiable risk factor for AD	[81]
Genetic factors	The CALM gene, which regulates the clearance of tau proteins, is linked to increased risk of AD, suggesting a genetic foundation for tau accumulation in the condition.	[82]
Social engagement	Social activities and cognitive engagement (e.g., reading, playing chess) are protective against AD, while factors like being widowed and physical illness increase risk.	[83]
Apolipoprotein E (APOE) gene	The presence of the APOE ε4 allele significantly increases the risk of developing sporadic Al	[84]
Cardiovascular health	Conditions such as diabetes and obesity are linked to a greater risk of AD. Poor cardiovascular health can reduce brain blood flow, contributing to cognitive decline.	[85]
Focal sepsis	Bacteria from focal infections, especially in the oral cavity, can cause systemic inflammation, potentially worsening neurodegenerative processes.	[86]

2.4 Comprehensive Diagnostic Approaches for Alzheimer's Disease

Traditionally, AD diagnosis relied on clinical criteria. However, research has shown that clinical symptoms may not always align with the underlying pathology. The current framework for diagnosing AD in clinical practice involves a comprehensive approach that includes cognitive assessments, ruling out other potential causes of dementia, neuroimaging, use of biomarkers, collaborative care models, and emerging computational methods [87, 88]. The diagnosis is based on a steady onset and evolving cognitive impairment, especially in memory. Early stages of AD typically do not show motor or sensory deficits [18]. Early detection is vital for effective intervention, as significant brain changes occur long before symptoms appear. The integration of technology and imaging is enhancing the accuracy and timeliness of diagnoses.

2.4.1 Clinical Assessment

Diagnosis often starts with interviews and cognitive assessments, focusing on memory and processing speed, which are sensitive indicators of early AD. Neuropsychological testing is commonly used to assess cognitive functions and identify impairments, facilitating the differentiation of AD from other dementia types. Tools like the Alzheimer's Disease Assessment Scale (ADAS) are crucial for enhancing diagnostic accuracy and are widely used in clinical trials [89]. Research has utilized audio recordings of interviews with Alzheimer's patients to analyze speech patterns, focusing on syntactic and lexical features of speech, which helps to distinguish between dementia and control groups [90]. Post-L-dopa administration hormonal assessments have been proposed as a potential diagnostic method.

Implementing collaborative care models in primary care settings has shown significant improvements in managing AD. Studies indicate that patients receiving collaborative care have better outcomes in behavioral and psychological symptoms compared to those receiving standard care [91].

While these advancements in diagnostic practices are promising, challenges remain in standardizing procedures and ensuring access to new biomarkers and technologies. This is particularly important for distinguishing AD from other dementias, as misdiagnosis can lead to inappropriate treatment strategies. The guidelines from the National Institute on Aging-Alzheimer's Association (NIA-AA) have significantly shaped clinical diagnoses, highlighting the importance of thorough assessments for both cognitively unimpaired and impaired individuals [92].

2.4.2 Neuroimaging Techniques

Advanced techniques for neuroimaging, such as MRI and PET scans, are crucial for visualizing brain associated with AD, including amyloid deposition and neurodegeneration [93]. These imaging biomarkers are instrumental in diagnosing AD and monitoring its progression. Specifically, amyloid PET and tau PET scans provide detailed visualization of pathological alterations in the brain, aiding in the differentiation of AD from other dementias [94]. MRI scans, particularly those focusing on the entorhinal cortex and hippocampus, are used to analyze brain structures, helping to identify atrophy and volumetric reductions indicative of AD [95]. These methods collectively enhance our ability to diagnose and track the disease progression. Various biomarkers are utilized for the detection of Alzheimer's disease.

2.4.3 Biomarkers

2.4.3.1 Cerebrospinal Fluid (CSF) Biomarkers

Cerebrospinal fluid proteomics biomarkers are gaining recognition for their potential in the early detection and diagnosis of AD. The analysis of CSF targets specific proteins such as amyloid-beta, total, and phosphorylated tau. Abnormal levels of these proteins are linked to Alzheimer's pathology and can aid in diagnosis. Measuring these biomarkers in CSF has become common practice in clinical settings. Elevated levels of p-tau and reduced levels of Aβ42 are particularly indicative of AD [96]. CSF proteomics not only helps in diagnosing Alzheimer's but also provides valuable prognostic information about disease progression. Monitoring changes in biomarker levels over time can help predict the onset of cognitive decline [97]. Recent studies are focused on enhancing the sensitivity and specificity of CSF proteomics assays. Overall, CSF proteomics biomarkers offer a great potential for the early detection and ongoing monitoring of AD, improving patient management and treatment strategies [98].

2.4.3.2 Genetic Biomarkers

Genetic biomarkers are essential for understanding the risk, progression, and potential treatment strategies for AD. Key genetic factors, such as the APOE ε4 allele, BDNF, KIBRA, and Spondin-1, are associated with a higher likelihood of developing AD.

Apolipoprotein E: The APOE gene is a major risk factor for AD. The ε4 allele is related to higher amyloid deposition and an increased risk of developing AD [99], as well as an earlier onset. Genotyping for the APOE allele can provide insights into an individual's risk profile, though it does not predict the disease's occurrence in any specific individual [100].

Brain-Derived Neurotrophic Factor (BDNF): Variants in the BDNF gene, especially the Val66Met polymorphism, are linked to cognitive decline and hippocampal atrophy in those at risk for AD. BDNF is crucial for neuronal survival and synaptic plasticity, which are vital for memory and understanding [101, 102].

KIBRA: This memory-associated protein is implicated in cognitive function. Genetic variations in the KIBRA gene may influence memory performance and the risk of AD. Abnormalities in KIBRA are associated with synaptic dysfunction and cognitive impairment [103].

Additional Genes: Other genes, such as CUB and Sushi multiple domains 1 (CSMD1) klotho (KL), and Spondin-1 (SPON1), have been identified as potential genetic biomarkers related to cognitive changes and AD risk [104]. Understanding these genetic biomarkers can help develop risk profiles for individuals, which is crucial for preclinical trials and assessing therapeutic efficacy. Overall, genetic biomarkers are vital for advancing our understanding of AD, providing insights into its etiology, progression, and potential therapeutic targets.

2.4.3.3 Blood and Urine Biomarkers

Blood and urine biomarkers are becoming promising, noninvasive, and accessible options for the early detection and diagnosis of AD. Plasma levels of amyloid-beta (Aβ42 and Aβ40) have been confirmed as reliable biomarkers for AD. Notably, increasing levels of Aβ42 over time are strongly linked to an elevated risk of developing the disease [105]. Research indicates that these levels change with age and are associated with the onset of cognitive impairment. Additionally, analyzing specific proteins and metabolites in urine can provide valuable insights into the disease. Overall, blood and urine biomarkers represent a significant breakthrough in Alzheimer's research, offering the potential to enhance early diagnosis and monitoring of the disease.

Amyloid Blood Biomarkers Researchers have discovered specific blood biomarkers that can help detect Alzheimer's disease. These biomarkers are associated with the amyloid plaques found in the brain, which are a hallmark of Alzheimer's pathogenesis [106].

2.4.3.4 Blood Lipid Biomarkers

Blood lipid biomarkers offer a noninvasive way to detect early-stage Alzheimer's disease, enabling timely intervention and management. Lipids are vital for various biological functions, including cell signaling and maintaining membrane integrity in the central nervous system. Research has focused on specific lipids like cholesterol, ceramides, and sphingomyelin for their potential links to Alzheimer's [107]. Changes in these lipid levels in the blood may indicate a higher risk of developing disease, its progression, and cognitive decline [108]. Overall, blood lipid biomarkers are a promising research area with the potential to improve our understanding of Alzheimer's mechanisms and enhance diagnostic capabilities.

Emerging biomarkers like glial fibrillary acidic protein (GFAP) are showing great promise for the detection of early stage AD. Higher levels of GFAP in the blood can act as reliable indicators, helping doctors intervene sooner and manage the disease more effectively [109].

Early evidence suggests that the biomarker neurogranin (Ng) could help differentiate AD from other neurodegenerative conditions. Elevated levels of Ng in CSF seem to be an early indicator of AD-related impairment. However, its effectiveness as a biomarker appears to decrease after the onset of dementia [110].

Current research has identified potential genetic variants linked to Alzheimer's disease using whole-genome sequencing. This approach can help predict disease progression. However, further validation of these genetic variants is necessary, and challenges such as data complexity, sample quality, and variant interpretation must be addressed. Additional research is essential to explore their implications in clinical settings [111].

2.4.4 Technological Advances

Recent research suggests that integrating structural MRI, clinical, and genetic data with machine learning techniques can achieve high classification accuracy for AD and MCI [112]. Advanced machine learning algorithms, such as Support Vector Machines (SVMs), have been utilized to predict the progression from MCI to AD by analyzing various features to enhance diagnostic precision. The use of a support vector classifier (SVC) in conjunction with neuropsychological testing represents a novel approach in the differential diagnosis of dementia. Additionally, innovative models like tensorized contrastive principal component analysis (T-cPCA) have been developed to predict the progression of Alzheimer's and identify its clinical subtypes, effectively reducing the confounding effects of natural aging [113]. Techniques such as the All-Pairs method, which compares temporal data points from patient records, and neural networks have also shown promise in predicting the progression of Alzheimer's disease.

2.4.4.1 Machine Learning Models

Recent studies have employed machine learning to classify AD neuropathology using clinical features and MRI data, achieving significant accuracy in distinguishing between affected and unaffected individuals. Clinically available features, such as the APOE genotype and cortical thickness measurements from MRI scans, which include Alzheimer's disease meta-regions of interest (meta-ROIs), can classify neuropathological status confirmed postmortem with 84.03% accuracy and 70.14% for in vivo amyloid positivity status [114]. These results suggest the potential utility of AD meta-ROIs in diagnosing AD in living persons.

2.4.4.2 Deep Learning Approaches

An Innovative deep learning-based approach called clustering-constrained-attention multiple instance learning (CLAM) was used to diagnose Alzheimer's disease and other tauopathies, including corticobasal degeneration, globular glial tauopathy, Pick's disease, and progressive supranuclear palsy. These techniques have been developed to analyze whole slide images of tau immunohistochemistry, enhancing diagnostic accuracy for AD and related tauopathies. The findings support the feasibility of using deep learning approaches for the classification of tauopathies in clinical settings. The study encourages further investigation into clinicopathological correlation studies, which could enhance understanding and treatment of these neurodegenerative disorders [115]. While these advancements improve diagnostic capabilities, the reliance on postmortem examination remains critical, as it provides definitive confirmation of AD pathology, underscoring the need for continued research in this area [116].

2.4.5 Diagnostic Procedure: Initial Screening and Evaluation Framework

A comprehensive diagnostic workup for AD significantly enhances patient outcomes and informs treatment plans by integrating advanced technologies and biomarkers. This includes standardized cognitive tests to evaluate memory and problem-solving skills, detailed patient history, and thorough physical and neurological exams [117, 118]. Monitoring behavioral and psychological symptoms, such as depression and anxiety, is essential. Advanced biomarker analysis, including blood tests for phosphorylated tau (p-tau217) and amyloid-β ratios, along with neuroimaging techniques like MRI and PET scans, help identify brain changes associated with AD [119]. New diagnostic criteria emphasize less invasive methods such as nasal samples for detecting tau proteins and amyloid-beta peptides. Additionally, occupational therapy plays a crucial role in maintaining patient independence and improving quality of life. This multifaceted approach ensures early and accurate diagnosis, leading to tailored treatment plans that address the specific needs of each patient.

2.4.6 Distinguishing Alzheimer's Disease from Other Cognitive Decline Conditions

Differentiating AD from other cognitive impairments is essential for precise diagnosis and effective treatment. Research underscores unique cognitive profiles and mechanisms that aid in this differentiation. AD patients typically show marked episodic memory deficits, especially in recognition tasks, which are less severe in Parkinson's disease dementia [120]. Conditions like diabetes and hypertension are significant risk factors for cognitive decline, complicating the clinical picture and potentially mimicking AD symptoms [121]. In contrast, cognitive impairment in major depressive disorder is linked with altered functional connectivity in the frontotemporal cortex, differing from AD's neurodegenerative patterns [122]. Cerebrovascular dysfunction is common in both AD and vascular cognitive impairment (VCI), but the underlying mechanisms differ: AD involves neurodegeneration, while VCI is linked to vascular health [123]. VCI often coexists with AD, complicating diagnosis, but primarily arises from vascular pathology rather than amyloid-β accumulation seen in AD [124]. Additionally, altered bile acid profiles have been linked to cognitive decline in AD, suggesting a potential biomarker for differentiation [125]. However, the overlap in symptoms and the presence of other health issues can make it difficult to distinguish between them, highlighting the need for thorough and comprehensive assessments to ensure an accurate diagnosis.

2.4.6.1 Clinical Considerations

Cognitive impairment is a common issue among elderly patients and requires thorough evaluation to differentiate it from normal aging and other medical conditions. Primary care clinicians must first rule out potentially reversible causes of cognitive impairment. Laboratory tests are essential for excluding other dementia causes, although they do not confirm AD. Assessing activities of daily living (ADLs) is crucial, as functional decline is often associated with cognitive disability in AD [126]. The Dementia Apraxia Test (DATE) has shown an 87% accuracy in distinguishing early AD from psychiatric conditions, particularly through limb praxis assessments [127]. Additionally, neuropsychological testing, when combined with machine learning algorithms, has achieved an overall accuracy of 82% in identifying AD among patients, increasing to 89% in early-stage cases. These methods underscore the importance of comprehensive assessments in ensuring accurate diagnosis [128] (Fig. 2.5).

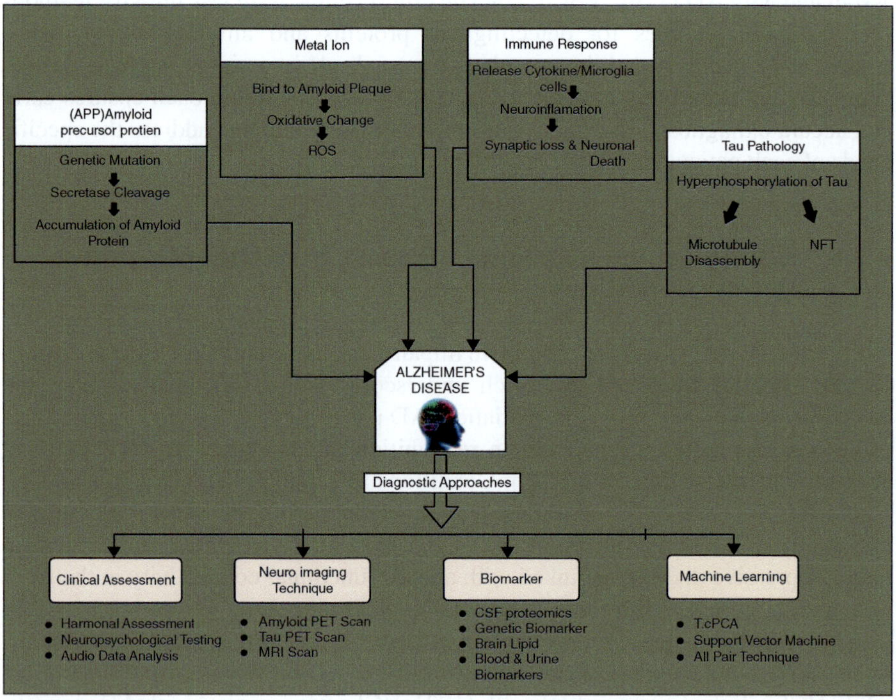

Fig. 2.5 Alzheimer's diagnosis and pathophysiology: from amyloid beta to tau pathology and neuronal loss

2.5 Treatment of Alzheimer's Disease

Currently, there are several pharmacological treatments available for Alzheimer's disease (AD), each targeting different aspects of the condition. FDA-approved medications like cholinesterase inhibitors (Donepezil, Rivastigmine, Galantamine) and memantine help manage symptoms by enhancing cholinergic function and regulating glutamate activity, respectively. Emerging treatments include gene therapy, which explores genetic modifications to treat or prevent AD, and theranostics, which combines therapy and diagnostics for personalized treatment. Immunotherapy leverages the immune system to target AD-related proteins, while peptidomimetics involve designing peptide-like molecules to interfere with disease processes [4]. Metal chelators aim to reduce metal ions that may contribute to AD pathology. Additionally, lifestyle interventions such as probiotics, prebiotics, and exercise are being investigated for their potential to support brain health. These diverse approaches reflect the ongoing efforts to find more effective treatments for AD, offering hope for better management and improved quality of life for patients [4]. Non-pharmacological treatments such as cognitive stimulation therapy, physical exercise, music therapy, art therapy, and reminiscence therapy are currently being utilized to improve cognitive function and enhance the quality of life for Alzheimer's disease (AD) patients. These therapies offer unique benefits, including engaging patients in activities to improve memory and problem-solving skills, boosting brain health through physical activity, evoking memories and reducing agitation with music, encouraging creative expression to reduce anxiety, and discussing past experiences to improve mood and cognitive function [129]. Preliminary data suggest improvements in outcomes like the Mini-Mental State Examination (MMSE), activities of daily living (ADL), and the Alzheimer's Disease Assessment Scale-cognitive section (ADAS-cog) [89, 91] (Fig. 2.6).

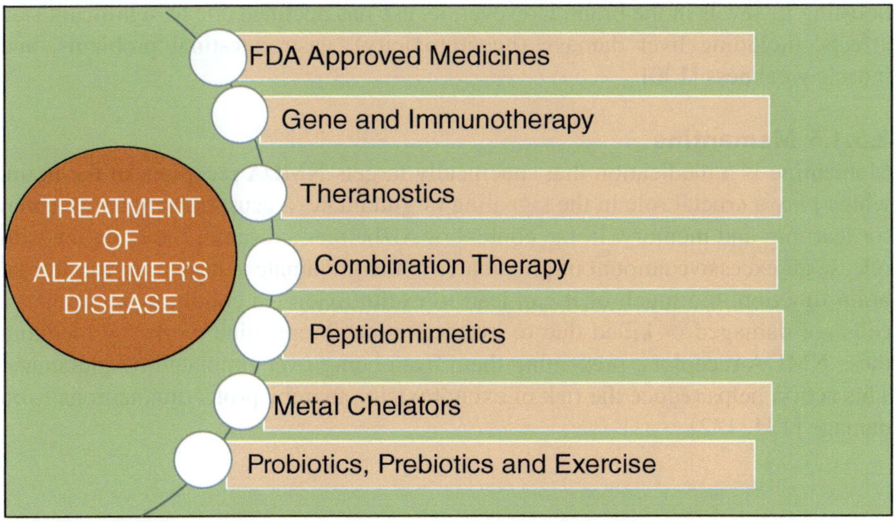

Fig. 2.6 Managing Alzheimer's: integrating approved medications, probiotics, and regular exercise for comprehensive care

2.5.1 FDA-Approved Medicines for Alzheimer's Disease

Cholinesterase inhibitors play a vital role in the treatment of Alzheimer's disease. These medications work by inhibiting the enzyme acetylcholinesterase, which breaks down acetylcholine, a neurotransmitter crucial for memory and learning. In individuals with Alzheimer's, acetylcholine levels are often reduced due to the degeneration of neurons. By blocking acetylcholinesterase, AChEIs increase the availability of acetylcholine in the brain, thereby enhancing communication between nerve cells. This improvement can lead to better cognitive functions such as memory, attention, and mental clarity. Although these drugs do not cure Alzheimer's, they can provide symptomatic relief and potentially slow cognitive decline. Here are some common AChEIs:

2.5.1.1 Donepezil

This medication is frequently prescribed for mild to moderate Alzheimer's disease. It can help improve cognitive function and slow the progression of symptoms.

2.5.1.2 Rivastigmine

Available in both oral and patch forms, rivastigmine is used for mild to moderate Alzheimer's and also for dementia associated with Parkinson's disease [130].

2.5.1.3 Galantamine

In addition to inhibiting acetylcholinesterase, galantamine also modulates nicotinic receptors, which can further enhance cognitive function [130].

2.5.1.4 Tacrine

Tacrine was one of the first cholinesterase inhibitors approved for treating Alzheimer's disease. It helps by preventing the breakdown of acetylcholine, thereby boosting its levels in the brain. However, its use has declined due to significant side effects, including liver damage (hepatotoxicity), gastrointestinal problems, and muscle weakness [130].

2.5.1.5 Memantine

Memantine is a medication that specifically targets NMDA receptors in the brain, which play a crucial role in the signaling of glutamate, a neurotransmitter essential for learning and memory. In the context of Alzheimer's disease, the brain tends to release an excessive amount of glutamate. While glutamate is necessary for normal brain function, too much of it can lead to excitotoxicity, a condition where nerve cells are damaged or killed due to overactivation. Memantine works by blocking these NMDA receptors, preventing them from being overstimulated by glutamate. This action helps reduce the risk of excitotoxicity, thereby protecting neurons from damage [131, 132].

2.5.2 Gene Therapy

Gene therapy is an innovative approach that aims to tackle diseases at their genetic roots by repairing faulty DNA, genes, or proteins. Recent advancements in the use of recombinant adeno-associated viruses (rAAVs) have opened up new possibilities for treating AD. In a study involving patients with early-stage Alzheimer's disease, researchers tested the effects of nerve growth factor (NGF) gene therapy. The findings were encouraging, revealing that the therapy promoted neuron growth and axonal sprouting. Remarkably, these positive effects persisted for a decade [133]. Another study utilized a modified virus to deliver the PGC1-alpha gene to mice, which led to improved memory and a reduction in amyloid plaques [131]. Furthermore, scientists discovered a technique to convert the APOE4 gene, which increases the risk of Alzheimer's, into the protective APOE3 gene, potentially lowering the risk of developing the disease [134].

2.5.3 Theranostics in Alzheimer's Disease

Theranostics is an innovative technology that merges therapeutic and diagnostic techniques to revolutionize the treatment of AD. This approach employs gold nanorods equipped with Aβ inhibitors to detect and destroy amyloid-beta aggregates using near-infrared radiation, enabling continuous monitoring of disease progression. Phenothiazine derivatives act as fluorescent probes, identifying and preventing Aβ plaque aggregation in mice, while charged molecules like DBA-SLOH bind to Aβ aggregates, inhibiting their formation. A specialized blood-brain barrier shuttle system delivers antibodies for Aβ immunotherapy and facilitates PET diagnosis, ensuring targeted treatment and precise monitoring. Nanoprobes with chelators detect excess copper ions, which contribute to amyloid plaque formation, and inhibit their accumulation. Iron oxide nanoparticles are used to detect amyloid plaques through MRI and release drugs to prevent oxidative damage. Quantum dots (QDs) offer simultaneous imaging, diagnosis, and therapy, providing a multifaceted approach to AD management. Graphene QDs have shown promise in reducing amyloid plaques and inflammation in mice, while low-level laser therapy, despite its limited brain penetration, helps reduce inflammation. These advanced strategies collectively represent a significant leap forward in the diagnosis and treatment of Alzheimer's disease [99, 100].

2.5.4 Immunotherapy

Amyloid-beta immunotherapy employs several mechanisms to combat Alzheimer's disease. Antibodies can neutralize and dissolve Aβ plaques, stimulate microglia to phagocytose these plaques, and prevent amyloid seeds from propagating. Additionally, some methods focus on unbundling plaques directly and facilitating the removal of Aβ from the brain to the plasma [135, 136]. Immunotherapy strategies for Alzheimer's disease are broadly categorized into active and passive immunization. Active immunization involves the administration of Aβ-42 or Aβ fragments to stimulate the body's immune response. This approach aims to generate an immune response that targets and clears Aβ plaques. On the other hand, passive immunization involves the use of monoclonal antibodies (mAb) such as bapineuzumab and solanezumab. These antibodies are designed to bind to Aβ and promote its clearance. However, these treatments can sometimes cause side effects, including amyloid-related imaging abnormalities (ARIA) [136, 137].

Newer engineered antibodies, such as BAN2401 and Aducanumab, have been developed to target specific forms of Aβ with fewer side effects. Aducanumab, in particular, is a monoclonal antibody that targets beta-amyloid plaques in the brain. It was conditionally approved by the FDA for its potential to modify the disease rather than just alleviate symptoms [136, 137].

2.5.5 Combination Therapy

In the treatment of AD, a combination of cholinesterase inhibitors and memantine is often employed to address symptoms more effectively. This combination is particularly useful for patients with moderate to severe AD as it targets different mechanisms involved in the disease. Donepezil, a ChEI, and memantine, an NMDA receptor antagonist, are commonly used together to enhance cognitive function and slow disease progression [138].

For treatments targeting tau pathology, a combination of anti-tau agents is used. These drugs work by inhibiting tau phosphorylation and reducing tau aggregation, which are key processes in the formation of neurofibrillary tangles, a hallmark of AD. However, Phase III trials with valproate in moderate AD showed no significant slowing of cognitive decline or delay in the onset of psychosis and agitation. Moreover, patients experienced toxic effects and greater brain volume loss, highlighting the need for safer and more effective treatments. Neurotrophic agents are also being explored for their potential in treating mild cognitive impairment and AD. Reduced neurotrophic signaling, such as that involving insulin-like growth factor I (IGF-I) and brain-derived neurotrophic factor, leads to neuron degeneration and cognitive decline. Drugs that boost these neurotrophic pathways are promising for AD treatment[5]. Another promising approach involves the use of PBT2, a copper/zinc ionophore, in combination with ChEIs. Chronic exposure to metals has been linked to AD, and PBT2 acts as a metal-chelating agent, reducing the toxicity of protein-metal complexes like those formed by Aβ oligomers with zinc and copper [138]. Early phase II clinical trials have shown that combining PBT2 with ChEIs such as galantamine, donepezil, or rivastigmine enhances executive functions in a dose-dependent manner and significantly lowers cerebrospinal fluid Aβ42 levels compared to ChEIs alone. While these findings are promising, further research is needed to confirm their efficacy and safety. Additionally, a combination of noradrenaline reuptake inhibitors and ChEIs or NMDA receptor antagonists is being studied for its potential benefits in AD. Intense noradrenergic dysfunction has been observed in AD, and atomoxetine, a noradrenaline reuptake inhibitor, has been studied in combination with ChEIs (galantamine, donepezil, or rivastigmine) or memantine in mild to moderately severe AD [130, 131].

2.5.6 Peptidomimetics

Peptidomimetics are innovative protein analogs designed to mimic the structure and function of natural proteins, often achieving similar or even superior biological effects. These peptide inhibitors can be derived from amyloid-beta sequences and are engineered to target phosphorylated tau proteins and human β-secretase, which are key players in the pathology of Alzheimer's disease.

Peptidomimetics can be categorized into two main types: those composed of natural amino acids and those incorporating modified amino acids. For tau and Aβ aggregates, the primary recognition sites are the central hydrophobic cores. For tau, these cores are represented by the sequences VQIVYK or VQIINK, while for Aβ, the core sequence is KLVFF. Peptidomimetics bind to these hydrophobic regions, effectively preventing further aggregation and, in some cases, dissolving pre-formed aggregates [139, 140]. When targeting Aβ aggregates, peptidomimetics can focus on either the central hydrophobic core or the C-terminus of the peptide. By using natural or modified amino acids, these peptidomimetics can be tailored to enhance their binding affinity and specificity, thereby improving their therapeutic potential. This approach not only helps in preventing the formation of new aggregates but also aids in breaking down existing ones, offering a dual mechanism of action against AD [140, 141].

2.5.7 Metal Chelators

The interaction of metals with amyloid-beta peptides leads to the generation of ROS and disrupts metal ion balance, significantly contributing to the pathogenesis of AD. Metal chelators, which are small, low-charge molecules capable of crossing the blood-brain barrier, play a crucial role in breaking the Aβ-metal complex, thereby restoring metal ion balance and reducing neurotoxicity. In AD, there is a notable increase in the levels of copper, iron, and zinc in the brain. Chelators such as Deferoxamine (DFO) and Deferiprone specifically target iron, helping to treat memory impairment and prevent the production of ROS. Copper-targeting chelators include PDTC, pyridine derivatives, and BTSC. Clioquinol, a bidentate ligand, has shown effectiveness in treating neurodegenerative diseases by balancing metal ions and dissolving Aβ-Cu2+/Zn2+ aggregates. Additionally, hybrid multifunctional modulators (HMMs) like TGR86 inhibit Aβ aggregation and prevent mitochondrial damage by binding to specific regions of Aβ, thereby increasing cell viability. Given that ROS is a major factor in AD, metal chelators represent a promising future therapy [142, 143].

2.5.8 Probiotics, Prebiotics, and Exercise

Changes in gut microbiota can contribute to AD by affecting neurotransmitter and immune responses through the brain-gut axis. Studies on both animals and humans have explored the impact of prebiotics and probiotics on AD. Probiotics, supported by prebiotics, promote the growth of beneficial gut microbes like *Lactobacillus* and *Bifidobacterium.* Another study combined probiotics, omega-3 fatty acids, vitamins, and exercise, showing that this regimen could slow AD progression and ease symptoms. Human studies also showed that probiotics, including various *Lactobacillus* and *Bifidobacterium* strains, along with vitamin D, activated the immune system. Probiotic supplements should be mild to avoid disrupting gut microbiota, which can trigger the immune system and cause neuroinflammation [144–147].

2.5.9 Voice as a Biomarker Using AI

Alzheimer's disease profoundly impacts language abilities, making communication increasingly difficult as the disease progresses. In the early stages, individuals often experience difficulty in finding the right words, leading to frequent pauses and substitutions. As the disease advances to the intermediate stage, their vocabulary becomes more restricted, and their ability to engage in everyday conversations diminishes. In the advanced stages, speech is often reduced to a few words or phrases, and responses become very limited. Research highlights the importance of analyzing speech and emotional cues for the early diagnosis of AD. For instance, a study demonstrated that speech analysis could effectively distinguish between different types of dementia [148]. This study involved collecting speech samples from patients and control groups through semi-structured interviews where participants described a picture. These recordings were then analyzed using a machine learning algorithm, specifically a multilayer perceptron, which achieved an impressive 88% accuracy in differentiating between dementia types. These recordings were then analyzed using a machine learning algorithm, specifically a multilayer perceptron, which achieved an impressive 88% accuracy in differentiating between dementia types [149, 150] (Fig. 2.7).

2.5.10 Non-pharmacological Treatment

Non-pharmacological treatments are increasingly being recognized for their potential to improve cognitive function and enhance the quality of life for patients with AD. These treatments encompass a variety of approaches, each offering unique benefits. Cognitive stimulation therapy, for instance, engages patients in activities designed to improve memory and problem-solving skills. Physical exercise is another crucial intervention, as it not only boosts overall brain health but also helps

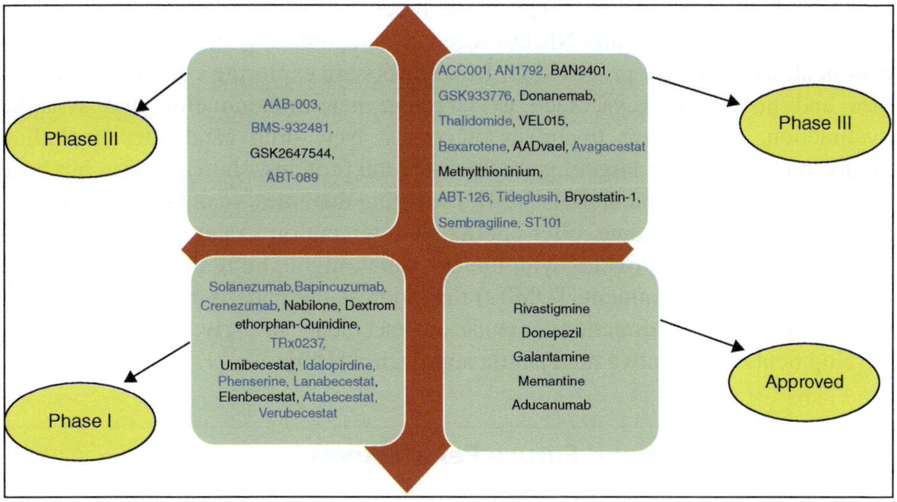

Fig. 2.7 Drugs tested at different clinical trial phases. Drugs marked in black signify those still in development, while those marked in blue are being phased out

[129] in maintaining physical fitness, which is vital for daily functioning. Music therapy has shown promise in evoking memories and reducing agitation [151]. By listening to or participating in music, patients can experience emotional and cognitive benefits. Art therapy encourages creative expression, which can significantly reduce anxiety and provide a sense of accomplishment. Reminiscence therapy involves discussing past experiences, which can improve mood and cognitive function by helping patients reconnect with their personal history. Preliminary data from various studies suggest that these therapies can lead to improvements in several key outcomes, such as the Mini-Mental State Examination (MMSE), activities of daily living (ADL), and the Alzheimer's Disease Assessment Scale-cognitive section (ADAS-cog). Similarly, cognitive stimulation programs have been shown to maintain cognitive function and improve self-reported quality of life in patients with mild to moderate AD. Non-pharmacological interventions focusing on cognitive aspects offer promising avenues for improving gait and reducing fall risk. Cognitive training exercises aimed at enhancing attention, executive function, and memory can lead to better motor control and stability. For instance, interventions that incorporate dual-task training, where patients practice walking while engaging in cognitive tasks, have shown potential in improving gait performance and reducing fall incidents [152]. Cognitive processes such as attention, executive function, and working memory play a vital role in maintaining stable and coordinated gait. When these cognitive functions are compromised, as seen in Alzheimer's disease, gait disturbances and instability become more pronounced. Research indicates that dual-task gait assessments, where individuals walk while performing a cognitive task, can reveal significant insights into the relationship between cognitive decline and gait performance [153].

Traditional treatments for AD primarily focus on symptom management through cholinesterase inhibitors and NMDA receptor antagonists, which have limited success in slowing disease progression. Researchers are exploring combination therapies and novel pathways, such as targeting neuroinflammation and synaptic dysfunction, to improve treatment outcomes. New drug candidates, including monoclonal antibodies targeting amyloid and tau proteins, show promise but have faced challenges in clinical trials [154]. Emerging therapies, such as stem cell treatments and herbal medicine, offer alternative intervention strategies. Current pharmacotherapy mainly provides symptomatic relief, highlighting the urgent need for disease-modifying treatments (DMTs) that address the underlying mechanisms of AD. Future research directions include advancements in early detection and the development of innovative therapeutic approaches [155].

2.6 Conclusion and Future Perspectives

In conclusion, Alzheimer's disease remains one of the most challenging neurodegenerative disorders, characterized by complex pathophysiology and multifaceted diagnostic processes. Despite significant advancements in understanding the disease mechanisms, including the roles of amyloid plaques, neurofibrillary tangles, neuronal degeneration, and neuroinflammation, effective therapies to halt or reverse disease progression are still elusive. The amyloid hypothesis, while dominant, has not yet yielded successful treatments, highlighting the need for continued research into alternative pathways and therapeutic targets.

The integration of advanced imaging techniques, such as MRI and PET scans, along with emerging biomarkers, has improved our ability to diagnose AD earlier and more accurately. However, distinguishing AD from other dementias remains a critical challenge, necessitating comprehensive and standardized diagnostic approaches. Additionally, lifestyle interventions, such as physical exercise, cognitive training, and dietary modifications, are being investigated for their potential to support brain health and mitigate the progression of AD. As we deepen our understanding of both genetic and environmental risk factors, as well as the differences and similarities between early-onset and late-onset forms of the disease, we move closer to developing effective interventions. The ongoing research and collaborative efforts in the scientific community offer hope that 1 day we will find a way to significantly alter the course of Alzheimer's disease, improving the quality of life for millions affected by this devastating condition.

References

1. Dubois B, Von Arnim CAF, Burnie N, Bozeat S, Cummings J. Biomarkers in Alzheimer's disease: role in early and differential diagnosis and recognition of atypical variants. Alzheimers Res Ther. 2023;15(1):175.
2. Ahmad FB, Cisewski JA, Xu J, Anderson RN. Provisional mortality data — United States. MMWR Morb Mortal Wkly Rep. 2022;71:597.

3. Ahmad FB, Cisewski JA, Anderson RN. Mortality in the United States — provisional data, 2023. MMWR Morb Mortal Wkly Rep. 2024;73(31):677–81.
4. Gustavsson A, Norton N, Fast T, Frölich L, Georges J, Holzapfel D, et al. Global estimates on the number of persons across the Alzheimer's disease continuum. Alzheimers Dement. 2023;19(2):658–70.
5. Aisen PS, Cummings J, Jack CR, Morris JC, Sperling R, Frölich L, et al. On the path to 2025: understanding the Alzheimer's disease continuum. Alzheimers Res Ther. 2017;9(1):60.
6. Tahami Monfared AA, Byrnes MJ, White LA, Zhang Q. Alzheimer's disease: epidemiology and clinical progression. Neurol Ther. 2022;11(2):553–69.
7. Li X, Feng X, Sun X, Hou N, Han F, Liu Y. Global, regional, and national burden of Alzheimer's disease and other dementias, 1990–2019. Front Aging Neurosci. 2022;14:937486.
8. Global status report on the public health response to dementia. Geneva: World Health Organization; 2021.
9. Nichols E, Steinmetz JD, Vollset SE, Fukutaki K, Chalek J, Abd-Allah F, et al. Estimation of the global prevalence of dementia in 2019 and forecasted prevalence in 2050: an analysis for the Global Burden of Disease Study 2019. Lancet Public Health. 2022;7(2):e105–25.
10. Ranson JM, Rittman T, Hayat S, Brayne C, Jessen F, Blennow K, et al. Modifiable risk factors for dementia and dementia risk profiling. A user manual for Brain Health Services—part 2 of 6. Alzheimers Res Ther. 2021;13(1):169.
11. Jones A, Ali MU, Kenny M, Mayhew A, Mokashi V, He H, et al. Potentially modifiable risk factors for dementia and mild cognitive impairment: an umbrella review and meta-analysis. Dement Geriatr Cogn Disord. 2024;53(2):91–106.
12. Zhang XX, Tian Y, Wang ZT, Ma YH, Tan L, Yu JT. The epidemiology of Alzheimer's disease modifiable risk factors and prevention. J Prev Alzheimers Dis. 2021;8:313–21.
13. Crous-Bou M, Minguillón C, Gramunt N, Molinuevo JL. Alzheimer's disease prevention: from risk factors to early intervention. Alzheimers Res Ther. 2017;9(1):71.
14. Silva MVF, Loures CDMG, Alves LCV, De Souza LC, Borges KBG, Carvalho MDG. Alzheimer's disease: risk factors and potentially protective measures. J Biomed Sci. 2019;26(1):33.
15. Nicolas G, Acuña-Hidalgo R, Keogh MJ, Quenez O, Steehouwer M, Lelieveld S, et al. Somatic variants in autosomal dominant genes are a rare cause of sporadic Alzheimer's disease. Alzheimers Dement. 2018;14(12):1632–9.
16. Hoogmartens J, Cacace R, Van Broeckhoven C. Insight into the genetic etiology of Alzheimer's disease: a comprehensive review of the role of rare variants. Alzheimers Dement Diagn Assess Dis Monit. 2021;13(1):e12155.
17. Salehi A, Ashford JW, Mufson EJ. The link between Alzheimer's disease and down syndrome. A historical perspective. Curr Alzheimer Res. 2016;13(1):2–6.
18. McKhann G, Drachman D, Folstein M, Katzman R, Price D, Stadlan EM. Clinical diagnosis of Alzheimer's disease: report of the NINCDS-ADRDA work group under the auspices of Department of Health and Human Services Task Force on Alzheimer's disease. Neurology. 1984;34(7):939–44.
19. Blair LJ, Frauen HD, Zhang B, Nordhues BA, Bijan S, Lin YC, et al. Tau depletion prevents progressive blood-brain barrier damage in a mouse model of tauopathy. Acta Neuropathol Commun. 2015;3:8.
20. Afsar A, Chacon Castro MDC, Soladogun AS, Zhang L. Recent development in the understanding of molecular and cellular mechanisms underlying the Etiopathogenesis of Alzheimer's disease. Int J Mol Sci. 2023;24(8):7258.
21. Lei P, Ayton S, Bush AI. The essential elements of Alzheimer's disease. J Biol Chem [Internet]. 2021 [cited 2024 Sep 10];296. Available from: https://pubmed.ncbi.nlm.nih.gov/33219130/.
22. Strimbu K, Tavel JA. What are biomarkers? Curr Opin HIV AIDS. 2010;5(6):463–6.
23. Shaikh UA, Shinde SP, Shaikh UA, Shinde SP. Alzheimer disease. World J Biol Pharm Health Sci [Internet]. 2024 [cited 2024 Sep 8];18(2). Available from: https://wjbphs.com/content/alzheimer-disease.

24. Alzheimer A, Stelzmann RA, Schnitzlein HN, Murtagh FR. An English translation of Alzheimer's 1907 paper, "Uber eine eigenartige Erkankung der Hirnrinde". Clin Anat. 1995;8(6):429–31.
25. Yen SH, Dickson DW, Crowe A, Butler M, Shelanski ML. Alzheimer's neurofibrillary tangles contain unique epitopes and epitopes in common with the heat-stable microtubule associated proteins tau and MAP2. Am J Pathol. 1987;126(1):81–91.
26. Bierer LM, Hof PR, Purohit DP, Carlin L, Schmeidler J, Davis KL, et al. Neocortical neurofibrillary tangles correlate with dementia severity in Alzheimer's disease. Arch Neurol. 1995;52(1):81–8.
27. Arriagada PV, Growdon JH, Hedley-Whyte ET, Hyman BT. Neurofibrillary tangles but not senile plaques parallel duration and severity of Alzheimer's disease. Neurology. 1992;42(3 Pt 1):631–9.
28. Masters CL, Multhaup G, Simms G, Pottgiesser J, Martins RN, Beyreuther K. Neuronal origin of a cerebral amyloid: neurofibrillary tangles of Alzheimer's disease contain the same protein as the amyloid of plaque cores and blood vessels. EMBO J. 1985;4(11):2757–63.
29. Wolf DS, Gearing M, Snowdon DA, Mori H, Markesbery WR, Mirra SS. Progression of regional neuropathology in Alzheimer disease and normal elderly: findings from the Nun study. Alzheimer Dis Assoc Disord. 1999;13(4):226–31.
30. Morris JC, Storandt M, McKeel DW, Rubin EH, Price JL, Grant EA, et al. Cerebral amyloid deposition and diffuse plaques in "normal" aging: evidence for presymptomatic and very mild Alzheimer's disease. Neurology. 1996;46(3):707–19.
31. Liu W, Wong A, Law ACK, Mok VCT. Cerebrovascular disease, amyloid plaques, and dementia. Stroke. 2015;46(5):1402–7.
32. Yang J, Wong A, Wang Z, Liu W, Au L, Xiong Y, et al. Risk factors for incident dementia after stroke and transient ischemic attack. Alzheimers Dement. 2015;11(1):16–23.
33. Kahn J, Anderton BH, Probst A, Ulrich J, Esiri MM. Immunohistological study of granulovacuolar degeneration using monoclonal antibodies to neurofilaments. J Neurol Neurosurg Psychiatry. 1985;48(9):924–6.
34. Ball MJ, Lo P. Granulovacuolar degeneration in the ageing brain and in dementia. J Neuropathol Exp Neurol. 1977;36(3):474–87.
35. Iqbal K, Liu F, Gong CX, Alonso A, Grundke-Iqbal I. Mechanisms of tau-induced neurodegeneration. Acta Neuropathol (Berl). 2009;118:53–69.
36. Galloway PG, Perry G, Gambetti P. Hirano body filaments contain actin and actin-associated proteins. J Neuropathol Exp Neurol. 1987;46(2):185–99.
37. Hirano A. Hirano bodies and related neuronal inclusions. Neuropathol Appl Neurobiol. 1994;20(1):3–11.
38. Jauregui GV. Translational neuroscience and aging program (TAP). Role of amyloid precursor protein in astrocytes in Alzheimer's disease. Alzheimers Dement [Internet]. 2023 [cited 2024 Sep 14];19(S1):e063218. Available from: https://onlinelibrary.wiley.com/doi/abs/10.1002/alz.063218.
39. Bellaver B, Povala G, Ferreira PCL, Ferrari-Souza JP, Leffa DT, Lussier FZ, et al. Astrocyte reactivity influences amyloid-β effects on tau pathology in preclinical Alzheimer's disease. Nat Med. 2023;29(7):1775–81.
40. Doher N, Davoudi V, Magaki S, Townley RA, Haeri M, Vinters HV. Illustrated neuropathologic diagnosis of Alzheimer's disease. Neurol Int. 2023;15(3):857–67.
41. Scheff SW, Price DA. Synaptic pathology in Alzheimer's disease: a review of ultrastructural studies. Neurobiol Aging. 2003;24(8):1029–46.
42. Scheff SW, Price DA, Sparks DL. Quantitative assessment of possible age-related change in synaptic numbers in the human frontal cortex. Neurobiol Aging. 2001;22(3):355–65.
43. Fisher RA, Miners JS, Love S. Pathological changes within the cerebral vasculature in Alzheimer's disease: new perspectives. Brain Pathol Zurich Switz. 2022;32(6):e13061.
44. Terry RD. Interrelations among the lesions of normal and abnormal aging of the brain. Prog Brain Res. 1986;70:41–8.

45. Haroutunian V, Schnaider-Beeri M, Schmeidler J, Wysocki M, Purohit DP, Perl DP, et al. Role of the neuropathology of Alzheimer disease in dementia in the oldest-old. Arch Neurol [Internet] 2008 [cited 2024 Sep 24];65(9):1211–7. Available from: https://www.ncbi.nlm.nih.gov/pmc/articles/PMC3071251/.
46. DeTure MA, Dickson DW. The neuropathological diagnosis of Alzheimer's disease. Mol Neurodegener. 2019;14(1):32.
47. Lardelli M. An alternative view of familial Alzheimer's disease genetics. J Alzheimers Dis [Internet]. 2023 [cited 2024 Sep 10];96(1):13–39. Available from: https://content.iospress.com/articles/journal-of-alzheimers-disease/jad230313.
48. Devkota S, Zhou R, Nagarajan V, Maesako M, Do H, Noorani A, et al. Familial Alzheimer mutations stabilize synaptotoxic γ-secretase-substrate complexes. Cell Rep [Internet]. 2024 [cited 2024 Sep 10];43(2). Available from: https://www.cell.com/cell-reports/abstract/S2211-1247(24)00089-5.
49. Davidson DS, Kraus JA, Montgomery JM, Lemkul JA. Effects of familial Alzheimer's disease mutations on the folding free energy and dipole-dipole interactions of the amyloid β-peptide [Internet]. bioRxiv; 2022 [cited 2024 Sep 10]. p. 2022.05.19.492750. Available from: https://www.biorxiv.org/content/10.1101/2022.05.19.492750v2.
50. Orobets KS, Karamyshev AL. Amyloid precursor protein and Alzheimer's disease. Int J Mol Sci [Internet]. 2023 [cited 2024 Sep 10];24(19):14794. Available from: https://www.mdpi.com/1422-0067/24/19/14794.
51. Milward EA, Papadopoulos R, Fuller SJ, Moir RD, Small D, Beyreuther K, et al. The amyloid protein precursor of Alzheimer's disease is a mediator of the effects of nerve growth factor on neurite outgrowth. Neuron. 1992;9(1):129–37.
52. Moya KL, Benowitz LI, Schneider GE, Allinquant B. The amyloid precursor protein is developmentally regulated and correlated with synaptogenesis. Dev Biol. 1994;161(2):597–603.
53. Koo EH, Sisodia SS, Archer DR, Martin LJ, Weidemann A, Beyreuther K, et al. Precursor of amyloid protein in Alzheimer disease undergoes fast anterograde axonal transport. Proc Natl Acad Sci USA. 1990;87(4):1561–5.
54. Sharma R, Abubakar MD, Bisht P, Rachamalla M, Kumar A, Murti K, et al. Arsenic exposure and amyloid precursor protein processing: a focus on Alzheimer's disease. Curr Mol Pharmacol. 2023;17:e18761429272806.
55. Hu D, Dong X, Wang Q, Liu M, Luo S, Meng Z, et al. PCP4 promotes Alzheimer's disease pathogenesis by affecting amyloid-β protein precursor processing. J Alzheimers Dis JAD. 2023;94(2):737–50.
56. Simmons LK, May PC, Tomaselli KJ, Rydel RE, Fuson KS, Brigham EF, et al. Secondary structure of amyloid beta peptide correlates with neurotoxic activity in vitro. Mol Pharmacol. 1994;45(3):373–9.
57. Schubert D, Behl C, Lesley R, Brack A, Dargusch R, Sagara Y, et al. Amyloid peptides are toxic via a common oxidative mechanism. Proc Natl Acad Sci U S A [Internet]. 1995 [cited 2024 Sep 15];92(6):1989–93. Available from: https://www.ncbi.nlm.nih.gov/pmc/articles/PMC42408/.
58. Behl C, Davis JB, Lesley R, Schubert D. Hydrogen peroxide mediates amyloid beta protein toxicity. Cell. 1994;77(6):817–27.
59. Behl C, Sagara Y. Mechanism of amyloid beta protein induced neuronal cell death: current concepts and future perspectives. J Neural Transm Suppl. 1997;49:125–34.
60. Zhaliazka K, Kurouski D. Elucidation of molecular mechanisms by which amyloid β1-42 fibrils exert cell toxicity. Biochim Biophys Acta Mol Cell Biol Lipids. 2024;1869(6):159510.
61. Banerjee S, Banerjee S. Amyloid Beta-mediated neurovascular toxicity in Alzheimer's disease. Methods Mol Biol. 2024;2761:355–72.
62. Chen Y, Gu Y, Li Y, Li GL, Chai R, Li W, et al. Generation of mature and functional hair cells by co-expression of Gfi1, Pou4f3, and Atoh1 in the postnatal mouse cochlea. Cell Rep. 2021;35(3):109016.
63. Forloni G. Beta-amyloid neurotoxicity. Funct Neurol. 1993;8(3):211–25.

64. Keske E, Tunali NE. Ameliorating amyloid beta oligomer toxicity with NAC. Alzheimers Dement [Internet]. 2023 [cited 2024 Sep 15];19(S13):e077820. Available from: https://onlinelibrary.wiley.com/doi/abs/10.1002/alz.077820.
65. Montresor S, Pigazzini ML, Baskaran S, Sleiman M, Adhikari G, Basilicata L, et al. HSP110 is a modulator of amyloid beta (Aβ) aggregation and proteotoxicity. J Neurochem. 2024;169:e16214.
66. van Eersel J, Przybyla M, Genoud S, Fath T, Ke YD, Ittner LM. Amyloid-β precursor protein facilitates spreading of tau pathology in Alzheimer models. Alzheimers Dement [Internet]. 2023 [cited 2024 Sep 14];19(S24):e082576. Available from: https://onlinelibrary.wiley.com/doi/abs/10.1002/alz.082576.
67. Goedert M. Tau protein and the neurofibrillary pathology of Alzheimer's disease. Ann N Y Acad Sci. 1996;777:121–31.
68. Spillantini MG, Goedert M. Tau protein pathology in neurodegenerative diseases. Trends Neurosci. 1998;21(10):428–33.
69. Singh M, Ali H, Renuka Jyothi S, Kaur I, Kumar S, Sharma N, et al. Tau proteins and senescent cells: targeting aging pathways in Alzheimer's disease. Brain Res. 2024;1844:149165.
70. Granholm AC, Hamlett ED. The role of tau pathology in Alzheimer's disease and down syndrome. J Clin Med. 2024;13(5):1338.
71. Zhou LT, Liu D, Kang HC, Lu L, Huang HZ, Ai WQ, et al. Tau pathology epigenetically remodels the neuron-glial cross-talk in Alzheimer's disease. Sci Adv. 2023;9(16):eabq7105.
72. Paidlewar M, Kumari S, Dhapola R, Sharma P, HariKrishnaReddy D. Unveiling the role of astrogliosis in Alzheimer's disease pathology: insights into mechanisms and therapeutic approaches. Int Immunopharmacol. 2024;141:112940.
73. Beckman D, Diniz GB, Ott S, Hobson B, Chaudhari AJ, Muller S, et al. Temporal progression of tau pathology and neuroinflammation in a rhesus monkey model of Alzheimer's disease. Alzheimers Dement. 2024;20(8):5198–219.
74. Zhang S, Gu B, Zhen K, Du L, Lv Y, Yu L. Effects of exercise on brain-derived neurotrophic factor in Alzheimer's disease models: a systematic review and meta-analysis. Arch Gerontol Geriatr. 2024;126:105538.
75. Qiu H, Yang P, Wang L. DCNeT: a disease comorbidity network-based temporal deep learning framework to predict cardiovascular risk in patients with mental disorders. Expert Syst Appl [Internet]. 2024 [cited 2024 Sep 16];254:124312. Available from: https://www.sciencedirect.com/science/article/pii/S0957417424011783.
76. Farooqui AA. Metabolic syndrome as a risk factor for Alzheimer disease. In: Farooqui AA, editor. Metabolic syndrome: an important risk factor for stroke, Alzheimer disease, and depression [Internet]. New York: Springer. 2013 [cited 2024 Sep 16]. p. 281–341. Available https://doi.org/10.1007/978-1-4614-7318-3_8.
77. Park YW, Zhu S, Palaniappan L, Heshka S, Carnethon MR, Heymsfield SB. The metabolic syndrome: prevalence and associated risk factor findings in the US population from the Third National Health and Nutrition Examination Survey, 1988–1994. Arch Intern Med. 2003;163(4):427–36.
78. Graham IM, Daly LE, Refsum HM, Robinson K, Brattström LE, Ueland PM, et al. Plasma homocysteine as a risk factor for vascular disease. The European Concerted Action Project. JAMA. 1997;277(22):1775–81.
79. Stampfer MJ, Hennekens CH, Manson JE, Colditz GA, Rosner B, Willett WC. Vitamin E consumption and the risk of coronary disease in women. N Engl J Med. 1993;328(20):1444–9.
80. Turk KW, Knobel MD, Nothern A, Friedman G, Noah H, Campbell B, et al. An Interprofessional team for disease-modifying therapy in Alzheimer disease implementation. Neurol Clin Pract. 2024;14(6):e200346.
81. Ávila-Villanueva M, Gómez-Ramírez J, Ávila J, Fernández-Blázquez MA. Loneliness as risk factor for Alzheimer's disease. Curr Aging Sci. 2022;15(3):293–6.
82. Wood H. Alzheimer disease: Alzheimer disease risk factor CALM modulates tau turnover. Nat Rev Neurol. 2014;10(11):610.

83. Wang PN, Wang SJ, Hong CJ, Liu TT, Fuh JL, Chi CW, et al. Risk factors for Alzheimer's disease: a case-control study. Neuroepidemiology. 1997;16(5):234–40.
84. Daviglus ML, Plassman BL, Pirzada A, Bell CC, Bowen PE, Burke JR, et al. Risk factors and preventive interventions for Alzheimer disease: state of the science. Arch Neurol. 2011;68(9):1185–90.
85. Blackford JM, Willius FA, Haines SB. Operative risk in cardiac disease. J Am Med Assoc [Internet]. 1917 [cited 2024 Sep 16];LXIX(24):2011–4. Available from: https://doi.org/10.1001/jama.1917.02590510003003.
86. Tilley H. Discussion on focal sepsis as a factor in disease. Proc R Soc Med. 1926;19(Gen Rep):26.
87. Petersen RC. Detecting Alzheimer disease clinically: how early can we go? Neurology. 2022;98(15):607–8.
88. Witkowska-Płusa U, Dubielis A, Płusa T. Recognizing Alzheimer's disease in daily practice. Pol Merkur Lek Organ Pol Tow Lek. 2022;50(298):253–6.
89. Mohs RC, Knopman D, Petersen RC, Ferris SH, Ernesto C, Grundman M, et al. Development of cognitive instruments for use in clinical trials of antidementia drugs: additions to the Alzheimer's disease assessment scale that broaden its scope. The Alzheimer's Disease Cooperative Study. Alzheimer Dis Assoc Disord. 1997;11(Suppl 2): S13–21.
90. Orimaye SO, Wong JSM, Golden KJ, Wong CP, Soyiri IN. Predicting probable Alzheimer's disease using linguistic deficits and biomarkers. BMC Bioinform. 2017;18(1):34.
91. Callahan CM, Boustani MA, Unverzagt FW, Austrom MG, Damush TM, Perkins AJ, et al. Effectiveness of collaborative care for older adults with Alzheimer disease in primary care: a randomized controlled trial. JAMA. 2006;295(18):2148–57.
92. Bieger A, Brum WS, Borelli WV, Therriault J, De Bastiani MA, Moreira AG, et al. Influence of different diagnostic criteria on Alzheimer disease clinical research. Neurology. 2024;103(5):e209753.
93. Johnson KA, Fox NC, Sperling RA, Klunk WE. Brain imaging in Alzheimer disease. Cold Spring Harb Perspect Med. 2012;2(4):a006213.
94. Fantoni ER, Chalkidou A, O' Brien JT, Farrar G, Hammers A. A systematic review and aggregated analysis on the impact of amyloid PET brain imaging on the diagnosis, diagnostic confidence, and management of patients being evaluated for Alzheimer's disease. J Alzheimers Dis JAD. 2018;63(2):783–96.
95. Ferreira LK, Busatto GF. Neuroimaging in Alzheimer's disease: current role in clinical practice and potential future applications. Clinics. 2011;66(Suppl 1):19–24.
96. Hansson O, Seibyl J, Stomrud E, Zetterberg H, Trojanowski JQ, Bittner T, et al. CSF biomarkers of Alzheimer's disease concord with amyloid-β PET and predict clinical progression: a study of fully automated immunoassays in BioFINDER and ADNI cohorts. Alzheimers Dement. 2018;14(11):1470–81.
97. Lashley T, Schott JM, Weston P, Murray CE, Wellington H, Keshavan A, et al. Molecular biomarkers of Alzheimer's disease: progress and prospects. Dis Model Mech. 2018;11(5):dmm031781.
98. Satake Y, Kanemoto H, Yoshiyama K, Nakahama R, Matsunaga K, Shimosegawa E, et al. Case report: usefulness of biomarkers for Alzheimer's disease in two cases with very-late-onset schizophrenia-like psychosis. Front Psychiatry. 2021;12:742659.
99. Poirier J, Davignon J, Bouthillier D, Kogan S, Bertrand P, Gauthier S. Apolipoprotein E polymorphism and Alzheimer's disease. Lancet Lond Engl. 1993;342(8873):697–9.
100. Wolk DA, Dickerson BC, Alzheimer's Disease Neuroimaging Initiative. Apolipoprotein E (APOE) genotype has dissociable effects on memory and attentional-executive network function in Alzheimer's disease. Proc Natl Acad Sci USA. 2010;107(22):10256–61.
101. Lim YY, Villemagne VL, Laws SM, Ames D, Pietrzak RH, Ellis KA, et al. BDNF Val66Met, Aβ amyloid, and cognitive decline in preclinical Alzheimer's disease. Neurobiol Aging. 2013;34(11):2457–64.

102. Lim YY, Villemagne VL, Laws SM, Ames D, Pietrzak RH, Ellis KA, et al. Effect of BDNF Val66Met on memory decline and hippocampal atrophy in prodromal Alzheimer's disease: a preliminary study. PLoS One. 2014;9(1):e86498.
103. Johannsen S, Duning K, Pavenstädt H, Kremerskothen J, Boeckers TM. Temporal-spatial expression and novel biochemical properties of the memory-related protein KIBRA. Neuroscience. 2008;155(4):1165–73.
104. Porter T, Burnham SC, Savage G, Lim YY, Maruff P, Milicic L, et al. A polygenic risk score derived from episodic memory weighted genetic variants is associated with cognitive decline in preclinical Alzheimer's disease. Front Aging Neurosci [Internet] 2018 [cited 2024 Sep 24];10. Available from: https://www.frontiersin.org/journals/aging-neuroscience/articles/10.3389/fnagi.2018.00423/full.
105. Schupf N, Tang MX, Fukuyama H, Manly J, Andrews H, Mehta P, et al. Peripheral Abeta subspecies as risk biomarkers of Alzheimer's disease. Proc Natl Acad Sci USA. 2008;105(37):14052–7.
106. Nabers A, Perna L, Lange J, Mons U, Schartner J, Güldenhaupt J, et al. Amyloid blood biomarker detects Alzheimer's disease. EMBO Mol Med. 2018;10(5):e8763.
107. Beal E. Lipid biomarkers for early-stage Alzheimer disease. Nat Rev Neurol. 2011;7(9):474.
108. Zarrouk A, Debbabi M, Bezine M, Karym EM, Badreddine A, Rouaud O, et al. Lipid biomarkers in Alzheimer's disease. Curr Alzheimer Res. 2018;15(4):303–12.
109. Ozcelikay-Akyildiz G, Karadurmus L, Cetinkaya A, Uludag İ, Ozcan B, Unal MA, et al. The evaluation of clinical applications for the detection of the Alzheimer's disease biomarker GFAP. Crit Rev Anal Chem [Internet] 2024 [cited 2024 Sep 17]. Available from: https://www.tandfonline.com/doi/full/10.1080/10408347.2024.2393874.
110. Jurasova V, Andel R, Katonova A, Veverova K, Zuntychova T, Horakova H, Vyhnalek M, Kolarova T, Matoska V, Blennow K, Hort J. CSF neurogranin levels as a biomarker in Alzheimer's disease and frontotemporal lobar degeneration: a cross-sectional analysis. Alzheimers Res Ther [Internet]. 2024 [cited 2024 Sep 8];16(1). Available from: https://pubmed.ncbi.nlm.nih.gov/39242539/.
111. Shi R, Fan KH, Lopez OL, Feingold E, Kamboh MI. A whole-genome-sequencing study to identify genetic variants associated with Alzheimer's disease progression. Alzheimers Dement [Internet]. 2022 [cited 2024 Sep 10];18(S4):e064739. Available from: https://onlinelibrary.wiley.com/doi/abs/10.1002/alz.064739.
112. Qiang YR, Zhang SW, Li JN, Li Y, Zhou QY, Alzheimer's Disease Neuroimaging Initiative. Diagnosis of Alzheimer's disease by joining dual attention CNN and MLP based on structural MRIs, clinical and genetic data. Artif Intell Med. 2023;145:102678.
113. Han T, Peng Y, Du Y, Li Y, Wang Y, Sun W, et al. Mining Alzheimer's disease clinical data: reducing effects of natural aging for predicting progression and identifying subtypes. Front Neurosci. 2024;18:1388391.
114. Zhuang X, Cordes D, Bender AR, Nandy R, Oh EC, Kinney J, et al. Classifying Alzheimer's disease neuropathology using clinical and MRI measurements. J Alzheimers Dis [Internet] 2024 [cited 2024 Sep 8];100(3):843–62. Available from: https://content.iospress.com/articles/journal-of-alzheimers-disease/jad231321.
115. Kim M, Sekiya H, Yao G, Martin NB, Castanedes-Casey M, Dickson DW, et al. Diagnosis of Alzheimer disease and tauopathies on whole-slide histopathology images using a weakly supervised deep learning algorithm. Lab Investig J Tech Methods Pathol. 2023;103(6):100127.
116. Cullinane PW, Wrigley S, Bezerra Parmera J, Valerio F, Millner TO, Shaw K, et al. Pathology of neurodegenerative disease for the general neurologist. Pract Neurol. 2024;24(3):188–99.
117. Hein C, Sourdet S, Piau A, Villars H, Nourhashemi F, Vellas B. Key points of the follow-up plan in the care of Alzheimer's disease patients. Rev Med Interne. 2011;32(3):154–8.

118. Jack CR. Alzheimer disease: new concepts on its neurobiology and the clinical role imaging will play. Radiology. 2012;263(2):344–61.
119. Palmqvist S, Tideman P, Mattsson-Carlgren N, Schindler SE, Smith R, Ossenkoppele R, et al. Blood biomarkers to detect Alzheimer disease in primary care and secondary care. JAMA. 2024;332:1245–57.
120. Hildebrandt H, Fink F, Kastrup A, Haupts M, Eling P. Cognitive profiles of patients with mild cognitive impairment or dementia in Alzheimer's or Parkinson's disease. Dement Geriatr Cogn Disord Extra. 2013;3(1):102–12.
121. Khan H, Farhana F, Mostafa F, Rafiq A, Nizia EW, Razzaq R, et al. Comparative study of risk factors associated with normal cognition and cognitive impairment in rural west elderly Texans. J Alzheimers Dis Rep. 2024;8(1):1133–51.
122. Lu X, Zhang Y, Zhong S, Lai S, Yan S, Song X, et al. Cognitive impairment in major depressive disorder with non-suicidal self-injury: association with the functional connectivity of frontotemporal cortex. J Psychiatr Res. 2024;177:219–27.
123. van der Flier WM, Skoog I, Schneider JA, Pantoni L, Mok V, Chen CLH, et al. Vascular cognitive impairment. Nat Rev Dis Primers. 2018;4:18003.
124. Toth P, Tarantini S, Csiszar A, Ungvari Z. Functional vascular contributions to cognitive impairment and dementia: mechanisms and consequences of cerebral autoregulatory dysfunction, endothelial impairment, and neurovascular uncoupling in aging. Am J Physiol Heart Circ Physiol. 2017;312(1):H1–20.
125. MahmoudianDehkordi S, Arnold M, Nho K, Ahmad S, Jia W, Xie G, et al. Altered bile acid profile associates with cognitive impairment in Alzheimer's disease-an emerging role for gut microbiome. Alzheimers Dement. 2019;15(1):76–92.
126. Womack CL, Perkins A, Arnold JM. Cognitive impairment in the primary care clinic. Prim Care. 2024;51(2):233–51.
127. Yliranta A, Karjalainen VL, Nuorva J, Ahmasalo R, Jehkonen M. Apraxia testing to distinguish early Alzheimer's disease from psychiatric causes of cognitive impairment. Clin Neuropsychol. 2023;37(8):1629–50.
128. Gurevich P, Stuke H, Kastrup A, Stuke H, Hildebrandt H. Neuropsychological testing and machine learning distinguish Alzheimer's disease from other causes for cognitive impairment. Front Aging Neurosci. 2017;9:114.
129. Wang LY, Pei J, Zhan YJ, Cai YW. Overview of meta-analyses of five non-pharmacological interventions for Alzheimer's disease. Front Aging Neurosci. 2020;12:594432.
130. Hansen RA, Gartlehner G, Webb AP, Morgan LC, Moore CG, Jonas DE. Efficacy and safety of donepezil, galantamine, and rivastigmine for the treatment of Alzheimer's disease: a systematic review and meta-analysis. Clin Interv Aging. 2008;3:211.
131. Yaghmaei E, Lu H, Ehwerhemuepha L, Zheng J, Danioko S, Rezaie A, et al. Combined use of donepezil and Memantine increases the probability of five-year survival of Alzheimer's disease patients. Commun Med. 2024;4(1):99.
132. Atri A, Hendrix SB, Pejović V, Hofbauer RK, Edwards J, Molinuevo JL, et al. Cumulative, additive benefits of memantine-donepezil combination over component monotherapies in moderate to severe Alzheimer's dementia: a pooled area under the curve analysis. Alzheimers Res Ther. 2015;7(1):28.
133. Tuszynski MH, Yang JH, Barba D, Bakay RA, Pay MM, et al. Nerve growth factor gene therapy activates neuronal responses in Alzheimer's disease. JAMA Neurol [Internet] 2015 [cited 2024 Sep 23];72(10):1139–47. Available from: https://www.ncbi.nlm.nih.gov/pmc/articles/PMC4944824/.
134. Folch J, Busquets O, Ettcheto M, Sánchez-López E, Castro-Torres RD, Verdaguer E, et al. Memantine for the treatment of dementia: a review on its current and future applications. J Alzheimers Dis. 2018;62(3):1223–40.
135. Taylor X, Clark IM, Fitzgerald GJ, Oluoch H, Hole JT, DeMattos RB, et al. Amyloid-β (Aβ) immunotherapy induced microhemorrhages are associated with activated perivascu-

lar macrophages and peripheral monocyte recruitment in Alzheimer's disease mice. Mol Neurodegener. 2023;18(1):59.
136. Yadollahikhales G, Rojas JC. Anti-amyloid immunotherapies for Alzheimer's disease: a 2023 clinical update. Neurotherapeutics. 2023;20(4):914–31.
137. Winblad B, Graf A, Riviere ME, Andreasen N, Ryan JM. Active immunotherapy options for Alzheimer's disease. Alzheimers Res Ther. 2014;6(1):7.
138. Guo T, Zhang D, Zeng Y, Huang TY, Xu H, Zhao Y. Molecular and cellular mechanisms underlying the pathogenesis of Alzheimer's disease. Mol Neurodegener. 2020;15(1):40.
139. Vassar R, Kovacs DM, Yan R, Wong PC. The β-secretase enzyme BACE in health and Alzheimer's disease: regulation, cell biology, function, and therapeutic potential. J Neurosci. 2009;29(41):12787–94.
140. Liu N, Haziyihan A, Zhao W, Chen Y, Chao H. Trajectory of brain-derived amyloid beta in Alzheimer's disease: where is it coming from and where is it going? Transl Neurodegener. 2024;13(1):42.
141. Serneels L, T'Syen D, Perez-Benito L, Theys T, Holt MG, De Strooper B. Modeling the β-secretase cleavage site and humanizing amyloid-beta precursor protein in rat and mouse to study Alzheimer's disease. Mol Neurodegener. 2020;15(1):60.
142. Rajasekhar K, Mehta K, Govindaraju T. Hybrid multifunctional modulators inhibit multifaceted Aβ toxicity and prevent mitochondrial damage. ACS Chem Neurosci. 2018;9(6):1432–40.
143. Khan S, Barve KH, Kumar MS. Recent advancements in pathogenesis, diagnostics and treatment of Alzheimer's disease. Curr Neuropharmacol. 2020;18(11):1106–25.
144. Wiatrak B, Balon K, Jawień P, Bednarz D, Jęśkowiak I, Szeląg A. The role of the microbiota-gut-brain axis in the development of Alzheimer's disease. Int J Mol Sci. 2022;23(9):4862.
145. Dissanayaka DMS, Jayasena V, Rainey-Smith SR, Martins RN, Fernando WMADB. The role of diet and gut microbiota in Alzheimer's disease. Nutrients. 2024;16(3):412.
146. Varesi A, Pierella E, Romeo M, Piccini GB, Alfano C, Bjørklund G, et al. The potential role of gut microbiota in Alzheimer's disease: from diagnosis to treatment. Nutrients. 2022;14(3):668.
147. Bostanciklioğlu M. The role of gut microbiota in pathogenesis of Alzheimer's disease. J Appl Microbiol. 2019;127(4):954–67.
148. Jarrold W, Peintner B, Wilkins D, Vergryi D, Richey C, Gorno-Tempini ML, et al. Aided diagnosis of dementia type through computer-based analysis of spontaneous speech. In: Resnik P, Resnik R, Mitchell M, editors. Proceedings of the workshop on computational linguistics and clinical psychology: from linguistic signal to clinical reality [Internet]. Baltimore: Association for Computational Linguistics; 2014 [cited 2024 Sep 23]. p. 27–37. Available from: https://aclanthology.org/W14-3204.
149. Vigo I, Coelho L, Reis S. Speech- and language-based classification of Alzheimer's disease: a systematic review. Bioengineering. 2022;9(1):27.
150. Yang Q, Li X, Ding X, Xu F, Ling Z. Deep learning-based speech analysis for Alzheimer's disease detection: a literature review. Alzheimers Res Ther. 2022;14(1):186.
151. Zucchella C, Sinforiani E, Tamburin S, Federico A, Mantovani E, Bernini S, et al. The multidisciplinary approach to Alzheimer's disease and dementia. A narrative review of non-pharmacological treatment. Front Neurol. 2018;9:1058.
152. Segev-Jacubovski O, Herman T, Yogev-Seligmann G, Mirelman A, Giladi N, Hausdorff JM. The interplay between gait, falls and cognition: can cognitive therapy reduce fall risk? Expert Rev Neurother. 2011;11(7):1057–75.
153. Allali G, Verghese J. Management of gait changes and fall risk in MCI and dementia. Curr Treat Options Neurol. 2017;19(9):29.
154. Purba JS. Potential implication of treatments for Alzheimer's disease: current and future. Medicinus [Internet] 2023 [cited 2024 Sep 19];36(1):3–10. Available from: https://medicinus.co/journal/article/view/medicinus-vol36-no1-3-10.
155. Yu TW, Lane HY, Lin CH. Novel therapeutic approaches for Alzheimer's disease: an updated review. Int J Mol Sci. 2021;22(15):8208.

Parkinson's Disease: Epidemiology, Pathophysiology, Diagnosis, and Treatment

3

Anushri Sharma, Agina Nair, and Dharini Dedhia

3.1 Introduction

Parkinson's disease (PD) is a degenerative condition impacting the nervous system and areas of the body governed by nerve activity [1]. While symptoms and disease progression are different from one person to another, for some people, the changes take up to 20 years to become evident. Some discover that the illness advances more swiftly. After a diagnosis, motor fluctuations may become problematic 5–10 years later [1, 2]. The disease is most common in older adults; yet, only 4% of cases are identified before the age of 50. More than 10 million people worldwide are affected by Parkinson's disease. Men have a 50% higher risk of developing Parkinson's disease compared to women. Approximately one million people in the United States are afflicted with Parkinson's disease (PD). The expectation is for this number to rise to 1.2 million by the year 2030. Parkinson's is the second most common disease after Alzheimer's neurological illness. Over 10 million people globally are affected by Parkinson's disease [2].

3.2 Classification

There are three main types of Parkinson's disease.

A. Sharma (✉)
Department of Biochemistry and Genetics, Barkatullah University, Bhopal, India

A. Nair
Department of Biotechnology, K.J Somaiya College of Science & Commerce, University of Mumbai, Mumbai, India

D. Dedhia
Department of Molecular Medicine and Stem Cell Technologies, Jain University, Bangalore, India

N. Puranik (ed.), *Proteostasis*, https://doi.org/10.1007/978-981-96-6202-9_3

3.2.1 Primary (Idiopathic) Parkinsonism

Also known as Juvenile Parkinson. The neurodegenerative process that causes idiopathic Parkinson's disease is linked to the development of protein-containing inclusion bodies in specific types of neurons in the human nervous system [2, 3]. The process begins in the upper brain stem in the dorsal motor nucleus of the vagus nerve and ascends through sensitive parts of the medulla oblongata, pontine tegmentum, midbrain, and basal forebrain before reaching the cerebral cortex [2]. Over time, different components of the autonomic, limbic, and motor systems experience a significant decline. All interconnected subcortical grays and cortical regions are susceptible [3].

3.2.2 Secondary Parkinsonism (Acquired)

Neurological conditions that are often triggered by brain tumors, toxins, or medications are encompassed in this form of Parkinsonism. Symptoms resembling those found in Parkinson's disease may improve with treatment, depending on the condition [4]. Other secondary causes of Parkinson's are Wilson's disease (metabolic disease), hypothyroidism (endocrine disease), Whipple's disease (infectious disease), and repetitive head trauma [5].

3.2.3 Atypical Parkinsonism

Also called plus syndromes which can include Multiple system atrophy (MSA), i.e., progressive neurodegenerative disorder that typically occurs in adults and is characterized by a range of symptoms [6]; progressive supranuclear palsy which is neurological disorder with unknown cause that shows a range of symptoms including difficulty walking, early balance issues, stiffness in the body's core, slow movement, lack of coordination, sluggish eye movements leading to limited eye movement, speech difficulties, and problems with decision-making [7, 8]; and dementia with Lewy bodies in which the illness is defined by different combinations of worsening cognitive impairment, Parkinson's-like symptoms, changing mental state, sensitivity to certain medications, and seeing things that are not there [9]. Some advanced symptoms tend to occur within 1–3 years, which typically progress faster as compared to PD.

3.3 Phases of Parkinson's

There are five phases of PD based on the level of clinical disability. On this scale, Phases 1 and 2 indicate early-stage, 2 and 3 indicate mid-stage, and 4 and 5 indicate advanced-stage PD [10].

3.3.1 Phase One

Mildest form of Parkinson's is Stage 1. Initially, the individual encounters mild symptoms that typically do not disrupt their daily routine. Tremor and other movement symptoms are present on just one side of the body. Walking, posture, and facial expressions undergo changes [2, 11]. At this point, prescribed drugs can effectively decrease and reduce symptoms.

3.3.2 Phase Two

Average form of Parkinson's phase II. The symptoms are much more noticeable as compared to phase I, such as stiffness, and others impact both sides of the body. Issues with walking and bad posture may be noticeable. Everyday responsibilities have become harder and more time-consuming [2, 12].

3.3.3 Phase Three

Considered mid-stage and known as a significant critical moment in the development of illness. The movement becomes slower. The person is restricted to daily activities but somehow lives an independent life [13].

3.3.4 Phase Four

Symptoms are fully and severely developed. Movement may require a walker or walking device assistance. The person needs help in daily activities and requires complete assistance [11].

3.3.5 Phase Five

It is the most evolved stage in which stiffness in legs can make standing or walking impossible. People in this stage require assistance to stand without falling. It can also lead to hallucination, delusions, and confusion. All-time care is required for all activities [14].

3.4 Symptoms

The symptoms of Parkinson's can be different for everyone. In Stage I, it remains mild and unnoticed. It can be divided into movement and non-movement symptoms. Dopamine is primarily in charge of regulating movement and emotional reactions. It is a chemical messenger or neurotransmitter. Patients with Parkinson's disease have cells that produce dopamine and have been harmed; as it progresses, more cells are lost in which brain eventually runs out of significant dopamine production capacity. This moves increasingly challenging [15].

Different movement and non-movement symptoms

S. no.	Movement symptoms		Non-movement symptoms	
1.	Primary	1. Bradykinesia 2. Tremor 3. Rigidity	Mental health	Anxiety Apathy Depression
2.	Secondary	1. Dysarthria 2. Dyskineasia 3. Dystonia 4. Freezing 5. Hypophomia 6. Shuffling gait	General well-being	Breathing Fatigue Pain Sleep Disorder Speech Swelling issue
3.			Thinking change	Cognitive change Dementia Hallucination
4.			Digestive issue	Constipation Nausea Weight management

3.5 Causes

Parkinson's disease involves the gradual breakdown or loss of certain brain cells called neurons. The reason behind numerous Parkinson's symptoms is the depletion of brain neurons responsible for producing dopamine. Lowered dopamine levels cause abnormal brain activity, exacerbating both motor impairments and other symptoms of Parkinson's disease [16].

The cause of this is still unknown; however, various factors are observed to have a significant impact, such as:

1. Age—Biggest risk factor developing Parkinson's in advanced stage. The onset is 65 years.
2. Genetic—Several genes are known to elevate the likelihood of developing PD, with studies showing their strong impact on disease occurrence at an individual or familial level [17].
3. Environmental factors—Most delayed PD cases are caused by environmental exposures, either on their own or in combination with genetic factors. Even though the likelihood is small, being exposed to certain toxins or environmental factors could increase the risk of developing Parkinson's disease later on [18]. There are a few risk factors, which include age, heredity, sex (male are more affected than female), and exposure to toxins.

3.6 Epidemiology

The epidemiology shows notable differences according to the period, place, age, sex, and ethnicity. In India, there is a low prevalence of PD, except for a small Parsi community, where Bharucha et al. came across a significant frequency. No genetic alterations have been found as of yet, even though familial cases and early onset of PD have been recorded in India. Currently affecting around one million Americans, PD is a degenerative, long-term neurological disorder. The number of people worldwide estimated to have Parkinson's disease has nearly doubled in the past 25 years, reaching over 8.5 million in 2019. Disability and mortality related to Parkinson's disease are increasing faster than any other neurological disorder. As per recent projections, PD led to 329,000 fatalities in 2019, reflecting a surge of over 100% compared to 2000, along with 5.8 million disability-adjusted life years, which is 8.1% higher than in 2000 [19].

3.6.1 Prevalence and Incidence

The prevalence of PD varies by country. Parkinson's disease is more common in Europe and North America than it is in West Africa and Asia. Across the globe, India has the lowest generalized incidence of Parkinson's disease. One cannot generalize about the prevalence of this disease among Indians as it varies by state and ethnic group. However, the prevalence of PD is observed to be more common in older people and men. It is challenging to directly compare estimates of the prevalence of Parkinson's disease due to the vast disparities in the methodologies used in epidemiological studies and the abundance of research on the illness. It is widely acknowledged that the condition affects 1% of people over 60 [20], but its frequency in unselected populations ranges from 1 to 2 per 1000 [21]. Parkinson's disease is uncommon before the age of 50 but can reach a prevalence of 4% in older age groups [22]. There are comparably fewer PD incidence studies than prevalence studies. Reports on the disease's occurrence vary widely from one another. Methodological variations, particularly those about case identification and diagnostic criteria application, are most likely liable for this. Incidences per 100,000 people vary from less than 10 to more than 20 per year. Underdiagnosis of Parkinson's disease (PD), especially in older adults, may have an impact on incidence studies [23].

3.6.2 Risk Factors Affecting PD

Several researchers have conducted epidemiological research on Parkinson's disease throughout the world. The prevailing belief is that both genetics and surroundings contribute to this outcome, most cases of PD, with a complex etiology. A head injury and exposure to harmful chemicals may raise the risk of Parkinson's, although some lifestyle choices may reduce it. Hereditary factors may mitigate the effects of environmental exposures. While specific mutations in certain genes are responsible for causing identifiable changes in Parkinson's, approximately 5–10% of cases, most cases lack these mutations. Additionally, the majority of Parkinson's-associated genetic mutations have incomplete insight, which suggests that other genetic or environmental factors contribute to the disease. A study that looked at how often twins have the same condition found that only 30% of Parkinson's disease risk is due to genetics, with the rest being caused by behavior and environment [24].

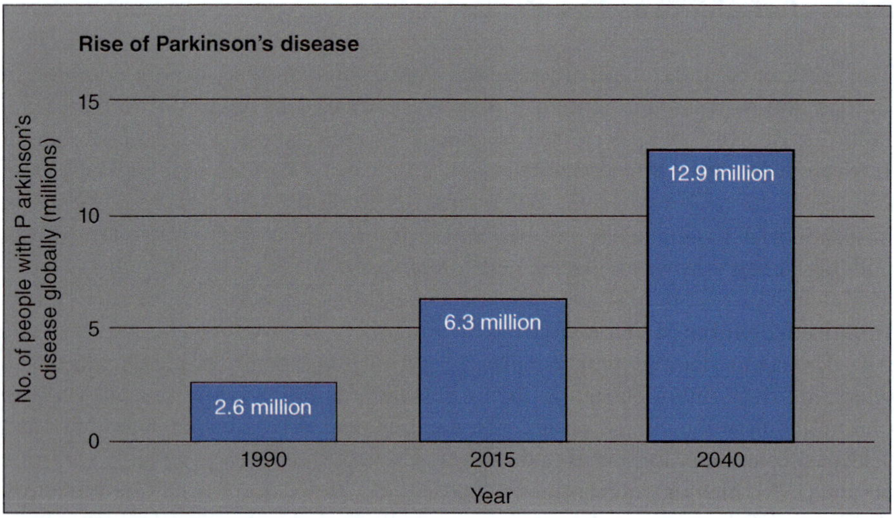

Graph shows rise of Parkinson's disease globally

Geographic differences cause the frequency distribution of Parkinson's disease. Different environmental factors are linked to PD in addition to geographical variations. The use of pesticides has been associated with Parkinson's disease. According to research conducted by Gupta et al. in 2014, many sociodemographic factors, including occupation, education, religion, and dietary habits, are unrelated to Parkinson's disease (PD) [25]. However, drinking water and being exposed to chemicals are linked to Parkinson's disease. Several additional factors have been studied in connection to Parkinson's disease. One significant prospective cohort study found that physically active men had a lower risk of PD. A particular risk-averse personality type, anxiety, depression, and olfactory abnormalities have all been linked to a higher chance of developing Parkinson's in the future [20].

Numerous population-based and follow-up studies indicate a correlation between Parkinson's and several medical conditions, including central obesity, hypertension, diabetes mellitus, and cognitive deficits. Several people with Parkinson's will continue to rise along with the span of the condition as the population ages and life expectancy rises, increasing in cases with advanced Parkinson's disease. Primary preventive measures based on the main cause of Parkinson's, along with more efficient medicines than those on the market today, are needed to tackle this burden. More incidence and prevalence studies are required to investigate time patterns and the variables influencing them, particularly for regions with limited data.

3.7 Pathophysiology

Parkinson's disease (PD) is a progressive neurodegenerative disorder characterized by the loss of dopamine-producing neurons in the substantia nigra, a region of the brain responsible for controlling movement. This loss of dopamine leads to a variety of motor symptoms, including tremor, rigidity, bradykinesia, and postural instability. However, PD also involves non-motor symptoms such as cognitive impairment, mood disorders, and autonomic dysfunction.

3.7.1 Neuropathology of Parkinson's Disease

A key characteristic of Parkinson's disease is the degeneration of dopaminergic neurons in the substantia nigra pars compacta, which is a crucial component of the basal ganglia system [26]. The loss of these neurons leads to decreased dopamine levels in the striatum, which is responsible for the characteristic motor deficits seen in patients. Dopamine is essential for coordinating voluntary movement, and its deficiency disrupts the balance within basal ganglia circuits, resulting in the classical manifestations of PD.

3.7.1.1 Lewy Bodies

Lewy bodies are pathological hallmark features of PD, primarily composed of aggregated alpha-synuclein, a protein implicated in synaptic function and neuronal health. The formation of Lewy bodies disrupts cellular homeostasis and contributes to neurodegeneration. In PD, the misfolding and accumulation of alpha-synuclein lead to the aggregation of these proteins within neurons, forming the characteristic inclusions that define the disease [27]. Lewy bodies are closely linked to neuronal cell death, especially in dopaminergic neurons of the substantia nigra. Research indicates that the accumulation of alpha-synuclein disrupts normal cellular processes and initiates a cascade of neuroinflammatory responses, which can worsen neuronal damage and create a harmful cycle that accelerates the progression of Parkinson's disease. Additionally, Lewy bodies may impair mitochondrial function, resulting in increased oxidative stress, a crucial factor in neuronal injury and death

[28]. Studies have demonstrated that the distribution of Lewy bodies is correlated with the severity of motor symptoms in patients with Parkinson's disease. Their presence in other brain regions, such as the cortex and limbic system, is linked to non-motor symptoms, including cognitive decline and mood disorders [29]. Studies have shown that the distribution of Lewy bodies correlates with the severity of motor symptoms in PD patients. Their presence in other brain regions, such as the cortex and limbic system, is associated with non-motor symptoms, including cognitive decline and mood disorders [30]. Targeted therapy development requires an understanding of Lewy bodies' function in the pathophysiology of Parkinson's disease. Researchers aim to improve outcomes for individuals affected by the disease and lessen the negative impact of these abnormal inclusions by investigating ways to either increase or avoid alpha-synuclein aggregation.

3.7.1.2 Neuroinflammation

A major factor in the pathophysiology of Parkinson's disease is neuroinflammation, which contributes to the destruction of dopaminergic neurons in the substantia nigra. Neuroinflammation in Parkinson's disease (PD) is characterized by the activation of microglia, the immune cells that reside in the brain. Upon activation, these cells release pro-inflammatory cytokines such as TNF-α and interleukin-1β (IL-1β), which can exacerbate neuronal injury and accelerate the process of neurodegeneration [28].

Activation of Microglia

The central nervous system's (CNS) resident immune cells, or microglia, are essential for maintaining homeostasis and responding to damage. The loss of dopaminergic neurons in PD triggers microglial activation. During this activation, the shape of microglia changes, and pro-inflammatory cytokines such as interleukin-1β (IL-1β), tumor necrosis factor-alpha (TNF-α), and interleukin-6 (IL-6) are produced in greater amounts [28]. These cytokines contribute to a neurotoxic environment that exacerbates neuronal damage and promotes further degeneration.

Cytokine Release and Neuronal Injury

The pro-inflammatory cytokines released by activated microglia can have detrimental effects on nearby neurons. Elevated levels of TNF-α, for instance, have been shown to induce apoptosis in dopaminergic neurons [31]. Similarly, IL-1β can enhance the excitotoxicity of glutamate, leading to increased calcium influx and further neuronal death [32]. This cycle of inflammation and neurodegeneration creates a feedback loop that accelerates disease progression.

The Role of ROS

Activated microglia contribute to the inflammatory response by releasing reactive oxygen species (ROS) in addition to cytokines. While ROS may be necessary for normal cellular signaling, excessive ROS production can lead to oxidative stress, which is particularly harmful to dopaminergic neurons [28]. The oxidative stress associated with neuroinflammation contributes to mitochondrial dysfunction and the accumulation of misfolded proteins, such as alpha-synuclein, further aggravating neuronal injury.

3.7.2 Mitochondrial Dysfunction

Mitochondrial dysfunction is a significant factor in the pathophysiology of Parkinson's disease (PD). Mitochondria are vital for energy production, and their impairment results in reduced ATP levels in dopaminergic neurons, particularly in the substantia nigra. This energy deficit contributes to neuronal degeneration [33]. Additionally, dysfunctional mitochondria increase the production of reactive oxygen species (ROS), resulting in oxidative stress, which further damages cellular components like lipids, proteins, and DNA [34]. Research has shown that mutations in genes such as PINK1 and PARKIN, which play a role in mitochondrial quality control, worsen these issues, underscoring the genetic basis of mitochondrial dysfunction in Parkinson's disease [35]. The interplay between mitochondrial impairment and oxidative stress creates a vicious cycle that accelerates neurodegeneration, making these pathways potential targets for therapeutic interventions aimed at slowing disease progression [36].

3.7.3 Neurotransmitter Imbalance

The main neurotransmitter impacted in PD is dopamine; however, other neurotransmitter systems are also involved. The loss of dopaminergic neurons results in an imbalance between dopamine and other neurotransmitters, particularly acetylcholine and glutamate.

3.7.3.1 Dopaminergic Pathways

Dopamine plays a vital role in regulating the activity of the direct and indirect pathways of the basal ganglia. The direct pathway promotes movement, while the indirect pathway suppresses it. In Parkinson's disease, the degeneration of dopaminergic neurons leads to reduced activation of the direct pathway and heightened activity of the indirect pathway. This imbalance contributes to the hallmark motor symptoms of PD, including bradykinesia, rigidity, and tremors [37].

3.7.3.2 Acetylcholine and Glutamate

In addition to dopamine, acetylcholine and glutamate are important contributors to the pathophysiology of Parkinson's disease. The loss of dopaminergic neurons results in increased cholinergic activity in the striatum, which exacerbates the motor symptoms seen in patients. Moreover, glutamate excitotoxicity, a process in which excessive glutamate causes neuronal damage, may also play a role in neurodegeneration in PD [38].

3.7.4 Genetic Factors

Genetic factors play a significant role in the risk of developing Parkinson's disease (PD). Although most cases are sporadic, genetic contributions are important in both familial and idiopathic forms of the disease. Understanding these genetic influences sheds light on the underlying mechanisms of neurodegeneration and suggests potential avenues for targeted therapies. Several genes have been implicated, including mutations in alpha-synuclein (SNCA), Leucine-rich repeat kinase 2 (LRRK2), and DJ-1 (PARK7).

3.7.4.1 SNCA (Alpha-Synuclein)

The SNCA gene, which encodes the protein alpha-synuclein, is key to the pathophysiology of PD. Mutations and duplications of this gene are associated with familial forms of the disease, particularly early-onset PD. A defining feature of PD is the formation of Lewy bodies, which are intracellular aggregates primarily made up of misfolded alpha-synuclein [27]. These aggregates disrupt normal cellular functions and contribute to neuronal death. Furthermore, research has shown that the presence of aggregated alpha-synuclein can trigger neuroinflammatory processes that exacerbate neurodegeneration [39].

3.7.4.2 LRRK2

Leucine-Rich Repeat Kinase 2 (LRRK2) is another important gene associated with both familial and sporadic Parkinson's disease. Mutations in the LRRK2 gene result in heightened kinase activity, which is connected to synaptic dysfunction and neuroinflammation [40]. LRRK2 mutations have been found to alter neuronal signaling pathways, resulting in the vulnerability of dopaminergic neurons. These mutations can also promote abnormal cellular responses to stress, contributing to the disease's progression [41].

3.7.4.3 PARK7 (DJ-1)

The PARK7 gene encodes the protein DJ-1, which is essential for protecting neurons from oxidative stress. Mutations in DJ-1 can impair its protective function, increasing the vulnerability of dopaminergic neurons to oxidative damage [42]. While mutations in PARK7 are relatively rare, they provide insights into how mitochondrial dysfunction and oxidative stress contribute to neuronal degeneration in PD.

3.7.4.4 PINK1 and PRKN (Parkin)

Both PINK1 and PRKN are involved in mitochondrial quality control. Mutations in PINK1 impair the cellular mechanisms that protect against mitochondrial dysfunction, which is a key feature of PD pathophysiology [43]. PRKN mutations disrupt the ubiquitin-proteasome system, essential for degrading misfolded proteins. This impairment results in the accumulation of damaged proteins, further exacerbating cellular stress and neuronal loss [39].

3.7.5 Environmental Factors

Parkinson's disease (PD) is largely influenced by environmental variables as well. According to epidemiological research, there is a link between specific exposures and a higher chance of contracting the illness. When these environmental influences are combined with genetic predispositions, they may trigger the onset of PD in susceptible individuals.

3.7.5.1 Toxins

Exposure to environmental toxins, such as pesticides and heavy metals, has been linked to an increased risk of Parkinson's disease. For instance, paraquat, a herbicide, has been shown to cause dopaminergic neurodegeneration in animal models, providing evidence for the hypothesis that environmental toxins can trigger PD [44].

3.7.5.2 Head Trauma

There is increasing evidence that head trauma may elevate the risk of developing PD. Studies indicate that individuals with a history of traumatic brain injury experience a higher incidence of PD later in life. Although the exact mechanisms are not fully understood, they may involve neuroinflammation and oxidative stress [41, 44].

3.7.6 Impaired Protein Homeostasis

Maintaining protein homeostasis, or proteostasis, is essential for neuronal health. In Parkinson's disease, the pathways responsible for protein degradation—such as the ubiquitin-proteasome system (UPS) and autophagy—are frequently disrupted. This impairment leads to the accumulation of misfolded proteins like alpha-synuclein, which further contributes to neurodegeneration [30].

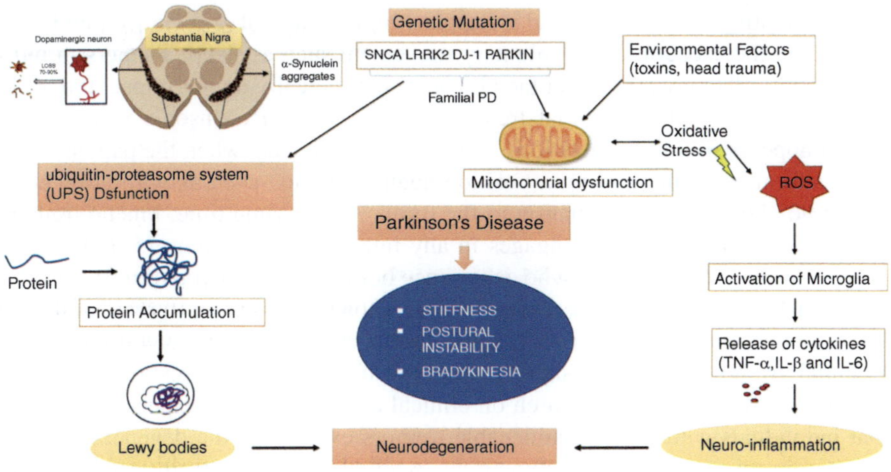

Overall pathophysiology of Parkinson's disease

3.8 Diagnosis

Making an accurate diagnosis of Parkinson's is complicated. Diagnosis of Parkinson's involves a combination of clinical evaluation, genetic history, neurological examination, and medical imaging.

3.8.1 Clinical Evaluation

Parkinson's disease has four essential characteristics [45].

3.8.1.1 Bradykinesia

The primary clinical symptom of Parkinson's disease (PD) is a decrease in movement speed; however, it is also noticeable in other conditions, such as depression. Bradykinesia is a signature feature of basal ganglia dysfunction [46]. Common initial indications include a decreased speed in completing everyday activities, as well as delayed responses to movements [47]. It also includes drooling due to impaired swelling, movement, gesture, loss of facial expression, and decreased blinking [48]. Patients are usually required to perform rapid evaluations in order to determine their condition, repetitive, alternating hand movement, and slowness and decreasing hand grips [46]. Despite the absence of a distinct clarification for bradykinesia, the cardinal Parkinson's disease symptom seems to be most closely associated with the severity of dopamine insufficiency [49]. Neuroimaging studies such as PET demonstrated that in the accumbens–caudate complex and striatum reduced absorption of F-fluorodopa is correlated with the severity of bradykinesia [50].

3.8.1.2 Tremor

Tremor disorders are difficult to diagnose. The most frequently seen of PD tremor is called resting tremor (RT), and it happens in the range of 4–6 Hz [51] and vanishes upon the execution of a voluntary movement such as sleep. Tremors in individuals with PD may also affect facial feature but typically do not impact the neck, head, or voice, unlike essential tremors [52]. In most Parkinson's patients, the tremor appears simultaneously [53]. Postural tremor occurs when the patient experiences tremors while holding a position against gravity, such as with arms positioned 90° horizontally in relation to the trunk. At the same time, kinetic tremors happen when the individual engages in any intentional movement [54]. The next group consists of PD patients who experience both postural/kinetic tremor episodes at higher frequencies than the resting tremor, known as essential tremor, alongside RT episodes. Numerous research studies support this idea because episodes of essential tremor can occur simultaneously with episodes of resting tremor in Parkinson's disease [55]. Research on clinical and pathological aspects has shown that individuals with PD and noticeable tremors experience the degeneration of a specific group of mid-brain (A8) neurons, while this region remains unaffected in PD patients who do not have tremors [56].

3.8.1.3 Rigidity

It is defined by heightened defiance, especially when linked to a tremor, which is evident during any passive movement of a limb (such as bending, straightening, or rotating a joint). It can happen close by (such as neck, shoulders, and hips) or further away (such as wrists and ankles) [57].

3.8.1.4 Postural Instability

Postural instability caused by the absence of postural reflexes typically appears in the advanced phases of PD, often following the emergence of other symptoms. It, combined with freezing of walking pattern, is the primary reason for falls and greatly increases the likelihood of hip fractures [58].

3.8.2 Medical History

The doctor reviews the patient's medical or genetic history and any family history of neurological disorders.

3.8.3 Neurological Examination

This may include a test of balance, coordination, and reflexes. The doctor may look for specific signs such as tremors, change in gait function, balance, and different movements of the body.

3.8.4 Effect of Medicine After Examination

A medication is prescribed (levodopa or Carbidopa), which may help to confirm the diagnosis [59].

3.8.5 Medical Imaging

While a single test cannot diagnose Parkinson's, different imaging techniques such as MRI (Routine MRI clinics can detect early signs of Parkinson's; also MRI is able to find the difference between idiopathic Parkinson's from Parkinsonian Syndromes) [60]. Dopamine transporter scans (DAT) and Positron emission tomography (PET) have higher sensitivity than MRI and can help rule out other conditions.

3.9 Treatment

The treatment of Parkinson's focuses on controlling symptoms and enhancing the quality of life because there is no treatment available at present. These are the primary methods, which include medical and nonmedical approaches as follows:

3.9.1 Medical Approach

3.9.1.1 Levodopa

Individuals with Parkinson's disease will eventually require medication known as levodopa. The reason behind the main movement symptoms in PD is the reduction of dopamine in the striatum caused by the death of DA neurons in the SNpc [61]. Giving levodopa instead of dopamine in the brain was a significant advancement in treating Parkinson's, as a result, several other options for dopamine-based treatments have been found. Levodopa is widely seen as the most effective treatment and is necessary for nearly all patients throughout their illness [62]. Motor fluctuations and dyskinesias make the long-term use of levodopa challenging. The reasons for these motor complications remain unknown. One possible explanation for the occurrence is the combination of pre- and postsynaptic mechanisms that result in abnormal pulsating activation of dopamine receptors in the striatum, triggering different negative neuronal reactions [63]. Enhancing Levodopa bioavailability can be achieved through creating better oral formulations like sustained release options or by exploring new methods of administration such as intestinal infusion, transcutaneous delivery through mini pumps, or inhalation. RYTARY/IPX066 is a novel levodopa-carbidopa (LD/CD) oral formulation combining the immediate-release and extended-release LD/CD [64, 65].

3.9.1.2 Dopamine Agonists

The primary focus of dopamine receptors is on the D2 receptor group. The first drugs in this drug family were derived from ergoline. Cardiac and pulmonary safety concerns were raised by ergoline drugs, so the agents currently used are all non-ergoline drugs. DA provide a more stable stimulation of striatal dopamine receptors compared to levodopa, which can significantly lower the likelihood of motor complications when used as the first treatment option [66].

Nausea, daytime sleepiness, and edema are linked to both levodopa and DA, but they tend to cause more frequent adverse effects. Dopamine agonists can lead to restrain problems and hallucinations caused by drugs are especially common in older individuals with cognitive impairments; therefore, they should not be used in populations at risk.

3.9.1.3 MAO B Inhibitor

Blocking MAO B results in elevated amounts of dopamine within the synapse and improves symptom relief. Rasagiline, an irreversible MAO B inhibitor, is commonly used as an additional treatment in patients experiencing motor fluctuations [67]. Treatment with MAO B inhibitors led to a notable decrease in the amount of levodopa needed and a reduced occurrence of dyskinesias [68]. Safinamide is an MAOB inhibitor that can be reversed and has properties that block glutamate. Safinamide provides better management of movement symptoms and enhances patients' quality of life with advanced Parkinson's [69].

3.9.1.4 COMT Inhibitors

Presently, medications are equipped to block the peripheral breakdown of dopamine, ultimately boosting the effectiveness of the drug. Tolcapone's use is limited because of its adverse reactions. Entocapone, a safer option, is now on the market but has lower effectiveness [70].

FDA-approved drugs for the treatment of Parkinson's disease [71]

S. no.	Drug type	Drug
1	Dopamine precursors	Levodopa
2	Dopamine agonists	Pramipexole, Ropinirole, Lisuride
3	Anticholinergies	Benztropine, Trihexyphenidyl
4	Others	Amantadine

3.9.2 Surgical Approach

Surgery for PD can have a profound impact on the lives of numerous cases. Continuous assessment and revisions are essential for reaching optimal results. Deep brain stimulation of the subthalamic nucleus or globus pallidus interna is a well-known treatment for individuals with movement problems [72]. DBS involves inserting electrodes into particular parts of the brain, like the subthalamic nucleus or globus pallidus interna. These electrodes provide electric pulses to control irregular brain function [73]. DBS is usually advised for patients with advanced Parkinson's disease who have not given a satisfactory response to medication or are dealing with significant side effects. DBS can be reversed and customized to accommodate changes in the disease's advancement.

3.9.3 Nonmedical Approach

Therapy such as physical therapy improves motor and balance, occupational therapy helps in daily living activities, speech therapy helps in speaking and swallowing trouble, exercise helps in regular movement and function of the body, and nutrition (balanced diet) helps in overall health condition.

Emotional and psychological support plays a most important role in well-being and coping with the disease.

3.10 Conclusion and Future Perspective

This neurodegerative condition that has significant impacts on movement and day-to-day activities. Despite the availability of effective management strategies such as medications, surgery, and supportive therapies that can improve symptoms and quality of life, the lack of a cure emphasizes the importance of ongoing research and innovation. Understanding the symptoms, treatments, and abiding progression of disease is crucial for effectively managing cases. Significant advancements have been achieved in comprehending the neuropathology of Parkinson disease. In the future, numerous hopeful pathways could transform the Parkinson's disease treatment scene such as advancement in medicine, technological innovation, stem cell research, holistic approach, and community and support network, which plays an essential part in improving the well-being of individuals suffering from Parkinson's disease.

References

1. https://www.mayoclinic.org/diseases-conditions/parkinsons-disease/symptoms-causes/syc-20376055
2. Homepage. Parkinson's Foundation. 2024. Available at: https://www.parkinson.org/
3. Braak H, Braak E. Pathoanatomy of Parkinson's disease. J Neurol. 2000;247:II3–10.
4. Braak H, Rüb U, Gai WP, Del Tredici K. Idiopathic Parkinson's disease: possible routes by which vulnerable neuronal types may be subject to neuroinvasion by an unknown pathogen. J Neural Transm. 2003;110:517–36.
5. Schulz-Schaeffer WJ. Is cell death primary or secondary in the pathophysiology of idiopathic Parkinson's disease. Biomolecules. 2015;5(3):1467–79.
6. Scholz SW, Bras J. Genetics underlying atypical parkinsonism and related neurodegenerative disorders. Int J Mol Sci. 2015;16(10):24629–55.
7. Wenning GK, Geser F, Krismer F, Seppi K, Duerr S, Boesch S, Köllensperger M, Goebel G, Pfeiffer KP, Barone P, Pellecchia MT. The natural history of multiple system atrophy: a prospective European cohort study. Lancet Neurol. 2013;12(3):264–74.
8. Williams DR, Lees AJ. Progressive supranuclear palsy: clinicopathological concepts and diagnostic challenges. Lancet Neurol. 2009;8(3):270–9.
9. Respondek G, Stamelou M, Kurz C, Ferguson LW, Rajput A, Chiu WZ, Van Swieten JC, Troakes C, Al Sarraj S, Gelpi E, Gaig C. The phenotypic spectrum of progressive supranuclear palsy: a retrospective multicenter study of 100 definite cases. Mov Disord. 2014;29(14):1758–66.
10. Lippa CF, Duda JE, Grossman M, Hurtig HI, Aarsland D, Boeve BF, Brooks DJ, Dickson DW, Dubois B, Emre M, Fahn S. DLB and PDD boundary issues: diagnosis, treatment, molecular pathology, and biomarkers. Neurology. 2007;68(11):812–9.
11. Poewe W. Global scales to stage disability in PD: the Hoehn and Yahr scale. In: Rating scales Parkinsons disease; 2012. p. 115–22. https://doi.org/10.1093/med/9780199783106.003.0258.
12. Goetz CG, Poewe W, Rascol O, Sampaio C, Stebbins GT, Counsell C, Giladi N, Holloway RG, Moore CG, Wenning GK, Yahr MD. Movement Disorder Society Task Force report on the Hoehn and Yahr staging scale: status and recommendations the Movement Disorder Society Task Force on rating scales for Parkinson's disease. Mov Disord. 2004;19(9):1020–8.
13. https://www.healthline.com/health/parkinsons/stages#stage-2
14. Poewe W, Mahlknecht P. The clinical progression of Parkinson's disease. Parkinsonism & related disorders. 2009;15:S28–32.

15. Jellinger KA. Neuropathobiology of non-motor symptoms in Parkinson disease. J Neural Transm. 2015;122:1429–40.
16. Wood-Kaczmar A, Gandhi S, Wood NW. Understanding the molecular causes of Parkinson's disease. Trends Mol Med. 2006;12(11):521–8.
17. Deng H, Wang P, Jankovic J. The genetics of Parkinson disease. Ageing Res Rev. 2018;42:72–85.
18. Chen H, Ritz B. The search for environmental causes of Parkinson's disease: moving forward. J Parkinsons Dis. 2018;8(s1):S9–17.
19. https://www.who.int/news-room/fact-sheets/detail/parkinson-disease
20. De Lau LM, Breteler MM. Epidemiology of Parkinson's disease. Lancet Neurol. 2006;5(6):525–35.
21. Von Campenhausen S, Bornschein B, Wick R, Bötzel K, Sampaio C, Poewe W, Oertel W, Siebert U, Berger K, Dodel R. Prevalence and incidence of Parkinson's disease in Europe. European neuropsychopharmacol. 2005;15(4):473–90.
22. De Rijk MD, Launer LJ, Berger K, Breteler MM, Dartigues JF, Baldereschi M, Fratiglioni L, Lobo A, Martinez-Lage J, Trenkwalder C, Hofman A. Prevalence of Parkinson's disease in Europe: a collaborative study of population-based cohorts. Neurologic Diseases in the Elderly Research Group. Neurology. 2000;54(11 Suppl 5):S21–3.
23. Tysnes OB, Storstein A. Epidemiology of Parkinson's disease. J Neural Transm. 2017;124:901–5.
24. Simon DK, Tanner CM, Brundin P. Parkinson disease epidemiology, pathology, genetics, and pathophysiology. Clin Geriatr Med. 2020;36(1, 1):–2.
25. Verma AK, Raj J, Sharma V, Singh TB, Srivastava S, Srivastava R. Epidemiology and associated risk factors of Parkinson's disease among the north Indian population. Clin Epidemiol Global Health. 2017;5(1):8–13.
26. Damier P, Hirsch EC, Agid Y, Graybiel A. The substantia nigra of the human brain: II. Patterns of loss of dopamine-containing neurons in Parkinson's disease. Brain. 1999;122(8):1437–48.
27. Spillantini MG, Schmidt ML, Lee VM, Trojanowski JQ, Jakes R, Goedert M. α-Synuclein in Lewy bodies. Nature. 1997;388(6645):839–40.
28. Hirsch EC, Vyas S, Hornykiewicz O. Neuroinflammation in Parkinson disease: a key player in the disease pathogenesis. J Neurochem. 2012;120(5):829–43.
29. Mazzulli JR, et al. Alpha-synuclein clearance and synaptic function. Nature. 2011;497(7448):345–51.
30. Braak H, Del Tredici K, Rüb U, De Vos RA, Steur EN, Braak E. Staging of brain pathology related to sporadic Parkinson's disease. Neurobiol Aging. 2003;24(2):197–211.
31. Tremblay MÈ, Lowery RL, Majewska AK. Microglial interactions with synapses are modulated by visual experience. PLoS Biol. 2010;8(11):e1000527.
32. Kummer KK, Zeidler M, Kalpachidou T, Kress M. Role of IL-6 in the regulation of neuronal development, survival and function. Cytokine. 2021;144:155582.
33. Obeso JA, Rodríguez-Oroz MC, Rodríguez M, Arbizu J, Giménez-Amaya JM. The basal ganglia and disorders of movement: pathophysiological mechanisms. Physiology. 2002;17(2):51–5.
34. Gressens P, Hagberg H. Perinatal brain damage: mechanisms and neuroprotective strategies. J Indian Institute Sci. 2012;92(4):465–84.
35. Martin I, Dawson VL, Dawson TM. Recent advances in the genetics of Parkinson's disease. Ann Rev Genomics Human Genet. 2011;12(1):301–25.
36. Xiong Y, Dawson TM, Dawson VL. Models of LRRK2-associated Parkinson's disease. Leucine-Rich Repeat Kinase. 2017;2(LRRK2):163–91. Klein C, Westenberger A. Genetics of Parkinson's disease. Cold Spring Harbor Perspect Med. 2012;2(1):a008888
37. Dolgacheva LP, Berezhnov AV, Fedotova EI, Zinchenko VP, Abramov AY. Role of DJ-1 in the mechanism of pathogenesis of Parkinson's disease. J Bioenerg Biomembr. 2019;51:175–88.
38. Valente EM, Abou-Sleiman PM, Caputo V, Muqit MM, Harvey K, Gispert S, Ali Z, Del Turco D, Bentivoglio AR, Healy DG, Albanese A. Hereditary early-onset Parkinson's disease caused by mutations in PINK1. Science. 2004;304(5674):1158–60.

39. Islam MS, Azim F, Saju H, Zargaran A, Shirzad M, Kamal M, Fatema K, Rehman S, Azad MM, Ebrahimi-Barough S. Pesticides and Parkinson's disease: current and future perspective. J Chem Neuroanat. 2021;115:101966.
40. Brett BL, Gardner RC, Godbout J, Dams-O'Connor K, Keene CD. Traumatic brain injury and risk of neurodegenerative disorder. Biol Psychiatry. 2022;91(5):498–507.
41. Schapira AH. Mitochondria in the aetiology and pathogenesis of Parkinson's disease. Lancet Neurol. 2008;7(1):97–109.
42. Hauser DN, Hastings TG. Mitochondrial dysfunction and oxidative stress in Parkinson's disease and monogenic parkinsonism. Neurobiol Disease. 2013;51:35–42.
43. Moon HE, Paek SH. Mitochondrial dysfunction in Parkinson's disease. Exp Neurobiol. 2015;24(2):103.
44. López-Doménech G, Covill-Cooke C, Howden JH, Birsa N, Morfill C, Brandon NJ, Kittler JT. Miro ubiquitination is critical for efficient damage-induced PINK1/Parkin-mediated mitophagy. BioRxiv. 2018:414664. https://doi.org/10.1101/414664.
45. Jankovic J. Parkinson's disease: clinical features and diagnosis. J Neurol Neurosurg Psychiatry. 2008;79(4):368–76.
46. Berardelli A, Rothwell JC, Thompson PD, Hallett M. Pathophysiology of bradykinesia in Parkinson's disease. Brain. 2001;124(11):2131–46.
47. Giovannoni G, Van Schalkwyk J, Fritz VU, Lees AJ. Bradykinesia akinesia inco-ordination test (BRAIN TEST): an objective computerised assessment of upper limb motor function. J Neurol Neurosurg Psychiatry. 1999;67(5):624–9.
48. Bagheri H, Damase-Michel C, Lapeyre-Mestre M, Cismondo S, O'Connell D, Senard JM, Rascol O, Montastruc JL. A study of salivary secretion in Parkinson's disease. Clin Neuropharmacol. 1999;22(4):213–5.
49. Vingerhoets FJ, Schulzer M, Calne DB, Snow BJ. Which clinical sign of Parkinson's disease best reflects the nigrostriatal lesion? Ann Neurol. 1997;41(1):58–64.
50. Lozza C, Marie RM, Baron JC. The metabolic substrates of bradykinesia and tremor in uncomplicated Parkinson's disease. Neuroimage. 2002;17(2):688–99.
51. Lyons KE, Pahwa R. Handbook of essential tremor and other tremor disorders. CRC Press; 2005.
52. Shulman LM, Singer C, Bean JA, Weiner WJ. Internal tremor in patients with Parkinson's disease. Mov Disord. 1996;11(1):3–7.
53. Wenzelburger R, Raethjen J, Löffler K, Stolze H, Illert M, Deuschl G. Kinetic tremor in a reach-to-grasp movement in Parkinson's disease. Mov Disord. 2000;15(6):1084–94.
54. Deuschl G, Raethjen J, Baron R, Lindemann M, Wilms H, Krack P. The pathophysiology of parkinsonian tremor: a review. J Neurol. 2000;247:V33–48.
55. Hristova A, Lyons K, Tröster AI, Pahwa R, Wilkinson SB, Koller WC. Effect and time course of deep brain stimulation of the globus pallidus and subthalamus on motor features of Parkinson's disease. Clin Neuropharmacol. 2000;23(4):208–11.
56. Burgess JG, Warwick K, Ruiz V, Gasson MN, Aziz TZ, Brittain JS, Stein J. Identifying tremor-related characteristics of basal ganglia nuclei during movement in the Parkinsonian patient. Parkinsonism Relat Disord. 2010;16(10):671–5.
57. Broussolle E, Krack P, Thobois S, Xie-Brustolin J, Pollak P, Goetz CG. Contribution of Jules Froment to the study of parkinsonian rigidity. Mov Disord. 2007;22(7):909–14.
58. Williams DR, Watt HC, Lees AJ. Predictors of falls and fractures in bradykinetic rigid syndromes: a retrospective study. J Neurol Neurosurg Psychiatry. 2006;77(4):468–73.
59. Tambasco N, Romoli M, Calabresi P. Levodopa in Parkinson's disease: current status and future developments. Curr Neuropharmacol. 2018;16(8):1239–52.
60. Samson E, Noseworthy MD. A review of diagnostic imaging approaches to assessing Parkinson's disease. Brain Disord. 2022;6:100037.
61. Beitz JM. Parkinson's disease: a review. Front Biosci (Schol Ed). 2014;6(1):65–74.
62. LeWitt PA, Fahn S. Levodopa therapy for Parkinson disease: a look backward and forward. Neurology. 2016;86(14_supplement_1):S3–12.

63. Olanow CW, Obeso JA, Stocchi F. Continuous dopamine-receptor treatment of Parkinson's disease: scientific rationale and clinical implications. Lancet Neurol. 2006;5(8):677–87.
64. Stocchi F, Hsu A, Khanna S, Ellenbogen A, Mahler A, Liang G, Dillmann U, Rubens R, Kell S, Gupta S. Comparison of IPX066 with carbidopa–levodopa plus entacapone in advanced PD patients. Parkinsonism Relat Disord. 2014;20(12):1335–40.
65. Dhall R, Kreitzman DL. Advances in levodopa therapy for Parkinson disease: review of RYTARY (carbidopa and levodopa) clinical efficacy and safety. Neurology. 2016;86(14_supplement_1):S13–24.
66. Frankel JP, Lees AJ, Kempster PA, Stern GM. Subcutaneous apomorphine in the treatment of Parkinson's disease. J Neurol Neurosurg Psychiatry. 1990;53(2):96–101.
67. Jankovic J, Poewe W. Therapies in Parkinson's disease. Curr Opin Neurol. 2012;25(4):433–47.
68. Cereda E, Cilia R, Canesi M, Tesei S, Mariani CB, Zecchinelli AL, Pezzoli G. Efficacy of rasagiline and selegiline in Parkinson's disease: a head-to-head 3-year retrospective case–control study. J Neurol. 2017;264(6):1254–63.
69. deSouza RM, Schapira A. Safinamide for the treatment of Parkinson's disease. Exp Opin Pharmacother. 2017;18(9):937–43.
70. Learmonth DA, Vieira-Coelho MA, Benes J, Alves PC, Borges N, Freitas AP, Soares-da-Silva P. Synthesis of 1-(3, 4-dihydroxy-5-nitrophenyl)-2-phenyl-ethanone and derivatives as potent and long-acting peripheral inhibitors of catechol-O-methyltransferase. J Med Chem. 2002;45(3):685–95.
71. Aggarwal L, Cervo FA. Therapeutic management of late-stage Parkinson's disease: review of the movement disorder society's guidelines. Ann Longterm Care. 2011;19(12):42–6.
72. Kalia SK, Sankar T, Lozano AM. Deep brain stimulation for Parkinson's disease and other movement disorders. Curr Opin Neurol. 2013;26(4):374–80.
73. Bronstein JM, Tagliati M, Alterman RL, Lozano AM, Volkmann J, Stefani A, Horak FB, Okun MS, Foote KD, Krack P, Pahwa R. Deep brain stimulation for Parkinson disease: an expert consensus and review of key issues. Arch Neurol. 2011;68(2):165.

A Proteasome System and Impairment in Ubiquitin-Proteasome System in Neurodegeneration and Neuroprotection

4

Shiv Kumar Yadav, Manish Kumar Maurya, Chhavi Tyagi, Riddhi Rai, Binu Bhat, and Mukul Das

4.1 Introduction

Neurodegenerative diseases are a broad group of neurological disorders distinguished by the gradual decline of neurons in the central nervous system. Neurodegeneration is recognized as a key pathological change in many brain-related disorders. Despite ongoing efforts by contemporary science to develop medical or surgical treatments, results have not been promising, and neurological conditions like Alzheimer's and dementia are still major clinical concerns, especially in the elderly population [35, 67]. A wide range of conditions that differ significantly in their underlying pathogenic mechanisms and clinical manifestations have been included in the category of neurodegenerative disorders [19]. One of the basic pathological hallmarks of neurodegenerative diseases is the buildup of certain proteins in the brain that are often used for disease classification and diagnosis. In many neurodegenerative diseases, protein aggregates are formed in several brain regions that align with the clinical signs and support their importance in disease pathology [18, 19, 67]. Neurodegenerative proteinopathies are categorized according to the primary protein accumulation, and these categories include tauopathies, α-synucleinopathies, tau-linked frontotemporal dementia [FTDtau], cerebral amyloidosis, prion diseases, trinucleotide repeat disorders, ferritinopathy, and neuroserpinopathy [29]. The buildup of ubiquitinated and aggregated proteins in many neurodegenerative diseases suggests disrupted proteostasis [12, 29, 58]. Protein homeostasis is regulated through two primary cellular mechanisms such as the ubiquitin-proteasome system (UPS) and autophagy-lysosome pathway (ALP). Proteostasis maintains the brain's proteome in a stable balance despite its dynamic nature. For optimum functioning of proteins, the proteostasis network maintains proper protein structure, localization, posttranslational modifications (PTMs), and protein–protein interactions [47]. The interruption of the ubiquitin-signaling

S. K. Yadav (✉) · M. K. Maurya · C. Tyagi · R. Rai · B. Bhat · M. Das
Shriram Institute for Industrial Research, Delhi, India

N. Puranik (ed.), *Proteostasis*, https://doi.org/10.1007/978-981-96-6202-9_4

pathways can result in protein mislocalization, DNA repair defects, oxidative stress, mitochondrial dysfunction, and various other harmful effects, which contribute to the commencement of neurodegenerative diseases [36].

Drug discovery and development for neurodegenerative diseases is a high-risk area, with only 12% of drugs in clinical trials receiving approval, and even these drugs mainly alleviate symptoms without averting disease progression. This low success rate of drugs is mainly attributed to the complex and diverse nature of neurodegenerative disorders [1]. The present-day management of neurodegenerative diseases, such as dopaminergic therapies for Parkinson's, cholinesterase inhibitors for Alzheimer's, and glutamate blockers for amyotrophic lateral sclerosis (ALS), are causing significant side effects. Moreover, current therapies are associated with many adverse drug reactions and are the fourth leading cause of death in the United States. Therefore, new treatment approaches that can act at the cellular level to alter disease pathogenesis instead of just managing symptoms like targeted protein degradation are urgently required. In this context, there is an increasing interest in targeting UPS and deubiquitinases (DUBs) for potential therapeutic interventions for neurodegenerative disorders. As a result, the UPS has gained recognition as a promising therapeutic target, with DUBs attracting significant interest due to their critical role in regulating protein stability and function [41, 48]. This chapter aims to explore the involvement of the UPS in neuronal function and its implication in neurodegenerative diseases with a particular focus on Alzheimer's disease (AD).

4.2 Ubiquitin-Proteasome System

In eukaryotes, the proteasome is the main protein-degrading complex present in both the cytosol and the nucleus. The proteasome is essential for numerous cellular functions, including protein homeostasis, stress response, and the regulation of critical processes such as cell division, DNA replication, transcription, and signal transduction [5]. In eukaryotic cells, UPS is necessary for the non-lysosomal degradation and clearance of short-lived, misfolded, mislocated, mutant, and damaged proteins [57]. The UPS is a multisubunit complex essential for maintaining the proper function of the nervous system. Dysfunction in its components has been associated with neurodegenerative diseases [61]. The majority of proteins in humans are degraded by the UPS, and the enzymes that are involved in this process are normally significant regulators of disease pathogenesis and desirable therapeutic targets. Among these, deubiquitinating enzymes (DUBs) have been identified as promising therapeutic targets for neurodegenerative diseases [52]. The cells tag proteins for proteasomal breakdown with ubiquitin. This enzyme-mediated process involves ubiquitin-activating enzymes (E1), ubiquitin-conjugating enzymes (E2), and ubiquitin ligases (E3) [51]. Ubiquitin can also modify protein function, interaction, or localization within the cell. Subsequently, the tagged proteins are recognized and degraded by 26S proteasomes, producing small peptides and amino acids that can be reused to produce new proteins, while the ubiquitin molecules are recycled back into the system and participate in the clearance of other abnormal proteins [48, 51].

Dysfunction of the UPS impairs the clearance of damaged and misfolded proteins, leading to their accumulation within cells, which subsequently disrupts cellular homeostasis. These pathological protein accumulations interfere with essential cellular processes and ultimately trigger cellular dysfunction and death [54]. UPS malfunction has been linked to the pathogenesis of several neurodegenerative disorders, including AD and Parkinson's disease (PD) [26, 54, 69].

4.3 The 26S Proteasome

The proteasome is a multisubunit protein complex found in both the nucleus and cytoplasm. It is crucial in eliminating the polyubiquitinated, oxidized, and misfolded proteins from the cell [21]. There are two types of proteasomes in eukaryotes: the 20S proteasome, which is a small-sized proteasome that is responsible for the degradation of unfolded proteins; and a 26S proteasome, which degrades the folded proteins in an ATP-independent or ATP-dependent manner [31, 40]. The structure of the 26S proteasome is made up of a 20S core particle (20S-CP) and one or two, i.e., additional 19S regulatory particles (19S-RP) [68]. The 19S regulatory particle is complicated and is composed of 19 distinct subunits and organized as a lid and base, and it mediates the recognition of polyubiquitinated targets, their deubiquitination and unfolding together with the translocation to the proteolytic cavity for degradation [15, 21]. The lid of the 19S-RP is made up of nine subunits (Rpn3, Rpn5-9, Rpn11, Rpn12, and Rpn15), and the base consists of ten subunits (Rpt1-6, Rpn10, Rpn1, Rpn2, and Rpn13) [68]. Rpn10/S5a and Rpn13 (also known as hRpn1, ADRM1, or ARM1) are also essential for the process of target protein degradation as ubiquitin receptors [39]. Moreover, the Rpt ring (Rpt1-6) possesses three HbYX motifs that interact with regulatory particles in order to open the N-terminal gate of the α-ring of the 20S-CP to allow substrate entry. The 20S-CP complex is formed by two outer α-rings and two inner β-rings which form a total of 28 subunits. Out of these, the two outer α-rings (α1-α7) consist of 14 subunits which form a central opening through which polypeptides in unfolded form can pass into the protease forming a cavity. Similarly, the two inner β-rings (β1-β7) also have 14 subunits, which form the proteolytic chamber, the site of protein breakdown. All of them have six proteolytic sites in each β-ring, which are known to primarily consist of, 2 β1 (trypsin-like site), 2 β2 (caspase-like site), and 2 β5 (chymotrypsin-like site) protease subunit which is required for degrading the substrate protein [21, 39, 68]. The fundamental molecular architecture of the 26S proteasome is shown in Fig. 4.1.

4.4 Ubiquitin-Proteasome System in Neurodegenerative Diseases

Neurodegenerative diseases are marked by the buildup of numerous proteins that would typically be degraded through two primary degradation pathways, such as UPS and ALP [36]. The extended lifespan of neuronal cells and their inability to

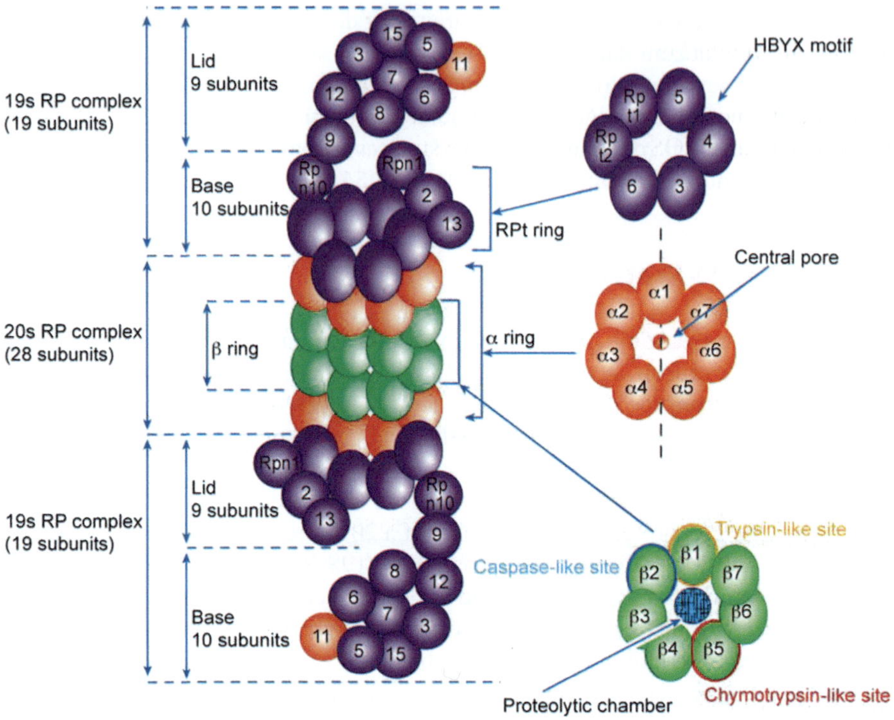

Fig. 4.1 Molecular structure of the 26S proteasome. (This figure is reproduced from Gadhave et al. [21]. Published by John Wiley & Sons Ltd. and Foundation for Cellular and Molecular Medicine). This is an open-access article under the terms of the Creative Commons Attribution License, which permits use, distribution, and reproduction in any medium, provided the original work is properly cited

divide make them more susceptible to the toxic effects of accumulated damaged organelles or misfolded proteins. Therefore, neurons depend on quality control mechanisms, and the failure of these systems is a key feature of aging, adversely affecting neuronal health in neurodegenerative diseases [3, 42]. A common feature of many neurodegenerative diseases is the accumulation of misfolded proteins in specific areas of the brain, indicating significant dysfunction in cellular protein degradation processes [13]. In AD, these aggregates typically comprise β-amyloid (Aβ) and hyperphosphorylated tau; in PD, α-synuclein; in Huntington's disease (HD), huntingtin (htt); and in ALS, superoxide dismutase 1 (SOD1) or TAR DNA binding protein-43 (TDP-43). These aggregates, especially extracellular Aβ plaques, frequently contain ubiquitin and enzymes involved in their attachment to substrates, indicating that abnormalities in proteostasis play a crucial role in disease pathogenesis [13, 56]. The development of many neurodegenerative diseases is linked to, and at least partially driven by the downregulation of UPS [14, 23]. Aging is an important risk factor contributing to reduced UPS activity in degenerating brains. Numerous studies have shown that proteasomal function declines with age,

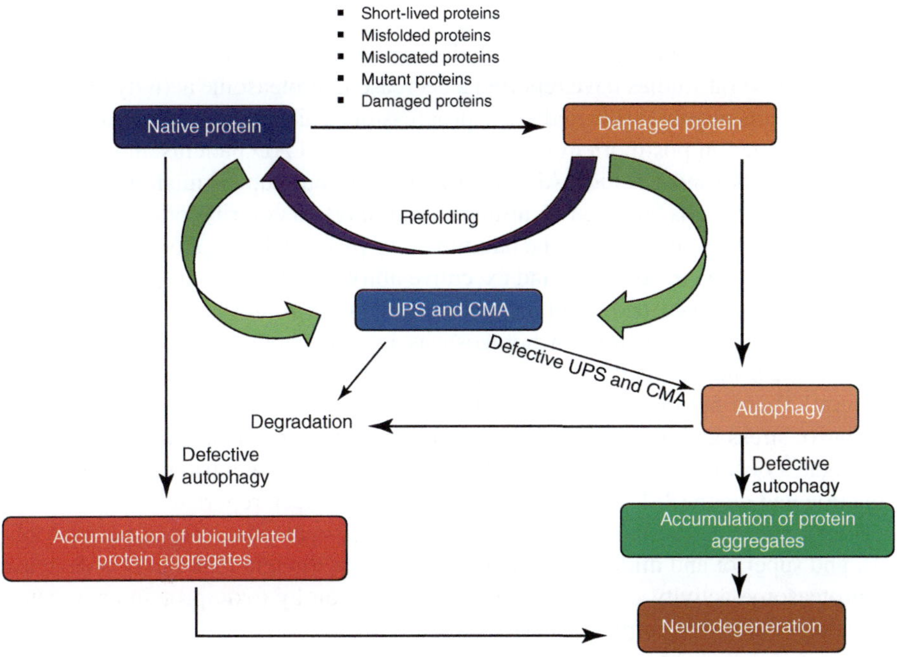

Fig. 4.2 General outline of protein quality control and neurodegeneration. This figure is adapted and modified from Bedford et al. [5]. (UPS: Ubiquitin-Proteasome System, CMA: Chaperone-Mediated Autophagy)

resulting in a weakened ability to break down misfolded proteins, promoting the formation of pathological protein aggregates in the brain [13, 32, 62]. Aggregated proteins represent an additional risk factor that can impair the function of UPS components, including the proteasome [13]. While most aggregated and misfolded proteins in the human proteome are cleared by cellular protein quality control systems, certain native and mutant proteins that tend to form β-sheet-enriched oligomers resist degradation by all known proteolytic pathways. These resistant proteins can accumulate, forming inclusion bodies or extracellular plaques. This buildup of protease-resistant proteins is a key mechanism in protein misfolding disorders, including neurodegenerative diseases [13]. Figure 4.2 shows the general outline of protein quality control in the cell.

4.5 Ubiquitination Modification of Important Proteins in Alzheimer's Disease

Alzheimer's disease is the well-known and most common cause of dementia. There are two categories of medications for the management of moderate to severe AD symptoms. These medications include cholinesterase inhibitors (Rivastigmine, Donepezil, and Galantamine) and N-methyl-D-aspartate (NMDA) receptor antagonist (Memantine). These medications help in alleviating AD symptoms by either improving neurotransmitter function or regulating glutamate activity in the brain [8]. Several studies have reported a decrease in proteasome activity in AD and highlighted its potential role in disease progression. This reduction has been consistently observed in postmortem brain tissue analyses of AD patients and in various animal models used to study AD [2, 10, 31]. The buildup of misfolded proteins within neuronal cells has been identified as an important contributor to neurodegeneration. AD is characterized by the buildup of Aβ plaques, hyperphosphorylated tau, and a slow decline in cognitive and executive abilities. The UPS plays an important role in protein homeostasis, and its dysfunction and ferroptosis are associated with AD progression. When protein homeostasis is compromised, misfolded proteins can no longer be efficiently removed from the cell, leading to their accumulation in the cell. These protein aggregates disrupt normal cellular functions by triggering oxidative stress and interfering with intracellular signaling. This buildup of protein aggregates becomes toxic to neuronal cells and ultimately contributes to their dysfunction and death [38]. Proteasome dysfunction has been reported in several brain regions, including the parahippocampal gyrus, hippocampus, inferior parietal lobule, and superior and middle temporal gyri [31]. It is thought that in AD, a decrease in proteasome activity occurs due to direct inhibition by toxic protein aggregates, especially Aβ and hyperphosphorylated tau [16].

4.5.1 Ubiquitination of Amyloid Precursor Protein

Numerous studies have made significant progress in uncovering the important molecules involved in the processing of β-amyloid precursor protein (APP). β-amyloid is a peptide derived from APP through enzymatic cleavage by β-secretase and γ-secretase. After β-secretase 1 (BACE1) cleaves APP in its ectodomain, γ-secretase processes the transmembrane domain of the carboxy-terminal fragments, leading to the release of Aβ peptides into the extracellular space. It has been suggested that distinct cholesterol levels affect several steps in the production of Aβ, including the degradation of APP and the actions of secretase enzymes that process it [63]. Posttranslational modifications of APP are crucial in regulating both formation and breakdown of Aβ peptides. Among these modifications, dysregulation of the ubiquitin modification process of APP has been recognized as a key factor in pathological effects caused by the Aβ accumulation in AD [17, 60]. The degradation of APP is mainly carried out by two E3 ubiquitin ligases, namely, HRD1 and FBL2, by the UPS pathway [30, 66]. By facilitating the ubiquitination of APP, HRD1, and FBL2 enhance its proteasome-dependent degradation, thereby lowering the levels of Aβ in the cell [38, 66]. Several studies have revealed that the accumulation of Aβ in neurons leads to a reduction in the activity of proteasomes and deubiquitinating enzymes. Removing Aβ can significantly reduce the early stages of tau pathology. However, Aβ buildup can impair proteasome function and encourage tau accumulation. Additionally, both mutant and wild-type APP in neural cells are known to influence the inhibition of downstream proteases [9, 48].

4.5.2 Ubiquitination of Tau

Tau is a soluble neuronal protein that mostly builds up in axons and plays a crucial role in maintaining the stability of microtubules. Moreover, it can undergo acetylation that influences its structural integrity and potential pathological alterations [37]. Neuronal degeneration and cognitive impairment are closely associated with abnormal tau protein accumulation and aggregation [38]. Numerous PTM affects tau polymerization and degradation, and ubiquitination plays a major role in guiding tau into the degradation pathway. Phosphorylated tau can be targeted for degradation through ubiquitination, leading to its breakdown by the proteasome and lysosome. Deubiquitinases like ubiquitin-specific protease 10 (USP10), which is widely expressed in the brain and promotes tau aggregation, can reverse ubiquitination. In AD, tau protein contains the highest number of ubiquitination sites among all ubiquitinated proteins, suggesting the significant role of UPS in tau regulation. The ubiquitination of tau protein acts as a vital mechanism for its degradation and clearance, preventing the accumulation of toxic protein aggregates [38]. In vivo, the E3 ubiquitin ligase CHIP (C-terminus of HSC70 Interacting Protein) tightly regulates the tau degradation through the UPS [49]. CHIP can bind to tau protein and help mark it for destruction, promoting its breakdown through a process called ubiquitination, which involves specific chaperone proteins [28]. Tau has many sites where it can interact with CHIP, allowing it to be recognized and processed for degradation due to its flexible structure [49]. The CHIP helps in maintaining the protein homeostasis and prevents the formation of toxic tau aggregates by facilitating the ubiquitination of tau. The clearance of tau is decreased when CHIP activity is impaired, which results in the accumulation of neurofibrillary tangles and subsequently associated with neurodegeneration in AD. Numerous studies reported that together with CHIP, other E3 ubiquitin ligases such as Parkin, Hrd1, and TRAF6 can also facilitate the degradation of phosphorylated tau and carry out the complete ubiquitination process through the UPS pathway [46, 50]. Some important E3 ligases involved in the ubiquitination of AD-associated proteins are shown in Table 4.1.

4.5.3 Ubiquitination of Proteins Associated with Ferroptosis in AD

Ferroptosis is a distinct type of regulated cell death caused by lipid peroxidation that is dependent on iron, which results in membrane instability and oxidative damage to the cell. Numerous studies have advocated that ferroptosis is a significant cause of neuronal loss and subsequent dysfunction of the central nervous system in multiple neurodegenerative diseases [43].

Several studies suggest that disruptions in iron homeostasis play a crucial role in the pathology of AD [38, 43]. Magnetic resonance imaging (MRI) scans of AD patients have revealed excessive iron accumulation in diverse regions of the brain such as the hippocampus, cortex, and basal ganglia and the disruption of these brain

Table 4.1 List of important E3 ligases involved in ubiquitination of AD-associated proteins and neurodegeneration

S.N.	E3 ligases	Molecular target	Functions	References
1	CHIP	Tau	It regulates the ubiquitin-dependent degradation of tau protein by binding to tau and facilitating its ubiquitination and subsequently promoting its clearance with the help of chaperone proteins	[28, 38]
2	HRD1	APP	It mediates APP degradation through the UPS pathway, promoting its ubiquitination and decreasing β-amyloid levels	[30]
3	FBL2	APP	It targets APP on the cell surface as well as inside the cell, promoting its ubiquitination and reducing the production of Aβ	[38, 66]
4.	Nedd4-2	DMT1	It selectively binds to DMT1 through Ndfip1 protein and facilitates its ubiquitination and degradation, which downregulates its expression and subsequently lowering intracellular iron accumulation	[20, 38]
5	Nedd4-1	AMPA receptors	NEDD4-1 targets AMPARs for ubiquitination and degradation, which results in synaptic dysfunction	[55]
6	Parkin	Aβ	It helps in reducing the intracellular Aβ(1–42) peptide levels and preventing its accumulation	[34]
7	TRAF6	Tau	It helps in the ubiquitination of tau	[4]

areas is strongly linked to AD pathology [43, 45]. The development of many neurodegenerative diseases has been linked with the disruptions in the intracellular transport of iron and other heavy metals, which is largely mediated by divalent metal transporter 1 (DMT1). DMT1 plays a significant role in regulating metal ion homeostasis by facilitating the uptake and intracellular distribution of divalent cations [27]. Ubiquitination is a crucial PTM that targets proteins for degradation by UPS using specific E3 ligases. Moreover, it also plays a vital role in ferroptosis and protein degradation [64]. The degradation of DMT1 is mostly controlled by ubiquitination, which is essential for preventing neuronal apoptosis and iron accumulation [43]. The Nedd4 family interacting protein 1 (Ndfip1) plays a significant role in adjusting iron imbalances, reducing Aβ secretion, and protecting neurons from iron-induced damage by regulating the degradation of DMT1. Therefore, Ndfip1 could be a promising AD prevention and treatment molecular target [59].

4.6 Ubiquitin Signaling as a Therapeutic Target in Neurodegeneration

Advancements in biotechnology are anticipated to play a significant role in neurological research, particularly in the development of targeted therapeutic approaches for neurodegenerative diseases. However, currently, no treatments are available that address the root cause of neurodegeneration to halt or slow the disease progression.

With the global prevalence of neurodegenerative diseases rising in an aging population, there is an urgent necessity for new therapeutic strategies. Although drugs that target the ubiquitin system are being tested in clinical settings as possible cancer treatments, their application in neurodegenerative diseases remains confined to pre-clinical research. The wide variety and specificity of over 600 E3 enzymes present a promising opportunity for precisely targeting disease-related pathways [25]. The components of ubiquitin signaling serve as direct therapeutic targets, leveraging ubiquitin-dependent degradation is also gaining recognition as a promising strategy for treating neurodegeneration. Over the past decade, targeted protein degradation (TPD) methods have evolved as an innovative approach to eliminate neurotoxic proteins from cells. This method uses heterobifunctional peptides or small molecules that bind both the target protein and the UPS or autophagy machinery, guiding this protein toward breakdown [56]. The most well-known TPD technologies are proteolysis targeting chimeras (PROTACs), which function by recruiting an E3 ubiquitin ligase to promote Lys48-linked polyubiquitination and subsequent proteasomal degradation of a target protein. PROTACs, a next-generation therapeutic molecule, are made up of two different components that bind distinctly to the target protein and the E3 ubiquitin ligase, coupled by a linker molecule. It forms the stable ternary complex by carrying the target protein and the E3 ligase together, which triggers the ubiquitination and subsequent proteasomal degradation of the target protein [7, 24]. Peptide-based PROTACs have been designed to selectively target and degrade the pathological aggregates of tau and α-synuclein in AD and PD, respectively [11, 44, 53]. In 2024, ARV-102, the first PROTAC protein degrader designed for neurological disorders, entered into phase I clinical trials. This orally bioavailable, brain-penetrant PROTAC specifically targets the kinase LRRK2 and has the potential to treat PD and progressive supranuclear palsy [22].

TPD is an emerging therapeutic method that enables the elimination of disease-causing proteins that have been challenging to target with traditional small molecules [7]. It represents a revolutionary change in small-molecule drug discovery and therapeutic development by providing new methods for modifying protein levels, increasing treatment options for previously inaccessible targets. One important benefit of PROTAC technology is its ability to degrade proteins that lack active sites, making them inaccessible to traditional small-molecule inhibitors. Unlike conventional inhibitors, which require direct binding to an active site to modulate protein function, PROTACs function by recruiting target proteins to the UPS and subsequently their selective degradation. PROTACs can target a wider range of disease-associated proteins related to neurodegenerative diseases. For instance, in tauopathies such as AD, progressive supranuclear palsy, and frontotemporal dementia small-molecule inhibitors are not effective but PROTACs can be effective [7, 65]. Notably, a tau-kelch-like ECH-associated protein 1 (KEAP1) chimeric peptide PROTAC molecule and a VHL-tau PROTAC have shown that the UPS triggers tau turnover and the degradability of tau species [44, 64]. With the significant unmet medical requirements posed by tauopathies and the appropriateness of the PROTAC strategy for addressing neurodegenerative diseases, this field is expected to achieve key milestones in the next decade, potentially leading to breakthrough medicines

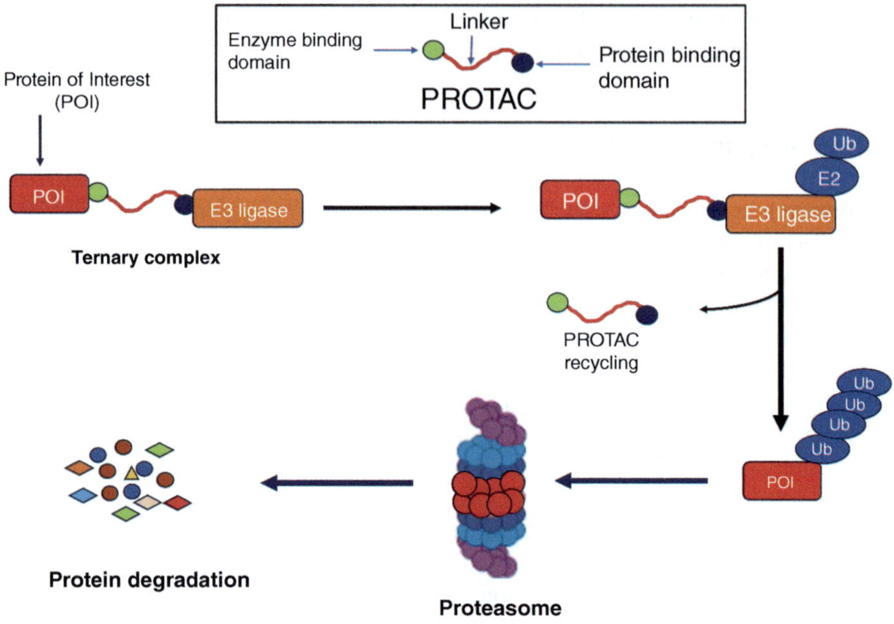

Fig. 4.3 Overview of proteolysis targeting chimeras (PROTAC) inducing protein degradation. A PROTAC is a small molecule consisting of three components: two active domains and a linker molecule. One domain binds to the target protein and at the same time the other domain attaches to an E3 ligase enzyme, bringing the two proteins together, which leads to the ubiquitination and subsequent clearance of the target protein by the proteasome

[7]. Moreover, PROTAC can be more effective compared to current therapies for treating other neurological targets, such as α-synuclein, a protein found in the brain of Parkinson's disease patients [33]. Small molecule PROTACs have more blood-brain barrier (BBB) permeability and bioavailability compared to peptide PROTACs. Further, it can also be appropriately designed to differentiate the disease-associated misfolded proteins from the physiological form of protein [56]. A general overview of PROTAC-induced protein degradation is shown in Fig. 4.3. Considering the important role of ubiquitin signaling pathway in numerous aspects of neuronal survival and function, targeting this pathway to either augment the removal of abnormal protein aggregates or address specific vulnerabilities driving neuronal degeneration could lead to an innovative treatment modality to slow or stop the progress of neurodegenerative diseases.

4.7 Conclusion

The UPS plays a significant role in regulating proteolysis within the cell. The ubiquitylated proteins are degraded by the 26S proteasome and through autophagy, which helps in maintaining neuronal homeostasis. The disruption in the function of these protein degradation pathways leads to neurodegeneration. The pharmacological manipulation of proteasome may provide an appropriate method for treating neurodegenerative disorders. There is an increasing curiosity in targeting UPS and deubiquitinases to develop more effective treatments for neurodegenerative diseases. Recent developments in this research area demonstrated that TPD could become a significant therapeutic modality. The TPD through PROTACs presents a potential new therapeutic approach for tackling disease-causing proteins that are difficult to target with traditional small molecules. The PROTACs's ability to cross the BBB and modulate the ubiquitin signaling cascade emphasizes their potential for revolutionizing the treatment of neurodegenerative diseases.

Acknowledgments The authors are thankful to the Director and Management of the Shriram Institute for Industrial Research, Delhi, for providing their support and encouragement in the preparation of this chapter, which has been assigned the number SRI-MS#20241010-01.

References

1. Akhtar A, Andleeb A, Waris TS, Bazzar M, Moradi AR, Awan NR, Yar M. Neurodegenerative diseases and effective drug delivery: a review of challenges and novel therapeutics. J Control Release. 2021;330:1152–67. https://doi.org/10.1016/j.jconrel.2020.11.021.
2. Almeida CG, Takahashi RH, Gouras GK. B-Amyloid accumulation impairs multivesicular body sorting by inhibiting the ubiquitin-proteasome system. J Neurosci. 2006;26(16):4277–88. https://doi.org/10.1523/JNEUROSCI.5078-05.2006.
3. Alves-Rodrigues A, Gregori L, Figueiredo-Pereira ME. Ubiquitin, cellular inclusions and their role in neurodegeneration. Trends Neurosci. 1998;21(12):516–20. https://doi.org/10.1016/s0166-2236(98)01276-4.
4. Babu JR, Geetha T, Wooten MW. Sequestosome 1/p62 shuttles polyubiquitinated tau for proteasomal degradation. J Neurochem. 2005;94(1):192–203. https://doi.org/10.1111/j.1471-4159.2005.03181.x.
5. Bard JA, Goodall EA, Greene ER, Jonsson E, Dong KC, Martin A. Structure and function of the 26S proteasome. Annu Rev Biochem. 2018;87(1):697–724. https://doi.org/10.1146/annurev-biochem-062917-011931.
6. Bedford L, Lowe J, Dick LR, Mayer RJ, Brownell JE. Ubiquitin-like protein conjugation and the ubiquitin-proteasome system as drug targets. Nat Rev Drug Discov. 2011;10(1):29–46. https://doi.org/10.1038/nrd3321.
7. Békés M, Langley DR, Crews CM. PROTAC targeted protein degraders: the past is prologue. Nat Rev Drug Discov. 2022;21(3):181–200. https://doi.org/10.1038/s41573-021-00371-6.
8. Briggs R, Kennelly SP, O'Neill D. Drug treatments in Alzheimer's disease. Clin Med. 2016;16(3):247–53. https://doi.org/10.7861/clinmedicine.16-3-247.
9. Budd Haeberlein S, O'Gorman J, Chiao P, Bussiere T, von Rosenstiel P, Tian Y, et al. Clinical development of aducanumab, an anti-abeta human monoclonal antibody being investigated for the treatment of early Alzheimer's disease. J Prev Alzheimers Dis. 2017;4:255–63. https://doi.org/10.14283/jpad.2017.39.
10. Chocron ES, Munkácsy E, Kim HS, Karpowicz P, Jiang N, Van Skike CE, DeRosa N, Banh A, Palavicini J, Wityk P, Kalinowski L, Galvan V, Osmulski P, Jankowska E, Gaczynska M, Pickering AM. Genetic and pharmacologic proteasome augmentation ameliorates Alzheimer's-like pathology in mouse and fly APP overexpression models. Sci Adv. 2022;8(23):eabk2252. https://doi.org/10.1126/sciadv.abk2252.

11. Chu TT, Gao N, Li QQ, Chen PG, Yang XF, Chen YX, Zhao YF, Li YM. Specific knockdown of endogenous tau protein by peptide-directed ubiquitin-proteasome degradation. Cell Chem Biol. 2016;23(4):453–61. https://doi.org/10.1016/j.chembiol.2016.02.016.
12. Ciechanover A, Brundin P. The ubiquitin proteasome system in neurodegenerative diseases: sometimes the chicken, sometimes the egg. Neuron. 2003;40(2):427–46. https://doi.org/10.1016/s0896-6273(03)00606-8.
13. Ciechanover A, Kwon YT. Degradation of misfolded proteins in neurodegenerative diseases: therapeutic targets and strategies. Exp Mol Med. 2015;47(3):e147. https://doi.org/10.1038/emm.2014.117.
14. Connell CM, Shaw BA, Holmes SB, Foster NL. Caregivers' attitudes toward their family members' participation in Alzheimer disease research: implications for recruitment and retention. Alzheimer Dis Assoc Disord. 2001;15(3):137–45. https://doi.org/10.1097/00002093-200107000-00005.
15. da Fonseca PC, Morris EP. Structure of the human 26S proteasome: subunit radial displacements open the gate into the proteolytic core. J Biol Chem. 2008;283(34):23305–14. https://doi.org/10.1074/jbc.M802716200.
16. Davidson K, Pickering AM. The proteasome: a key modulator of nervous system function, brain aging, and neurodegenerative disease. Front Cell Dev Biol. 2023;11:1124907. https://doi.org/10.3389/fcell.2023.1124907.
17. Dawkins E, Small DH. Insights into the physiological function of the β-amyloid precursor protein: beyond Alzheimer's disease. J Neurochem. 2014;129(5):756–69. https://doi.org/10.1111/jnc.12675.
18. Dugger BN, Dickson DW. Pathology of neurodegenerative diseases. Cold Spring Harb Perspect Biol. 2017;9(7):a028035. https://doi.org/10.1101/cshperspect.a028035.
19. Erkkinen MG, Kim MO, Geschwind MD. Clinical neurology and epidemiology of the major neurodegenerative diseases. Cold Spring Harb Perspect Biol. 2018;10(4):a033118. https://doi.org/10.1101/cshperspect.a033118.
20. Foot NJ, Dalton HE, Shearwin-Whyatt LM, Dorstyn L, Tan SS, Yang B, Kumar S. Regulation of the divalent metal ion transporter DMT1 and iron homeostasis by a ubiquitin-dependent mechanism involving Ndfips and WWP2. Blood. 2008;112(10):4268–75. https://doi.org/10.1182/blood-2008-04-150953.
21. Gadhave K, Bolshette N, Ahire A, Pardeshi R, Thakur K, Trandafir C, Balea M. The ubiquitin proteasomal system: a potential target for the management of Alzheimer's disease. J Cell Mol Med. 2016;20(7):1392–407. https://doi.org/10.1111/jcmm.12817.
22. Gregory JA, Hickey CM, Chavez J, Cacace AM. New therapies on the horizon: targeted protein degradation in neuroscience. Cell Chem Biol. 2024;31(9):1688–98. https://doi.org/10.1016/j.chembiol.2024.08.010.
23. Hegde AN, Upadhya SC. Role of ubiquitin–proteasome-mediated proteolysis in nervous system disease. Biochim Biophys Acta. 2011;1809(2):128–40. https://doi.org/10.1016/j.bbagrm.2010.07.006.
24. Hsia O, Hinterndorfer M, Cowan AD, Iso K, Ishida T, Sundaramoorthy R, Nakasone MA, Imrichova H, Schätz C, Rukavina A, Husnjak K, Wegner M, Correa-Sáez A, Craigon C, Casement R, Maniaci C, Testa A, Kaulich M, Dikic I, Winter GE, Ciulli A. Targeted protein degradation via intramolecular bivalent glues. Nature. 2024;627(8002):204–11. https://doi.org/10.1038/s41586-024-07089-6.
25. Huang X, Dixit VM. Drugging the undruggables: exploring the ubiquitin system for drug development. Cell Res. 2016;26(4):484–98. https://doi.org/10.1038/cr.2016.31.
26. Hyttinen JM, Amadio M, Viiri J, Pascale A, Salminen A, Kaarniranta K. Clearance of misfolded and aggregated proteins by aggrephagy and implications for aggregation diseases. Ageing Res Rev. 2014;18:16–28. https://doi.org/10.1016/j.arr.2014.07.002.
27. Ingrassia R, Garavaglia B, Memo M. DMT1 expression and iron levels at the crossroads between aging and neurodegeneration. Front Neurosci. 2019;13:575. https://doi.org/10.3389/fnins.2019.00575.

28. Joshi V, Amanullah A, Upadhyay A, Mishra R, Kumar A, Mishra A. A decade of boon or burden: what has the CHIP ever done for cellular protein quality control mechanism implicated in neurodegeneration and aging? Front Mol Neurosci. 2016;9:93. https://doi.org/10.3389/fnmol.2016.00093.
29. Jucker M, Walker LC. Self-propagation of pathogenic protein aggregates in neurodegenerative diseases. Nature. 2013;501(7465):45–51. https://doi.org/10.1038/nature12481.
30. Kaneko M, Koike H, Saito R, Kitamura Y, Okuma Y, Nomura Y. Loss of HRD1-mediated protein degradation causes amyloid precursor protein accumulation and amyloid-β generation. J Neurosci. 2010;30(11):3924–32. https://doi.org/10.1523/JNEUROSCI.2422-09.2010.
31. Keller JN, Hanni KB, Markesbery WR. Impaired proteasome function in Alzheimer's disease. J Neurochem. 2000a;75(1):436–9. https://doi.org/10.1046/j.1471-4159.2000.0750436.x.
32. Keller JN, Huang FA, Markesbery WR. Decreased levels of proteasome activity and proteasome expression in aging spinal cord. Neuroscience. 2000b;98(1):149–56. https://doi.org/10.1016/s0306-4522(00)00067-1.
33. Kotzbauer PT, Trojanowski JQ, Lee VMY. Lewy body pathology in Alzheimer's disease. J Mol Neurosci. 2001;17:225–32. https://doi.org/10.1385/jmn:17:2:225.
34. Kumar P, Pradhan K, Karunya R, Ambasta RK, Querfurth HW. Cross-functional E3 ligases Parkin and C-terminus Hsp70-interacting protein in neurodegenerative disorders. J Neurochem. 2012;120(3):350–70. https://doi.org/10.1111/j.1471-4159.2011.07588.x.
35. Lamptey RN, Chaulagain B, Trivedi R, Gothwal A, Layek B, Singh J. A review of the common neurodegenerative disorders: current therapeutic approaches and the potential role of nanotherapeutics. Int J Mol Sci. 2022;23(3):1851. https://doi.org/10.3390/ijms23031851.
36. Le Guerroué F, Youle RJ. Ubiquitin signaling in neurodegenerative diseases: an autophagy and proteasome perspective. Cell Death Differ. 2021;28(2):439–54. https://doi.org/10.1038/s41418-020-00667-x.
37. Lee MJ, Lee JH, Rubinsztein DC. Tau degradation: the ubiquitin–proteasome system versus the autophagy-lysosome system. Prog Neurobiol. 2013;105:49–59. https://doi.org/10.1016/j.pneurobio.2013.03.001.
38. Lin N, Gao XY, Li X, Chu WM. Involvement of ubiquitination in Alzheimer's disease. Front Neurol. 2024;15:1459678. https://doi.org/10.3389/fneur.2024.1459678.
39. Liu CW, Jacobson AD. Functions of the 19S complex in proteasomal degradation. Trends Biochem Sci. 2013;38(2):103–10. https://doi.org/10.1016/j.tibs.2012.11.009.
40. Liu CW, Li X, Thompson D, Wooding K, Chang TL, Tang Z, DeMartino GN. ATP binding and ATP hydrolysis play distinct roles in the function of 26S proteasome. Mol Cell. 2006;24(1):39–50. https://doi.org/10.1016/j.molcel.2006.08.025.
41. Liu P, Chen Z, Guo Y, He Q, Pan C. Recent advances in small molecule inhibitors of deubiquitinating enzymes. Eur J Med Chem. 2025;287:117324. https://doi.org/10.1016/j.ejmech.2025.117324.
42. Lopez-Otin C, Blasco MA, Partridge L, Serrano M, Kroemer G. The hallmarks of aging. Cell. 2013;153:1194–217. https://doi.org/10.1016/j.cell.2013.05.039.
43. Lu L, Jifu C, Xia J, Wang J. E3 ligases and DUBs target ferroptosis: a potential therapeutic strategy for neurodegenerative diseases. Biomed Pharmacother. 2024;175:116753. https://doi.org/10.1016/j.biopha.2024.116753.
44. Lu M, Liu T, Jiao Q, Ji J, Tao M, Liu Y, You Q, Jiang Z. Discovery of a Keap1-dependent peptide PROTAC to knockdown Tau by ubiquitination-proteasome degradation pathway. Eur J Med Chem. 2018;146:251–9. https://doi.org/10.1016/j.ejmech.2018.01.063.
45. Ma H, Dong Y, Chu Y, Guo Y, Li L. The mechanisms of ferroptosis and its role in Alzheimer's disease. Front Mol Biosci. 2022;9:965064. https://doi.org/10.3389/fmolb.2022.965064.
46. Mazanetz MP, Fischer PM. Untangling tau hyperphosphorylation in drug design for neurodegenerative diseases. Nat Rev Drug Discov. 2007;6(6):464–79. https://doi.org/10.1038/nrd2111.
47. McClatchy DB. Quantitative analysis of proteostasis networks: are there applications for dementia? Expert Rev Proteomics. 2024;21(1–3):1–3. https://doi.org/10.1080/14789450.2024.2320807.

48. Momtaz S, Memariani Z, El-Senduny FF, Sanadgol N, Golab F, Katebi M, Abdolghaffari A, Farzaei H, Abdollahi M. Targeting ubiquitin-proteasome pathway by natural products: novel therapeutic strategy for treatment of neurodegenerative diseases. Front Physiol. 2020;11:361. https://doi.org/10.3389/fphys.2020.00361.
49. Munari F, Mollica L, Valente C, Parolini F, Kachoie EA, Arrigoni G, Assfalg M. Structural basis for chaperone-independent ubiquitination of tau protein by its E3 ligase CHIP. Angew Chem Int Ed. 2022;61(15):e202112374. https://doi.org/10.1002/anie.202112374.
50. Oddo S, Vasilevko V, Caccamo A, Kitazawa M, Cribbs DH, LaFerla FM. Reduction of soluble Aβ and tau, but not soluble Aβ alone, ameliorates cognitive decline in transgenic mice with plaques and tangles. J Biol Chem. 2006;281(51):39413–23. https://doi.org/10.1074/jbc.M608485200.
51. Pickart CM. Ubiquitin in chains. Trends Biochem Sci. 2000;25(11):544–8. https://doi.org/10.1016/s0968-0004(00)01681-9.
52. Qin B, Chen X, Wang F, Wang Y. DUBs in Alzheimer's disease: mechanisms and therapeutic implications. Cell Death Discov. 2024;10(1):475. https://doi.org/10.1038/s41420-024-02237-3.
53. Qu J, Ren X, Xue F, He Y, Zhang R, Zheng Y, Huang H, Wang W, Zhang J. Specific knockdown of α-synuclein by peptide-directed proteasome degradation rescued its associated neurotoxicity. Cell Chem Biol. 2020;27(6):751–62. https://doi.org/10.1016/j.chembiol.2020.03.010.
54. Rao G, Croft B, Teng C, Awasthi V. Ubiquitin-proteasome system in neurodegenerative disorders. J Drug Metab Toxicol. 2015;6(4):187. https://doi.org/10.4172/2157-7609.1000187.
55. Rodrigues EM, Scudder SL, Goo MS, Patrick GN. Aβ-induced synaptic alterations require the E3 ubiquitin ligase Nedd4-1. J Neurosci. 2016;36(5):1590–5. https://doi.org/10.1523/JNEUROSCI.2964-15.2016.
56. Schmidt MF, Gan ZY, Komander D, Dewson G. Ubiquitin signalling in neurodegeneration: mechanisms and therapeutic opportunities. Cell Death Differ. 2021;28(2):570–90. https://doi.org/10.1038/s41418-020-00706-7.
57. Sherman MY, Goldberg AL. Cellular defenses against unfolded proteins: a cell biologist thinks about neurodegenerative diseases. Neuron. 2001;29(1):15–32. https://doi.org/10.1016/s0896-6273(01)00177-5.
58. Taylor JP, Hardy J, Fischbeck KH. Toxic proteins in neurodegenerative disease. Science. 2002;296(5575):1991–5. https://doi.org/10.1126/science.1067122.
59. Tian J, Zheng W, Li XL, Cui YH, Wang ZY. Lower expression of Ndfip1 is associated with Alzheimer disease pathogenesis through decreasing DMT1 degradation and increasing iron influx. Front Aging Neurosci. 2018;10:165. https://doi.org/10.3389/fnagi.2018.00165.
60. Tsatsanis A, Dickens S, Kwok JC, Wong BX, Duce JA. Post translational modulation of β-amyloid precursor protein trafficking to the cell surface alters neuronal iron homeostasis. Neurochem Res. 2019;44:1367–74. https://doi.org/10.1007/s11064-019-02747-y.
61. Türker F, Cook EK, Margolis SS. The proteasome and its role in the nervous system. Cell Chem Biol. 2021;28(7):903–17. https://doi.org/10.1016/j.chembiol.2021.04.003.
62. Tydlacka S, Wang CE, Wang X, Li S, Li XJ. Differential activities of the ubiquitin–proteasome system in neurons versus glia may account for the preferential accumulation of misfolded proteins in neurons. J Neurosci. 2008;28(49):13285–95. https://doi.org/10.1523/JNEUROSCI.4393-08.2008.
63. Wang C, Shou Y, Pan J, Du Y, Liu C, Wang H. The relationship between cholesterol level and Alzheimer's disease-associated APP proteolysis/Aβ metabolism. Nutr Neurosci. 2019;22(7):453–63. https://doi.org/10.1080/1028415X.2017.1416942.
64. Wang X, Wang Y, Li Z, Qin J, Wang P. Regulation of ferroptosis pathway by ubiquitination. Front Cell Dev Biol. 2021;9:699304. https://doi.org/10.3389/fcell.2021.699304.
65. Wang Y, Mandelkow E. Tau in physiology and pathology. Nat Rev Neurosci. 2016;17(1):22–35. https://doi.org/10.1038/nrn.2015.1.
66. Watanabe T, Hikichi Y, Willuweit A, Shintani Y, Horiguchi T. FBL2 regulates amyloid precursor protein (APP) metabolism by promoting ubiquitination-dependent APP degradation and inhibition of APP endocytosis. J Neurosci. 2012;32(10):3352–65. https://doi.org/10.1523/JNEUROSCI.5659-11.2012.

67. Wilson DM, Cookson MR, Van Den Bosch L, Zetterberg H, Holtzman DM, Dewachter I. Hallmarks of neurodegenerative diseases. Cell. 2023;186(4):693–714. https://doi.org/10.1016/j.cell.2022.12.032.
68. Xie Y. Structure, assembly and homeostatic regulation of the 26S proteasome. J Mol Cell Biol. 2010;2(6):308–17. https://doi.org/10.1093/jmcb/mjq030.
69. Zheng C, Geetha T, Babu JR. Failure of ubiquitin proteasome system: risk for neurodegenerative diseases. Neurodegener Dis. 2015;14(4):161–75. https://doi.org/10.1159/000367694.

Protein Misfolding and Cell Death in Neurodegenerative Diseases: Consequences and Strategies

5

Nidhi Saxena and Sagarkumar Joshi

5.1 Introduction

Neurodegenerative diseases (NDDs) refer to a class of disorders, listed together to reflect their shared and common feature of loss of neuronal structure or function leading eventually to neuronal cell death in a progressive manner. NDDs create a global public health problem related to the aging population and result in severe physical, cognitive, and economic burdens on individuals, families, and healthcare systems. Well-studied and characterized examples of neurodegenerative disorders are Alzheimer's disease (AD), Parkinson's disease (PD), Huntington's disease (HD), and amyotrophic lateral sclerosis (ALS). While having diverse clinical presentations, perhaps the most prevalent pathological characterization of many NDDs is the presence of misfolded proteins that aggregate into toxic forms that induce cellular dysfunction and death [23]. An understanding of the processes of protein misfolding and aggregation will be essential for the development of acute and chronic therapeutic strategies to ameliorate these devastating disorders (Fig. 5.1).

Neurodegenerative diseases encompass very diverse disorders that share the common aspect of degenerative neuronal tissue or cell loss. Neurodegenerative diseases can include the early onset familial disorders, or can be broadly categorized as a sporadic disorder, which could also be genetic in the mean for that individual [30]. Disorders can be categorized by the brain regions involved, types of neurons involved, and specific clinical symptoms observed at presentation, namely, the following disorders may present in similar fashion: AD—primarily involves the

N. Saxena
Department of Microbiology, Atmiya University, Rajkot, Gujarat, India

S. Joshi (✉)
Department of Microbiology and Biotechnology, Gujarat University, Ahmedabad, Gujarat, India

Department of Microbiology, Darshan Institute of Science, Darshan University, Rajkot, Gujarat, India

N. Puranik (ed.), *Proteostasis*, https://doi.org/10.1007/978-981-96-6202-9_5

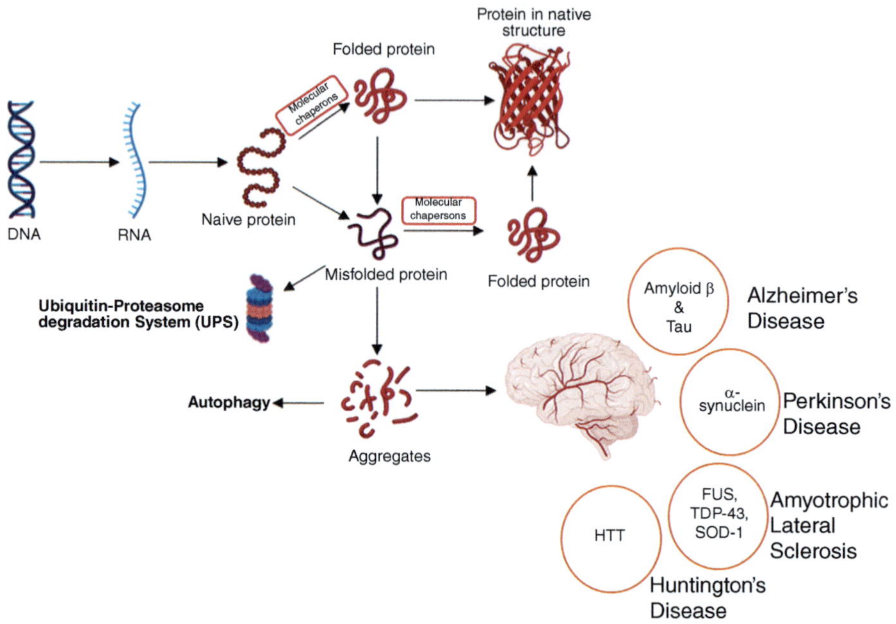

Fig. 5.1 Overview of protein misfolding and neurodegenerative diseases

coordinated activities of the hippocampus followed by cortical degeneration (memory), PD—primarily involves degeneration of the substantia nigra of the brainstem (motor); HD—as the name implies, involves neurological degeneration of the basal ganglia of the brain and subsequently produces a motor movement disorder, psychiatric symptoms, and cognitive declines, etc. ALS generally overlaps into both the "AD" and "PD" category in that it involves the loss of lower motor neuron function (spinal cord and brainstem) and further demonstrates upper and lower motor neuron signs (the corticospinal tract and frontal lobe implications and other craniofacial signs) [50]. All neurodegenerative diseases may have a latent period for weeks to decades, depending on the rate of cellular process decline, but generally have a slow pathway of disease progress over time, where motor, cognitive, and behavioral consequences may be evident [50]. The pathophysiology of degenerative changes, regardless of the label attached (to an NDD), tends to be a course that spans years and can be influenced by corelated and modifiable factors (genetic and environmental/lifestyle) that are under control.

More recent research has also shed light on the cellular and molecular events, especially those involving protein homeostasis, or "proteostasis," that dictate the life-or-death fate of neurons under conditions of stress related to aging and disease.

5.2 Folding and Misfolding of Proteins

Proteins are fundamental cellular components, mediating a multitude of functions including performance and activity; however, to serve these functions with specificity, proteins require both folding catalysts as well as molecular chaperones, forming a critical role in the expression and formation of functional protein structures under strictly regulated conditions. On the other hand, misfolding of a protein may give rise to either a toxic or inactive form. The neurodegenerative diseases are notable for the aggregation into fibrils or plaques of misfolded proteins [43]. These aggregates prevent the breakdown of protein and mitochondrial function. Apart from this, the possibility for the abnormally folded proteins to evade quality control mechanisms also contributes to the progressive nature of neurodegenerative disorders. Causes range from aging and external stresses to genetic mutants.

5.2.1 Mechanism of Protein Folding

5.2.1.1 Role of Cellular Physiology in Protein Folding

Proteins are the most basic cellular macromolecules performing an enormous range of functions, from providing structural support to catalyzing metabolic processes. From this perspective, it would mean that the amino acid sequences of proteins have to fold within the cytosol and endoplasmic reticulum of cells into a precise three-dimensional (3D) form; it needs to be a multistage phenomenon so that homeostasis, and cellular function, is not adversely affected. One of the quality control systems developed by the cell for monitoring and regulation folding to preserve proteostasis is in its endoplasmic reticulum unfolded protein response (UPR) [10].

5.2.1.2 Chaperone-Mediated Protein Folding

Molecular chaperones are crucial components of protein folding because they enable immature or misfolded proteins to reach and keep their correct conformation. Chaperones are grouped into proteins such as peptidyl-prolyl isomerases, heat-shock proteins, and protein disulfide isomerases [10]. Some folding operations that they support include:

(a) *Chaperones Provide Help in Folding*: Chaperones bind to partially or electrochemically denatured polypeptide to prevent wrong interactions that lead to aggregation [6].
(b) *Refolding of Misfolded Proteins*: Chaperones are also capable of rearranging or refolding existing misfolded proteins from states that resulted due to oxidative damage or heat shock as these can be considered types of cellular stress [22].
(c) *The Proteostasis Network*: Chaperones direct misfolded proteins to degradation pathways, including autophagy-lysosome-mediated degradation of protein aggregates or the ubiquitin-proteasome system (UPS) when refolding is not feasible [2].

5.2.2 Protein Misfolding

If a protein cannot correctly fold into its overall three-dimensional structure, it may misfold and adopt a dangerous or dysfunctional conformation [38]. Protein misfolding is caused by various mechanisms such as:

(a) *Folding Pathway Errors*—wrong folding can randomly happen creating the intermediates that are misfolded [1].
(b) *Loss-of-Function Mutations*—mutations changing the amino acid sequence of proteins may also lead to alternative protein shapes and hence destabilize misfolded back to their native conformation [5].
(c) *Posttranslational Modifications*—can block defect-causing posttranslational modifications that interfere with folding of a typically properly folded protein and lead to amyloidogenic propensity [35]. Such modified form include phosphorylation, glycosylation, and oxidization.
(d) *Hydrophobic Regions in Proteins*—ones that are normally tucked inside their folded, 3D native state—become exposed when those proteins misfold. Aggregates may develop as a result of interactions between these exposed areas and other misfolded proteins. The accumulation of misfolded proteins can lead to cellular toxicity and the advancement of illness as they can evade cellular quality control systems [12].

5.2.2.1 Triggers for Misfolding

Protein misfolding can be brought on by a number of causes, which can lead to the development and progression of NDDs:

(a) *Genetic Mutations*: Some of the genes can have inherited mutation; such mutations will yield proteins with the intrinsic higher tendency to misfold and be aggregated. Mutations in the APP gene (i.e., amyloid precursor protein) are linked to an early-onset form of Alzheimer's disease (AD) [18], while mutations in the SNCA gene (i.e., α-synuclein encoding gene) are noted for familial Parkinson's disease (PD) [11].
(b) *Extrinsic Factors*: Several environmental agents like metals, poisons, and reactive oxygen species (ROS) can provoke protein misfolding and inflict loss-of-function or gain-of-toxicity operations. Tau and α-synuclein proteins are critical in the pathophysiology of Alzheimer's disease (AD) and Parkinson's disease (PD), respectively, and their aggregation has been shown to be related to pesticides and heavy metals [17].
(c) *Aging*: For all NDDs, aging is a major risk factor. With aging, the efficacy of the cellular quality control systems weakens, and the build-up of oxidative damage rises. Over time, this increases the likelihood that proteins may misfold and aggregate [39].

5.3 Types of Protein Aggregates

The aggregates of proteins can have different forms, and their toxicity can vary [45].

(a) *Oligomers*, small soluble aggregates that disrupt the cell membranes and synaptic functions.
(b) *Fibrils*, when the monomers of a misfolded protein polymerize into long insoluble aggregates.
(c) *Amyloids* are highly ordered due to the cross-beta sheet structure. Amyloid may be responsible for the intracellular inclusions of PD and ALS, as well as extracellular plaques in AD.

5.4 Consequences of Protein Misfolding in Neurodegenerative Disease

The misfolding of proteins has now been proven to be a major pathogenic component in an enormous group of neurodegenerative disorders. In every type of the diseases, it was established that the misfolding and aggregation of certain proteins caused the malfunctioning and death of neurons.

5.4.1 Alzheimer's Disease (AD)

Alzheimer's is one of the main types of dementia, which affects memory and other cognitive processes [41]. It is characterized by two major pathological features as extracellular deposition of amyloid-beta (Aβ) plaques [21] and intra-neuronal formation of neurofibrillary tangles, composed of hyperphosphorylated tau protein. Aβ is derived from an amyloid peptide produced by proteolytic cleavage of a precursor protein that accumulates in the brain. They are found as insoluble fractions and cause toxicity as oligomers, fibrils, or plaques [21]. In neurons, the tau protein complexes with microtubules to stabilize them, but when it is hyperphosphorylated these complexes dwindle while the unbound tau builds up and then forms neurofibrillary tangles (NFT) [3]. This disruption causes damage to synaptic communication and leads to neuronal loss. Toxicity is mainly caused by membrane disruption, oxidative stress, and inflammatory responses mediated by the activation of microglia. Hyperphosphorylation of tau contributes to cell death by interfering with axonal transport and enhancing synaptic dysfunction, while interaction of Aβ with tau aggregates aggravates neurodegeneration [18].

5.4.2 Parkinson's Disease (PD)

Parkinson's disease is a disorder of movement that meets significantly increased criteria, namely, tremor, rigidity, bradykinesia, and postural instability.

Pathologically, the hallmark in PD is intracellular α-synuclein-containing Lewy bodies in dopaminergic neurons of the substantia nigra [40]. In PD, α-synuclein misfolds, forms oligomers, and eventually aggregates into fibrils and Lewy bodies, where it induces neuron toxicity through the alteration of synaptic function. These misfolded forms of α-synuclein are transmitted from neuron to neuron in a prion-like fashion, thereby propagating the disease pathologically. The signature feature of PD is the loss of mitochondrial function, and this leads to increased generation of ROS with cellular component damage. Oxidative stress is further enhanced by the direct impact of α-synuclein oligomers on mitochondrial dynamics, thereby resulting in neuronal damage [16].

5.4.3 Huntington's Disease (HD)

It is caused by the mutation of huntington gene (HTT) by the expansion of CAG trinucleotide repeats, which produces the toxic oligomers and intracellular inclusions [4]. These inclusions cause anomalies to normal cellular functions such as transcriptional regulation, mitochondrial functions, and axonal transport, and thus contribute to neurodegeneration. One of the main mechanisms associated with HD pathology is excitotoxicity through the overactivation of glutamate receptors. Mutated HTT (mHTT) protein causes dysfunction in glutamate transporter function, so that glutamate levels increases outside the neurons, thereby resulting in overactivation of N-methyl-D-aspartate (NMDA) receptors. All these have led to selective neurodegeneration in the striatum and cortex through various mechanisms, including but not limited to calcium influx, mitochondrial dysfunction, and activation of apoptotic pathways, leading eventually to neuronal death [4].

5.4.4 Amyotrophic Lateral Sclerosis (ALS)

ALS is a neurodegenerative disease in which motor neurons are lost from the brain and spinal cord. There could be many factors to cause this disease such as muscle weakness, muscle atrophy, or muscle paralysis. Conversely, there is still a direct link to disease for some of the proteins that aggregate (98–100), inhibiting events such as RNA processes, protein translation, and autophagy like TDP-43, FUS, and SOD1 [33]. Here, they are mislocalized from the nucleus into the cytoplasm to eventually form insoluble aggregates. The ALS protein aggregates, most importantly, exert mitochondrial dysfunction, oxidative stress, and affected RNA metabolism. This leads to mitochondrial dynamics and most often results in energy deficits and increased production of reactive oxygen species (ROS) [37]. In addition, it interferes with cellular pathways and stress response as well as the proteostasis mechanisms and causes the death of motor neurons [19].

5.5 Protein Misfolding and Cell Death

Apoptosis Membrane blebbing and cell shrinkage are resultant of the protein misfolding and are responsible for DNA fragmentation and cell death. For instance, Aβ oligomers that induce Alzheimer's disease cause permeabilization of the outer mitochondrial membrane, leading to further leakage of cytochrome c, as this enhances activation of caspases and induction of intrinsic pathways for apoptosis [32]. Aggregated, misfolded SOD1 for ALS was shown to induce apoptosis through caspase-8 activation and death receptors as well [36].

Necroptosis When mixed lineage kinase domain-like protein (MLKL) or receptor-interacting protein kinases 1 (RIPK 1) and RIPK 3 are activated, they form a complex that leads to necroptosis. But if they are blocked or saturated, necroptosis pathway is inhibited and causes cell death in neurodegenerative diseases [29]. Recent work on α-synuclein aggregates in Parkinson's and tau clumps in Alzheimer's may trigger the death of neurons and forward progression of disease [13, 49].

5.6 Therapeutic Strategies to Target Protein Misfolding

In particular, neurodegenerative diseases (NDDs) such as Alzheimer's disease (AD), Parkinson's disease (PD), Huntington's disease (HD), and amyotrophic lateral sclerosis (ALS) are well-known proteinopathies caused by the misfolding and aggregation of proteins due to an exacerbated cellular stress response that results in loss of function and cell death.

5.6.1 Small Molecule Inhibitors

They have been made in such a way to target the aggregation-prone proteins head-on or the native conformations in a stabilized manner so as to stop the misfolding and aggregating. β-secretase inhibitors have been made to help prevent the accumulation of Aβ peptides by inhibiting the enzymes that are responsible for the generation of the aggregated peptides and through designing counter-proteins for such a disease as AD [48]. The α-synuclein aggregation inhibitors have given hope to save people as they have further reduced the production of the toxic and the fatal aggregates and the neurotoxic that result from them in animal models of PD [9]. Compounds that are small have been categorized to be able to prevent misfolding by stabilizing the proteins in their native conformation.

5.6.2 Regulation of Molecular Chaperones

Molecular chaperones such as endogenous and pharmacological chaperones are required to correct folding of proteins to prevent aggregation of misfolded proteins inside the tissues. Heat-shock proteins (HSPs) are type of chaperones that may maintain protein homeostasis by increasing the endogenous activity. Indeed, it has been demonstrated that overexpression of HSP70 decreases cellular toxicity through inhibiting polyglutamine-expanded huntingtin aggregation in HD models [8]. Pharmacological chaperones are small molecules that stabilize the native forms of some proteins, preventing their misfolding and aggregation. For instance, ambroxol has been shown to act as a pharmacological chaperone of glucocerebrosidase, enhancing its activity in both Gaucher disease and Parkinson's disease related to the induction of proper folding of the protein and a reduction in α-synuclein aggregation [9].

5.6.3 Proteostasis Regulators

It is controlled by proteostasis regulators that include the autophagy-lysosome pathway and the ubiquitin-proteasome system (UPS) as a means of enhancing cellular capabilities in respect of the maintenance of protein homeostasis. In models of AD, PD, and HD, drugs like rapamycin and derivatives or rapalogs inhibit the mTOR signaling pathway to elicit induction of autophagy [31]. Protease activators have also proved to promote the removal of the neurotoxic proteins associated with other NDDs [47].

5.6.4 Immunotherapy Approaches

Immunotherapy utilizes the body's immune system, as in the case of Monoclonal Antibodies, to flush out the clumps of the protein that had got misfolded. They are engineered to target specifically to neutralize or otherwise reduce the toxic protein clumps. One good example is aducanumab, drug that targets Aβ aggregates in AD and increases clearance, and it slows down the rate of cognitive decline in clinical trials [42].

5.7 Neuroprotective Strategies

Neuroprotective therapy targets reductions in neuronal damage and loss seen as typical variations of NDDs by indicating to protect neurons and the function of the neurological system. These approaches all intervene at the level of molecular mechanisms underlying neuronal dysfunction, i.e., oxidative stress, mitochondrial damage, inflammation, and protein misfolding.

5.7.1 Natural Products for Neuroprotection

Efforts have been made to find the natural products of plants and other organisms since times ago to save neurons damage during aging as they show the neuroprotective activity. These are just a few of the numerous compounds that could be used in the treatment of NDDs, given their antioxidant, anti-inflammatory, and antiapoptotic properties.

Polyphenols They are a family of natural-origin bioactive compounds in fruits, vegetables, and even beverages like teas and wines. Their antioxidant activity has been reported as very strong because they scavenge free radicals and reduce oxidative stress, which may well be responsible for most of the neuronal damage actually manifested in NDDs. There is a lot of research on the presence of resveratrol, which is a polyphenol that is present in red wine. It is tested on models of both AD and PD. Activation of the SIRT1 pathway increases the function of mitochondria, activates autophagy, and prevents the formation of amyloid-β in models of AD [15].

There is also a prospect for curcumin as an amyloid plaque-reducing agent in AD, and also as tau aggregation inhibitor and in preventing loss of dopaminergic neurons in models of PD [7].

Flavonoids Flavonoids are yet another category of phytochemicals encompassing the ability to act as neuroprotective agents. These phytochemicals also regulated signaling mechanisms associated with oxidative stress and inflammation, cardinal features of NDD pathology. Of the flavonoids in green tea, epigallocatechin gallate is claimed as inhibitory to amyloid-β protein (principal cause of Alzheimer's disease) and on α-synuclein protein (formed in abnormal state, causes Parkinson's disease). It also inhibits neuroinflammation and pro-inflammatory cytokines, which is an important anti-inflammatory mechanism of action [26]. There is more than just primary prevention of onset; Quercetin acts as an agent to neutralize oxidative stress and trigger autophagy—all in the hope that the misfolded proteins which caused NDDs will not aggregate [14].

Other Bioactive Compounds Ginseng is rich in ginsenosides while *Bacopa monnieri* produces bacosides; all these bioactive compounds established to possess neuroprotective activities in diverse models of neurodegenerations. Oxidative stress reduction, increased synaptic plasticity, and neuronal survival have been demonstrated by these compounds [27, 52].

5.7.2 Mechanisms of Neuroprotection by Natural Products

Several mechanisms are proposed in which natural products may provide protection to neurons:

Oxidative Activity Natural products are scavengers of reactive oxygen species (ROS), thereby preventing oxidative damage to neuronal membranes, proteins, and DNA [27].

Anti-Inflammatory Activity The compound was found to be microglia and astrocytes activator inhibitor and the central mediators of neuro-inflammatory in NDDs. Thus, there is a reduction of the release of pro-inflammatory cytokines such as TNF-α and IL-6 that may cause neuronal damage [52]. Most of the natural compounds have been known as inducers of autophagy, a cellular degradation mechanism that removes misfolded proteins. Maintained by the action of degradative pathways, it sets up proteostasis and ensures neuronal survival through the clearing of protein aggregates [31].

Gene Therapy Gene therapy should become a promising strategy to tackle NDDs at the very roots by tackling the genetic cause of protein misfolding leading to cellular dysfunction. With scientific breakthroughs like the CRISPR-Cas9 gene editing tool and RNA interference (RNAi), it has been possible to intervene precisely with this kind of disease at its very core or affect the impact of disease-causing genes.

RNA Interference (RNAi) RNAi is the process in which, specific mRNA get degraded by siRNA molecules, subsequently resulting in no production of proteins, which cause diseases. This is the exact approach that is useful in diseases like Huntington's in which a toxic protein is expressed from HTT gene and RNAi targets it, reducing mutant huntingtin levels which improves motor function in HD animal models [20].

Gene Editing with CRISPR-Cas9 This latest advancement in DNA modifications is a direct result of the CRISPR-Cas9 gene-editing tool. In NDDs, mutations inducing protein misfolding would be prevented from being expressed or counteracted after having been secreted by inhibiting their ensuing pathological consequences. CRISPR/Cas9 genome editing of the AD risk factor, apo-lipoprotein E4 (APOE4), has been shown to suppress amyloid β production and aggregation [51, 53].

Gene Transfer of Neuroprotective Genes Neurons are also cured by the transfer of neuroprotective genes to neurons via gene therapy. For instance, through the transfection of Brain-Derived Neurotrophic Factor (BDNF) genes in neurons, there was observed to be an increase in survival of neurons, synapse plasticity, and potentially neurogenesis. BDNF gene therapy shows some of the highest promise in both AD and PD models; therefore, it enhances cognitive and behavioral performance [28].

Mechanisms of Neuroprotection by Gene Therapy The gene-therapeutic strategy leads to neuroprotection simply by reducing the expression of a toxic protein. In fact, both the RNAi and CRISPR-Cas9 can lower or remove the misfolded protein then existing in NDDs. Neurotrophic factors such as BDNF, which enhance survival pathways and synaptic function, would favor neuronal longevity [28, 34, 46]. Similarly, gene therapy arranges for correction of the genetic mutations; hence, normal cellular processes are restored and neurodegeneration can be halted.

5.8 Future Perspectives and Challenges

For many years, research has been ongoing to create a potential neurodegenerative disease (NDD) cure. Understanding protein misfolding, and how it contributes to diseases progression has led to many discoveries, but still also presents many challenges. Most of these are due to in part the restrictions within current therapeutic approaches; predict new combinatorial treatment possibilities and underscore the critical importance of using a personalized medicine approach to drug-discovery for NDDs.

Because of blood-brain barrier, the efficacy and specificity of existing therapies such as small molecule inhibitors and monoclonal antibodies [44] are often restricted. Delivering drugs with nanoparticles and focused ultrasound is equally difficult [24, 25]. In addition, the heterogeneous nature of disease that complicates the path to universal therapies means that some mechanisms underpinning protein aggregate formation may not deal with other aspects of disease pathology [41].

5.9 Conclusion

Protein misfolding is a frequent feature of neurodegenerative diseases such as Alzheimer's disease, Parkinson's disease, and Huntington's disease that ends with neuronal death. All those diseases disrupt cellular homeostasis, so it includes neuroinflammation and oxidative stress that implicates in mitochondrial dysfunction. This has culminated in the identification of new therapeutic approaches that can prevent or slow down disease progression. Key neuoprotective natural compounds include polyphenols and flavonoids, while gene therapy strategies, including the inhibition of misfolded protein production and genetic mutations correction, are breakthrough (i.e., RNA interference, CRISPR-Cas9 gene editing). Such an approach brings attention to the need for multiple-targeted strategies in order to stop and balance out damage as cells are oddly complex structures.

Additional studies will be necessary in their optimization with the goal of developing new therapies. It is the power of integrating discoveries in molecular biology with therapeutic advances that holds the potential to delay — and perhaps someday even reverse — the slow-motion cascade of neurodegeneration that steals hope from those living under a life sentence in this invisible prison.

References

1. Anfinsen CB. Principles that govern the folding of protein chains. Science. 1973;181(4096):223–30. https://doi.org/10.1126/science.181.4096.223.
2. Balch WE, Morimoto RI, Dillin A, Kelly JW. Adapting Proteostasis for disease intervention. Science. 2008;319(5865):916–9. https://doi.org/10.1126/science.1141448.
3. Ballatore C, Lee VM-Y, Trojanowski JQ. Tau-mediated neurodegeneration in Alzheimer's disease and related disorders. Nat Rev Neurosci. 2007;8(9):663–72. https://doi.org/10.1038/nrn2194.
4. Bates GP, Dorsey R, Gusella JF, Hayden MR, Kay C, Leavitt BR, Nance M, et al. Huntington disease. Nat Rev Dis Prim. 2015;1(1):15005. https://doi.org/10.1038/nrdp.2015.5.
5. Bucciantini M, Giannoni E, Chiti F, Baroni F, Formigli L, Zurdo J, Taddei N, Ramponi G, Dobson CM, Stefani M. Inherent toxicity of aggregates implies a common mechanism for protein Misfolding diseases. Nature. 2002;416(6880):507–11. https://doi.org/10.1038/416507a.
6. Bukau B, Weissman J, Horwich A. Molecular chaperones and protein quality control. Cell. 2006;125(3):443–51. https://doi.org/10.1016/j.cell.2006.04.014.
7. Chandrasekaran V, Hediyal TA, Anand N, Kendaganna PH, Gorantla VR, Mahalakshmi AM, Ghanekar RK, Yang J, Sakharkar MK, Chidambaram SB. Polyphenols, autophagy and neurodegenerative diseases: a review. Biomol Ther. 2023;13(8):1196. https://doi.org/10.3390/biom13081196.
8. Ciechanover A, Kwon YT. Degradation of misfolded proteins in neurodegenerative diseases: therapeutic targets and strategies. Exp Mol Med. 2015;47(3):e147. https://doi.org/10.1038/emm.2014.117.
9. Colucci F, Avenali M, De Micco R, Poli MF, Cerri S, Stanziano M, Bacila A, et al. Ambroxol as a disease-modifying treatment to reduce the risk of cognitive impairment in GBA -associated Parkinson's disease: a multicentre, randomised, double-blind, placebo-controlled, phase II trial. The AMBITIOUS Study Protocol. BMJ Neurol Open. 2023;5(2):e000535. https://doi.org/10.1136/bmjno-2023-000535.
10. Dobson CM. Protein folding and Misfolding. Nature. 2003;426(6968):884–90. https://doi.org/10.1038/nature02261.
11. Eisele YS, Monteiro C, Fearns C, Encalada SE, Luke Wiseman R, Powers ET, Kelly JW. Targeting protein aggregation for the treatment of degenerative diseases. Nat Rev Drug Discov. 2015;14(11):759–80. https://doi.org/10.1038/nrd4593.
12. Eisenberg D, Jucker M. The amyloid state of proteins in human diseases. Cell. 2012;148(6):1188–203. https://doi.org/10.1016/j.cell.2012.02.022.
13. Exner N, Lutz AK, Haass C, Winklhofer KF. Mitochondrial dysfunction in Parkinson's disease: molecular mechanisms and pathophysiological consequences: mitochondrial dysfunction in Parkinson's disease. EMBO J. 2012;31(14):3038–62. https://doi.org/10.1038/emboj.2012.170.
14. Faysal M, Dehbia Z, Zehravi M, Sweilam SH, Akiful Haque M, Kumar KP, Chakole RD, et al. Flavonoids as potential therapeutics against neurodegenerative disorders: unlocking the prospects. Neurochem Res. 2024;49(8):1926–44. https://doi.org/10.1007/s11064-024-04177-x.
15. Floyd JA, Galperin A, Ratner BD. Drug encapsulated polymeric microspheres for intracranial tumor therapy: a review of the literature. Adv Drug Deliv Rev. 2015;91(August):23–37. https://doi.org/10.1016/j.addr.2015.04.008.
16. Friedman JH. Dementia with Lewy bodies and Parkinson disease dementia: it is the same disease! Parkinsonism Relat Disord. 2018;46(January):S6–9. https://doi.org/10.1016/j.parkreldis.2017.07.013.
17. Gitler AD, Chesi A, Geddie ML, Strathearn KE, Hamamichi S, Hill KJ, Caldwell KA, et al. a-Synuclein is part of a diverse and highly conserved interaction network that includes PARK9 and manganese toxicity. Nat Genet. 2009;41(3):308–15. https://doi.org/10.1038/ng.300.
18. Goedert M, Eisenberg DS, Anthony Crowther R. Propagation of tau aggregates and neurodegeneration. Annu Rev Neurosci. 2017;40(1):189–210. https://doi.org/10.1146/annurev-neuro-072116-031153.

19. Grosskreutz J, Van Den Bosch L, Keller BU. Calcium dysregulation in amyotrophic lateral sclerosis. Cell Calcium. 2010;47(2):165–74. https://doi.org/10.1016/j.ceca.2009.12.002.
20. Ha M, Narry Kim V. Regulation of microRNA biogenesis. Nat Rev Mol Cell Biol. 2014;15(8):509–24. https://doi.org/10.1038/nrm3838.
21. Hardy J, Selkoe DJ. The amyloid hypothesis of Alzheimer's disease: progress and problems on the road to therapeutics. Science. 2002;297(5580):353–6. https://doi.org/10.1126/science.1072994.
22. Hartl FU, Hayer-Hartl M. Molecular chaperones in the cytosol: from nascent chain to folded protein. Science. 2002;295(5561):1852–8. https://doi.org/10.1126/science.1068408.
23. Hetz C, Saxena S. ER stress and the unfolded protein response in neurodegeneration. Nat Rev Neurol. 2017;13(8):477–91. https://doi.org/10.1038/nrneurol.2017.99.
24. Hsu C-Y, Rheima AM, Kadhim MM, Ahmed NN, Mohammed SH, Abbas FH, Abed ZT, et al. An overview of nanoparticles in drug delivery: properties and applications. S Afr J Chem Eng. 2023;46(October):233–70. https://doi.org/10.1016/j.sajce.2023.08.009.
25. Huang L, Huang X-H, Yang X, Jia-Qin H, Zhu Y-Z, Yan P-Y, Xie Y. Novel Nano-drug delivery system for natural products and their application. Pharmacol Res. 2024;201(March):107100. https://doi.org/10.1016/j.phrs.2024.107100.
26. Hussain G, Zhang L, Rasul A, Anwar H, Sohail M, Razzaq A, Aziz N, Shabbir A, Ali M, Sun T. Role of plant-derived flavonoids and their mechanism in attenuation of Alzheimer's and Parkinson's diseases: an update of recent data. Molecules. 2018;23(4):814. https://doi.org/10.3390/molecules23040814.
27. Jeyasri R, Muthuramalingam P, Suba V, Ramesh M, Chen J-T. Bacopa Monnieri and their bioactive compounds inferred multi-target treatment strategy for neurological diseases: a cheminformatics and system pharmacology approach. Biomol Ther. 2020;10(4):536. https://doi.org/10.3390/biom10040536.
28. Lee MH, Kang S, Um K-H, Lee SW, Hwang H, Baek K, Choi JW. Brain-targeted delivery of neuroprotective survival gene minimizing hematopoietic cell contamination: implications for Parkinson's disease treatment. J Transl Med. 2024;22(1):53. https://doi.org/10.1186/s12967-023-04816-x.
29. Liddelow SA, Guttenplan KA, Clarke LE, Bennett FC, Bohlen CJ, Schirmer L, Bennett ML, et al. Neurotoxic reactive astrocytes are induced by activated microglia. Nature. 2017;541(7638):481–7. https://doi.org/10.1038/nature21029.
30. Lindholm D, Wootz H, Korhonen L. ER stress and neurodegenerative diseases. Cell Death Differ. 2006;13(3):385–92. https://doi.org/10.1038/sj.cdd.4401778.
31. Menzies FM, Fleming A, Rubinsztein DC. Compromised autophagy and neurodegenerative diseases. Nat Rev Neurosci. 2015;16(6):345–57. https://doi.org/10.1038/nrn3961.
32. Morris M, Maeda S, Vossel K, Mucke L. The many faces of tau. Neuron. 2011;70(3):410–26. https://doi.org/10.1016/j.neuron.2011.04.009.
33. Neumann M, Sampathu DM, Kwong LK, Truax AC, Micsenyi MC, Chou TT, Bruce J, et al. Ubiquitinated TDP-43 in frontotemporal lobar degeneration and amyotrophic lateral sclerosis. Science. 2006;314(5796):130–3. https://doi.org/10.1126/science.1134108.
34. Parambi DG, Thomas KS, Alharbi RK, Harilal S, Batiha GE-S, Cruz-Martins N, Magdy O, Musa A, Panda DS, Mathew B. Gene therapy approach with an emphasis on growth factors: theoretical and clinical outcomes in neurodegenerative diseases. Mol Neurobiol. 2022;59(1):191–233. https://doi.org/10.1007/s12035-021-02555-y.
35. Prusiner SB. Biology and genetics of prions causing neurodegeneration. Annu Rev Genet. 2013;47(1):601–23. https://doi.org/10.1146/annurev-genet-110711-155524.
36. Raoul C, Estévez AG, Nishimune H, Cleveland DW, deLapeyrière O, Henderson CE, Haase G, Pettmann B. Motoneuron death triggered by a specific pathway downstream of Fas. Neuron. 2002;35(6):1067–83. https://doi.org/10.1016/S0896-6273(02)00905-4.
37. Renton AE, Chiò A, Traynor BJ. State of play in amyotrophic lateral sclerosis genetics. Nat Neurosci. 2014;17(1):17–23. https://doi.org/10.1038/nn.3584.
38. Ross CA, Poirier MA. Protein aggregation and neurodegenerative disease. Nat Med. 2004;10(S7):S10–7. https://doi.org/10.1038/nm1066.

39. Rubinsztein DC. The roles of intracellular protein-degradation pathways in neurodegeneration. Nature. 2006;443(7113):780–6. https://doi.org/10.1038/nature05291.
40. Schapira AH. Mitochondria in the Aetiology and pathogenesis of Parkinson's disease. Lancet Neurol. 2008;7(1):97–109. https://doi.org/10.1016/S1474-4422(07)70327-7.
41. Selkoe DJ, Hardy J. The amyloid hypothesis of Alzheimer's disease at 25 years. EMBO Mol Med. 2016;8(6):595–608. https://doi.org/10.15252/emmm.201606210.
42. Sevigny J, Chiao P, Bussière T, Weinreb PH, Williams L, Maier M, Dunstan R, et al. The antibody Aducanumab reduces Aβ plaques in Alzheimer's disease. Nature. 2016;537(7618):50–6. https://doi.org/10.1038/nature19323.
43. Soto C. Unfolding the role of protein Misfolding in neurodegenerative diseases. Nat Rev Neurosci. 2003;4(1):49–60. https://doi.org/10.1038/nrn1007.
44. Sperling RA, Aisen PS, Beckett LA, Bennett DA, Craft S, Fagan AM, Iwatsubo T, et al. Toward defining the preclinical stages of Alzheimer's disease: recommendations from the National Institute on Aging-Alzheimer's Association Workgroups on diagnostic guidelines for Alzheimer's disease. Alzheimers Dement. 2011;7(3):280–92. https://doi.org/10.1016/j.jalz.2011.03.003.
45. Taylor JP, Hardy J, Fischbeck KH. Toxic proteins in neurodegenerative disease. Science. 2002;296(5575):1991–5. https://doi.org/10.1126/science.1067122.
46. Valdés P, Schneider BL. Gene therapy: a promising approach for neuroprotection in Parkinson's disease? Front Neuroanat. 2016;10(December) https://doi.org/10.3389/fnana.2016.00123.
47. van Oosten-Hawle P, Porter RS, Morimoto RI. Regulation of organismal Proteostasis by transcellular chaperone signaling. Cell. 2013;153(6):1366–78. https://doi.org/10.1016/j.cell.2013.05.015.
48. Vassar R, Bennett BD, Babu-Khan S, Kahn S, Mendiaz EA, Denis P, Teplow DB, et al. b-Secretase cleavage of Alzheimer's amyloid precursor protein by the transmembrane aspartic protease BACE. Science. 1999;286(5440):735–41. https://doi.org/10.1126/science.286.5440.735.
49. Volpicelli-Daley LA, Luk KC, Patel TP, Tanik SA, Riddle DM, Stieber A, Meaney DF, Trojanowski JQ, Lee VM-Y. Exogenous α-Synuclein fibrils induce Lewy body pathology leading to synaptic dysfunction and neuron death. Neuron. 2011;72(1):57–71. https://doi.org/10.1016/j.neuron.2011.08.033.
50. Wang X. Selective neuronal vulnerability to oxidative stress in the brain. Front Aging Neurosci. 2010;2 https://doi.org/10.3389/fnagi.2010.00012.
51. Yang S, Chang R, Yang H, Zhao T, Hong Y, Kong HE, Sun X, et al. CRISPR/Cas9-mediated gene editing ameliorates neurotoxicity in mouse model of Huntington's disease. J Clin Invest. 2017;127(7):2719–24. https://doi.org/10.1172/JCI92087.
52. Yenisetti SC, Manjunath MJ, Muralidhara. Neuropharmacological properties of Withania Somnifera – Indian ginseng: an overview on experimental evidence with emphasis on clinical trials and patents. Recent Pat CNS Drug Discov. 2016;10(2):204. https://doi.org/10.2174/1574889810666160615014106.
53. Zhao D, Badur MG, Luebeck J, Magaña JH, Birmingham A, Sasik R, Ahn CS, Ideker T, Metallo CM, Mali P. Combinatorial CRISPR-Cas9 metabolic screens reveal critical redox control points dependent on the KEAP1-NRF2 regulatory Axis. Mol Cell. 2018;69(4):699–708.e7. https://doi.org/10.1016/j.molcel.2018.01.017.

Proteostasis Disturbances and Inflammation in Neurodegenerative Diseases

6

Shivani Malviya, Rupali Yadav, Neha Arya, Rajesh Singh, Priyanka V. Kashyap, and Ashok Kumar

6.1 Introduction

Proteostasis, or protein homeostasis, is the cellular process that regulates a delicate balance between protein synthesis, folding, and degradation. It is a multiphase process crucial for the maintenance of cellular health, particularly within the neurons, as they have limited regenerative capacity [27]. Therefore, dysregulated proteostasis greatly affects neuronal function, and it is one of the major hallmarks of various neurodegenerative disorders. Additionally, aging triggers various cellular transformations that significantly affect protein homeostasis. As individuals age, the efficiency of proteostatic mechanisms such as the ubiquitin-proteasome system (UBS) and autophagy declines [30]. This decline in cellular homeostasis leads to inefficiency in regulating protein folding and clearance, contributing to the accumulation of damaged proteins [6]. The accumulation of such proteins invokes cellular stress, including endoplasmic reticulum (ER) stress, mitochondrial stress, and ultimately leads to the activation of inflammation [11]. For example, the accumulation of amyloid-β in Alzheimer's disease (AD) and α-synuclein in Parkinson's disease (PD) explains how failures in protein homeostasis can instigate neuroinflammatory responses leading to neuronal death [4, 41]. The inflammatory environment not only aggravates proteostatic imbalances but also disrupts neuronal functions, contributing to cognitive decline and motor deficits [57]. Thus, the intricate relationship

S. Malviya · N. Arya
Department of Translational Medicine, All India Institute of Medical Sciences, Bhopal, India

R. Yadav · A. Kumar (✉)
Department of Biochemistry, All India Institute of Medical Sciences, Bhopal, India
e-mail: ashok.biochemistry@aiimsbhopal.edu.in

R. Singh
Department of Human Genetics, Banaras Hindu University, Varanasi, India

P. V. Kashyap
Department of Neurology, All India Institute of Medical Sciences, Bhopal, India

N. Puranik (ed.), *Proteostasis*, https://doi.org/10.1007/978-981-96-6202-9_6

between aging, proteostasis, and inflammation is important for understanding the biology of neurological disorders. Enhanced levels of inflammatory cytokines such as interleukin-1 beta (IL-1β) and tumor necrosis factor-alpha (TNF-α) have been well documented in the aging brain and their levels further elevate in case of neurodegenerative disorders [17]. Along with cytokines, activation of glial cells like microglia and astrocytes further instigates the overall process and plays a pivotal role in the progression of neurodegenerative diseases such as Alzheimer's disease and Parkinson's disease [9]. Additionally, chronic inflammation can disrupt proteostasis, creating a feedback loop that exacerbates neuronal dysfunction and death [45].

Proteostatic stress induces an immune response. Besides cell-mediated immune response, the components of innate immunity are also involved in neuroinflammation. Inflammasomes are cytosolic multimeric complexes present in several cells, including immune cells, neurons, microglia, and astrocytes. Inflammasomes recognize pathogen-associated molecular patterns (PAMPs) and death-associated molecular patterns (DAMPs) [54]. One of such complexes, nucleotide-binding oligomerization domain leucine-rich repeat-containing protein (NLRP) 3, is a well-studied inflammasome, and its activation has been implicated in the onset and progression of several neurodegenerative diseases. Multiple studies based on cellular and animal models have demonstrated that protein aggregation induces glial cells, which release IL-1β and activate NLRP3 inflammasome pathway [13, 44].

In this chapter, we discuss the mechanism of proteostasis, particularly protein degradation pathways. Then, we deliberate the multifaceted relationship between proteostasis disturbances and inflammation including inflammasome pathways, and their implication in neurodegenerative disorders [54]. Furthermore, we highlight the potential therapeutic avenues that target these intertwined processes, aiming to perturb or mitigate the progression of neurodegenerative diseases.

6.2 Proteostasis in the Central Nervous System

Many environmental cues and pathological conditions can dysregulate the proteostasis. Heat shock, oxidative stress, proteasome dysfunction, mutations, and aging-related changes are just a few of the extracellular and intracellular stresses that can disrupt proteostasis [40]. In the upcoming section, we will emphasize on the mechanisms of the protein degradation by ubiquitin-protease system and autophagy.

As with other organs, maintaining proteostasis is necessary for the CNS to operate correctly and provide suitable stress responses. The vulnerability of CNS proteome to impaired proteostasis is attributed to its distinctive structural features, particularly their long processes and synapses, which are located far from the soma. The complexity of neuronal proteostasis is increased by the fact that CNS proteomes are extremely dynamic because of both constitutive and inducible protein and peptide secretion from both neuronal and glial cells [73].

6.2.1 Mechanisms of Protein Control

6.2.1.1 Ubiquitin-Proteasome System

Proteasomes and autophagy mediate the breakdown of intracellular proteins and the recycling of component amino acids in eukaryotic cells (Fig. 6.1). Ubiquitin-proteasome system (UPS) is responsible for approximately 80% of intracellular proteolysis [39]. The 76-residue evolutionarily conserved protein ubiquitin (Ub) first labels the UPS substrates (target proteins), after which the proteasome recognizes, unfolds, and proteolyze them. Protein ubiquitination is one of the most common post-translational modifications. A cascade of three enzymes, an E1 ubiquitin-activating enzyme, an E2 ubiquitin-conjugating enzyme, and an E3 ubiquitin ligase, is necessary for protein ubiquitination. Ubiquitin activation entails the creation of a thioester bond between an E1 active site cysteine and the C-terminal glycine residue of free Ub. The E1 enzyme then transfers the active ubiquitin to the E2 enzyme. The E3 ubiquitin ligase then interacts with the Ub–E2 complex and moves Ub to a lysine residue of the substrate protein through its ε-amino group [32]. Additional ubiquitins can be conjugated to any of the seven lysine residues of Ub to further modify the generated ubiquitin–protein complex. A polyubiquitin (polyUb)

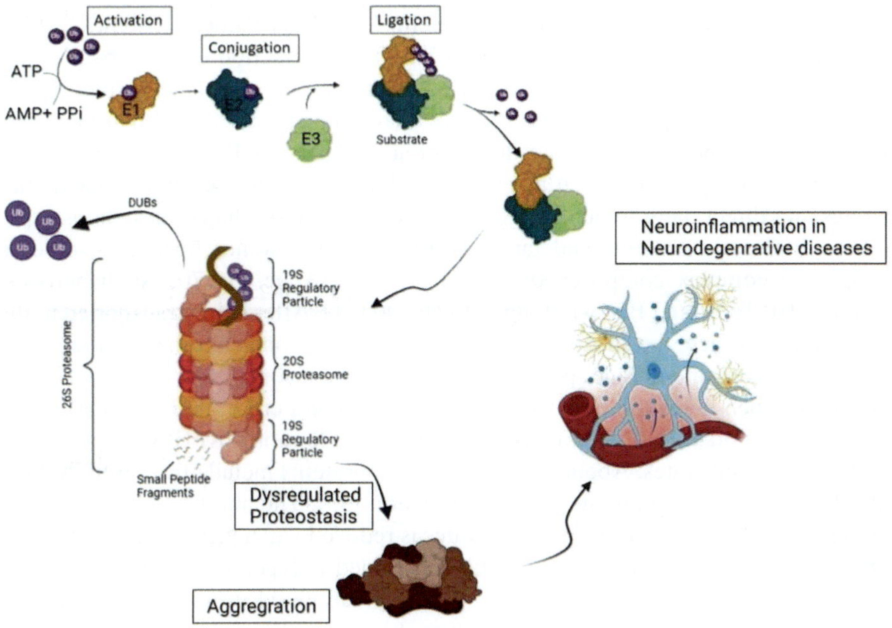

Fig. 6.1 Defects in protein control systems (proteosomal degradation or autophagy) causes dysregulated proteostasis, leading to protein aggregation, resulting in neuroinflammation, and thereby contributing to neurodegenerative diseases. (Made using Biorender.com) The subunits of the 19S cap recognize and deubiquitinate the poly-ubiquitinated proteins, which are then unfolded and sent to the 20S core for ATP-dependent degradation that produces short peptides (3–23 amino acid residues) [9, 48]

chain's topology and conjugated ubiquitin count predicts the fate of the altered proteins [59]. The 26S proteasome targets proteins with a Lys48-linked polyUb chain for degradation [14]. The multisubunit structure known as an active proteasome is made up of a 20S core and 19S caps [22].

6.2.1.2 Autophagy

Autophagy is a process by which cellular components, whole organelles, or intracellular pathogens are eliminated by lysosome-mediated degradation pathway [50]. Autophagy plays a vital role in various physiological and pathological processes such as tumor suppression, cell death, development, aging, defense against microbes, antigen presentation, adaptation to nutrient starvation, and the removal of extra or damaged organelles and proteins [21]. Lysosomes, single membrane-bound organelles, play a central role in this process. They contain a wide range of acid hydrolases, including endo and exo-proteases. Based on the trafficking of cargo molecules to lysosomes, autophagy is classified into three main categories; chaperone-mediated autophagy (CMA), microautophagy, and macroautophagy (often called autophagy). The processes of macroautophagy and CMA lead to degradation of misfolded and aggregated proteins [22]. The process involves the endocytosis and sequestration of cargo, its transport to lysosomes, and degradation followed by the utilization of degradation products [34]. The cytoplasmic components sequester into a double membrane vesicular organelle known as autophagosomes, which fuses with lysosome for further degradation. The production of autophagosomes and its fusion with lysosomes is regulated by numerous autophagic protein complexes [9]. The ubiquitinated substrates which majorly constitute misfolded proteins are resistant to proteasome degradation and degraded via macroautophagy. Various adaptor proteins, such as p62 or NRB1, recognize the ubiquitinated substrates and direct them to autophagosomes [48]. In CMA, chaperones choose specific cytoplasmic proteins and target them to the lysosomes [50]. It involves a chaperone complex composed of constitutive chaperone, Hsc70, co-chaperones; Hsp40, HIP, HOP, and BAG-1 along with Hsp90 [21]. After being transported to the lysosome's surface, the substrate attached to the chaperone complex interacts with the single-span membrane protein LAMP-2A. Lysosomal Hsc-70 (lys-Hsc70) chaperone then helps the targeted substrate unfold and cross the lysosomal membrane for degradation [27]. A consensus pentapeptide motif, KFERQ is present in all the CMA substrates. About 30% of cytosolic proteins include this motif, but not all of them are susceptible to CMA degradation [33]. The ability of chaperones to aid in the folding of their original substrates is reduced when they are trapped by the abnormally folded proteins. As a result, misfolded polypeptides that are prone to aggregation and folding intermediates progressively build up inside the cell [30]. Even though the proteostatic balance may be restored by a number of survival mechanisms, such as up-regulating the chaperone machinery and allowing the UPS and autophagy to degrade misfolded proteins, it is still a major challenge for neurons to maintain such equilibrium under stressful conditions or during the aging process.

6.3 Protein Misfolding and Aggregation in Neurodegenerative Diseases

As mentioned before, following the protein synthesis, if misfolded proteins, either due to defective protein-folding machinery or aberrant posttranslational modifications (PTMs), are not removed properly, it may lead to toxic aggregate formation [68]. Chaperones, UPS, or lysosomes are responsible for the removal of misfolded and aggregated proteins and maintain proteostasis [68]. Beta amyloids are present in the form of amorphous aggregates or fibrillar structures. These are resistant to degradation and results in impairment of the refolding/degradation process, leading to neurodegeneration [30]. In addition, such aggregates impede the entry of other substrates intended for destruction by binding to proteasomes and autophagosomes and result in a progressive decline in proteostasis [19]. For example, diminished proteasomal activity has been reported in cortical areas of Alzheimer's patients due to the accumulation of extracellular amyloid beta (Aβ) plaques [31, 53]. Additionally, tau proteins get hyperphosphorylated and form aggregates in Alzheimer's disease. Similarly, α-synuclein and Huntingtin aggregate intracellularly, impairing proteasomal action to exacerbate neuropathy in Parkinson's and Huntington's disease, respectively [16, 46, 66]. The conformational folding of proteins is aided by chaperones, and misfolded proteins are redirected toward lysosomes and proteasomes for destruction. Chaperones also assist ubiquitin ligases in identifying misfolded targets in order to deliver them to lysosomal membranes for embedding [9, 29]. Thus, the expression of chaperones is tightly regulated in neurons. Altered expression of chaperones has been reported during aging or neurodegenerative disorders; some of them are upregulated to protect against proteostatic stress while others are downregulated as a result of disease insults [61]. Moreover, a conserved proteostatic process called ER-associated protein degradation (ERAD) is also activated due to the prolonged production of misfolded proteins as a result of erroneous protein synthesis leading to ER stress and the activation of unfolded protein response (UPR) [55]. Neurodegeneration occurs when the buildup of pathogenic proteins exceeds the ERAD capacity [7].

6.4 Immunoproteasomes in Neurodegenerative Diseases

Immunoproteasome is a variant of canonical proteasome responsible for antigen processing and degradation of proteins inside the immune cells. This proteolytic complex slightly differs from the conventional proteasomes by the presence of more efficient catalytic subunits in the 20S core unit. The 20S core unit of the conventional proteasome is composed of two pairs of outer α rings with seven subunits, two pairs of inner β rings with seven β units, and two 19S regulatory subunits located at either end of the core subunits. Three beta subunits, i.e., β1, β2, and β3 of proteosome, are responsible for proteolytic activity [42]. In contrast these catalytic subunits are replaced by iβ1 (LMP-2, large multifunctional peptidase-2, or PSMB-9, proteasome subunit beta type-9), iβ2 (LMP-10 or MECL-1, multicatalytic

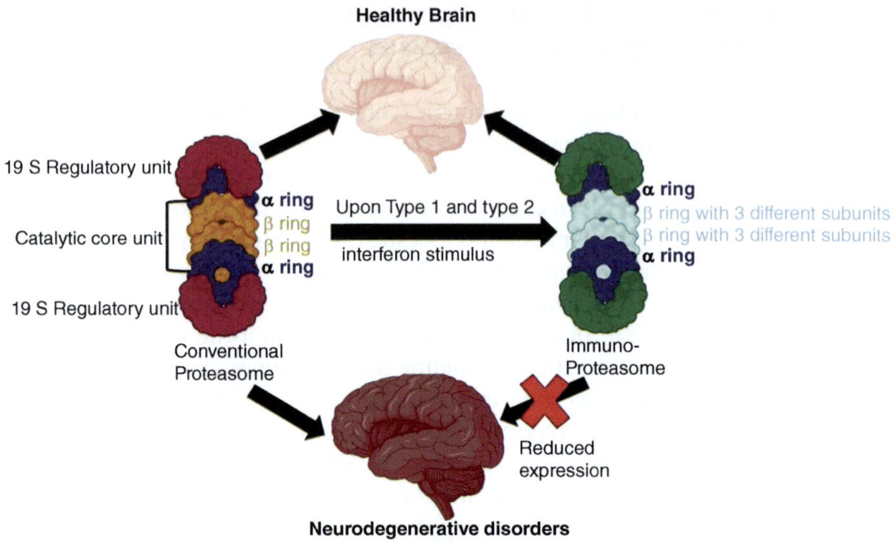

Fig. 6.2 Structural components of conventional proteasome and immunoproteasome. Its expression is either absent or reduced in neurodegenerative disorders. (Made using Biorender.com)

endopeptidase complex like-1), and iβ5 (LMP-7 or PSMB-10) in immunoproteasome as depicted in Fig. 6.2 [23, 49]. This efficient and modified version of the proteasome is responsible for carrying out the regular functions that the conventional proteasome does, along with the degradation of viral proteins and antigen presentation. In addition to immune cells, immunoproteasome assembly can also be induced upon Type 1 and Type 2 interferon stimulation [8, 42].

Recent studies demonstrated the instrumental role of immunoproteasome in regulating brain function during aging. Enhanced poly-ubiquitination of proteins has been reported in triple-knockout mice lacking immunoproteasome, which led to the development of epileptic seizures [42]. This finding was further corroborated by data from human epileptic patients showing the reduced expression of immunoproteasome. Moreover, mice lacking one of the core subunits, LMP7, have been reported to have impaired proteotoxic stress-mediated microglial function [8]. The cellular mechanism of how immunoproteasome protects the brain function is still elusive at the moment. Future studies are required to establish the fact that immunoproteasome can be a novel target for the treatment of chronic neurodegenerative diseases such as Alzheimer's, Parkinson, or epilepsy.

6.5 Neuroinflammation in Neurodegenerative Diseases

Neuroinflammation has emerged as a crucial factor in the pathogenesis of various neurodegenerative diseases, including Alzheimer's disease, Parkinson's disease, and multiple sclerosis [2, 24, 73]. In the upcoming section, we will describe various inflammatory responses in the brain, the roles of microglia and astrocytes, and how these processes contribute to neuronal dysfunction.

Neuroinflammation refers to the brain's immune response to injury or disease, characterized by the activation of glial cells and the release of inflammatory mediators [73]. Microglia and astrocytes are the primary responders to brain injury. Initially, their activation can promote repair and homeostasis in neurons, but in the chronic stage, it leads to a sustained inflammatory environment detrimental to neuronal health [58]. Chronic microglial activation has been seen in Alzheimer's disease, promoting neurodegeneration through the release of pro-inflammatory cytokines [20, 58]. Persistent inflammation is often observed in neurodegenerative diseases, contributing to a vicious cycle of neuronal damage and further inflammatory responses [20, 58]. Inflammatory mediators play a pivotal role in neuroinflammation, and the key players are cytokines, chemokines, and reactive oxygen species (ROS). Pro-inflammatory cytokines such as TNF-α, IL-1β, and IL-6 are critical in regulating the inflammatory response, and their elevated levels are frequently found in the brains of patients with neurodegenerative diseases [17, 26]. Enhanced expression of cytokine levels is well correlated with disease severity in Alzheimer's patients [49]. Dysregulation of the chemokines, such as CCL2 and CXCL10, are involved in the recruitment of immune cells to inflammation sites, and it can further exacerbate the pathology of neurodegenerative diseases [71]. Furthermore, the generation of reactive oxygen species (ROS) as a result of inflammatory responses leads to oxidative stress, thereby damaging the neurons [38] ROS can also trigger apoptotic pathways in neurons, leading to neurodegeneration [38].

Neuroinflammation can also compromise the integrity of the blood-brain barrier (BBB), allowing harmful substances to enter the brain and exacerbate neuronal injury [64]. The inflammatory cytokines can also disrupt the tight junctions of endothelial cells, leading to increased permeability and subsequent neuroinflammation [15].

6.5.1 Role of Microglia: Activation and Effects on Central Nervous System

Microglia, the resident immune cells of the brain, play a dual role in neuroinflammation, as their activation can lead to beneficial and detrimental outcomes for neuronal health [5, 71]. Activated microglia can adopt a pro-inflammatory phenotype, releasing cytokines and chemokines that promote inflammation and neurodegeneration [57]. Conversely, microglia can exhibit neuroprotective functions, such as phagocytosing debris and promoting repair mechanisms [72]. The balance between these opposing roles is critical in determining neuronal fate, and its dysregulation can tip the scale toward neurodegeneration [58]. Microglia has also been reported to interact with protein aggregates, such as amyloid-β and tau, in Alzheimer's disease [75].

6.5.1.1 Phagocytosis

Numerous brain conditions are linked to excessive microglial phagocytosis of neurons and synapses [5]. Microglia are capable of clearing the protein aggregates by phagocytosis, but their phagocytic efficacy may diminish with age or in chronic neurodegenerative diseases [62]. Microglial phagocytosis of dead or dying neurons further inhibits the release of pro-inflammatory and/or harmful intracellular components [51]. Nevertheless, microglia have been shown to phagocytose living neurons in certain circumstances, like inflammation, causing their death [18].

6.5.1.2 Secretion of Inflammatory Mediators

The activation of CNS resident glial cells, such as microglia and astrocytes, infiltrated peripheral immune cells, including macrophages, neutrophils, natural killer cells, and lymphocytes, through a compromised blood-brain barrier (BBB) are the primary source of secretory inflammatory mediators governing various CNS diseases [35]. These cells are implicated in the generation of inflammatory mediators, including cytokines, chemokines, and matrix metalloproteinases [12, 70]. In response to aggregates, microglia may also secrete pro-inflammatory mediators, potentially leading to further neuronal damage [5]. The interaction between microglia and protein aggregates can result in a feedback loop of inflammation and neurodegeneration [70].

6.5.2 Role of Astrocytes: Responses to Proteostatic Disturbances

Astrocytes are vital for preserving brain homeostasis and responding to neuroinflammation. These cells undergo reactive astrogliosis in response to injury or disease [43]. Astrogliosis involves morphological and functional changes, such as increased expression of glial fibrillary acidic protein (GFAP) and altered release of neurotrophic factors [43]. Depending on the context, reactive astrogliosis can either support neuronal survival or contribute to neurotoxicity [28]. Moreover, astrocytes play a critical role in clearing excess neurotransmitters, and disruptions in astrocytic function can lead to excitotoxicity and neuronal dysfunction [28, 43, 45].

6.5.2.1 Contribution to the Inflammatory Milieu and Neuronal Dysfunction

Astrocytes significantly contribute to the inflammatory milieu in neurodegenerative diseases. Activated astrocytes release various cytokines and chemokines, amplifying the inflammatory response and potentially leading to neurotoxicity [43]. The inflammatory mediators released by astrocytes can alter synaptic function and promote neuronal apoptosis, further contributing to disease progression [12].

6.6 Cross Talk Between Proteostasis and Inflammation

As mentioned in previous sections, neuroinflammation and dysregulated proteostasis are two important factors that are common among multiple neurodegenerative disorders. A cross talk between these two processes is one of the significant determinants of disease progression, as shown in Fig. 6.3 [74]. The accumulation of misfolded or aggregated proteins owing to dysregulated protein homeostasis induces cellular stress, including ER, mitochondrial stress and triggers inflammatory responses in neurons, leading to the activation of microglia. Thus, neuroinflammation is a result of excessive immune responses in neurons due to the cumulative effect of protein misfolding, aggregation, accumulation, or other alterations [69]. The resulting inflammation can further impair protein homeostasis, creating a feedback loop and worsening the protein homeostasis.

Contrary to this, another school of thought exists that highlights the role of neuroinflammation in the onset of neurodegenerative disorders. A substantial amount of data indicates the activation of glial cells results in the release of inflammatory cytokines and suggests the major role of inflammation in the beginning and development of neurodegeneration [74]. Following neuroinflammation, the growing burden of protein accumulation in the neurons leads to ER stress and mitochondrial

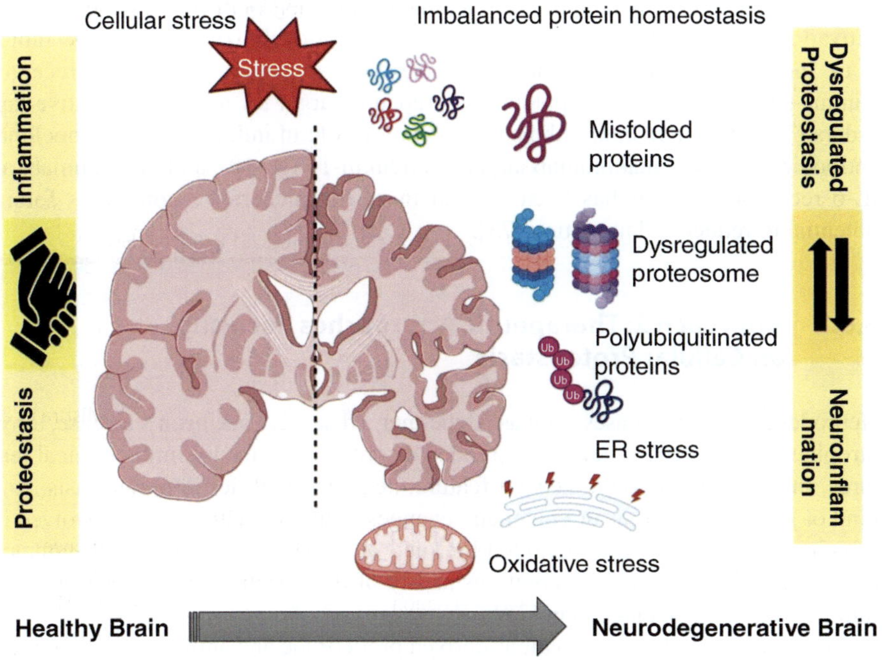

Fig. 6.3 Pictorial representation highlighting the cellular events that occur during the onset and progression of neurodegenerative disorder and the cross-talk between proteostasis and neuroinflammation. (Made using Biorender.com) Therapeutic Approaches and Future Directions

dysfunction, thereby creating oxidative stress, which finally causes neurodegeneration. In support of these findings, genetic mutations in multiple immunomodulatory genes have been reported that are associated with several neurodegenerative disorders. The bidirectional relationship between these two processes in neurons is detrimental to healthy brain function and the onset of neurodegenerative disease. Thus, treatment strategies targeting proteostasis or neuroinflammation may be beneficial.

The following section discusses the therapeutic approaches targeting cellular proteostasis machinery or inflammatory pathways to reduce neuroinflammation. Although these approaches may not cure the disease, they can slow their progression by preserving neuronal integrity and reducing inflammation.

6.6.1 Approach 1: Targeting Neuroinflammatory Pathways

It has been reported that the use of anti-inflammatory drugs may prevent or slow down the progression of neurodegenerative diseases [24]. Multiple non-steroidal anti-inflammatory drugs (NSAIDs) like ibuprofen, diclofenac, and naproxen are currently being used to reduce inflammation in neurodegenerative diseases. Prednisone is currently used to treat multiple sclerosis for controlling inflammation; however, due to potential adverse effects, its long-term use is limited [67]. As discussed earlier in this chapter, neuroinflammation plays a significant role in the pathophysiology of Alzheimer's disease and other degenerative illnesses via the activation of glial cells. Thus, the drugs targeting non-cyclooxygenase (COX) inhibitory mechanisms and activation of peroxisome proliferator-activated receptor gamma (PPARγ) might be a potential therapeutic target for neurodegenerative disorders [73, 76]. Since interleukins play a central role in inflammation, monoclonal antibodies such as canakinumab target interleukin-1β (IL-1β), and Tocilizumab, an IL-6 receptor inhibitor has been studied in neurodegenerative conditions for its potential to reduce inflammation [73].

6.6.2 Approach 2: Therapeutic Approaches Focusing on Cellular Proteostasis

Neurodegenerative diseases such as Alzheimer's disease, Parkinson's disease, amyotrophic lateral sclerosis, and prion-related disorders, despite dissimilar clinical features, share one common molecular feature, i.e., abnormalities in proteostasis. Any kind of genetic mutations or epigenetic changes in the regulatory genes involved in protein folding, trafficking, degradation, and autophagy can lead to dysregulated proteostasis. Therapeutic interventions at any of these steps can have a significant impact on disease management [37].

Multiple studies based on patient-derived brain tissue and animal models suggest ER stress is a common hallmark of most neurological disorders and highlight the role of eIF2α phosphorylation blockers as a therapeutic target [25]. Furthermore, numerous proteins that are deposited as a result of disrupted proteostasis begin to show up in biofluids and may be useful as biomarkers for the early detection of neurodegenerative disease.

In diseases like Alzheimer's, monoclonal antibodies such as anti-amyloid antibodies (aducanumab) have received approval from the US Food and Drug Administration (FDA). Lecanemab, a humanized monoclonal antibody having binding affinity to amyloid- β protofibrils, has shown promising effects in the reduction in brain amyloid levels [1].

However, it requires long-term clinical trials and further elaborative research to determine its efficacy and safety.

6.6.3 Effectiveness and Limitations of Current Therapies

For most neurodegenerative diseases, the approved treatments either provide only symptomatic relief or delay the progression, but none of them cure or stop the progression of the disease. As mentioned above, various drugs, small molecules, or antibodies are currently being used; however, much more research is required to delineate their effectiveness (Table 6.1).

6.6.4 Emerging Therapies: Potential Targets for Neuroinflammation and Proteostasis

Combining different therapeutic strategies, such as using chaperones with autophagy enhancers or proteasome activators, may provide a more comprehensive approach to manage proteostasis-related disorders. Recent studies have revealed

Table 6.1 List of drugs, molecules, or antibodies targeting proteostasis and neuroinflammation in neurodegenerative disorders

S. no.	Drugs	Mechanism of action	Targets
1	Lecanemab	FDA-approved anti-amyloid-β (anti-Aβ) monoclonal antibody	Proteostasis
2	Aducanumab	FDA-approved anti-amyloid-β (anti-Aβ) monoclonal antibody	Proteostasis
3	GSK2606414	Inhibitor of the kinase domain of PERK, targeting eIF2α-P-mediated translational repression	Proteostasis
4	ISRIB	Inhibits ATF4 expression targeting eIF2α-P-mediated translational repression	Proteostasis
5	Ibuprofen	Nonsteroidal anti-inflammatory drugs (NSAIDs)	Neuroinflammation
6	Diclofenac	Nonsteroidal anti-inflammatory drugs (NSAIDs)	Neuroinflammation
7	Naproxen	Nonsteroidal anti-inflammatory drugs (NSAIDs)	Neuroinflammation
8	Prednisone	Neuroactive steroid	Neuroinflammation
9	Canakinumab	Interleukin-1β (IL-1β) monoclonal antibody, reduce inflammation	Neuroinflammation
10	Tocilizumab	IL-6 receptor inhibitor, reduce inflammation	Neuroinflammation

that the oral administration of GSK2606414, an inhibitor of the PERK/eIF2α pathway, provides neuroprotection to prion-infected mice [56, 60]. Drugs that modulate microglial activation, such as P2X7 receptor antagonists, are also being researched for their ability to reduce excessive neuroinflammation [3].

6.6.4.1 Gene Therapy

The process of replacement or modification of defective genes to prevent or cure the disease is known as gene therapy. Gene therapy is a promising field to research for neurodegenerative illnesses, wherein most conventional medications have not been effective. Finding medications that are safe and able to cross the blood-brain barrier is one of the biggest obstacles in the clinical implementation of these medicines. Studies are being conducted on various neurological genetic disorders, such as Huntington's disease, frontotemporal dementia, and spinal muscular atrophy [10, 52]. Virus-based approach is currently being investigated for utilizing these viruses (Adeno-associated viruses) as a vehicle for targeted gene delivery [65].

Nanomedicine The major key challenge in utilizing pharmacological therapy for neurodegenerative disorders is to achieve optimal drug concentration within the central nervous system (CNS) due to the BBB, which hinders drug transport. Nanoparticles, or NPs, have recently gained interest as potential medicine carriers due to their ability to cross the BBB. Nanoparticles can be designed and modified to specifically target activated microglia and deliver anti-inflammatory drugs. Recent investigations demonstrated that Tau protein polymerization can be inhibited using protein-capped iron oxide and cadmium sulfide nanoparticles [63]. In a rat model, solid-lipid particles containing an HDAC inhibitor, nicotinamide has shown a significant reduction in cognitive impairment associated with Alzheimer's disease by reducing Tau protein phosphorylation [36]. Furthermore, the use of NPs as vectors for delivering genetic material, such as siRNA or CRISPR/Cas9, has been reported to modify the expression of genes related to neurological disorders or neurotrophic factors [47].

Acknowledgments We acknowledge the ICMR (Indian Council of Medical Research), Govt. of India for funding (Grant No. 5/4-5/3/6(Neuro)/2022-NCD1).

References

1. Ameen TB, Kashif SN, Abbas SMI, Babar K, Ali SMS, Raheem A. Unraveling Alzheimer's: the promise of aducanumab, lecanemab, and donanemab. Egypt J Neurol Psychiatry Neurosurg. 2024;60:72. https://doi.org/10.1186/s41983-024-00845-5.
2. Bai B, Wang X, Li Y, Chen P-C, Yu K, Dey KK, Yarbro JM, Han X, Lutz BM, Rao S, Jiao Y, Sifford JM, Han J, Wang M, Tan H, Shaw TI, Cho J-H, Zhou S, Wang H, Niu M, Mancieri A, Messler KA, Sun X, Wu Z, Pagala V, High AA, Bi W, Zhang H, Chi H, Haroutunian V, Zhang B, Beach TG, Yu G, Peng J. Deep multilayer brain proteomics identifies molecular net-

works in Alzheimer's disease progression. Neuron. 2020;106:700. https://doi.org/10.1016/j.neuron.2020.04.031.

3. Bianchi C, Alvarez-Castelao B, Sebastián-Serrano Á, Di Lauro C, Soria-Tobar L, Nicke A, Engel T, Díaz-Hernández M. P2X7 receptor inhibition ameliorates ubiquitin–proteasome system dysfunction associated with Alzheimer's disease. Alzheimers Res Ther. 2023;15:105. https://doi.org/10.1186/s13195-023-01258-x.
4. Bruni AC, Bernardi L, Gabelli C. From beta amyloid to altered proteostasis in Alzheimer's disease. Ageing Res Rev. 2020;64:101126. https://doi.org/10.1016/j.arr.2020.101126.
5. Butler CA, Popescu AS, Kitchener EJA, Allendorf DH, Puigdellívol M, Brown GC. Microglial phagocytosis of neurons in neurodegeneration, and its regulation. J Neurochem. 2021;158:621–39. https://doi.org/10.1111/jnc.15327.
6. Calamini B, Morimoto RI. Protein homeostasis as a therapeutic target for diseases of protein conformation. Curr Top Med Chem. 2012;12:2623–40. https://doi.org/10.2174/156802661212220014.
7. Carroll EC, Marqusee S. Site-specific ubiquitination: deconstructing the degradation tag. Curr Opin Struct Biol. 2022;73:102345. https://doi.org/10.1016/j.sbi.2022.102345.
8. Çetin G, Studencka-Turski M, Venz S, Schormann E, Junker H, Hammer E, Völker U, Ebstein F, Krüger E. Immunoproteasomes control activation of innate immune signaling and microglial function. Front Immunol. 2022;13:982786. https://doi.org/10.3389/fimmu.2022.982786.
9. Ciechanover A, Kwon YT. Degradation of misfolded proteins in neurodegenerative diseases: therapeutic targets and strategies. Exp Mol Med. 2015;47:e147. https://doi.org/10.1038/emm.2014.117.
10. Ciurea AV, Mohan AG, Covache-Busuioc R-A, Costin H-P, Glavan L-A, Corlatescu A-D, Saceleanu VM. Unraveling molecular and genetic insights into neurodegenerative diseases: advances in understanding Alzheimer's, Parkinson's, and Huntington's diseases and amyotrophic lateral sclerosis. Int J Mol Sci. 2023;24:10809. https://doi.org/10.3390/ijms241310809.
11. Díaz-Villanueva J, Díaz-Molina R, García-González V. Protein folding and mechanisms of Proteostasis. Int J Mol Sci. 2015;16:17193–230. https://doi.org/10.3390/ijms160817193.
12. Fakhoury M. Role of immunity and inflammation in the pathophysiology of neurodegenerative diseases. Neurodegener Dis. 2015;15:63–9. https://doi.org/10.1159/000369933.
13. Fekete C, Vastagh C, Dénes Á, Hrabovszky E, Nyiri G, Kalló I, Liposits Z, Sárvári M. Chronic amyloid β oligomer infusion evokes sustained inflammation and microglial changes in the rat hippocampus via NLRP3. Neuroscience. 2019;405:35–46. https://doi.org/10.1016/j.neuroscience.2018.02.046.
14. Finley D. Recognition and processing of ubiquitin-protein conjugates by the proteasome. Annu Rev Biochem. 2009;78:477–513. https://doi.org/10.1146/annurev.biochem.78.081507.101607.
15. Fonseca AC, Oliveira CR, Pereira CF, Cardoso SM. Loss of proteostasis induced by amyloid beta peptide in brain endothelial cells. Biochim Biophys Acta BBA Mol Cell Res. 2014;1843:1150–61. https://doi.org/10.1016/j.bbamcr.2014.02.016.
16. Franco-Iborra S, Plaza-Zabala A, Montpeyo M, Sebastian D, Vila M, Martinez-Vicente M. Mutant HTT (huntingtin) impairs mitophagy in a cellular model of Huntington disease. Autophagy. 2021;17:672–89. https://doi.org/10.1080/15548627.2020.1728096.
17. Frankola KA, Greig NH, Luo W, Tweedie D. Targeting TNF-α to elucidate and ameliorate neuroinflammation in neurodegenerative diseases. CNS Neurol Disord Drug Targets. 2011;10:391–403. https://doi.org/10.2174/187152711794653751.
18. Gabandé-Rodríguez E, Keane L, Capasso M. Microglial phagocytosis in aging and Alzheimer's disease. J Neurosci Res. 2020;98:284–98. https://doi.org/10.1002/jnr.24419.
19. Giandomenico SL, Alvarez-Castelao B, Schuman EM. Proteostatic regulation in neuronal compartments. Trends Neurosci. 2022;45:41–52. https://doi.org/10.1016/j.tins.2021.08.002.
20. Glass CK, Saijo K, Winner B, Marchetto MC, Gage FH. Mechanisms underlying inflammation in neurodegeneration. Cell. 2010;140:918–34. https://doi.org/10.1016/j.cell.2010.02.016.
21. Glick D, Barth S, Macleod KF. Autophagy: cellular and molecular mechanisms. J Pathol. 2010;221:3–12. https://doi.org/10.1002/path.2697.

22. Glickman MH, Ciechanover A. The ubiquitin-proteasome proteolytic pathway: destruction for the sake of construction. Physiol Rev. 2002;82:373–428. https://doi.org/10.1152/physrev.00027.2001.
23. Griffin TA, Nandi D, Cruz M, Fehling HJ, Kaer LV, Monaco JJ, Colbert RA. Immunoproteasome assembly: cooperative Incorporation of Interferon γ (IFN-γ)– inducible subunits. J Exp Med. 1998;187:97–104. https://doi.org/10.1084/jem.187.1.97.
24. Wilms H, Zecca L, Rosenstiel P, Sievers J, Deuschl G, Lucius R. Inflammation in Parkinsons diseases and other neurodegenerative diseases: cause and therapeutic implications. Curr Pharm Des. 2007;13:1925–8. https://doi.org/10.2174/138161207780858429.
25. Halliday M, Radford H, Zents KAM, Molloy C, Moreno JA, Verity NC, Smith E, Ortori CA, Barrett DA, Bushell M, Mallucci GR. Repurposed drugs targeting eIF2α-P-mediated translational repression prevent neurodegeneration in mice. Brain. 2017;140:1768–83. https://doi.org/10.1093/brain/awx074.
26. Hefting LL, D'Este E, Arvedsen E, Benned-Jensen T, Rasmussen HB. Multiple domains in the Kv7.3 C-terminus can regulate localization to the axon initial segment. Front Cell Neurosci. 2020;14:10. https://doi.org/10.3389/fncel.2020.00010.
27. Hoppe T, Cohen E. Organismal protein homeostasis mechanisms. Genetics. 2020;215:889–901. https://doi.org/10.1534/genetics.120.301283.
28. Jiwaji Z, Hardingham GE. Good, bad, and neglectful: astrocyte changes in neurodegenerative disease. Free Radic Biol Med. 2022;182:93–9. https://doi.org/10.1016/j.freeradbiomed.2022.02.020.
29. Johnston HE, Samant RS. Alternative systems for misfolded protein clearance: life beyond the proteasome. FEBS J. 2021;288:4464–87. https://doi.org/10.1111/febs.15617.
30. Kaushik S, Cuervo AM. Chaperone-mediated autophagy: a unique way to enter the lysosome world. Trends Cell Biol. 2012;22:407–17. https://doi.org/10.1016/j.tcb.2012.05.006.
31. Keck S, Nitsch R, Grune T, Ullrich O. Proteasome inhibition by paired helical filament-tau in brains of patients with Alzheimer's disease. J Neurochem. 2003;85:115–22. https://doi.org/10.1046/j.1471-4159.2003.01642.x.
32. Kim W, Bennett EJ, Huttlin EL, Guo A, Li J, Possemato A, Sowa ME, Rad R, Rush J, Comb MJ, Harper JW, Gygi SP. Systematic and quantitative assessment of the ubiquitin-modified proteome. Mol Cell. 2011;44:325–40. https://doi.org/10.1016/j.molcel.2011.08.025.
33. Klionsky DJ, Cregg JM, Dunn WA Jr, Emr SD, Sakai Y, Sandoval IV, Sibirny A, Subramani S, Thumm M, Veenhuis M, Ohsumi Y. A unified nomenclature for yeast autophagy-related genes. Dev Cell. 2003;5:539–45. https://doi.org/10.1016/S1534-5807(03)00296-X.
34. Klionsky DJ, Emr SD. Autophagy as a regulated pathway of cellular degradation. Science. 2000;290:1717–21. https://doi.org/10.1126/science.290.5497.1717.
35. Koronyo-Hamaoui M, Gaire BP, Frautschy SA, Alvarez JI. Editorial: role of inflammation in neurodegenerative diseases. Front Immunol. 2022;13:958487. https://doi.org/10.3389/fimmu.2022.958487.
36. Krsek A, Baticic L. Nanotechnology-driven therapeutic innovations in neurodegenerative disorders: a focus on Alzheimer's and Parkinson's disease. Future Pharmacol. 2024;4:352–79. https://doi.org/10.3390/futurepharmacol4020020.
37. Kulkarni A, Preeti K, Tryphena KP, Srivastava S, Singh SB, Khatri DK. Proteostasis in Parkinson's disease: recent development and possible implication in diagnosis and therapeutics. Ageing Res Rev. 2023;84:101816. https://doi.org/10.1016/j.arr.2022.101816.
38. Kurtishi A, Rosen B, Patil KS, Alves GW, Møller SG. Cellular proteostasis in neurodegeneration. Mol Neurobiol. 2019;56:3676–89. https://doi.org/10.1007/s12035-018-1334-z.
39. Labbadia J, Morimoto RI. The biology of Proteostasis in aging and disease. Annu Rev Biochem. 2015;84:435–64. https://doi.org/10.1146/annurev-biochem-060614-033955.
40. Laskowska E, Kuczyńska-Wiśnik D, Lipińska B. Proteomic analysis of protein homeostasis and aggregation. J Proteome. 2019;198:98–112. https://doi.org/10.1016/j.jprot.2018.12.003.
41. Lehtonen Š, Sonninen T-M, Wojciechowski S, Goldsteins G, Koistinaho J. Dysfunction of cellular Proteostasis in Parkinson's disease. Front Neurosci. 2019;13:457. https://doi.org/10.3389/fnins.2019.00457.

42. Leister H, Krause FF, Gil B, Prus R, Prus I, Hellhund-Zingel A, Mitra M, Da Rosa Gerbatin R, Delanty N, Beausang A, Brett FM, Farrell MA, Cryan J, O'Brien DF, Henshall DC, Helmprobst F, Pagenstecher A, Steinhoff U, Visekruna A, Engel T. Immunoproteasome deficiency results in age-dependent development of epilepsy. Brain Commun. 2024;6:fcae017. https://doi.org/10.1093/braincomms/fcae017.
43. Liddelow SA, Guttenplan KA, Clarke LE, Bennett FC, Bohlen CJ, Schirmer L, Bennett ML, Münch AE, Chung W-S, Peterson TC, Wilton DK, Frouin A, Napier BA, Panicker N, Kumar M, Buckwalter MS, Rowitch DH, Dawson VL, Dawson TM, Stevens B, Barres BA. Neurotoxic reactive astrocytes are induced by activated microglia. Nature. 2017a;541:481–7. https://doi.org/10.1038/nature21029.
44. Liu RT, Wang A, To E, Gao J, Cao S, Cui JZ, Matsubara JA. Vinpocetine inhibits amyloid-beta induced activation of NF-κB, NLRP3 inflammasome and cytokine production in retinal pigment epithelial cells. Exp Eye Res. 2014;127:49–58. https://doi.org/10.1016/j.exer.2014.07.003.
45. Liu W, Tang Y, Feng J. Cross talk between activation of microglia and astrocytes in pathological conditions in the central nervous system. Life Sci. 2011;89:141–6. https://doi.org/10.1016/j.lfs.2011.05.011.
46. Liu X, Hebron ML, Mulki S, Wang C, Lekah E, Ferrante D, Shi W, Kurd-Misto B, Moussa C. Ubiquitin specific protease 13 regulates tau accumulation and clearance in models of Alzheimer's disease 1. J Alzheimers Dis. 2019;72:425–41. https://doi.org/10.3233/JAD-190635.
47. Mani S, Jindal D, Singh M. Gene therapy, a potential therapeutic tool for neurological and neuropsychiatric disorders: applications, challenges and future perspective. Curr Gene Ther. 2023;23:20–40. https://doi.org/10.2174/1566523222666220328142427.
48. Maria Cuervo A. Autophagy: in sickness and in health. Trends Cell Biol. 2004;14:70–7. https://doi.org/10.1016/j.tcb.2003.12.002.
49. Mishto M, Bellavista E, Santoro A, Stolzing A, Ligorio C, Nacmias B, Spazzafumo L, Chiappelli M, Licastro F, Sorbi S, Pession A, Ohm T, Grune T, Franceschi C. Immunoproteasome and LMP2 polymorphism in aged and Alzheimer's disease brains. Neurobiol Aging. 2006;27:54–66. https://doi.org/10.1016/j.neurobiolaging.2004.12.004.
50. Mizushima N. Autophagy: process and function. Genes Dev. 2007;21:2861–73. https://doi.org/10.1101/gad.1599207.
51. Neher JJ, Neniskyte U, Brown GC. Primary phagocytosis of neurons by inflamed microglia: potential roles in neurodegeneration. Front Pharmacol. 2012;3 https://doi.org/10.3389/fphar.2012.00027.
52. Ng ASL, Rademakers R, Miller BL. Frontotemporal dementia: a bridge between dementia and neuromuscular disease. Ann N Y Acad Sci. 2015;1338:71–93. https://doi.org/10.1111/nyas.12638.
53. Oh S, Hong HS, Hwang E, Sim HJ, Lee W, Shin SJ, Mook-Jung I. Amyloid peptide attenuates the proteasome activity in neuronal cells. Mech Ageing Dev. 2005;126:1292–9. https://doi.org/10.1016/j.mad.2005.07.006.
54. Papendorf JJ, Krüger E, Ebstein F. Proteostasis perturbations and their roles in causing sterile inflammation and autoinflammatory diseases. Cells. 2022;11:1422. https://doi.org/10.3390/cells11091422.
55. Park S-K, Arslan F, Kanneganti V, Barmada SJ, Purushothaman P, Verma SC, Liebman SW. Overexpression of a conserved HSP40 chaperone reduces toxicity of several neurodegenerative disease proteins. Prion. 2018;12:16–22. https://doi.org/10.1080/19336896.2017.1423185.
56. Radford H, Moreno JA, Verity N, Halliday M, Mallucci GR. PERK inhibition prevents tau-mediated neurodegeneration in a mouse model of frontotemporal dementia. Acta Neuropathol (Berl). 2015;130:633–42. https://doi.org/10.1007/s00401-015-1487-z.
57. Ransohoff RM. How neuroinflammation contributes to neurodegeneration. Science. 2016;353:777–83. https://doi.org/10.1126/science.aag2590.
58. Ransohoff RM, Cardona AE. The myeloid cells of the central nervous system parenchyma. Nature. 2010;468:253–62. https://doi.org/10.1038/nature09615.

59. Ravid T, Hochstrasser M. Diversity of degradation signals in the ubiquitin–proteasome system. Nat Rev Mol Cell Biol. 2008;9:679–89. https://doi.org/10.1038/nrm2468.
60. Rozpędek-Kamińska W, Siwecka N, Wawrzynkiewicz A, Wojtczak R, Pytel D, Diehl JA, Majsterek I. The PERK-dependent molecular mechanisms as a novel therapeutic target for neurodegenerative diseases. Int J Mol Sci. 2020;21:2108. https://doi.org/10.3390/ijms21062108.
61. Shorter J. The mammalian Disaggregase machinery: Hsp110 synergizes with Hsp70 and Hsp40 to catalyze protein disaggregation and reactivation in a cell-free system. PLoS One. 2011;6:e26319. https://doi.org/10.1371/journal.pone.0026319.
62. Sokolowski JD, Mandell JW. Phagocytic clearance in neurodegeneration. Am J Pathol. 2011;178:1416–28. https://doi.org/10.1016/j.ajpath.2010.12.051.
63. Sonawane SK, Ahmad A, Chinnathambi S. Protein-capped metal nanoparticles inhibit tau aggregation in Alzheimer's disease. ACS Omega. 2019;4:12833–40. https://doi.org/10.1021/acsomega.9b01411.
64. Sonninen T-M, Goldsteins G, Laham-Karam N, Koistinaho J, Lehtonen Š. Proteostasis disturbances and inflammation in neurodegenerative diseases. Cells. 2020;9:2183. https://doi.org/10.3390/cells9102183.
65. Sun J, Roy S. Gene-based therapies for neurodegenerative diseases. Nat Neurosci. 2021;24:297–311. https://doi.org/10.1038/s41593-020-00778-1.
66. Suzuki G, Imura S, Hosokawa M, Katsumata R, Nonaka T, Hisanaga S-I, Saeki Y, Hasegawa M. a-synuclein strains that cause distinct pathologies differentially inhibit proteasome. eLife. 2020;9:e56825. https://doi.org/10.7554/eLife.56825.
67. Terzi M, Altun G, Şen S, Kocaman A, Kaplan AA, Yurt KK, Kaplan S. The use of non-steroidal anti-inflammatory drugs in neurological diseases. J Chem Neuroanat. 2018;87:12–24. https://doi.org/10.1016/j.jchemneu.2017.03.003.
68. Tseng C-S, Chao Y-W, Liu Y-H, Huang Y-S, Chao H-W. Dysregulated proteostasis network in neuronal diseases. Front Cell Dev Biol. 2023;11:1075215. https://doi.org/10.3389/fcell.2023.1075215.
69. Uddin MS, Yu WS, Lim LW. Exploring ER stress response in cellular aging and neuroinflammation in Alzheimer's disease. Ageing Res Rev. 2021;70:101417. https://doi.org/10.1016/j.arr.2021.101417.
70. Voet S, Srinivasan S, Lamkanfi M, Van Loo G. Inflammasomes in neuroinflammatory and neurodegenerative diseases. EMBO Mol Med. 2019;11:e10248. https://doi.org/10.15252/emmm.201810248.
71. Wendimu MY, Hooks SB. Microglia phenotypes in aging and neurodegenerative diseases. Cells. 2022;11:2091. https://doi.org/10.3390/cells11132091.
72. Wilson MR, Satapathy S, Vendruscolo M. Extracellular protein homeostasis in neurodegenerative diseases. Nat Rev Neurol. 2023;19:235–45. https://doi.org/10.1038/s41582-023-00786-2.
73. Wu S, Zhou Y. Monoclonal antibody precision therapy targeting inflammation for bipolar disorder: a narrative review. Ther Adv Psychopharmacol. 2024;14:20451253241227772. https://doi.org/10.1177/20451253241227772.
74. Zhang W, Xiao D, Mao Q, Xia H. Role of neuroinflammation in neurodegeneration development. Signal Transduct Target Ther. 2023;8:267. https://doi.org/10.1038/s41392-023-01486-5.
75. Zheng H, Koo EH. The amyloid precursor protein: beyond amyloid. Mol Neurodegener. 2006;1:5. https://doi.org/10.1186/1750-1326-1-5.
76. Zolezzi JM, Santos MJ, Bastías-Candia S, Pinto C, Godoy JA, Inestrosa NC. PPARS in the central nervous system: roles in neurodegeneration and neuroinflammation. Biol Rev. 2017;92:2046–69. https://doi.org/10.1111/brv.12320.

Proteostasis: Investigating Molecular Dynamics in Neurodegenerative Disorders

7

Dhiraj Kumar Singh, Ayesha Rahman, Rashmi Gupta, Amit Kumar Verma, Qulsum Akhter Mir, and Riyaz Ahmad Mir

7.1 Introduction

Proteostasis, short for "protein homeostasis," refers to the processes that cells use to maintain the balance and proper functioning of proteins. Just like homeostasis maintains internal conditions like temperature or pH, proteostasis ensures that proteins are correctly synthesized, folded, and degraded as needed. Protein homeostasis or "proteostasis" is the process that regulates proteins within the cell in order to maintain the health of both the cellular proteome and the organism itself.

A complicated network of interconnected mechanisms that affect a protein's destiny from production to breakdown is known as proteostasis. When one component is impacted, the others adapt to keep everything operating normally. Pathologies like diabetes, cancer, and Alzheimer's disease (AD) can arise when one or more of these proteostasis influences is disrupted [1].

Enzo offers an unparalleled range of ubiquitin/proteasome and heat shock protein (HSP) research options to support your proteostatis studies. Our research into protein synthesis, modification, and degradation continues to yield state-of-the-art instruments for analyzing important biological processes in the field of proteostasis. First-to-market immunoassays, cell-based assays, and chemical libraries for autophagy demonstrate this dedication by assisting researchers in comprehending the cellular regulation of protein turnover and its connection to illness [2].

Here is a breakdown of the key components involved in proteostasis:

1. *Protein Synthesis*: Proteins are synthesized in cells based on genetic instructions. This involves transcription of DNA to mRNA and translation of mRNA into

D. K. Singh · A. Rahman · R. Gupta · Q. A. Mir · R. A. Mir (✉)
Department of Biochemistry, All India Institute of Medical Sciences, New Delhi, India
e-mail: riyazmir@aiims.edu

A. K. Verma
Department of Biosciences, Jamia Millia Islamia, New Delhi, India

N. Puranik (ed.), *Proteostasis*, https://doi.org/10.1007/978-981-96-6202-9_7

amino acid chains by ribosomes. Reactive oxygen species (ROS) buildup and the energetic state of the mitochondria, known as mitochondrial membrane potential, are commonly measured as indicators of drug-induced cellular stress.

2. *Protein Folding*: Newly synthesized proteins need to fold into specific three-dimensional shapes to be functional. Molecular chaperones assist in this folding process, ensuring that proteins achieve and maintain their correct structures [3].
3. *Quality Control*: Proteins that fail to fold correctly or are damaged are identified by quality control mechanisms. These mechanisms often involve chaperones and other proteins that help refold or target misfolded proteins for degradation.
4. *Protein Degradation*: Proteins that are no longer needed or are damaged are broken down and removed from the cell. This process is crucial for clearing out faulty proteins and regulating protein levels. More than 500 proteases that have been found in the genome make up a broad family of enzymes that are essential for numerous essential functions. Major contributors to protein homeostasis, or proteostasis, are lysosomal proteases (including cathepsins) and the proteasome, which make up the degradative processes for protein turnover, misfolded, and damaged proteins. On the other hand, it is becoming more widely acknowledged that specific and restricted proteolysis (e.g., MMPs, ADAMs, caspases, and DPPIV) is a significant regulatory mechanism that can either activate or deactivate a variety of signaling targets, such as other proteases, growth factors, and peptides. The ubiquitin-proteasome system (UPS) and autophagy are key systems for protein degradation. In proteomic studies of biological systems, PTMs, protein synthesis, and folding are crucial. Beyond translation, assessment of protein homeostasis (proteostasis) includes chemical treatment to return a protein to its native shape and molecular chaperone-assisted folding. Furthermore, to control appropriate localization and function, a wide range of chemical changes are used, including phosphorylation, ubiquitinylation, acetylation, glycosylation, SUMOylation, and methylation.
5. *Stress Responses*: Proteostasis is challenged by various stresses such as heat, toxins, or genetic mutations. Cells have stress response mechanisms, like the heat shock response (HSR), that help manage these challenges and restore proteostasis. Proteostasis must be maintained for cells to be healthy and operate properly. Protein aggregation and illnesses, including neurodegenerative conditions like Parkinson's and Alzheimer's, can result from disturbances in these mechanisms [4].

7.2 Proteostasis

One of the most prevalent types of macromolecules in the biological system are proteins, and homeostasis depends on the proper structures and activities of proteins. Proteostasis, also known as protein homeostasis, is the network of cell homeostasis that includes every aspect of a protein's life cycle, from synthesis to failure to achieve native structure, loss of native structure, aggregation, or fulfilling a functional role in cells, to degradation upon completion. Molecular chaperone proteins are essential for folding newly synthesized polypeptides, blocking undesirable protein-protein interactions and misfolding/aggregation, maintaining protein solubility, solubilizing aggregated proteins, and guiding misfolded or aggregated proteins toward proteolysis during translation at ribosomes. Autophagy and the ubiquitin proteasome system (UPS) are the two primary mechanisms of proteolysis in cells [5, 6].

Molecular chaperones are proteins that mediate folding and are essential to proteostasis. Misfolded proteins with exposed hydrophobic areas are recognized by chaperones like Hsp70 and Hsp90. Until correct folding is accomplished by ATP hydrolysis and cochaperone activity, they keep these proteins in a foldable form. Notably, misfolded proteins are also directed toward destruction by chaperones. Proteostatic failure results from aging-related reductions in the heat shock response's capacity to activate chaperones in response to proteotoxic stress. During translation at ribosomes, molecular chaperone proteins are necessary for folding newly synthesized polypeptides, preventing unwanted protein-protein interactions and misfolding/aggregation, preserving protein solubility, solubilizing aggregated proteins, and directing aggregated or misfolded proteins toward proteolysis. The two main processes of proteolysis in cells are autophagy and the ubiquitin proteasome system (UPS) [7].

A multisubunit proteolysis system known as the 26S proteasome links many small proteins, including ubiquitin, for degradation once they have fulfilled their cellular functional tasks or exhibit abnormal folding. On the other hand, the lysosome-dependent autophagic mechanism breaks down big protein aggregates and cellular organelles. The two primary pathways of cellular degradation are the autophagy-lysosomal pathway, which breaks down large aggregates of proteins and macromolecules, individual proteins, and whole organelles. The UPS breaks down both functional and dysfunctional proteins [8].

7.2.1 Protein Folding, Refolding, and Misfolding

The "energy landscape" theory provides an explanation for the intricate biomolecular self-assembly phenomenon that is protein folding. Cells generate molecular chaperones, which direct the forming polypeptide chain along the proper folding pathway and occasionally reverse misfolding and aggregation to aid in proper folding and avoid protein aggregation. Since proteins are dynamic and subject to both internal and external stress, chaperones collaborate with protein degradation systems as part of a broader proteostasis network to maintain protein balance within the cell. If a protein misfolds, cells implement various protective mechanisms. Chaperones attempt to refold the protein first, but if that fails, they can trigger cellular processes such as the ubiquitin-proteasome system (UPS), heat shock response (HSR), unfolded protein response (UPR), and endoplasmic reticulum-associated degradation (ERAD) to either correct or degrade the misfolded protein.

However, protein folding is inherently error-prone, as misfolding can occur during the search for stable interactions between amino acid residues. Although small proteins may fold efficiently in vitro, larger proteins often encounter aggregation issues due to off-pathway events. Misfolding and aggregation, which occur mainly due to nonspecific interactions between exposed hydrophobic regions, represent significant biological and medical challenges [4, 9]. Chaperones play a key role in quality control by binding to these hydrophobic regions in unfolded proteins to prevent nonspecific aggregation. Chaperones perform other roles in higher eukaryotes, such as adhering to preexisting aggregates to facilitate disintegration, hiding their hydrophobic surfaces, or enlarging them into larger assemblies, all of which lessen toxicity. When the chaperone mechanism fails, which frequently happens after pathogenic events that induce ER stress, unfolded proteins accumulate in the endoplasmic reticulum (ER) and trigger the unfolded protein response (UPR). The UPR maintains ER protein homeostasis through a variety of processes, including the ubiquitin-proteasome system (UPS) and other transcriptionally regulated pathways involved in protein folding, transport, and destruction. Misfolded proteins can arise in two ways. First, proteins may be correctly synthesized but adopt an alternative, stable conformation, leading to misfolding. Alternatively, genetic mutations can cause misfolding and dysfunction, where even a single erroneous amino acid can result in improper folding, leading to protein aggregation and potential cellular disaster. Mutations in the cell's protein-folding machinery can also result in protein misfolding diseases. While mutations in degradation-related proteins stop misfolded proteins from being broken down, allowing them to accumulate and form fibrils, mutations in chaperones facilitate the folding of proteins into alternate conformations more readily. Additionally, connected to cellular stresses such as inflammation, calcium imbalance, and dysfunctional mitochondria is protein misfolding. Protein aggregates have been found to be directly linked to more than 40 human diseases, including neurodegenerative conditions like Alzheimer's and Parkinson's disease. Due to the proteostasis network's decreased efficiency with age, many of these disorders are age-related. A crucial protein in these illnesses is frequently misfolded, eluding the systems responsible for folding and breaking down proteins. This leads to the formation of aggregates that eventually enlarge into fibrillar formations [10].

7.2.2 Ubiquitin-Proteasome System (UPS)

The ubiquitin-proteasome system (UPS), a critical proteolytic mechanism that degrades the majority of intracellular proteins, is necessary to maintain normal cellular activity. Eighty to ninety percent of all cellular proteins, including those that are short-lived, misfolded, damaged, and outdated, are broken down by the UPS, which is composed of more than 850 different components. Its central component, the barrel-shaped, four-ringed 20S proteasome, is where proteolysis takes place. While the two outer non-catalytic rings, known as α-rings (with subunits α1–α7), regulate substrate recognition, binding, and entry into the inner chamber, the two inner β-rings (β1–β7) contain proteolytic activities that break down proteins, such as chymotrypsin-like (β5), caspase-like (β1), and trypsin-like (β2) activities.

The 26S proteasome is created when the 19S regulatory particle attaches to the α-rings, expanding the variety of proteins that can be broken down. At least 18 subunits make up the 19S unit, including a lid that detects polyubiquitin signals and a base of six ATPases with chaperone-like roles. This enables ubiquitinated proteins to bind, unfold, and enter the 20S proteasome for destruction. Together, the three main enzymes E1 (ubiquitin-activating enzyme), E2 (ubiquitin-conjugating enzyme), and E3 (ubiquitin ligase) use the small protein ubiquitin to mark proteins that are going to be destroyed. A fourth enzyme, E4, can sometimes be used to prolong the polyubiquitin chain.

Proteases require a minimum of four ubiquitins to recognize them. Proteins that have been polyubiquitinated are taken to the 26S proteasome, where they are broken down into tiny peptides. After being released, ubiquitin molecules are recycled for later usage. The 26S proteasome is effective at breaking down polyubiquitinated proteins, but it has trouble breaking down oxidized proteins, which are not usually ubiquitinated. Proteins that are oxidatively modified can be selectively degraded without the need for ubiquitination, as these proteins often undergo conformational changes that expose hydrophobic amino acids. However, highly oxidized proteins tend to aggregate, which can hinder proteasomal degradation due to the size and structure of these aggregates [11].

7.2.3 Endoplasmic Reticulum-Associated Protein Degradation (ERAD)

Degrading freshly generated proteins that do not fold correctly is one of the proteasome's primary functions. The endoplasmic reticulum (ER) retains these proteins until they have fully matured. Even though appropriate protein folding is crucial for maintaining cellular homeostasis, a sizable percentage of freshly produced polypeptides fail to acquire their correct shape. Proteins that are unable to fold or assemble correctly are prone to ER-associated protein degradation (ERAD), which is supervised by ER quality control mechanisms. Misfolded proteins are identified and targeted as part of the intricate, multistep ERAD process, which is then followed by retrotranslocation, ubiquitination, and proteasome destruction. This procedure makes sure that misfolded proteins that could be dangerous are removed from the ER [12].

The recognition of ERAD substrates is mediated by various posttranslational modifications, and chaperones like BiP/GRP78, EDEM1, and lectins such as OS9 and XTP3B help recruit misfolded proteins to the ERAD complex, specifically the Sel1L-Hrd1 complex. De-mannosylation, or the removal of mannose residues from misfolded glycoproteins, plays a critical role in marking them for degradation. Substrates are retrotranslocated from the ER membrane to the cytoplasm after being chosen for ERAD. There, they are ubiquitinated by ER-resident E3 ligases and primed for proteasome degradation [13].

Luminal proteins must be fully retrotranslocated, while only certain domains of membrane proteins require transport. Following retrotranslocation, ubiquitinated substrates are quickly directed to the proteasome to prevent cytosolic aggregate formation. Cytosolic proteins that act as ubiquitin receptors guide these [14] substrates to the proteasome for degradation. ERAD operates at low levels under normal conditions, but its activity increases during proteotoxic stress. Neurofibrillary tangles (NFTs), unfolded protein response (UPR) activation, raised levels of GRP78/BiP, and amyloid-beta (Aβ) plaques are all signs of increased ER stress in the brains of AD patients. An accumulation of unfolded or misfolded proteins in the ER lumen reduces the quantity of free chaperones available, causing ER stress.

In order to lessen the amount of new proteins entering the ER, this triggers the UPR, which slows down protein synthesis. When folding attempts are unsuccessful, the UPR also increases the expression of genes linked to ERAD, which facilitates the breakdown of faulty proteins. The UPR can trigger cell death processes like autophagy or apoptosis when there is ongoing cellular stress [15].

7.2.4 Autophagy-Lysosome System

Apart from the proteasomal pathway, the lysosomal system is the primary "digestive mechanism" of the cell, with a virtually limitless potential for breakdown. Among the enzymes present in membrane-bound organelles known as lysosomes are hydrolases, proteases, lipases, peptidases, glycosidases, nucleases, phosphatases, and sulfatases.

They are necessary for the degradation of internal and extracellular macromolecules. Processes including receptor-mediated endocytosis, phagocytosis, and pinocytosis can all be used to engulf extracellular particles. Furthermore, autophagy is a crucial mechanism in cellular maintenance. It involves the transfer of large aggregates, extracellular materials, long-lived proteins, and organelles to lysosomes for breakdown. The term "autophagy," which translates to "self-eating," was first used in the 1960s following the discovery of double-membrane vesicles that capture proteins and intracellular organelles [16].

Targeting bigger protein clumps or membrane-associated proteins, the autophagic system is made up of over 500 components and is essential for cell survival and metabolic balance. It is a hunger response that recycles cellular components to assist in maintaining intracellular organelle homeostasis. Particularly in times of cellular stress, autophagy plays a critical role in eliminating harmful protein aggregates and damaged organelles. Numerous human illnesses, including lysosomal storage diseases, malignancies, and neurodegenerative disorders, have been connected to its dysregulation. All cell types undergo basal autophagy, which aids in controlling the size of the endoplasmic reticulum (ER) and the makeup of the cytoplasm. When autophagy is dysregulated, it can also cause type II apoptosis and contribute to the etiology of a number of illnesses. Autophagy comes in three primary forms: chaperone-mediated autophagy (CMA), microautophagy, and macroautophagy. While lysosomal degradation is the end result of all three routes, their methods vary [17].

7.2.5 Immunoproteasome

When exposed to oxidative stress and triggers, including lipopolysaccharide, interferons, and tumor necrosis factor-α, a specialized form of the proteasome known as the immunoproteasome can be formed. This is achieved by replacing the traditional catalytic subunits with immunoproteasomal subunits, such as LMP2 (β1i), MECL1 (β2i), and LMP7 (β5i). Moreover, 11S (PA28) regulators have the ability to bind to the 20S core unit of the proteasome. The primary role of the immunoproteasome is to digest oligopeptides for antigen presentation in the adaptive immune response. It is essential for immunity as well as the elimination of oxidized proteins.

The immunoproteasome and 20S proteasome enhance their ability to break down oxidized proteins under oxidative stress, in contrast to the 26S proteasome, which is ineffective at doing so.

In order to help cells deal with oxidative damage, the 26S proteasome disassembles during oxidative stress, increasing the availability of 20S proteasomes. The 20S proteasomes' capacity to break down oxidized proteins is increased when 11S regulators bind to them. There is mounting evidence that the immunoproteasome is more active and selective than the typical 20S proteasome in its ability to break-down oxidized proteins [18].

The 11S regulator further enhances the immunoproteasome's function in this regard, suggesting that it has a broader biological role beyond immune function. It has been shown that even low levels of hydrogen peroxide (H_2O_2) can trigger an increase in 20S proteasomes, immunoproteasomes, and 11S regulators, preparing cells to better withstand future oxidative stress. Additionally, enhanced immunoproteasome function has been linked to increased lifespan in certain species, likely due to its ability to maintain protein quality and remove damaged proteins. The immunoproteasome has also been associated with various diseases, including cancer, inflammatory disorders, and neurodegenerative diseases. For example, increased immunoproteasome expression has been observed in amyloid plaques in Alzheimer's disease, suggesting that cells attempt to clear protein aggregates, albeit unsuccessfully. In models of amyotrophic lateral sclerosis (ALS), increased immunoproteasome activity in astrocytes and microglia has been linked to better survival outcomes, indicating that the immunoproteasome may play a role in limiting disease progression [18].

Additionally, a number of illnesses, such as cancer, inflammatory conditions, and neurodegenerative diseases, have been linked to the immunoproteasome. For instance, Alzheimer's disease amyloid plaques have been shown to exhibit elevated immunoproteasome expression, indicating that cells make an ineffective attempt to remove protein clumps.

7.2.6 Neurodegenerative Disorders (NDDs)

Because they are essential for communication, neurons are necessary for the human brain to operate properly. Neurons are present all across the body, but the bulk of them start in the brain. The majority of neurons are produced by neural stem cells during childhood, but this production sharply declines as people age. Neurodegeneration, or the progressive loss of neurons, their structures, or their activities, is a critical health issue and a major contributing factor to the development of many brain ailments, even though neurons are not immortal. Neurodegeneration is associated with the buildup of altered protein variants in the brain, disruptions in neural networks, and synaptic dysfunction. A class of illnesses known as neurodegenerative diseases (NDs) is characterized by neurodegeneration [19] (Table 7.1).

Among the most common NDs are amyotrophic lateral sclerosis (ALS), Parkinson's disease, Alzheimer's disease, prion disease, motor neuron disease, Huntington's disease (HD), spinal muscular atrophy, and spinocerebellar ataxia. The progressive loss of neurons in the central nervous system (CNS) is a hallmark of neurodegenerative diseases, such as Alzheimer's, Parkinson's, and ALS. Some of the more common NDs include Parkinson's disease, Alzheimer's disease, prion disease, amyotrophic lateral sclerosis (ALS), motor neuron disease, Huntington's disease, spinal muscular atrophy, and spinocerebellar ataxia. The progressive loss of neurons in the central nervous system is a feature of neurodegenerative diseases like Alzheimer's, Parkinson's, and ALS [20].

Although there have been significant advancements in understanding these disorders, therapeutic strategies primarily focus on alleviating symptoms rather than addressing the underlying mechanisms of the diseases. This limitation is largely due to the blood-brain barrier (BBB), a highly selective barrier that prevents nearly 99% of therapeutic drugs from entering the brain while also protecting it from harmful substances. Consequently, the effectiveness of therapies targeting the mechanisms of pathogenesis is restricted.

7.3 Pathophysiology of Neurodegenerative Disorders

7.3.1 Alzheimer's Disease (AD)

The most prevalent type of dementia, Alzheimer's disease, is characterized by abnormal behavior, memory loss, and cognitive impairment. Amyloid-beta plaque buildup and tau protein neurofibrillary tangles in the brain are biochemical indicators of AD. Neuronal death results from the disruption of synaptic transmission caused by these abnormal protein aggregates. The condition mostly affects the brain's hippocampus and cerebral cortex, which are responsible for memory and higher order cognitive functions [21].

7.3.2 Parkinson's Disease (PD)

Parkinson's disease (PD), a degenerative neurological ailment, is characterized by tremors, muscle rigidity, uneven gait, and issues with balance and coordination. The

Table 7.1 Neurodegenerative disorder and its example with their mechanism of action

Neurodegenerative disorder	Therapeutic class	Examples	Mechanism of action	Challenges
Alzheimer's disease (AD)	Cholinesterase inhibitors	Donepezil, Rivastigmine, Galantamine	Prevent the breakdown of acetylcholine, enhancing cholinergic transmission, providing symptomatic cognitive benefits	Limited cognitive improvements, does not halt disease progression, BBB penetration concerns
	Glutamate regulators	Memantine	NMDA receptor antagonist that reduces excitotoxicity and neuronal damage, managing symptoms of moderate-to-severe AD	Primarily symptomatic, limited efficacy in reversing pathology
	Antibody targeting Aβ plaques	Aducanumab (Aduhelm)	IgG1 monoclonal antibody targets and removes amyloid-beta plaques, potentially modifying disease	High cost, controversial effectiveness, requires regular IV infusion
Parkinson's disease (PD)	Dopamine replacement therapy	Levodopa-Carbidopa	Replenishes dopamine in the substantia nigra, improving motor function	Long-term use leads to motor complications, no impact on disease progression
	Dopamine agonists	Pramipexole, Ropinirole, Rotigotine	Mimic dopamine action, improving motor symptoms	Side effects (e.g., impulse control issues), symptomatic management only
	MAO-B inhibitors	Selegiline, Rasagiline	Inhibits breakdown of dopamine, increasing dopamine levels in the brain	Symptomatic treatment, limited efficacy in advanced stages
	COMT inhibitors	Entacapone, Tolcapone	Inhibits catechol-O-methyltransferase, prolonging levodopa's effect	Associated with liver toxicity (Tolcapone), symptomatic relief only

(continued)

Table 7.1 (continued)

Neurodegenerative disorder	Therapeutic class	Examples	Mechanism of action	Challenges
Amyotrophic lateral sclerosis (ALS)	Glutamate modulators	Riluzole	Reduces glutamate toxicity, slightly extending survival	Limited survival
	Free radical scavengers	Edaravone	Reduces oxidative stress, slowing disease progression	Administered via IV, minimal impact on overall survival

largest risk factor for Parkinson's disease (PD) is aging, but environmental and genetic factors also play a part. Other factors, such as smoking, excessive coffee use, and exposure to environmental toxins, are known to influence the risk of developing Parkinson's disease, even if the precise mechanisms are yet unknown. The pathophysiology of Parkinson's disease includes frontal cortical atrophy and ventricular hypertrophy. However, the most obvious anatomical change is the loss of pigmentation in the locus coeruleus and substantia nigra pars compacta (SNpc) due to the degeneration of dopaminergic, neuromelanin-containing neurons. The nigrostriatal pathway is hampered by this significant neuron loss, which lowers dopamine levels in the striatum and, as a result, causes the characteristic movement signs of Parkinson's disease. Moreover, cell death in several regions, such as the raphe nuclei, locus coeruleus, pedunculopontine nucleus, dorsal motor nucleus of the vagus, hypothalamus, and olfactory bulb, as well as the nucleus basalis of Meynert, is the source of PD's non-motor symptoms.

Parkinson's disease (PD) progresses due to a number of factors, including α-synuclein misfolding and aggregation, mitochondrial failure, impaired protein clearance systems such as the ubiquitin-proteasome and autophagy-lysosome pathways, and neuroinflammation. Microscopic characteristics of Parkinson's disease (PD) include dystrophic neurites, sometimes called Lewy neurites, and Lewy bodies, which are abnormal cytoplasmic deposits containing α-synuclein. It is possible for these Lewy bodies to get phosphorylated and travel to different areas of the central nervous system. Like Alzheimer's disease, Parkinson's disease (PD) also involves protein misfolding, with tau protein being a common example. Neurofibrillary tangles (NFTs) arise as a result of tau hyperphosphorylation. Widespread amyloid-beta plaque and NFT deposits are also seen in certain PD patients [22].

7.3.3 Amyotrophic Lateral Sclerosis (ALS)

Lou Gehrig's disease, also known as amyotrophic lateral sclerosis (ALS), is a neurological illness that primarily affects motor neurons, which control voluntary muscle contractions. Motor neurons gradually deteriorate and die, causing muscle atrophy, weakness, and ultimately paralysis. When ALS affects both upper and lower motor neurons, the neuromuscular system slowly breaks down. The majority of individuals die from respiratory failure within 3–5 years of diagnosis, though occasionally behavioral and cognitive problems may also manifest. The impulses needed for muscular activity cannot reach their intended sites when ALS-related motor neuron degeneration occurs in the brain and spinal cord. Despite the fact that more than 30 genes have been linked to ALS, mutations in four key genes—C9orf72, TARDBP, SOD1, and FUS—account for more than 70% of cases. These genes generate proteins necessary for DNA repair, cellular homeostasis, mitochondrial functioning, and glial cell activity. These processes are thought to be malfunctioning, which results in the loss of motor neurons seen in ALS. ALS is characterized by protein clumps inside neurons, with TAR DNA-binding protein being the most prevalent in most instances. Other proteins, such as superoxide dismutase-1 and neurofilament, can also form aggregates. However, it is still unclear if these protein clumps cause or are a result of neuron damage [23].

7.4 Therapeutic Options for Neurodegenerative Disorders

The majority of the therapeutic options available for neurological disorders are symptomatic and palliative, despite decades of research. The main goal of approved drugs is to lessen symptoms rather than to stop or reverse the course of the illness.

7.4.1 Alzheimer's Disease

Cholinesterase inhibitors (donepezil, rivastigmine) and NMDA receptor antagonists (memantine) protect neurons against excitotoxicity by improving cholinergic transmission. However, these drugs only offer little clinical alleviation and ignore the fundamental reasons of the disease.

7.4.2 Antibody Targeting Amyloid-Beta (Aβ) Plaques

For moderate-to-severe AD cases, greater doses of donepezil (23 mg/day) were recently approved, either by itself or in conjunction with memantine. Rivastigmine is a cholinesterase inhibitor that is used for mild-to-moderate AD. It is unique among cholinesterase inhibitors in that it may be applied as a transdermal patch. The enzymes butyrylcholinesterase and acetylcholinesterase are both inhibited by it. In addition to inhibiting cholinesterase, galantamine functions as an allosteric modulator of nicotinic cholinergic receptors and is prescribed at doses of 16–24 mg/day for mild-to-moderate AD. Cholinesterase inhibitors are still the sole commonly accessible therapy option for AD despite the development of other medications [24].

This suggests that more antibody-based therapies could receive FDA approval for AD treatment in the near future.

7.4.3 Cholinesterase Inhibitors

Currently, the first-line treatment for Alzheimer's disease (AD) is cholinesterase inhibitors. In clinical practice, galantamine, rivastigmine, and donepezil are the main cholinesterase inhibitors. Cholinergic neurons and acetylcholine levels in the brain's cortical regions are significantly reduced in AD. Increasing the availability of acetylcholine in dementia patients has been demonstrated to halt cognitive deterioration. Cholinesterase inhibitors increase cholinergic activity and offer some cognitive advantages to patients by blocking the breakdown of acetylcholine. The FDA authorized tacrine as the first cholinesterase inhibitor in 1993; however, it was later withdrawn because of hepatotoxicity. Oral pills containing 5 or 10 mg of donepezil daily are indicated for mild-to-moderate AD. For moderate-to-severe AD cases, greater doses of donepezil (23 mg/day) were recently approved, either by itself or in conjunction with memantine. Rivastigmine is a cholinesterase inhibitor that is used for mild-to-moderate AD. It is unique among cholinesterase inhibitors in that it may be applied as a transdermal patch. The enzymes butyrylcholinesterase and acetylcholinesterase are both inhibited by it. In addition to inhibiting cholinesterase, galantamine functions as an allosteric modulator of nicotinic cholinergic receptors and is prescribed at doses of 16–24 mg/day for mild-to-moderate AD. Cholinesterase inhibitors are still the sole commonly accessible therapy option for AD despite the development of other medications.

The newly approved drug aducanumab has sparked debate regarding its effectiveness and high cost. However, the efficacy of cholinesterase inhibitors is limited, offering only modest improvements in cognitive function and being classified as symptomatic treatments rather than therapies that alter the disease's underlying pathology. There are also concerns about whether current medications can cross the blood-brain barrier (BBB) in sufficient doses to produce the desired pharmacological effects [25].

7.4.4 Glutamate Regulators

The main excitatory neurotransmitter in the brain is glutamate. Glutamate can lead to neuronal injury and neurodegeneration when it is overactivated in postsynaptic neurons, especially through NMDA receptors. However, there have been serious adverse effects from totally blocking NMDA receptors. A noncompetitive NMDA receptor antagonist called memantine was created to remedy this. By controlling NMDA receptor activity and shielding patients from the negative consequences of over-activation, it offers therapeutic advantages. In 2003, memantine was authorized for use in treating moderate-to-severe Alzheimer's disease (AD) at a daily dosage of 5–20 mg. Compared to a placebo, memantine has been demonstrated to enhance cognitive function when administered as a monotherapy. Additionally, clinical trials showed improved efficacy over monotherapy for up to a year when paired with acetylcholinesterase inhibitors [26].

However, like other drugs in this class, memantine does not address the underlying pathology of AD and is primarily used to manage symptoms.

7.4.5 Parkinson's Disease

Despite being the second most prevalent neurodegenerative disease, Parkinson's disease (PD) currently lacks a viable treatment that directly modifies the underlying pathogenesis. In order to relieve patients' symptoms, the majority of current therapy approaches concentrate on controlling both motor and non-motor symptoms. Restoring dopamine levels in the substantia nigra, a part of the brain where dopamine depletion takes place, is the primary method of treating Parkinson's disease. The combination of levodopa and carbidopa is the most often used treatment. Carbidopa stops levodopa from being broken down before it reaches the brain, while levodopa, a precursor to dopamine, helps enhance motor function by restoring dopamine levels. To further maintain levodopa's availability in the brain, medications such as entacapone and tolcapone also inhibit catechol-O-methyltransferase (COMT), which stops levodopa from being methylated and broken down. Dopamine agonists, such as apomorphine hydrochloride, pergolide, pramipexole, ropinirole, and rotigotine, mimic the effects of dopamine and are also used to treat Parkinson's disease. Monoamine oxidase (MAO) inhibitors, such as selegiline and rasagiline, help maintain steady dopamine levels in the brain by halting its oxidative degradation. Parkinson's disease (PD) is also associated with a number of non-motor symptoms (NMS), which are treated symptomatically. These symptoms include depression, psychosis, sleep problems, constipation, dementia, and olfactory abnormalities. To treat dementia symptoms, galantamine, donepezil, and rivastigmine are used. Psychosis is treated with medications like clozapine, quetiapine, and pimavanserin [27].

Melatonin or clonazepam is recommended for managing sleep disturbances. While levodopa remains the gold standard for treating motor symptoms, with MAO-B inhibitors and dopamine agonists offering additional options, advanced cases may require deep brain stimulation (DBS). However, despite these treatments, the progression of the disease remains unaltered.

7.4.6 Amyotrophic Lateral Sclerosis (ALS)

Frontotemporal dementia, behavioral problems, and cognitive decline are common symptoms of amyotrophic lateral sclerosis (ALS), a motor neuron disease that worsens with time. Most ALS patients pass away from respiratory failure within 3–5 years of the onset of symptoms. Currently, the FDA has approved two medications for ALS: edaravone and riluzole. Riluzole, a glutamate-receptor antagonist, was first approved in 1995 as an oral tablet containing 100 mg per day. Clinical trials have shown that riluzole extends the life of ALS patients by 3–4 months when compared to a placebo group. The free-radical scavenger edaravone was licensed in

2017 and administered intravenously at a dose of 60 mg daily to slow the disease's progression.

Despite these treatments, the therapeutic options for ALS remain limited, and neither drug significantly alters the course of the disease. Symptomatic management, including physical therapy, respiratory support, and nutritional guidance, plays a critical role in improving the quality of life for ALS patients, though no definitive disease-modifying therapies are available at this time [28].

7.4.7 Major Challenges of Drug Delivery

The blood-brain barrier (BBB) functions as a diffusion barrier to prevent various substances from the bloodstream from entering the brain, preserving homeostasis and normal brain function. Numerous brain cell types, such as neurons, astrocytes, tight junctions, brain microvascular endothelial cells, and basal membranes, form a tightly sealed brain capillary system. The lack of fenestrations in the endothelial cells of the brain capillaries restricts the movement of proteins and tiny molecules. These endothelial cells are also connected by inter-endothelial junctions, which form an uninterrupted barrier and prevent the passage of substances that are soluble in water. The astrocytes, pericytes, and basal lamina that surround endothelial cells further restrict the flow of drug molecules from the blood into the brain.

Efflux transporters in brain capillaries actively pump chemicals back into the circulation, strengthening the barrier. The permeability of the blood-brain barrier is regulated by protein complexes at the inter-endothelial junctions, including adherens, gap, and tight junctions. Certain physicochemical properties, such as charge, lipid solubility, surface activity, size, and molecular weight, determine whether a drug may cross the blood-brain barrier via the paracellular or transcellular pathways. Small chemicals that can passively diffuse through the blood-brain barrier include barbiturates, carbon dioxide, and ethanol. Transporters, including insulin, transferrin, and glucose transporter-1 (GLUT-1), are involved in receptor-mediated processes that move hydrophilic molecules like proteins and peptides. Furthermore, some medical disorders might weaken the BBB, allowing chemicals to seep into the brain [29].

Protein homeostasis, commonly known as proteostasis, is an essential process that ensures that proteins are synthesized, folded, and degraded correctly within cells. With aging, disruptions in proteostasis and the emergence of inflammation become increasingly prominent, especially in neurodegenerative diseases. The intricate relationship between protein misfolding and inflammation significantly contributes to the progression of age-related disorders.

7.4.8 The Proteostasis Network

The proteostasis network consists of various molecular mechanisms that work together to preserve the integrity of functional proteins both inside and outside of cells. Key components of this network include molecular chaperones, which facilitate the proper folding of newly synthesized proteins, as well as the ubiquitin-proteasome system and autophagy, which are responsible for degrading misfolded or damaged proteins.

7.4.8.1 Molecular Chaperones

These proteins play a crucial role in preventing protein aggregation by stabilizing unfolded or misfolded proteins and aiding in their proper refolding. Their importance is heightened in neurons, where there is a high rate of protein synthesis and turnover [30, 31].

7.4.8.2 Ubiquitin-Proteasome System (UPS)

This system identifies misfolded proteins for degradation through a process called ubiquitination, which allows these proteins to be processed by the proteasome. The effectiveness of the UPS is vital for maintaining proteostasis, especially in postmitotic neurons that do not divide and have limited capacity for regeneration.

7.4.8.3 Autophagy

This mechanism targets larger aggregates and damaged organelles, facilitating the turnover of cellular components. In neurons, autophagy plays a significant role in removing aggregated proteins that may disrupt cellular functions.

7.4.8.4 Inflammation as a Response to Cellular Stress

Inflammation is the body's biological response to harmful stimuli, including cellular stress and protein damage. When proteostasis is disrupted, conditions such as oxidative stress, mitochondrial dysfunction, and endoplasmic reticulum (ER) stress can occur, leading to increased protein misfolding. The buildup of misfolded proteins can overwhelm the cellular degradation systems, resulting in protein aggregation and the formation of toxic species.

7.4.8.5 Activation of Glial Cells

In the central nervous system (CNS), glial cells, including microglia and astrocytes, are crucial in responding to stress. When proteostasis fails, these cells can become activated and release inflammatory mediators such as cytokines and chemokines, which can further intensify inflammation.

7.4.8.6 Cytokine Release

Pro-inflammatory cytokines, such as tumor necrosis factor-alpha (TNF-α) and interleukin-6 (IL-6), can create a feedback loop that fosters chronic inflammation. This persistent inflammation can disrupt neuronal function, contributing to declines in cognitive and motor abilities [32].

7.5 Interplay Between Proteostatic Stress and Inflammation

The intricate relationship between proteostatic stress and inflammation has emerged as a key area of research in understanding the progression of diseases such as neurodegeneration, cancer, and metabolic disorders. Proteostasis, or protein homeostasis, is critical for maintaining cellular function, while inflammation represents the immune system's response to cellular dysfunction. The relationship between proteostatic stress and excessive immune responses plays a critical role in the development of neurodegenerative diseases, including Alzheimer's disease, Parkinson's disease, and amyotrophic lateral sclerosis (ALS) [33].

7.5.1 Proteostasis and Proteostatic Stress

Proteostasis is maintained by molecular chaperones, the ubiquitin-proteasome system (UPS), and autophagy, which together ensure proper protein folding and degradation. However, disruptions in these systems lead to proteostatic stress, causing the accumulation of misfolded proteins.

Misfolded proteins can be toxic and lead to cellular damage, triggering stress response pathways such as:

7.5.1.1 Unfolded Protein Response (UPR)

UPR is activated by ER stress, attempts to restore proteostasis by reducing protein synthesis and enhancing folding capacity. The UPR is an essential adaptive response to proteostatic stress that operates mainly in the ER. Under persistent stress conditions, however, the UPR can turn maladaptive, propagating inflammation:

7.5.1.2 IRE1α Activation

IRE1α is a protein sensor that is misfolded in the ER. When activated, it splices X-box binding protein 1 (XBP1) mRNA, producing a transcription factor that boosts ER chaperone synthesis. However, persistent activation of IRE1α draws TNF receptor-associated factor 2 (TRAF2), which activates NF-κB and triggers the JNK pathway, causing pro-inflammatory cytokines to be produced. This is particularly crucial for long-term diseases like type 2 diabetes and obesity, where inflammation triggered by ER stress exacerbates metabolic dysfunction and insulin resistance.

7.5.1.3 PERK Pathway and Inflammatory Crosstalk

Activation of PKR-like ER kinase (PERK) in response to ER stress leads to phosphorylation of eukaryotic initiation factor 2α (eIF2α), halting protein translation to reduce ER load. However, prolonged PERK activity also enhances inflammatory responses by activating transcription factor 4 (ATF4) and NF-κB, contributing to inflammation-driven tissue damage, particularly in neurodegenerative disorders.

7.5.1.4 Heat Shock Response (HSR)

Induced in the cytoplasm, this response upregulates heat shock proteins (HSPs) to assist in refolding proteins. The heat shock response (HSR) is a cytosolic mechanism that promotes proteostasis by upregulating molecular chaperones, such as heat shock proteins (HSPs), which help refold damaged proteins. Interestingly, the HSR also plays a role in modulating immune responses:

7.5.1.5 HSPs as Immunomodulators

Certain heat shock proteins, such as HSP70 and HSP90, are recognized as danger signals when released into the extracellular space during cell stress or damage. These extracellular HSPs can act as damage-associated molecular patterns (DAMPs), activating toll-like receptors (TLRs) and promoting the secretion of pro-inflammatory cytokines. This phenomenon links proteostatic stress directly to innate immune activation, particularly in chronic inflammatory conditions like arthritis and neurodegenerative diseases [34].

7.5.2 Inflammation and Immune Response

Inflammation is triggered by infection, injury, or cellular stress. It is mediated by *pro-inflammatory cytokines* (TNF-α, IL-1β, IL-6) and involves the recruitment of immune cells (macrophages, neutrophils). Inflammation can be either acute (beneficial) or chronic (detrimental). Chronic inflammation has been linked to various pathologies, including autoimmune diseases, cardiovascular diseases, and cancers.

7.5.3 Proteostatic Stress Activates Inflammatory Pathways

Research has demonstrated that proteostatic stress can directly activate inflammatory signaling.

Following are some *examples*:

7.5.3.1 UPR and Inflammation

Chronic ER stress activates transcription factors such as ATF6, PERK, and IRE1, leading to the production of inflammatory cytokines. In particular, IRE1 activates the JNK pathway, which is involved in pro-inflammatory responses.

7.5.3.2 NLRP3 Inflammasome Activation

Misfolded proteins, such as β-amyloid in Alzheimer's disease, activate the NLRP3 inflammasome, leading to the production of IL-1β and contributing to neuroinflammation.

7.5.3.3 Oxidative Stress and NF-κB

Oxidative stress caused by proteostatic stress induces NF-κB activation, a key regulator of inflammatory gene expression.

7.5.3.4 Inflammation Aggravates Proteostatic Stress

Conversely, chronic inflammation can worsen proteostatic stress by impairing the cell's proteostasis machinery:

7.5.3.5 Cytokine-Mediated Proteasome Dysfunction

Inflammatory cytokines such as TNF-α reduce proteasome activity, impeding the degradation of misfolded proteins.

7.5.3.6 Suppression of Heat Shock Response

Chronic inflammation downregulates HSR, reducing the ability to manage protein misfolding.

7.6 Disease Implications of Proteostatic Stress and Inflammation

The interplay between proteostatic stress and inflammation has been implicated in various diseases:

7.6.1 Neurodegenerative Diseases

In Alzheimer's and Parkinson's disease, protein aggregates (e.g., β-amyloid and α-synuclein) activate the inflammasome and induce chronic neuroinflammation, exacerbating proteostatic stress and neuronal damage.

7.6.2 Alzheimer's Disease

In Alzheimer's, the buildup of amyloid-beta plaques and tau tangles leads to neuronal death. The resulting inflammatory response can further disrupt proteostasis by activating microglia, which release inflammatory factors detrimental to neuronal health.

7.6.3 Amyloid-Beta Plaques

Aβ peptides are derived from the cleavage of amyloid precursor protein (APP). In AD, these peptides misfold and aggregate into extracellular plaques, which are toxic to neurons. The accumulation of Aβ is often linked to *endoplasmic reticulum (ER) stress* and impaired proteostasis, which overwhelm the cell's capacity to manage misfolded proteins.

7.6.4 Tau Tangles

Intracellular neurofibrillary tangles are created in AD as a result of aberrant hyperphosphorylation of tau, a microtubule-associated protein. These tangles lead to cellular malfunction and death by interfering with neuronal transport processes. Tau protein aggregation also results in abnormalities of cytoplasmic proteostasis, which exacerbates neuronal toxicity.

7.6.5 Parkinson's Disease

In Parkinson's disease, misfolded alpha-synuclein (α-syn) proteins aggregate to form Lewy bodies, which trigger an inflammatory response in the brain. This inflammation contributes to neuronal cell death and worsens proteostasis disturbances, leading to further aggregation and toxicity. In PD, the balance of protein synthesis, folding, trafficking, and degradation within cells is essential for maintaining cellular homeostasis. This balance is compromised due to the accumulation of misfolded α-syn, which overwhelms the cell's ability to manage protein quality control.

7.6.6 Alpha-Synuclein Misfolding and Aggregation

Alpha-synuclein (α-syn) is a small, soluble protein primarily found at presynaptic terminals, where it is involved in synaptic vesicle regulation. In PD, α-syn undergoes a conformational change, misfolds, and aggregates into oligomers and fibrils that form the core of Lewy bodies. Several factors contribute to the misfolding and aggregation of α-syn, including the following.

7.6.7 Mutations in the SNCA Gene

Mutations in the gene encoding α-syn, such as A53T or gene multiplications, increase the propensity of α-syn to misfold and aggregate.

7.6.8 Posttranslational Modifications

Phosphorylation at Ser129 and other modifications enhance α-syn aggregation and neurotoxicity.

7.6.9 Mitochondrial Dysfunction and Oxidative Stress

Impaired mitochondrial function in PD leads to increased oxidative stress, which promotes the aggregation of misfolded proteins.

7.6.10 Amyotrophic Lateral Sclerosis (ALS)

Motor neurons in the brain and spinal cord are impacted by the progressive neurodegenerative illness known as amyotrophic lateral sclerosis (ALS). The illness first appears as a progressive loss of muscle function, which progresses to respiratory failure and paralysis. Both familial and sporadic ALS can occur, with many gene mutations linked to family cases. One of these, mutations in the SOD1 gene, which codes for superoxide dismutase 1, is a known factor in the pathophysiology of ALS. Misfolded proteins, such as mutant SOD1, TDP-43, and FUS, accumulate in motor neurons to form toxic aggregates, which is a characteristic of ALS.

This misfolding induces cellular stress, disrupts protein homeostasis (proteostasis), and triggers a neuroinflammatory response through the activation of microglia and astrocytes. The resulting inflammation exacerbates neuronal damage, creating a vicious cycle that accelerates motor neuron degeneration. In ALS, misfolded protein aggregates such as mutant SOD1 trigger a robust neuroinflammatory response that significantly contributes to disease progression. This response is mediated by the activation of glial cells, including microglia and astrocytes, which release pro-inflammatory cytokines and toxic factors that exacerbate motor neuron damage [35].

7.6.10.1 Microglial Activation in ALS

The central nervous system's (CNS) principal immune cells, microglia, are in charge of preserving homeostasis and reacting to damage. In ALS, microglia become chronically activated in response to misfolded SOD1 and other toxic proteins. This activation results in:

Release of Pro-Inflammatory Cytokines

Activated microglia secrete cytokines such as TNF-α, IL-1β, and IL-6, which promote inflammation and contribute to motor neuron toxicity.

Oxidative and Nitrosative Stress

Microglia also produce high levels of ROS and reactive nitrogen species (RNS), which induce oxidative damage in motor neurons and exacerbate proteostatic stress by modifying and destabilizing cellular proteins.

While microglial activation is initially aimed at clearing toxic protein aggregates, chronic activation leads to a self-perpetuating cycle of inflammation that accelerates neurodegeneration.

7.6.10.2 Astrocyte Dysfunction in ALS

The most prevalent glial cells in the central nervous system, astrocytes, are also essential to the pathophysiology of ALS.

In healthy conditions, astrocytes support motor neurons by regulating the extracellular environment and providing metabolic support. However, in ALS, astrocytes become reactive in response to mutant SOD1 and other misfolded proteins, contributing to the inflammatory response.

7.6.11 Huntington's Disease (HD)

In Huntington's disease (HD), neurodegeneration is driven by the combined effects of proteostatic stress and chronic neuroinflammation. The accumulation of misfolded mutant huntingtin protein overwhelms the cell's protein quality control systems, such as the ubiquitin-proteasome system and autophagy, leading to toxic aggregates that disrupt neuronal function and survival. At the same time, neuroinflammation, marked by the overactivation of microglia and astrocytes, results in the release of pro-inflammatory cytokines, which exacerbate oxidative stress and neuronal damage. This interplay between proteostatic stress and inflammation accelerates neuronal death, particularly in the striatum and cortex, contributing to the hallmark motor, cognitive, and psychiatric symptoms of HD [36]. Targeting both processes may offer therapeutic avenues to slow disease progression.

1. *Cancer*: In certain cancers, proteostatic stress enhances tumorigenesis by creating a pro-inflammatory environment, while chronic inflammation promotes mutations and genomic instability.
2. *Metabolic Diseases*: In obesity and diabetes, ER stress promotes insulin resistance by triggering inflammatory pathways like JNK and NF-κB, leading to chronic metabolic dysfunction.

References

1. Enzo [Internet]. [cited 2024 Oct 5]. Proteostasis. Available from: https://www.enzo.com/basic-research/field-of-research/proteostasis/.
2. Kuechler ER, Rose A, Bolten M, Madero A, Kammoonah S, Colborne S, et al. Protein feature analysis of heat shock induced ubiquitination sites reveals preferential modification site localization. J Proteome. 2021;239:104182.
3. Liberek K, Lewandowska A, Ziętkiewicz S. Chaperones in control of protein disaggregation. EMBO J. 2008;27(2):328.

4. Díaz-Villanueva JF, Díaz-Molina R, García-González V. Protein folding and mechanisms of proteostasis. Int J Mol Sci. 2015;16(8):17193–230.
5. Amm I, Sommer T, Wolf DH. Protein quality control and elimination of protein waste: the role of the ubiquitin–proteasome system. Biochim Biophys Acta (BBA) – Mol Cell Res. 2014;1843(1):182–96.
6. Juarez-Navarro K, Ayala-Garcia VM, Ruiz-Baca E, Meneses-Morales I, Rios-Banuelos JL, Lopez-Rodriguez A. Assistance for folding of disease-causing plasma membrane proteins. Biomol Ther. 2020;10(5):728.
7. Taylor RC, Dillin A. Aging as an event of proteostasis collapse. Cold Spring Harb Perspect Biol. 2011;3(5):a004440.
8. Kim Y, Kim EK, Chey Y, Song MJ, Jang HH. Targeted protein degradation: principles and applications of the proteasome. Cells. 2023;12(14):1846.
9. Stefani M. Protein folding and misfolding on surfaces. Int J Mol Sci. 2008;9(12):2515.
10. Livneh I, Cohen-Kaplan V, Cohen-Rosenzweig C, Avni N, Ciechanover A. The life cycle of the 26S proteasome: from birth, through regulation and function, and onto its death. Cell Res. 2016;26(8):869–85.
11. Paci A, Liu PXH, Zhang L, Zhao R. The proteasome subunit Rpn8 interacts with the small nucleolar RNA protein (snoRNP) assembly protein Pih1 and mediates its ubiquitin-independent degradation in Saccharomyces cerevisiae. J Biol Chem. 2016;291(22):11761–75.
12. van der Goot AT, Pearce MMP, Leto DE, Shaler TA, Kopito RR. Redundant and antagonistic roles of XTP3B and OS9 in decoding glycan and non-glycan degrons in ER-associated degradation. Mol Cell. 2018;70(3):516.
13. Kim P, Scott MR, Meador-Woodruff JH. Abnormal expression of ER quality control and ER associated degradation proteins in the dorsolateral prefrontal cortex in schizophrenia. Schizophr Res. 2018;197:484.
14. Chakrabarti A, Chen AW, Varner JD. A review of the mammalian unfolded protein response. Biotechnol Bioeng. 2011;108(12):2777.
15. Kelsen SG. The unfolded protein response in chronic obstructive pulmonary disease. Ann Am Thorac Soc. 2016;13(Suppl 2):S138.
16. Cooper GM. Endocytosis. In: The cell: a molecular approach. 2nd ed [Internet]. Sinauer Associates; 2000 [cited 2024 Nov 3]. Available from: https://www.ncbi.nlm.nih.gov/books/NBK9831/.
17. Todde V, Veenhuis M, van der Klei IJ. Autophagy: principles and significance in health and disease. Biochim Biophys Acta (BBA) – Mol Basis Dis. 2009;1792(1):3–13.
18. Ben Haim L, Carrillo-de Sauvage MA, Ceyzériat K, Escartin C. Elusive roles for reactive astrocytes in neurodegenerative diseases. Front Cell Neurosci [Internet]. 2015 [cited 2024 Nov 3];9. Available from: https://www.frontiersin.org/journals/cellular-neuroscience/articles/10.3389/fncel.2015.00278/full.
19. Brain basics: the life and death of a neuron | National Institute of Neurological Disorders and Stroke [Internet]. [cited 2024 Nov 3]. Available from: https://www.ninds.nih.gov/health-information/public-education/brain-basics/brain-basics-life-and-death-neuron.
20. Lamptey RNL, Chaulagain B, Trivedi R, Gothwal A, Layek B, Singh J. A review of the common neurodegenerative disorders: current therapeutic approaches and the potential role of nanotherapeutics. Int J Mol Sci. 2022;23(3):1851.
21. National Institute on Aging [Internet]. What happens to the brain in Alzheimer's disease? 2024 [cited 2024 Nov 3]. Available from: https://www.nia.nih.gov/health/alzheimers-causes-and-risk-factors/what-happens-brain-alzheimers-disease.
22. Models of α-synuclein aggregation in Parkinson's disease | Acta Neuropathol Commun | Full Text [Internet]. [cited 2024 Nov 3]. Available from: https://actaneurocomms.biomedcentral.com/articles/10.1186/s40478-014-0176-9.
23. Blokhuis AM, Groen EJN, Koppers M, van den Berg LH, Pasterkamp RJ. Protein aggregation in amyotrophic lateral sclerosis. Acta Neuropathol. 2013;125(6):777.

24. Singh R, Sadiq NM. Cholinesterase Inhibitors. In: StatPearls [Internet]. Treasure Island: StatPearls Publishing; 2024 [cited 2024 Nov 3]. Available from: http://www.ncbi.nlm.nih.gov/books/NBK544336/.
25. Kandiah N, Pai MC, Senanarong V, Looi I, Ampil E, Park KW, et al. Rivastigmine: the advantages of dual inhibition of acetylcholinesterase and butyrylcholinesterase and its role in subcortical vascular dementia and Parkinson's disease dementia. Clin Interv Aging. 2017;12:697.
26. Thomas SJ, Grossberg GT. Memantine: a review of studies into its safety and efficacy in treating Alzheimer's disease and other dementias. Clin Interv Aging. 2009;4:367.
27. Tan YY, Jenner P, Chen SD. Monoamine oxidase-B inhibitors for the treatment of Parkinson's disease: past, present, and future. J Parkinsons Dis. 2022;12(2):477.
28. Hoxhaj P, Hastings N, Kachhadia MP, Gupta R, Sindhu U, Durve SA, et al. Exploring advancements in the treatment of amyotrophic lateral sclerosis: a comprehensive review of current modalities and future prospects. Cureus. 2023;15(9):e45489.
29. Dong X. Current strategies for brain drug delivery. Theranostics. 2018;8(6):1481.
30. Mir RA, Bele A, Mirza S, Srivastava S, Olou AA, Ammons SA, et al. A novel interaction of ecdysoneless (ECD) protein with R2TP complex component RUVBL1 is required for the functional role of ECD in cell cycle progression. Mol Cell Biol. 2016;36(6):886–99.
31. Bioinformatics analysis reveals PIH1D1 as an important prognostic marker in breast cancer. J Appl Biol Biotech. 2024. Online First. https://doi.org/10.7324/JABB.2024.189230.
32. Popko K, Gorska E, Stelmaszczyk-Emmel A, Plywaczewski R, Stoklosa A, Gorecka D, et al. Proinflammatory cytokines IL-6 and TNF-α and the development of inflammation in obese subjects. Eur J Med Res. 2010;15(Suppl 2):120.
33. Sonninen TM, Goldsteins G, Laham-Karam N, Koistinaho J, Lehtonen Š. Proteostasis disturbances and inflammation in neurodegenerative diseases. Cells. 2020;9(10):2183.
34. Roh JS, Sohn DH. Damage-associated molecular patterns in inflammatory diseases. Immune Netw. 2018;18(4):e27.
35. Sweeney P, Park H, Baumann M, Dunlop J, Frydman J, Kopito R, et al. Protein misfolding in neurodegenerative diseases: implications and strategies. Transl Neurodegener. 2017;6:6.
36. Li Y, Li YJ, Zhu ZQ. To re-examine the intersection of microglial activation and neuroinflammation in neurodegenerative diseases from the perspective of pyroptosis. Front Aging Neurosci [Internet]. 2023 [cited 2024 Nov 3];15. Available from: https://www.frontiersin.org/journals/aging-neuroscience/articles/10.3389/fnagi.2023.1284214/full.

Biomarkers for Proteostasis Disruption: An Overview

8

Deena Prakash and Divya Bisht

8.1 Introduction

Neurodegenerative disorders (NDDs) represent a complex and challenging group of diseases characterized by the progressive degeneration of nerve cells (neurons), leading to a gradual decline in cognitive, motor, and functional abilities. These conditions typically involve the gradual and irreversible loss of neuron function, which impacts various aspects of daily life and overall quality of life [61]. For example, Alzheimer's disease (AD), the most prevalent form of dementia, primarily affects memory, reasoning, and other cognitive functions [57]. As the disease advances, patients experience increasing difficulties with memory recall, orientation, and performing everyday tasks, ultimately leading to severe cognitive impairment and dependence on caregivers.

In neurodegenerative disorders, specific proteins, such as amyloid-beta and tau in Alzheimer's disease, have been extensively studied as potential biomarkers for proteostasis disruption [5]. Elevated levels of these proteins are often found in the brains of affected individuals and correlate with symptom severity. Different diseases exhibit characteristic patterns of proteostasis disruption [86]. For instance, in neurodegenerative diseases, such as Alzheimer's and Parkinson's, the accumulation of misfolded proteins such as amyloid-beta and alpha-synuclein is a hallmark feature [50]. In cancer, the proteostasis network is often hijacked to promote tumor growth by stabilizing oncoproteins or enhancing the degradation of tumor suppressors [26, 136]. Moreover, changes in the levels of chaperone proteins, such as heat shock proteins (HSPs), have been proposed as potential biomarkers, reflecting the cellular response to misfolded proteins. HSPs play a crucial role in assisting with protein folding and preventing aggregation; thus, alterations in their expression can indicate dysregulation in proteostasis [108].

D. Prakash (✉) · D. Bisht
Biochemistry Discipline, School of Sciences, Indira Gandhi National Open University, New Delhi, India

N. Puranik (ed.), *Proteostasis*, https://doi.org/10.1007/978-981-96-6202-9_8

Proteostasis, a term derived from the fusion of "protein" and "homeostasis," refers to the cellular processes responsible for the synthesis, folding, trafficking, and degradation of proteins, ensuring their functionality [60]. The maintenance of proteostasis is critical for cell survival and proper function, especially in highly specialized cells such as neurons. Within cells, intricate networks of molecular chaperones, proteases, and autophagy mechanisms ensure that proteins reach their proper conformation and the damaged or misfolded proteins are efficiently degraded [125]. The intricate balance required for maintaining proteostasis is mediated by various cellular signals and factors, highlighting the potential for identifying biomarkers that could signify proteostasis disruption. When proteostasis is disrupted, it can lead to a range of pathological conditions, including neurodegenerative diseases (NDDs), cancer, and age-related disorders. Neurodegenerative disorders, including Alzheimer's disease, Parkinson's disease, Huntington's disease, and amyotrophic lateral sclerosis (ALS), are often characterized by the accumulation of misfolded proteins, leading to cellular stress, inflammation, and ultimately neuronal death [119]. In recent years, the identification of reliable biomarkers for proteostasis disruption in neurodegenerative disorders has gained interest, as these biomarkers can provide insights into disease mechanisms, aid in early diagnosis, and serve as potential therapeutic targets [123].

The concept of proteostasis disruption in neurodegenerative disorders revolves around the failure of the cellular machinery responsible for maintaining protein integrity. Several factors contribute to this disruption, including genetic mutations, environmental toxins, and aging, resulting in increased levels of misfolded and aggregated proteins [112]. This abnormal accumulation triggers a series of cellular responses, including the activation of stress response pathways, such as the unfolded protein response (UPR) and autophagy, in a bid to restore proteostasis [37]. The cellular proteostasis network disruption occurs at three fundamental levels: i.e., during translational regulation including protein synthesis at the ribosomes; at the protein folding via chaperones and cochaperones; and, lastly, during the autophagy or cell death, protein degradation via UPS and endoplasmic reticulum (ER)-associated clearance [75, 125]. The ER is responsible for proper protein folding and posttranslational modifications. Unfolded protein response (UPR), which is based on the endoplasmic reticulum (ER), continuously monitors the quality of the proteins. The UPR attempts to restore proteostasis by enhancing protein folding and degradation, but chronic ER stress can lead to cell dysfunction and apoptosis [94, 96]. In AD, the UPS is often impaired, leading to the accumulation of ubiquitinated proteins and contributing to neurodegenerative processes. Reduced proteasome activity has been observed in AD brains, which correlates with increased levels of misfolded proteins and cellular stress [22, 87]. Inositol-requiring enzyme-1 (IRE1), activating transcription factor 6 (ATF6), and protein kinase RNA-like endoplasmic reticulum kinase (PERK) are the three UPR activators that are triggered by misfolding events in the ER [117]. By triggering the transcription of protein quality control (PQC) genes, these proteins furthermore mediate a variety of reactions. When a protein's folding is not possible, ER-associated protein degradation (ERAD) is used to remove the protein [20, 130]. In the quality control connected with the ER membrane, the misfolded protein clients are ubiquitinated by E3 ubiquitin ligases, including HRD1, gp78, and MARCH6, in preparation for proteasomal breakdown [76, 100]. Misfolded cytosolic proteins are frequently segregated at several perinuclear inclusion bodies. These protein inclusions are shown to be positive for many quality control elements and prevent misfolded elements from interacting with other healthy cellular proteins [16]. Protein aggregates that have undergone severe misfolding may be carried to the mitochondria and dealt with there by the "Mitochondria as Guardian in Cytosol" (MAGIC) mechanism [99]. In the process, misfolded cytosolic proteins that are prone to aggregation are broken down by mitochondrial matrix proteases, like Pim1 [132]. Furthermore, a process known as mitochondrial-associated degradation (MAD) can be used to label faulty mitochondrial proteins with ubiquitin by E3 ubiquitin ligases and remove them from the outer mitochondrial membrane via the Cdc48 complex in preparation for cytosolic proteasome breakdown [71, 98]. The irregular handling in these processes leads to the onset of various neurodegenerative diseases, aging, and cancers. However, when these adaptive mechanisms are overwhelmed, they may lead to neuronal dysfunction and death, further exacerbating the condition [83, 127].

This review discusses major proteostasis markers, including amyloid-beta peptides, tau protein, ubiquitin-proteasome system (UPS), chaperone proteins, endoplasmic reticulum (ER) stress, and apoptotic pathways. Understanding these markers provides insight into disease mechanisms and potential therapeutic strategies. Various protein markers for Alzheimer's and Parkinson disease are represented in Fig. 8.1.

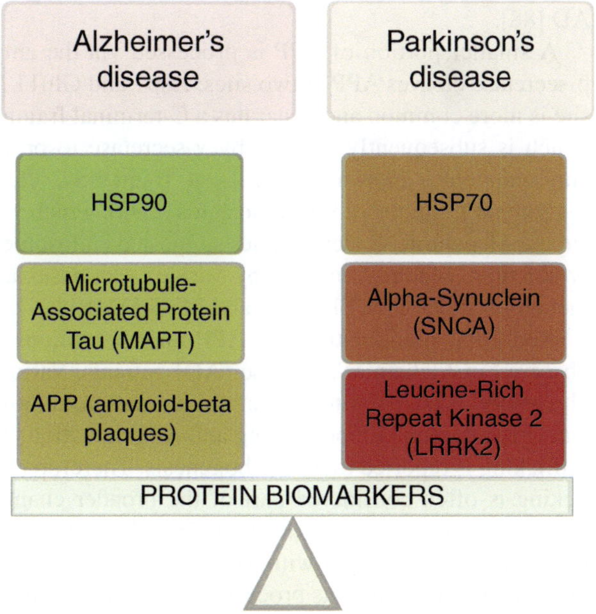

Fig. 8.1 Various protein biomarkers altered in neurodegenerative diseases

8.2 Amyloid Precursor Protein (APP)

Alzheimer's disease (AD) is a progressive neurodegenerative disorder characterized by dementia, impaired cognitive functions, and behavioral changes followed by brain atrophy. Pathologically, AD is caused due to the accumulation of amyloid beta plaques [12, 57].

The full-length APP is synthesized in the endoplasmic reticulum (ER) and then transported through the Golgi and trans-Golgi network (TGN), where it undergoes posttranslational modifications as part of its maturation process [46, 131]. Pathogenic mutation in the amyloid-β precursor protein (APP) gene and duplication in APP locus are also related to causing the early onset AD [44]. The APP gene is present on chromosome 21 and covers approximately 290,586 base pairs [138]. There are three isoforms of APP present in humans, i.e., APP695, APP751, and APP770 [148]. Among these, APP665 is predominantly present in the neural network and the remaining two are present in the non-neural cells. Amyloid precursor protein (APP) undergoes processing through both amyloidogenic and non-amyloidogenic pathways, mediated by α, β, and γ secretases [85, 149]. In the non-amyloidogenic pathway, which is the predominant route, α-secretase cleaves APP, resulting in an N-terminal soluble APP-alpha (sAPPα) and a C-terminal fragment of 83 amino acids (C83) [74]. The C83 fragment is further processed by γ-secretase to release a small 3 kDa peptide known as P3 and the APP intracellular domain (AICD). Additionally, β-site APP-cleaving enzyme 2 (BACE2), functioning as a θ-secretase, is involved in APP processing without producing amyloid-beta (Aβ). Amyloid-beta (Aβ) is a well-known proteolytic fragment of amyloid precursor protein (APP) [17, 145]. Mutations in the APP gene or in the genes encoding these secretases can lead to increased production of toxic Aβ, contributing to proteostasis disruption in AD [88].

A smaller portion of APP is processed via the amyloidogenic pathway, where β-secretase cleaves APP at two sites: Asp1 and Glu11 [148]. Cleavage at the Glu11 site is more common and generates a C-terminal fragment of 89 amino acids (C89), which is subsequently cleaved by γ-secretase to produce truncated Aβ in a non-amyloidogenic pathway [24]. Less frequently, cleavage at Asp1 produces a C-terminal fragment of 99 amino acids (C99), which γ-secretase processes to generate the Aβ peptide. Recent studies suggested endosomes are the main site for the Aβ production, which is ensured by various secretases and membrane-type 5 matrix metalloproteinase (MT1-MMP and MT5-MMP) [29].

Abnormal modifications of APP and disruptions in its trafficking have been observed in Alzheimer's disease (AD) patients. Various factors like glycosylation, phosphorylation, palmitoylation, ubiquitination, sumoylation, and sulfation are crucial in regulating APP processing and Aβ production [131]. The impairment of APP trafficking is reported to AD pathogenesis. Therefore, the dysregulation of APP trafficking is often considered part of the broader changes in its modifications. The abrupt regulation of kinases like GSK3, Protein kinase C, and phosphatases like PP1 and PP2A are linked with AD [9, 68]. The critical role of APP and its reduction to O-GlcNAcylation in its processing have also been identified in AD patients [93].

APP is primarily cleaved by α-secretase, generating non-amyloidogenic sAPPα. Any uncleaved APP in the plasma membrane is internalized into endosomes, where it can either be recycled back to the cell surface or sent to the lysosomal system for degradation [18]. Increased retention of APP in endosomes and lysosomes enhances amyloidogenic processing and Aβ production, as the acidic environment in these compartments is optimal for β-secretase activity [140]. The increased APP O-GlcNAcylation formation has promoted α secretase cleavage of APP while decreasing Aβ production. The O-GlcNAcase (OGN) inhibition increased the α cleavage with a significant reduction of β-cleavage of APP, attributing to the inhibition of APP endocytosis, which further leads to the increased level of APP in the plasma membrane [59]. Furthermore, the O-GlcNAcylation status of α-, β-, and γ-secretases may also influence the changes in APP processing induced by OGN inhibitors.

There are numerous reported studies that show that APP ubiquitination plays a role in inhibiting its processing and reducing Aβ production. The facilitation of APP ubiquitination by HRD1 causes APP degradation and leads to a decrease in Aβ generation [49]. However, preventing APP ubiquitination by replacing Lys649–651 with arginine residues disrupts APP sorting and increases Aβ production [81]. The ubiquitination of Lys651 is also crucial for the FBL2-induced reduction of Aβ levels [133]. Additionally, ubiquilin 1-mediated K63-linked polyubiquitination of APP slows its proteolytic processing, while a reduction in UBQLN1 levels is associated with increased Aβ production [46]. Other mutations have been studied in the AD, i.e., the APP gene on chromosome 21, the PSEN1 gene on chromosome 14, and the PSEN2 gene on chromosome 1 [64].

Posttranslational modifications of APP take place along the constitutive secretory pathway and also influence its trafficking suggesting a mutual regulation between APP modifications and trafficking [131]. These processes are tightly regulated to ensure APP's proper physiological functions and maintain its normal processing. Increasing evidence indicates that both APP modifications and trafficking significantly impact its processing and Aβ production. Abnormal APP modifications can lead to changes in trafficking and protein conformation, potentially shifting the cleavage preferences of various secretases, thereby disrupting APP processing and increasing Aβ generation [35]. Additionally, altered APP trafficking affects its interaction with different secretases, further influencing its processing and Aβ production. Therefore, understanding the regulation of APP modifications and trafficking is essential for developing new therapeutic strategies for Alzheimer's disease by targeting these processes.

8.3 Microtubule-Associated Protein Tau (MAPT)

The second key protein implicated in Alzheimer's disease (AD) is tau, a structural protein involved in stabilizing microtubules and facilitating axonal transport. In AD, tau becomes hyperphosphorylated, insoluble, and filamentous, forming the primary component of neurofibrillary tangles (NFTs) [54]. Tau inclusions are believed to contribute to AD progression, as they are found in brain regions with impaired function, and the formation of NFTs correlates with the duration and progression of the disease. Tau is an intrinsically disordered protein, and studies have shown that it can be degraded by the 20S proteasome without the need for ubiquitination [109]. While tau is expressed in the central and peripheral nervous systems, as well as in tissues such as the kidney, lungs, and testes, it is most abundant in neuronal axons, oligodendrocytes, and the somatodendritic compartments of neurons [82].

Microtubule formation and stability are critical for processes like neural development, cellular stabilization, and intracellular transport, and tau plays a critical role in these processes. During normal function, tau binds to the interface between α- and β-tubulin heterodimers through certain residues inside and surrounding its microtubule-binding repeats, helping to preserve the health of axons. The N-terminal acidic projection domain, proline-rich domain, microtubule binding domain, and C-terminal tail are the four primary domains that form tau [4]. While monomeric tau is naturally disordered and usually found unfolded, it can also adopt a "paperclip" shape in which the N-terminus and microtubule-binding domain are in close proximity to and interact with the C-terminus [38]. Aberrant aggregation may be encouraged by structural disruption brought on by interactions with other molecules, posttranslational changes, truncations, or mutations. Regions of the microtubule-binding domain are part of the "paired-helical filament (PHF) core," which is the basic structural component of tau aggregates [55]. Two hexapeptide motifs, PHF6* (275-VQIINK-280) and PHF6 (306-VQIVYK-311), with PHF6 being both required and sufficient for the formation of β-sheet aggregates in tau, are essential for aggregation within this core [78]. This relationship can be interfered with by mutations in these binding areas, which can increase the amount of free, unbound tau and destabilize microtubules.

Research into tau turnover suggests that both autophagy and the ubiquitin-proteasome system (UPS) are involved in its degradation, with different pathways targeting distinct tau pools [69]. For instance, hypoacetylated tau is primarily degraded by the UPS. Another hallmark of AD is the accumulation of advanced glycation end products (AGEs), which form insoluble protein aggregates linked to oxidative stress and inflammation [41]. AGEs have been shown to increase reactive oxygen species (ROS) production, which in turn activates pathways involved in APP processing and Aβ production. Additionally, AGEs are predominantly located in elevated levels of the immunoproteasome in amyloid plaques [48]. It has been demonstrated that AGEs stimulate a significant increase in immunoproteasome expression and activity through the activation of the receptor for AGEs (RAGE), triggering the phosphorylation of STAT1 via Jak1/2, which leads to the transcriptional upregulation of immunoproteasome subunits [89].

In 1996, it was found that some types of frontotemporal dementia (FTD) with parkinsonism associated with chromosome 17 (FTDP-17), currently known as FTDP-17t, are caused by autosomal dominant mutations in the MAPT gene, which produces the tau protein [67, 137]. This discovery verified that significant neurodegeneration in the central nervous system (CNS) can be caused solely by tau malfunction. More than 50 distinct pathogenic MAPT missense, silent, and intronic variants have been identified as of 2018 [115]. The buildup of filamentous tau inclusions in neurons and occasionally glia throughout the frontal and temporal lobes, which causes atrophy in these brain regions and ventricular enlargement, is the defining feature of FTDP-17t disease [126]. These MAPT mutations cause a variety of cognitive, behavioral, and motor impairments in affected individuals. Several central nervous system (CNS) disorders—including Alzheimer's disease (AD), Pick's disease (PiD), corticobasal degeneration (CBD), progressive supranuclear palsy (PSP), and argyrophilic grain disease—are marked by the abnormal accumulation of tau [62]. Therefore, targeting tau offers not only therapeutic potential for AD but also for a broader range of tauopathies.

8.4 Alpha-Synuclein (SNCA)

Alpha-synuclein (SNCA) was shown to be associated with Parkinson's disease (PD), and it established the first conclusive genetic linkage to PD [113]. Abnormal accumulations of alpha-synuclein, including Lewy bodies (LBs), Lewy neurites, and inclusions in glial cells, have been associated with various sporadic neurodegenerative disorders known as alpha-synucleinopathies such as idiopathic Parkinson's disease (PD), dementia with Lewy bodies (DLBs), and multiple system atrophy (MSA) [14]. The discovery of a genetic mutation associated with Parkinson's disease (PD) in certain families transformed the field and prompted a plethora of research investigations into the underlying genetics of the illness. In 1997, alpha-synuclein (SNCA) gene was one of the major genes associated with sporadic Parkinson's disease (PD), and it prompted researchers to swiftly create antibodies targeting alpha-synuclein [92]. It was discovered that alpha-synuclein is strongly present in Lewy bodies (LBs), particularly in the "halo" region of these inclusions. Today, the use of alpha-synuclein immunohistochemistry is considered for neuropathological evaluation of PD [2].

The SNCA gene produces a protein made up of 140 amino acids that remains structurally undefined in aqueous solutions, making it known as a "natively unfolded protein" [144]. Despite this, alpha-synuclein forms alpha-helical structures when it interacts with negatively charged lipids, such as phospholipids found in cell membranes. Over extended periods of incubation, it can also develop into beta-sheet-rich structures [103]. The protein is divided into three main regions: (1) the amino terminus (residues 1–60), which includes apolipoprotein lipid-binding motifs and is predicted to form amphiphilic helices, allowing alpha-helical structures to form upon membrane binding; (2) the central hydrophobic region (residues 61–95), known as the NAC (non-Aβ component), which enables beta-sheet formation; and (3) the carboxyl terminus, which carries a high negative charge and remains mostly unstructured [45].

Recent studies have shown a transient interaction between alpha-synuclein monomers and the negatively charged flexible C-terminal ends of fibrils, mediated by the positively charged N-terminus of the monomers [58]. These intermolecular interactions lead to the unfolding of the protein, revealing aggregation-prone regions. This unfolding is a critical step that facilitates protein aggregation, resulting in a rapid increase in alpha-synuclein and promoting the formation of new amyloids from monomers on the fibril surfaces, a phenomenon known as secondary nucleation. The oligomers formed during this secondary surface nucleation can disrupt cell membranes due to their lipophilic properties [129]. These actions result in early-stage raised baseline levels of cytosolic calcium, increased calcium influx upon stimulation, and delayed recovery of calcium levels in neurons. Disruption of neuronal calcium (Ca^{2+}) homeostasis results in increased ATP consumption to maintain the electrochemical gradient across the membrane, heightening neuronal vulnerability and affecting neurotransmitter release [32]. In the substantia nigra pars compacta (SNpc) dopaminergic neurons, slow Ca^{2+} oscillations initiate their characteristic slow tonic firing, leading to fluctuations in membrane potential. These oscillations also facilitate calcium entry into mitochondria, which stimulates oxidative phosphorylation and ATP production [118]. Increased mitochondrial calcium levels and enhanced oxidative phosphorylation further promote alpha-synuclein aggregation. Consequently, the interplay between calcium oscillations in dopaminergic neurons and pathological alpha-synuclein aggregates can trigger a toxic cascade that contributes to neurodegeneration [72]. Alpha-synuclein accumulation causes neuroinflammation, striatal neuron degeneration, and the activation of microglia [25].

Numerous posttranslational changes, including phosphorylation, have a major impact on alpha-synuclein folding and aggregation and are crucial in the development of synucleinopathies. Less than 5% of soluble, monomeric alpha-synuclein is phosphorylated in a physiologically normal state. On the other hand, phosphorylation at serine 129 and alpha-synuclein aggregation in Parkinson's disease (PD) and associated illnesses are strongly correlated, with around 90% of alpha-synuclein in Lewy bodies being phosphorylated. These illnesses have also been connected to the phosphorylation of other residues found in the protein's C-terminal region [14, 31, 122, 146]. The course of neuronal cell death in Parkinson's disease (PD) is known to be influenced by a number of inflammatory mediators, including nitric oxide (NO), reactive oxygen species (ROS), tumor necrosis factor-alpha (TNF-α), and interleukin-1 beta (IL-1β) produced by non-neuronal cells [120].

8.5 Leucine-Rich Repeat Kinase 2 (LRRK2)

Leucine-rich repeat kinase 2 (LRRK2) is a large multi-domain protein present in various tissues and involved in multiple biological processes. LRRK2 is majorly expressed in the olfactory bulb, striatum, cortex, hippocampal regions, midbrain, brainstem, and cerebellum of the brain [28]. Mutations in LRRK2 are identified as genetic risk factors for familial Parkinson's disease (PD) and may also play a role in the development of the more prevalent sporadic form of the disease. LRRK2 is a large protein with a mass of 286 kDa and 2527 amino acids [53]. It has several domains, each with a unique purpose. A GTPase, a COR (C-terminal of Roc) domain, and a serine/threonine kinase are encoded by the Ras of Complex proteins (ROC) domain found in the center region of LRRK2 [21]. A number of domains implicated in protein–protein interactions are also present in LRRK2, including the C-terminal WD40 domain, leucine-rich repeat region (LRR), ankyrin-like repeat region (ANK), and armadillo repeat region (ARM) [147]. Autophosphorylation activity appears in both the ROC and kinase domains of LRRK2. Ser-910, Ser-935, and more recently, Ser-1292 phosphorylation are thought to be markers of LRRK2 kinase activity [106, 121]. Given that LRRK2 is involved in physiological processes that are normal, it is possible that mutations in this protein could interfere with natural signaling pathways, particularly those considered to be connected to the pathology of Parkinson's disease.

According to recent research, LRRK2 phosphorylates MAPK scaffolding proteins and MAPKKs, which may activate c-Jun N-terminal kinase (JNK) downstream [101]. Alzheimer's disease is frequently accompanied by synapse degradation, which can be attributed to the disruption of the Wnt/β-catenin signaling pathway. This pathway is essential for mature neurons to operate properly [143]. LRRK2 interacts with sequestered β-catenin complexes and is recruited to the plasma membrane in response to Wnt ligand stimulation of cells. There, it serves as a scaffold to join low-density lipoprotein receptor-related protein 6 (LRP6), dishevelled oligomers, and β-catenin. Pathogenic LRRK2 mutations may lessen the scaffolding capacity of the protein, which would decrease Wnt signaling [8].

LRRK2 appears to intersect with mitochondrial pathways. LRRK2 interacts with dynamin-related protein 1 (DRP1), a regulator of mitochondrial fission. Phosphorylation at S616 is known to activate DRP1, promoting mitochondrial fission. Overexpression of wild-type LRRK2 has been shown to enhance DRP1-mediated mitochondrial fragmentation, a phenomenon also seen in the G2019S mutant [39]. Furthermore, hyperphosphorylation of S616 has been observed in cases of sporadic Parkinson's disease. LRRK2 also interacts with mitochondrial fusion regulators, including mitofusin 1 and 2 (MFN1/2) and optic atrophy 1 (OPA1) [1]. A recent study confirmed that PD patients with the G2019S mutation exhibit decreased OPA1 expression. Overall, the activation of LRRK2 compromises mitochondrial homeostasis [107].

8.6 Heat Shock Proteins (HSPs)

Molecular chaperones, such as heat shock proteins (HSPs), a highly conserved group, assist in protein folding and prevent aggregation. These proteins are named "heat shock" because their production is triggered by exposure to high temperatures [43]. However, other stressors besides heat can also cause protein denaturation, including oxidative stress, mitochondrial dysfunction, hypoxia, radiation, endoplasmic reticulum stress, and physical injury [15]. Under normal physiological conditions, the three mechanisms—chaperone-mediated protein refolding, ubiquitin-proteasome degradation, and lysosomal autophagy—are typically adequate to manage minor environmental fluctuations [6]. Over time, defense mechanisms can become exhausted, resulting in reduced heat shock protein expression, increased susceptibility to proteotoxic stress, and eventual cell death [105]. Certain regions of the brain may express higher levels of heat shock proteins than others, which can lead to uneven patterns of neurodegeneration and the distinctive distribution of protein inclusions typically seen in neurodegenerative diseases. These chaperones may become overwhelmed by the high load of misfolded proteins, reducing their ability to maintain proteostasis and contributing to the pathology of the disease [65].

Heat shock proteins are categorized into six major families based on their molecular weight [43]:

1. Small heat shock protein less than 34 kDa, such as αB-crystallin, Hsp27, and Hsp32
2. Hsp40 (35–54 kDa)
3. Hsp60 (55–64 kDa)
4. Hsp70 (65–80 kDa)
5. Hsp90 (81–99 kDa)
6. Hsp100 (greater than 100 kDa)

Among these, the Hsp70 family is the most extensively studied, consisting of both stress-inducible and constitutively expressed members. Hsp70 proteins bind to hydrophobic regions of unfolded proteins in an ATP-dependent process [114]. Structurally, they consist of a C-terminal substrate-binding domain, which recognizes the unstructured polypeptides, and an N-terminal ATPase domain, which enables the protein to switch between an ATP-bound, open state with low substrate affinity and an ADP-bound, closed state with high substrate affinity [80].

Hsp70 and Hsc70 work together in forming protein complexes. Hsp70 interacts with Hip, Hop, and Hsp90, aiding in the refolding of denatured proteins through an ATP-dependent process [30]. Hip, also known as ST13 (Hsp70-interacting protein), extends the half-life of ADP-bound Hsp70-substrate complexes. By stabilizing Hsp70 in its ADP-bound state, Hip enhances its affinity for substrate proteins [40]. For instance, Hip stabilizes the complex between Hsp70 and α-synuclein, which supports Hsp70's anti-amyloid function [97]. The reduced levels of Hip have been consistently observed in the blood of patients with Parkinson's disease.

Hsp70 proteins are crucial in blocking both caspase-dependent and caspase-independent pathways of apoptosis [102]. Hsp70's neutralize apoptosis-inducing factor (AIF), a mitochondrial intermembrane protein that migrates to the nucleus to trigger chromatin condensation and DNA fragmentation [95]. Additionally, the ATPase domain of Hsp70 can bind directly to the caspase recruitment domain of apoptotic protease activation factor 1 (Apaf1), thereby inhibiting Apaf1 oligomerization and preventing the recruitment of procaspase-9 to the apoptosome [63]. HSP70 also plays an immunological role by interacting with toll-like receptors (TLRs), which enable HSP70 to initiate cytokine production and enhance the innate immune response, while scavenger receptors facilitate the transport of HSP70-bound antigens to antigen-presenting cells, thereby stimulating the adaptive immune response [11].

In neurodegenerative diseases, Hsp70 has been found to promote the refolding or degradation of α-synuclein, reducing both its fibrillization and toxicity [52]. GRP78, a member of the Hsp70 family located in the endoplasmic reticulum, has been demonstrated to reduce α-synuclein toxicity in experimental Parkinson's disease models [33]. Increased expression of HSPA1A and HSPA1B has been observed in the substantia nigra in the progressive supranuclear palsy, and frontotemporal dementia with Parkinsonism [65]. Hsp70 has also been found within protein aggregates in Alzheimer's disease and Lewy bodies in Parkinson's disease, likely representing a natural cellular attempt to mitigate the toxicity of these accumulations [128]. Individuals with mild cognitive impairment have been observed to have higher levels of Hsp70 in their plasma [66, 111].

In models of Parkinson's disease, both CHIP and Hsp90 play a role in regulating the stability of leucine-rich repeat kinase 2 (LRRK2), a protein associated with an increased risk of developing the disorder due to mutations [90]. Additionally, the mitochondrial Hsp90 protein, tumor necrosis factor receptor-associated protein 1 (TRAP1), has been shown to alleviate α-synuclein toxicity [13]. Hsp90 levels are elevated in the brains of individuals with Parkinson's disease and Hsp90 is also present in Lewy bodies associated with this condition. In contrast, serum Hsp90 levels are reported to be lower in patients with Alzheimer's disease and Hsp90 expression has been found to be downregulated in the temporal lobe of those with the disease [10].

Another HSP, HSP27 is typically expressed at low levels in neurons, but its expression can increase in response to proteotoxic stress. In the context of neurodegenerative diseases, elevated levels of HSP27 in both neurons and glial cells are associated with the pathological accumulation of abnormal proteins [84]. HSP27 serves as a crucial mediator of survival responses to central nervous system (CNS) insults, primarily through its antioxidant activity and its ability to inhibit pathways that lead to cell death. HSP90 also affects neuroimmunological disorders. Scientists reported a significantly higher titer of IgG autoantibodies against HSP90 in the cerebrospinal fluid and serum of patients with Guillain-Barré syndrome (GBS) when compared to those with motor neuron disease. Dysregulation of HSP expression can exacerbate proteostasis disruption in neurodegenerative diseases [7].

8.7 C9orf72 Gene

The C9orf72 gene is situated on the short (p) arm of chromosome 9 and encodes a protein predominantly found in the cytoplasm of neurons and at presynaptic terminals. Typically, pedigrees exhibiting the mutated allele of C9orf72 displayed a range of affected phenotypes across generations, including amyotrophic lateral sclerosis (ALS) and frontotemporal dementia (FTD). Approximately 35% of patients with the C9orf72 hexanucleotide repeat expansion (HRE) are initially diagnosed with conditions other than ALS or FTD at the onset of their illness [56, 110].

Three different mechanisms have been suggested to contribute synergistically to the cellular dysfunction associated with C9orf72 disorders: haploinsufficiency (reduced expression of the HRE C9orf72 allele), sequestration of RNA-binding proteins (RBPs), and the toxic effects of RNA foci and dipeptide repeats (DPRs) [3, 142]. The hypermethylation in the promoter region of the HRE C9orf72 allele results in transcriptional silencing of the enlarged allele and a partial to whole absence of translation (haploinsufficiency) [73]. The HRE C9orf72 allele's transcription results in RNA-DNA hybrids (R-loops), which set off the DNA damage response and may have an impact on the expansion's stability as well as the somatic variability seen in the C9orf72 expansion. Furthermore, the long repeat RNA has a longer half-life than the wild-type RNA, which leads to the sequestration of RNA-binding proteins (RBPs) [36]. Both sense and antisense long repeat RNAs, along with RBPs, interact to create RNA foci. These RNA foci sequester RBPs, disrupting overall cellular translation activity. The DNA damage response can be further triggered by intra- and intermolecular interactions inside RNA foci, such as G-quadruplex forms [79].

8.8 Valosin-Containing Protein (VCP)

Valosin-containing protein (VCP) is a protein that is expressed throughout various tissues and belongs to the AAA+ (ATPases Associated with diverse cellular Activities) protein family [19, 125]. VCP is composed of four distinct domains: the N-terminal domain, two ATPase domains (D1 and D2), and the C-terminal domain [23]. Both ATPase domains are responsible for ATP hydrolysis, while the N- and C-terminal domains facilitate interactions with substrates, other partners, and cofactors, playing a crucial role in a range of cellular processes. These processes include protein degradation through the ubiquitin–proteasome system (UPS), apoptosis, and the maturation and initiation of autophagosomes. The ubiquitin-proteasome system (UPS) is a pathway responsible for degrading misfolded or damaged proteins via the proteasome with the assistance of ubiquitin-conjugating enzymes in the cytosol [69]. Upon binding to various interacting cofactors, VCP harnesses the energy from ATP hydrolysis occurring in its D2 ATPase domain to initiate conformational changes. These changes enable VCP to identify and extract substrates from a variety of cellular structures or protein complexes for degradation [139].

Mutations in VCP (such as K251A/K524A or E578Q) or the depletion of VCP can also induce apoptosis [141]. When VCP is mutated or knocked down, it disrupts the endoplasmic reticulum-associated degradation (ERAD) pathway, leading to the accumulation of polyubiquitinated proteins, endoplasmic reticulum (ER) stress, and abnormal ER expansion [70]. Additionally, the excessive buildup of polyubiquitinated proteins in the mitochondria, along with the impaired clearance of dysfunctional mitochondria, may further promote apoptosis by generating significant oxidative stress and activating caspases, both of which are key factors in the apoptotic process [116].

Patients with human Inclusion Body Myopathy with Early-Onset Paget Disease of Bone and/or Frontotemporal Dementia (IBMFPD) were the first to be found to carry six missense variants in VCP: R95G, R155H, R155P, R155C, R191Q, and

A232E [104, 134]. Many people believe that these mutations are the cause of diseases like ALS, HD, and IBMPFD [141]. Mutations in VCP, identified in familial ALS patients through whole-exome sequencing, are responsible for approximately 1% of all familial ALS cases [47]. Additional screening of VCP mutations in a cohort study suggested additional mutation, i.e., the R155H mutation, especially in IBMPFD [135]. The R155H mutation impairs autophagosome maturation and autophagic degradation, resulting in the accumulation of ubiquitinated inclusions that contain TDP-43. The degeneration of motor neurons accompanied by TDP-43 inclusions is a key pathological feature of ALS, suggesting that VCP mutations may contribute to the cellular toxicity associated with TDP-43 accumulation [51].

The amplification of CAG repeats in the gene that encodes the Huntingtin (Htt) protein, which translates into a polyglutamine (polyQ) stretch, causes Huntington's disease (HD), a genetic neurological condition. The presence of polyQ expansions in mutant Htt (mHtt) proteins leads to neuronal cell death in the brains of HD patients [124]. The K524A mutation within the D2 ATPase domain of VCP hinders its function in proteasomal protein degradation, resulting in the formation of mHtt aggregates [91]. Neurons containing Htt aggregates become more susceptible to cell death induced by endoplasmic reticulum (ER) stress. Research using an HD model has demonstrated that VCP interacts with mHtt in the mitochondria, triggering excessive mitophagy (a type of autophagy specific to mitochondria), which ultimately leads to neuronal cell death [27, 77]. Conversely, blocking the interaction between VCP and mHtt reduces neuronal cell death, indicating that VCP may play a significant role in the pathology of HD [34].

Additionally, oxidative stress and neuroinflammation are common in neurodegenerative disease and can exacerbate proteostasis dysfunction. Oxidative stress arises from an imbalance between reactive oxygen species (ROS) and the cell's antioxidant defenses, contributing to protein damage and aggregation. Neuroinflammation, driven by activated microglia and other immune cells, can further impair proteostasis by releasing inflammatory cytokines and promoting neuronal damage [42].

8.9 Conclusion

Proteostasis markers play a critical role in the pathology of neurodegenerative diseases. Disruptions in amyloid-beta processing, tau phosphorylation, UPS function, autophagy, chaperone activity, ER stress, and apoptotic pathways are central to disease mechanisms. Understanding these markers offers valuable insights into disease pathogenesis and highlights potential therapeutic targets. Ongoing research into restoring proteostasis and developing targeted therapies holds promise for improving outcomes in neurodegenerative diseases. These biomarkers not only enhance our understanding of the underlying mechanisms of diseases such as Alzheimer's, Huntington's, and amyotrophic lateral sclerosis but also provide potential targets for therapeutic intervention. Continued research into the identification and validation of these biomarkers will be crucial for developing effective strategies to restore proteostasis and mitigate the impact of protein aggregation-related disorders.

References

1. Arduíno DM, Esteves AR, Cardoso SM. Mitochondrial fusion/fission, transport and autophagy in Parkinson's disease: when mitochondria get nasty. Parkinson's Dis. 2011;2011:767230. https://doi.org/10.4061/2011/767230.
2. Baba M, Nakajo S, Tu PH, Tomita T, Nakaya K, Lee VM, Trojanowski JQ, Iwatsubo T. Aggregation of alpha-synuclein in Lewy bodies of sporadic Parkinson's disease and dementia with Lewy bodies. Am J Pathol. 1998;152(4):879–84.
3. Babić Leko M, Župunski V, Kirincich J, Smilović D, Hortobágyi T, Hof PR, Šimić G. Molecular mechanisms of neurodegeneration related to C9orf72 hexanucleotide repeat expansion. Behav Neurol. 2019;2019:2909168. https://doi.org/10.1155/2019/2909168.
4. Barbier P, Zejneli O, Martinho M, Lasorsa A, Belle V, Smet-Nocca C, Tsvetkov PO, Devred F, Landrieu I. Role of tau as a microtubule-associated protein: structural and functional aspects. Front Aging Neurosci. 2019;11:204. https://doi.org/10.3389/fnagi.2019.00204.
5. Barmaki H, Nourazarian A, Khaki-Khatibi F. Proteostasis and neurodegeneration: a closer look at autophagy in Alzheimer's disease. Front Aging Neurosci. 2023;15:1281338. https://doi.org/10.3389/fnagi.2023.1281338.
6. Bejarano E, Cuervo AM. Chaperone-mediated autophagy. Proc Am Thorac Soc. 2010;7(1):29–39. https://doi.org/10.1513/pats.200909-102JS.
7. Beretta G, Shala AL. Impact of heat shock proteins in neurodegeneration: possible therapeutical targets. Ann Neurosci. 2022;29(1):71–82. https://doi.org/10.1177/09727531211070528.
8. Berwick DC, Harvey K. LRRK2 functions as a Wnt signaling scaffold, bridging cytosolic proteins and membrane-localized LRP6. Hum Mol Genet. 2012;21(22):4966–79. https://doi.org/10.1093/hmg/dds342.
9. Beurel E, Grieco SF, Jope RS. Glycogen synthase kinase-3 (GSK3): regulation, actions, and diseases. Pharmacol Ther. 2015;148:114–31. https://doi.org/10.1016/j.pharmthera.2014.11.016.
10. Bohush A, Niewiadomska G, Weis S, Filipek A. HSP90 and its novel co-chaperones, SGT1 and CHP-1, in brain of patients with Parkinson's disease and dementia with Lewy bodies. J Parkinsons Dis. 2019;9(1):97–107. https://doi.org/10.3233/JPD-181443.
11. Borges TJ, Wieten L, van Herwijnen MJ, Broere F, van der Zee R, Bonorino C, van Eden W. The anti-inflammatory mechanisms of Hsp70. Front Immunol. 2012;3:95. https://doi.org/10.3389/fimmu.2012.00095.
12. Breijyeh Z, Karaman R. Comprehensive review on Alzheimer's disease: causes and treatment. Molecules (Basel, Switzerland). 2020;25(24):5789. https://doi.org/10.3390/molecules25245789.
13. Butler EK, Voigt A, Lutz AK, Toegel JP, Gerhardt E, Karsten P, Falkenburger B, Reinartz A, Winklhofer KF, Schulz JB. The mitochondrial chaperone protein TRAP1 mitigates α-synuclein toxicity. PLoS Genet. 2012;8(2):e1002488. https://doi.org/10.1371/journal.pgen.1002488.
14. Calabresi P, Mechelli A, Natale G, Volpicelli-Daley L, Di Lazzaro G, Ghiglieri V. Alpha-synuclein in Parkinson's disease and other synucleinopathies: from overt neurodegeneration back to early synaptic dysfunction. Cell Death Dis. 2023;14(3):176. https://doi.org/10.1038/s41419-023-05672-9.
15. Cao SS, Kaufman RJ. Endoplasmic reticulum stress and oxidative stress in cell fate decision and human disease. Antioxid Redox Signal. 2014;21(3):396–413. https://doi.org/10.1089/ars.2014.5851.
16. Chakrabarti O, Rane NS, Hegde RS. Cytosolic aggregates perturb the degradation of nontranslocated secretory and membrane proteins. Mol Biol Cell. 2011;22(10):1625–37. https://doi.org/10.1091/mbc.E10-07-0638.

17. Checler F, Afram E, Pardossi-Piquard R, Lauritzen I. Is γ-secretase a beneficial inactivating enzyme of the toxic APP C-terminal fragment C99? J Biol Chem. 2021;296:100489. https://doi.org/10.1016/j.jbc.2021.100489.
18. Choy RW, Cheng Z, Schekman R. Amyloid precursor protein (APP) traffics from the cell surface via endosomes for amyloid β (Aβ) production in the trans-Golgi network. Proc Natl Acad Sci USA. 2012;109(30):E2077–82. https://doi.org/10.1073/pnas.1208635109.
19. Chu S, Xie X, Payan C, Stochaj U. Valosin containing protein (VCP): initiator, modifier, and potential drug target for neurodegenerative diseases. Mol Neurodegener. 2023;18(1):52. https://doi.org/10.1186/s13024-023-00639-y.
20. Ciechanover A, Kwon YT. Protein quality control by molecular chaperones in neurodegeneration. Front Neurosci. 2017;11:185. https://doi.org/10.3389/fnins.2017.00185.
21. Civiero L, Dihanich S, Lewis PA, Greggio E. Genetic, structural, and molecular insights into the function of ras of complex proteins domains. Chem Biol. 2014;21(7):809–18. https://doi.org/10.1016/j.chembiol.2014.05.010.
22. Dantuma NP, Bott LC. The ubiquitin-proteasome system in neurodegenerative diseases: precipitating factor, yet part of the solution. Front Mol Neurosci. 2014;7:70. https://doi.org/10.3389/fnmol.2014.00070.
23. DeLaBarre B, Brunger AT. Complete structure of p97/valosin-containing protein reveals communication between nucleotide domains. Nat Struct Biol. 2003;10(10):856–63. https://doi.org/10.1038/nsb972.
24. Deng Y, Wang Z, Wang R, Zhang X, Zhang S, Wu Y, Staufenbiel M, Cai F, Song W. Amyloid-β protein (Aβ) Glu11 is the major β-secretase site of β-site amyloid-β precursor protein-cleaving enzyme 1(BACE1), and shifting the cleavage site to Aβ Asp1 contributes to Alzheimer pathogenesis. Eur J Neurosci. 2013;37(12):1962–9. https://doi.org/10.1111/ejn.12235.
25. Deyell JS, Sriparna M, Ying M, Mao X. The interplay between α-synuclein and microglia in α-synucleinopathies. Int J Mol Sci. 2023;24(3):2477. https://doi.org/10.3390/ijms24032477.
26. Duan S, Pagano M. Ubiquitin ligases in cancer: functions and clinical potentials. Cell Chem Biol. 2021;28(7):918–33. https://doi.org/10.1016/j.chembiol.2021.04.008.
27. Franco-Iborra S, Plaza-Zabala A, Montpeyo M, Sebastian D, Vila M, Martinez-Vicente M. Mutant HTT (huntingtin) impairs mitophagy in a cellular model of Huntington disease. Autophagy. 2021;17(3):672–89. https://doi.org/10.1080/15548627.2020.1728096.
28. Gandhi PN, Chen SG, Wilson-Delfosse AL. Leucine-rich repeat kinase 2 (LRRK2): a key player in the pathogenesis of Parkinson's disease. J Neurosci Res. 2009;87(6):1283–95. https://doi.org/10.1002/jnr.21949.
29. García-González L, Pilat D, Baranger K, Rivera S. Emerging alternative proteinases in APP metabolism and Alzheimer's disease pathogenesis: a focus on MT1-MMP and MT5-MMP. Front Aging Neurosci. 2019;11:244. https://doi.org/10.3389/fnagi.2019.00244.
30. Genest O, Wickner S, Doyle SM. Hsp90 and Hsp70 chaperones: collaborators in protein remodeling. J Biol Chem. 2019;294(6):2109–20. https://doi.org/10.1074/jbc.REV118.002806.
31. Ghanem SS, Majbour NK, Vaikath NN, Ardah MT, Erskine D, Jensen NM, Fayyad M, Sudhakaran IP, Vasili E, Melachroinou K, Abdi IY, Poggiolini I, Santos P, Dorn A, Carloni P, Vekrellis K, Attems J, McKeith I, Outeiro TF, Jensen PH, et al. α-Synuclein phosphorylation at serine 129 occurs after initial protein deposition and inhibits seeded fibril formation and toxicity. Proc Natl Acad Sci USA. 2022;119(15):e2109617119. https://doi.org/10.1073/pnas.2109617119.
32. Gleichmann M, Mattson MP. Neuronal calcium homeostasis and dysregulation. Antioxid Redox Signal. 2011;14(7):1261–73. https://doi.org/10.1089/ars.2010.3386.
33. Gorbatyuk MS, Shabashvili A, Chen W, Meyers C, Sullivan LF, Salganik M, Lin JH, Lewin AS, Muzyczka N, Gorbatyuk OS. Glucose regulated protein 78 diminishes α-synuclein neurotoxicity in a rat model of Parkinson disease. Mol Ther: J Am Soc Gene Ther. 2012;20(7):1327–37. https://doi.org/10.1038/mt.2012.28.
34. Guo X, Sun X, Hu D, Wang YJ, Fujioka H, Vyas R, Chakrapani S, Joshi AU, Luo Y, Mochly-Rosen D, Qi X. VCP recruitment to mitochondria causes mitophagy impairment and neuro-

degeneration in models of Huntington's disease. Nat Commun. 2016;7:12646. https://doi.org/10.1038/ncomms12646.
35. Haass C, Kaether C, Thinakaran G, Sisodia S. Trafficking and proteolytic processing of APP. Cold Spring Harb Perspect Med. 2012;2(5):a006270. https://doi.org/10.1101/cshperspect.a006270.
36. Haeusler AR, Donnelly CJ, Periz G, Simko EA, Shaw PG, Kim MS, Maragakis NJ, Troncoso JC, Pandey A, Sattler R, Rothstein JD, Wang J. C9orf72 nucleotide repeat structures initiate molecular cascades of disease. Nature. 2014;507(7491):195–200. https://doi.org/10.1038/nature13124.
37. Hetz C, Zhang K, Kaufman RJ. Mechanisms, regulation and functions of the unfolded protein response. Nat Rev Mol Cell Biol. 2020;21(8):421–38. https://doi.org/10.1038/s41580-020-0250-z.
38. Himmelstein DS, Ward SM, Lancia JK, Patterson KR, Binder LI. Tau as a therapeutic target in neurodegenerative disease. Pharmacol Ther. 2012;136(1):8–22. https://doi.org/10.1016/j.pharmthera.2012.07.001.
39. Ho DH, Je AR, Lee H, Son I, Kweon HS, Kim HG, Seol W. LRRK2 kinase activity induces mitochondrial fission in microglia via Drp1 and modulates neuroinflammation. Exp Neurobiol. 2018;27(3):171–80. https://doi.org/10.5607/en.2018.27.3.171.
40. Höhfeld J, Minami Y, Hartl FU. Hip, a novel cochaperone involved in the eukaryotic Hsc70/Hsp40 reaction cycle. Cell. 1995;83(4):589–98. https://doi.org/10.1016/0092-8674(95)90099-3.
41. Höhn A, Tramutola A, Cascella R. Proteostasis failure in neurodegenerative diseases: focus on oxidative stress. Oxidative Med Cell Longev. 2020;2020:5497046. https://doi.org/10.1155/2020/5497046.
42. Houldsworth A. Role of oxidative stress in neurodegenerative disorders: a review of reactive oxygen species and prevention by antioxidants. Brain Commun. 2024;6(1):fcad356. https://doi.org/10.1093/braincomms/fcad356.
43. Hu C, Yang J, Qi Z, Wu H, Wang B, Zou F, Mei H, Liu J, Wang W, Liu Q. Heat shock proteins: biological functions, pathological roles, and therapeutic opportunities. MedComm. 2022;3(3):e161. https://doi.org/10.1002/mco2.161.
44. Hung C, Fertan E, Livesey FJ, Klenerman D, Patani R. APP antisense oligonucleotides reduce amyloid-β aggregation and rescue endolysosomal dysfunction in Alzheimer's disease. Brain J Neurol. 2024;147(7):2325–33. https://doi.org/10.1093/brain/awae092.
45. Iyer A, Claessens MMAE. Disruptive membrane interactions of alpha-synuclein aggregates. Biochim Biophys Acta Proteins Proteomics. 2019;1867(5):468–82. https://doi.org/10.1016/j.bbapap.2018.10.006.
46. Jiang S, Li Y, Zhang X, Bu G, Xu H, Zhang YW. Trafficking regulation of proteins in Alzheimer's disease. Mol Neurodegener. 2014;9:6. https://doi.org/10.1186/1750-1326-9-6.
47. Johnson JO, Mandrioli J, Benatar M, Abramzon Y, Van Deerlin VM, Trojanowski JQ, Gibbs JR, Brunetti M, Gronka S, Wuu J, Ding J, McCluskey L, Martinez-Lage M, Falcone D, Hernandez DG, Arepalli S, Chong S, Schymick JC, Rothstein J, Landi F, et al. Exome sequencing reveals VCP mutations as a cause of familial ALS. Neuron. 2010;68(5):857–64. https://doi.org/10.1016/j.neuron.2010.11.036.
48. Johnston-Carey HK, Pomatto LC, Davies KJ. The immunoproteasome in oxidative stress, aging, and disease. Crit Rev Biochem Mol Biol. 2015;51(4):268–81. https://doi.org/10.3109/10409238.2016.1172554.
49. Kaneko M, Koike H, Saito R, Kitamura Y, Okuma Y, Nomura Y. Loss of HRD1-mediated protein degradation causes amyloid precursor protein accumulation and amyloid-beta generation. J Neurosci. 2010;30(11):3924–32. https://doi.org/10.1523/JNEUROSCI.2422-09.2010.
50. Khanam H, Ali A, Asif M, Shamsuzzaman. Neurodegenerative diseases linked to misfolded proteins and their therapeutic approaches: a review. Eur J Med Chem. 2016;124:1121–41. https://doi.org/10.1016/j.ejmech.2016.08.006.

51. Klickstein JA, Johnson MA, Antonoudiou P, Maguire J, Paulo JA, Gygi SP, Weihl C, Raman M. ALS-related p97 R155H mutation disrupts lysophagy in iPSC-derived motor neurons. Stem Cell Rep. 2024;19(3):366–82. https://doi.org/10.1016/j.stemcr.2024.01.002.
52. Klucken J, Shin Y, Masliah E, Hyman BT, McLean PJ. Hsp70 reduces alpha-synuclein aggregation and toxicity. J Biol Chem. 2004;279(24):25497–502. https://doi.org/10.1074/jbc.M400255200.
53. Kluss JH, Mamais A, Cookson MR. LRRK2 links genetic and sporadic Parkinson's disease. Biochem Soc Trans. 2019;47(2):651–61. https://doi.org/10.1042/BST20180462.
54. Kolarova M, García-Sierra F, Bartos A, Ricny J, Ripova D. Structure and pathology of tau protein in Alzheimer disease. Int J Alzheimers Dis. 2012;2012:731526. https://doi.org/10.1155/2012/731526.
55. Kontsekova E, Zilka N, Kovacech B, Skrabana R, Novak M. Identification of structural determinants on tau protein essential for its pathological function: novel therapeutic target for tau immunotherapy in Alzheimer's disease. Alzheimers Res Ther. 2014;6(4):45. https://doi.org/10.1186/alzrt277.
56. Kortazar-Zubizarreta I, Manero-Azua A, Afonso-Agüera J, Perez de Nanclares G. C9ORF72 gene GGGGCC hexanucleotide expansion: a high clinical variability from amyotrophic lateral sclerosis to frontotemporal dementia. J Pers Med. 2023;13(9):1396. https://doi.org/10.3390/jpm13091396.
57. Kumar A, Sidhu J, Lui F, et al. Alzheimer disease. [Updated 2024 Feb 12]. In: StatPearls [Internet]. Treasure Island: StatPearls Publishing; 2024. Available from: https://www.ncbi.nlm.nih.gov/books/NBK499922/.
58. Kumari P, Ghosh D, Vanas A, Fleischmann Y, Wiegand T, Jeschke G, Riek R, Eichmann C. Structural insights into α-synuclein monomer-fibril interactions. Proc Natl Acad Sci USA. 2021;118(10):e2012171118. https://doi.org/10.1073/pnas.2012171118.
59. Kwon OH, Cho YY, Lee JH, Chung S. O-GlcNAcylation inhibits endocytosis of amyloid precursor protein by decreasing its localization in lipid raft microdomains. Membranes. 2021;11(12):909. https://doi.org/10.3390/membranes11120909.
60. Labbadia J, Morimoto RI. The biology of proteostasis in aging and disease. Annu Rev Biochem. 2015;84:435–64. https://doi.org/10.1146/annurev-biochem-060614-033955.
61. Lamptey RNL, Chaulagain B, Trivedi R, Gothwal A, Layek B, Singh J. A review of the common neurodegenerative disorders: current therapeutic approaches and the potential role of nanotherapeutics. Int J Mol Sci. 2022;23(3):1851. https://doi.org/10.3390/ijms23031851.
62. Langerscheidt F, Wied T, Al Kabbani MA, van Eimeren T, Wunderlich G, Zempel H. Genetic forms of tauopathies: inherited causes and implications of Alzheimer's disease-like TAU pathology in primary and secondary tauopathies. J Neurol. 2024;271(6):2992–3018. https://doi.org/10.1007/s00415-024-12314-3.
63. Lanneau D, Brunet M, Frisan E, Solary E, Fontenay M, Garrido C. Heat shock proteins: essential proteins for apoptosis regulation. J Cell Mol Med. 2008;12(3):743–61. https://doi.org/10.1111/j.1582-4934.2008.00273.x.
64. Lanoiselée HM, Nicolas G, Wallon D, Rovelet-Lecrux A, Lacour M, Rousseau S, Richard AC, Pasquier F, Rollin-Sillaire A, Martinaud O, Quillard-Muraine M, de la Sayette V, Boutoleau-Bretonniere C, Etcharry-Bouyx F, Chauviré V, Sarazin M, le Ber I, Epelbaum S, Jonveaux T, Rouaud O, et al. Collaborators of the CNR-MAJ Project. APP, PSEN1, and PSEN2 mutations in early-onset Alzheimer disease: a genetic screening study of familial and sporadic cases. PLoS Med. 2017;14(3):e1002270. https://doi.org/10.1371/journal.pmed.1002270.
65. Leak RK. Heat shock proteins in neurodegenerative disorders and aging. J Cell Commun Signal. 2014;8(4):293–310. https://doi.org/10.1007/s12079-014-0243-9.
66. Lee KS, Chung JH, Oh BH, Hong CH. Increased plasma levels of heat shock protein 70 in patients with vascular mild cognitive impairment. Neurosci Lett. 2008;436(2):223–6. https://doi.org/10.1016/j.neulet.2008.03.025.
67. Lendon CL, Lynch T, Norton J, McKeel DW Jr, Busfield F, Craddock N, Chakraverty S, Gopalakrishnan G, Shears SD, Grimmett W, Wilhelmsen KC, Hansen L, Morris JC, Goate

AM. Hereditary dysphasic disinhibition dementia: a frontotemporal dementia linked to 17q21-22. Neurology. 1998;50:1546–55.
68. Leslie SN, Nairn AC. cAMP regulation of protein phosphatases PP1 and PP2A in brain. Biochim Biophys Acta – Mol Cell Res. 2019;1866(1):64–73. https://doi.org/10.1016/j.bbamcr.2018.09.006.
69. Li Y, Li S, Wu H. Ubiquitination-proteasome system (UPS) and autophagy two main protein degradation machineries in response to cell stress. Cells. 2022;11(5):851. https://doi.org/10.3390/cells11050851.
70. Liang CJ, Chang YC, Chang HC, Wang CK, Hung YC, Lin YE, Chan CC, Chen CH, Chang HY, Sang TK. Derlin-1 regulates mutant VCP-linked pathogenesis and endoplasmic reticulum stress-induced apoptosis. PLoS Genet. 2014;10(9):e1004675. https://doi.org/10.1371/journal.pgen.1004675.
71. Liao PC, Wolken DMA, Serrano E, Srivastava P, Pon LA. Mitochondria-associated degradation pathway (MAD) function beyond the outer membrane. Cell Rep. 2020;32(2):107902. https://doi.org/10.1016/j.celrep.2020.107902.
72. Lin KJ, Lin KL, Chen SD, Liou CW, Chuang YC, Lin HY, Lin TK. The overcrowded crossroads: mitochondria, alpha-synuclein, and the endo-lysosomal system interaction in Parkinson's disease. Int J Mol Sci. 2019;20(21):5312. https://doi.org/10.3390/ijms20215312.
73. Liu EY, Russ J, Wu K, Neal D, Suh E, McNally AG, Irwin DJ, Van Deerlin VM, Lee EB. C9orf72 hypermethylation protects against repeat expansion-associated pathology in ALS/FTD. Acta Neuropathol. 2014;128(4):525–41. https://doi.org/10.1007/s00401-014-1286-y.
74. Liu X, Liu Y, Ji S. Secretases related to amyloid precursor protein processing. Membranes. 2021;11(12):983. https://doi.org/10.3390/membranes11120983.
75. Lualdi M, Alberio T, Fasano M. Proteostasis and proteotoxicity in the network medicine era. Int J Mol Sci. 2020;21(17):6405. https://doi.org/10.3390/ijms21176405.
76. Luo H, Jiao Q, Shen C, Shao C, Xie J, Chen Y, Feng X, Zhang X. Unraveling the roles of endoplasmic reticulum-associated degradation in metabolic disorders. Front Endocrinol. 2023;14:1123769. https://doi.org/10.3389/fendo.2023.1123769.
77. Maity S, Komal P, Kumar V, Saxena A, Tungekar A, Chandrasekar V. Impact of ER stress and ER-mitochondrial crosstalk in Huntington's disease. Int J Mol Sci. 2022;23(2):780. https://doi.org/10.3390/ijms23020780.
78. Malhis M, Kaniyappan S, Aillaud I, Chandupatla RR, Ramirez LM, Zweckstetter M, Horn AHC, Mandelkow E, Sticht H, Funke SA. Potent Tau aggregation inhibitor D-peptides selected against Tau-repeat 2 using mirror image phage display. Chembiochem: Eur J Chem Biol. 2021;22(21):3049–59. https://doi.org/10.1002/cbic.202100287.
79. Malik I, Kelley CP, Wang ET, Todd PK. Molecular mechanisms underlying nucleotide repeat expansion disorders. Nat Rev Mol Cell Biol. 2021;22(9):589–607. https://doi.org/10.1038/s41580-021-00382-6.
80. Mayer MP, Bukau B. Hsp70 chaperones: cellular functions and molecular mechanism. Cell Mol Life Sci: CMLS. 2005;62(6):670–84. https://doi.org/10.1007/s00018-004-4464-6.
81. Morel E, Chamoun Z, Lasiecka ZM, Chan RB, Williamson RL, Vetanovetz C, et al. Phosphatidylinositol-3-phosphate regulates sorting and processing of amyloid precursor protein through the endosomal system. Nat Commun. 2013;4:2250. https://doi.org/10.1038/ncomms3250.
82. Morris M, Maeda S, Vossel K, Mucke L. The many faces of tau. Neuron. 2011;70(3):410–26. https://doi.org/10.1016/j.neuron.2011.04.009.
83. Moujalled D, Strasser A, Liddell JR. Molecular mechanisms of cell death in neurological diseases. Cell Death Differ. 2021;28(7):2029–44. https://doi.org/10.1038/s41418-021-00814-y.
84. Navarro-Zaragoza J, Cuenca-Bermejo L, Almela P, Laorden ML, Herrero MT. Could small heat shock protein HSP27 be a first-line target for preventing protein aggregation in Parkinson's disease? Int J Mol Sci. 2021;22(6):3038. https://doi.org/10.3390/ijms22063038.
85. O'Brien RJ, Wong PC. Amyloid precursor protein processing and Alzheimer's disease. Annu Rev Neurosci. 2011;34:185–204. https://doi.org/10.1146/annurev-neuro-061010-113613.

86. Ochneva A, Zorkina Y, Abramova O, Pavlova O, Ushakova V, Morozova A, Zubkov E, Pavlov K, Gurina O, Chekhonin V. Protein misfolding and aggregation in the brain: common pathogenetic pathways in neurodegenerative and mental disorders. Int J Mol Sci. 2022;23(22):14498. https://doi.org/10.3390/ijms232214498.
87. Oddo S. The ubiquitin-proteasome system in Alzheimer's disease. J Cell Mol Med. 2008;12(2):363–73. https://doi.org/10.1111/j.1582-4934.2008.00276.x.
88. Orobets KS, Karamyshev AL. Amyloid precursor protein and Alzheimer's disease. Int J Mol Sci. 2023;24(19):14794. https://doi.org/10.3390/ijms241914794.
89. Ott C, Jacobs K, Haucke E, Navarrete Santos A, Grune T, Simm A. Role of advanced glycation end products in cellular signaling. Redox Biol. 2014;2:411–29. https://doi.org/10.1016/j.redox.2013.12.016.
90. Pérez-Carrión MD, Posadas I, Solera J, Ceña V. LRRK2 and proteostasis in Parkinson's disease. Int J Mol Sci. 2022;23(12):6808. https://doi.org/10.3390/ijms23126808.
91. Poksay KS, Madden DT, Peter AK, Niazi K, Banwait S, Crippen D, Bredesen DE, Rao RV. Valosin-containing protein gene mutations: cellular phenotypes relevant to neurodegeneration. J Mol Neurosci: MN. 2011;44(2):91–102. https://doi.org/10.1007/s12031-010-9489-8.
92. Polymeropoulos MH, Lavedan C, Leroy E, Ide SE, Dehejia A, Dutra A, Pike B, Root H, Rubenstein J, Boyer R, Stenroos ES, Chandrasekharappa S, Athanassiadou A, Papapetropoulos T, Johnson WG, Lazzarini AM, Duvoisin RC, Di Iorio G, Golbe LI, Nussbaum RL. Mutation in the alpha-synuclein gene identified in families with Parkinson's disease. Science (New York, N.Y.). 1997;276(5321):2045–7. https://doi.org/10.1126/science.276.5321.2045.
93. Pratt MR, Vocadlo DJ. Understanding and exploiting the roles of O-GlcNAc in neurodegenerative diseases. J Biol Chem. 2023;299(12):105411. https://doi.org/10.1016/j.jbc.2023.105411.
94. Raffeiner M, Zhu S, González-Fuente M, Üstün S. Interplay between autophagy and proteasome during protein turnover. Trends Plant Sci. 2023;28(6):698–714. https://doi.org/10.1016/j.tplants.2023.01.013.
95. Ravagnan L, Gurbuxani S, Susin SA, Maisse C, Daugas E, Zamzami N, Mak T, Jäättelä M, Penninger JM, Garrido C, Kroemer G. Heat-shock protein 70 antagonizes apoptosis-inducing factor. Nat Cell Biol. 2001;3(9):839–43. https://doi.org/10.1038/ncb0901-839.
96. Read A, Schröder M. The unfolded protein response: an overview. Biology. 2021;10(5):384. https://doi.org/10.3390/biology10050384.
97. Roodveldt C, Bertoncini CW, Andersson A, van der Goot AT, Hsu ST, Fernández-Montesinos R, de Jong J, van Ham TJ, Nollen EA, Pozo D, Christodoulou J, Dobson CM. Chaperone proteostasis in Parkinson's disease: stabilization of the Hsp70/alpha-synuclein complex by Hip. EMBO J. 2009;28(23):3758–70. https://doi.org/10.1038/emboj.2009.298.
98. Ross JM, Olson L, Coppotelli G. Mitochondrial and ubiquitin proteasome system dysfunction in ageing and disease: two sides of the same coin? Int J Mol Sci. 2015;16(8):19458–76. https://doi.org/10.3390/ijms160819458.
99. Ruan L, Zhou C, Jin E, Kucharavy A, Zhang Y, Wen Z, Florens L, Li R. Cytosolic proteostasis through importing of misfolded proteins into mitochondria. Nature. 2017;543(7645):443–6. https://doi.org/10.1038/nature21695.
100. Ruggiano A, Foresti O, Carvalho P. Quality control: ER-associated degradation: protein quality control and beyond. J Cell Biol. 2014;204(6):869–79. https://doi.org/10.1083/jcb.201312042.
101. Rui Q, Ni H, Li D, Gao R, Chen G. The role of LRRK2 in neurodegeneration of Parkinson disease. Curr Neuropharmacol. 2018;16(9):1348–57. https://doi.org/10.2174/1570159X16666180222165418.
102. Sabirzhanov B, Stoica BA, Hanscom M, Piao CS, Faden AI. Over-expression of HSP70 attenuates caspase-dependent and caspase-independent pathways and inhibits neuronal apoptosis. J Neurochem. 2012;123(4):542–54. https://doi.org/10.1111/j.1471-4159.2012.07927.x.
103. Sarchione A, Marchand A, Taymans JM, Chartier-Harlin MC. Alpha-synuclein and lipids: the elephant in the room? Cells. 2021;10(9):2452. https://doi.org/10.3390/cells10092452.

104. Scarian E, Fiamingo G, Diamanti L, Palmieri I, Gagliardi S, Pansarasa O. The role of VCP mutations in the spectrum of amyotrophic lateral sclerosis-frontotemporal dementia. Front Neurol. 2022;13:841394. https://doi.org/10.3389/fneur.2022.841394.
105. Schroeder HT, De Lemos Muller CH, Heck TG, Krause M, Homem de Bittencourt PI. Heat shock response during the resolution of inflammation and its progressive suppression in chronic-degenerative inflammatory diseases. Cell Stress Chaperones. 2024;29(1):116–42. https://doi.org/10.1016/j.cstres.2024.01.002.
106. Sheng Z, Zhang S, Bustos D, Kleinheinz T, Le Pichon CE, Dominguez SL, Solanoy HO, Drummond J, Zhang X, Ding X, Cai F, Song Q, Li X, Yue Z, van der Brug MP, Burdick DJ, Gunzner-Toste J, Chen H, Liu X, Estrada AA, et al. Ser1292 autophosphorylation is an indicator of LRRK2 kinase activity and contributes to the cellular effects of PD mutations. Sci Transl Med. 2012;4(164):164ra161. https://doi.org/10.1126/scitranslmed.3004485.
107. Singh A, Zhi L, Zhang H. LRRK2 and mitochondria: recent advances and current views. Brain Res. 2019;1702:96–104. https://doi.org/10.1016/j.brainres.2018.06.010.
108. Singh MK, Shin Y, Ju S, Han S, Choe W, Yoon KS, Kim SS, Kang I. Heat shock response and heat shock proteins: current understanding and future opportunities in human diseases. Int J Mol Sci. 2024;25(8):4209. https://doi.org/10.3390/ijms25084209.
109. Sinsky J, Pichlerova K, Hanes J. Tau protein interaction partners and their roles in Alzheimer's disease and other tauopathies. Int J Mol Sci. 2021;22(17):9207. https://doi.org/10.3390/ijms22179207.
110. Smeyers J, Banchi EG, Latouche M. C9ORF72: what it is, what it does, and why it matters. Front Cell Neurosci. 2021;15:661447. https://doi.org/10.3389/fncel.2021.661447.
111. Son SJ, Lee KS, Chung JH, Chang KJ, Roh HW, Kim SH, Jin T, Back JH, Kim HJ, Lee Y, Choi SH, Noh JS, Lim KY, Chung YK, Hong CH, Oh BH. Increased plasma levels of heat shock protein 70 associated with subsequent clinical conversion to mild cognitive impairment in cognitively healthy elderly. PLoS One. 2015;10(3):e0119180. https://doi.org/10.1371/journal.pone.0119180.
112. Sonninen TM, Goldsteins G, Laham-Karam N, Koistinaho J, Lehtonen Š. Proteostasis disturbances and inflammation in neurodegenerative diseases. Cells. 2020;9(10):2183. https://doi.org/10.3390/cells9102183.
113. Stefanis L. α-Synuclein in Parkinson's disease. Cold Spring Harb Perspect Med. 2012;2(2):a009399. https://doi.org/10.1101/cshperspect.a009399.
114. Stetler RA, Gan Y, Zhang W, Liou AK, Gao Y, Cao G, Chen J. Heat shock proteins: cellular and molecular mechanisms in the central nervous system. Prog Neurobiol. 2010;92(2):184–211. https://doi.org/10.1016/j.pneurobio.2010.05.002.
115. Strang KH, Golde TE, Giasson BI. MAPT mutations, tauopathy, and mechanisms of neurodegeneration. Lab Investig J Tech Methods Pathol. 2019;99(7):912–28. https://doi.org/10.1038/s41374-019-0197-x.
116. Sun F, Kanthasamy A, Anantharam V, Kanthasamy AG. Mitochondrial accumulation of polyubiquitinated proteins and differential regulation of apoptosis by polyubiquitination sites Lys-48 and -63. J Cell Mol Med. 2009;13(8B):1632–43. https://doi.org/10.1111/j.1582-4934.2009.00775.x.
117. Sundaram A, Appathurai S, Plumb R, Mariappan M. Dynamic changes in complexes of IRE1α, PERK, and ATF6α during endoplasmic reticulum stress. Mol Biol Cell. 2018;29(11):1376–88. https://doi.org/10.1091/mbc.E17-10-0594.
118. Surmeier DJ. Determinants of dopaminergic neuron loss in Parkinson's disease. FEBS J. 2018;285(19):3657–68. https://doi.org/10.1111/febs.14607.
119. Sweeney P, Park H, Baumann M, Dunlop J, Frydman J, Kopito R, McCampbell A, Leblanc G, Venkateswaran A, Nurmi A, Hodgson R. Protein misfolding in neurodegenerative diseases: implications and strategies. Transl Neurodegener. 2017;6:6. https://doi.org/10.1186/s40035-017-0077-5.
120. Tansey MG, Wallings RL, Houser MC, Herrick MK, Keating CE, Joers V. Inflammation and immune dysfunction in Parkinson disease. Nat Rev Immunol. 2022;22(11):657–73. https://doi.org/10.1038/s41577-022-00684-6.

121. Taymans JM, Fell M, Greenamyre T, Hirst WD, Mamais A, Padmanabhan S, Peter I, Rideout H, Thaler A. Perspective on the current state of the LRRK2 field. NPJ Parkinson's Dis. 2023;9(1):104. https://doi.org/10.1038/s41531-023-00544-7.
122. Tenreiro S, Eckermann K, Outeiro TF. Protein phosphorylation in neurodegeneration: friend or foe? Front Mol Neurosci. 2014;7:42. https://doi.org/10.3389/fnmol.2014.00042.
123. Toader C, Dobrin N, Brehar FM, Popa C, Covache-Busuioc RA, Glavan LA, Costin HP, Bratu BG, Corlatescu AD, Popa AA, Ciurea AV. From recognition to remedy: the significance of biomarkers in neurodegenerative disease pathology. Int J Mol Sci. 2023;24(22):16119. https://doi.org/10.3390/ijms242216119.
124. Tong H, Yang T, Xu S, Li X, Liu L, Zhou G, Yang S, Yin S, Li XJ, Li S. Huntington's disease: complex pathogenesis and therapeutic strategies. Int J Mol Sci. 2024;25(7):3845. https://doi.org/10.3390/ijms25073845.
125. Tseng CS, Chao YW, Liu YH, Huang YS, Chao HW. Dysregulated proteostasis network in neuronal diseases. Front Cell Dev Biol. 2023;11:1075215. https://doi.org/10.3389/fcell.2023.1075215.
126. van Swieten J, Spillantini MG. Hereditary frontotemporal dementia caused by Tau gene mutations. Brain Pathol (Zurich, Switzerland). 2007;17(1):63–73. https://doi.org/10.1111/j.1750-3639.2007.00052.x.
127. Varela L, Garcia-Rendueles MER. Oncogenic pathways in neurodegenerative diseases. Int J Mol Sci. 2022;23(6):3223. https://doi.org/10.3390/ijms23063223.
128. Venediktov AA, Bushueva OY, Kudryavtseva VA, Kuzmin EA, Moiseeva AV, Baldycheva A, Meglinski I, Piavchenko GA. Closest horizons of Hsp70 engagement to manage neurodegeneration. Front Mol Neurosci. 2023;16:1230436. https://doi.org/10.3389/fnmol.2023.1230436.
129. Vidović M, Rikalovic MG. Alpha-synuclein aggregation pathway in Parkinson's disease: current status and novel therapeutic approaches. Cells. 2022;11(11):1732. https://doi.org/10.3390/cells11111732.
130. Wang X, Terpstra EJ. Ubiquitin receptors and protein quality control. J Mol Cell Cardiol. 2013;55:73–84. https://doi.org/10.1016/j.yjmcc.2012.09.012.
131. Wang X, Zhou X, Li G, Zhang Y, Wu Y, Song W. Modifications and trafficking of APP in the pathogenesis of Alzheimer's disease. Front Mol Neurosci. 2017;10:294. https://doi.org/10.3389/fnmol.2017.00294.
132. Wang Y, Ruan L, Zhu J, Zhang X, Chih-Chieh Chang A, Tomaszewski A, Li R. Metabolic regulation of misfolded protein import into mitochondria. bioRxiv: The Preprint Server for Biology. 2023:2023.03.29.534670. https://doi.org/10.1101/2023.03.29.534670.
133. Watanabe T, Hikichi Y, Willuweit A, Shintani Y, Horiguchi T. FBL2 regulates amyloid precursor protein (APP) metabolism by promoting ubiquitination-dependent APP degradation and inhibition of APP endocytosis. J Neurosci. 2012;32:3352–65. https://doi.org/10.1523/jneurosci.5659-11.2012.
134. Watts GD, Thomasova D, Ramdeen SK, Fulchiero EC, Mehta SG, Drachman DA, Weihl CC, Jamrozik Z, Kwiecinski H, Kaminska A, Kimonis VE. Novel VCP mutations in inclusion body myopathy associated with Paget disease of bone and frontotemporal dementia. Clin Genet. 2007;72(5):420–6. https://doi.org/10.1111/j.1399-0004.2007.00887.x.
135. Watts GD, Wymer J, Kovach MJ, Mehta SG, Mumm S, Darvish D, Pestronk A, Whyte MP, Kimonis VE. Inclusion body myopathy associated with Paget disease of bone and frontotemporal dementia is caused by mutant valosin-containing protein. Nat Genet. 2004;36(4):377–81. https://doi.org/10.1038/ng1332.
136. Wolska-Washer A, Smolewski P. Targeting protein degradation pathways in tumors: focusing on their role in hematological malignancies. Cancers. 2022;14(15):3778. https://doi.org/10.3390/cancers14153778.
137. Wszolek ZK, Tsuboi Y, Ghetti B, Pickering-Brown S, Baba Y, Cheshire WP. Frontotemporal dementia and parkinsonism linked to chromosome 17 (FTDP-17). Orphanet J Rare Dis. 2006;1:30. https://doi.org/10.1186/1750-1172-1-30.

138. Wu Y, Zhang S, Xu Q, Zou H, Zhou W, Cai F, Li T, Song W. Regulation of global gene expression and cell proliferation by APP. Sci Rep. 2016;6:22460. https://doi.org/10.1038/srep22460.
139. Xia D, Tang WK, Ye Y. Structure and function of the AAA+ ATPase p97/Cdc48p. Gene. 2016;583(1):64–77. https://doi.org/10.1016/j.gene.2016.02.042.
140. Ye X, Cai Q. Snapin-mediated BACE1 retrograde transport is essential for its degradation in lysosomes and regulation of APP processing in neurons. Cell Rep. 2014;6(1):24–31. https://doi.org/10.1016/j.celrep.2013.12.008.
141. Yeo BK, Yu SW. Valosin-containing protein (VCP): structure, functions, and implications in neurodegenerative diseases. Anim Cells Syst. 2016;20(6):303–9. https://doi.org/10.1080/19768354.2016.1259181.
142. Zampatti S, Peconi C, Campopiano R, Gambardella S, Caltagirone C, Giardina E. C9orf72-related neurodegenerative diseases: from clinical diagnosis to therapeutic strategies. Front Aging Neurosci. 2022;14:907122. https://doi.org/10.3389/fnagi.2022.907122.
143. Zhang CT, Wang J, Wang WY. Wnt signaling in synaptogenesis of Alzheimer's disease. Ibrain. 2023;9(3):316–25. https://doi.org/10.1002/ibra.12130.
144. Zhang G, Xia Y, Wan F, Ma K, Guo X, Kou L, Yin S, Han C, Liu L, Huang J, Xiong N, Wang T. New perspectives on roles of alpha-synuclein in Parkinson's disease. Front Aging Neurosci. 2018;10:370. https://doi.org/10.3389/fnagi.2018.00370.
145. Zhang H, Ma Q, Zhang YW, Xu H. Proteolytic processing of Alzheimer's β-amyloid precursor protein. J Neurochem. 2012;120(Suppl 1):9–21. https://doi.org/10.1111/j.1471-4159.2011.07519.x.
146. Zhang J, Li X, Li JD. The roles of post-translational modifications on α-synuclein in the pathogenesis of Parkinson's diseases. Front Neurosci. 2019;13:381. https://doi.org/10.3389/fnins.2019.00381.
147. Zhang X, Kortholt A. LRRK2 structure-based activation mechanism and pathogenesis. Biomolecules. 2023;13(4):612. https://doi.org/10.3390/biom13040612.
148. Zhang YW, Thompson R, Zhang H, Xu H. APP processing in Alzheimer's disease. Mol Brain. 2011;4:3. https://doi.org/10.1186/1756-6606-4-3.
149. Zhou ZD, Chan CH, Ma QH, Xu XH, Xiao ZC, Tan EK. The roles of amyloid precursor protein (APP) in neurogenesis: implications to pathogenesis and therapy of Alzheimer disease. Cell Adhes Migr. 2011;5(4):280–92. https://doi.org/10.4161/cam.5.4.16986.

Insights into Proteostasis System in Ageing and Neurodegenerative Disorders

9

Yashaswee Mishra, Rahul Kumar, and Ritesh Kumar Netam

9.1 Introduction

To grow, survive, and reproduce, organisms must maintain a homeostatic state. Malfunctions in this state result in abnormal physiological conditions, which, if they remain unchecked, could be lethal to the organism. To maintain a steady functional state, an organism must integrate the environmental and cellular signals. Homeostasis is a phenomenon that, by means of physiological and behavioral responses, maintains the predefined range of physiological parameters through feedback regulation in response to external and/or internal environment. In humans, sweating and shivering are the most common examples of homeostatic response to hot and cold environments, respectively. Such changes are transient and prevent cellular damage; however, a malfunction in homeostasis may result in hyperthermia or hypothermia, which can be fatal to the organism. The homeostatic system includes molecular sensors that sense and signal the organism to respond to the changes to maintain a stable state, failing which could be damaging to the organism. Proteins are one of the major components mediating the homeostatic response, and their turnover in the cellular milieu decides the normal cellular functioning. The term proteostasis is a portmanteau of protein and homeostasis. It is an organized network that regulates the production of correctly folded proteins and maintains their optimal level in a cell [221]. In various neurodegenerative disorders (NDDs), such as Alzheimer's disease (AD), Parkinson's disease (PD), and prion diseases (e.g., Creutzfeldt-Jakob disease), either over-accumulation or misfolding of proteins occurs.

Proteostasis is an interlinked network of pathways that generate and maintain a pool of essential cellular proteins through regulating their synthesis, proper folding,

Y. Mishra · R. K. Netam
Department of Physiology, All India Institute of Medical Sciences, New Delhi, India

R. Kumar (✉)
Department of Zoology, Sheodeni Sao College (Magadh University), Kaler, India

N. Puranik (ed.), *Proteostasis*, https://doi.org/10.1007/978-981-96-6202-9_9

and posttranslational modification, along with degradation and clearance of excess or misfolded proteins [107]. It is like demand and supply failing, to which the system shifts to a new level of set points of functioning, affecting homeostasis and leading to a diseased state. Proteostasis includes positive and negative feedback circuits to maintain the turnover of proteins and further bring them to their basal levels after a transient change in the system. Only correctly folded proteins with proper chemical modifications are able to perform the right biological functions. These modifications are performed during or after the protein is translated. A mutated transcript or fault in the folding of nascent polypeptide leads to the accumulation of non-functional misfolded proteins, which may be cytotoxic for the cells that are kept in check by the components of the proteostasis system. Cellular senescence during ageing and age-related neurodegenerative disorders are due to the failure of these proteostasis components [66].

9.2 Why Is Proteostasis Required?

The nascent polypeptide chains are linear in nature and are capable of self-assembly through which they attain a three-dimensional stochiometric conformation. They stabilize their shape by forming ionic and covalent interactions within their amino acid residues of side chains, which bring proteins their signature conformation like helical or sheet-like structures. A biologically functional protein must be correctly folded to attain a stochiometric conformation. However, they can attain a confirmational state of minimal energy through an alternative folding scheme which may render the proteins nonfunctional. Such confirmations result in production of biologically inactive proteins which are insoluble in nature [77]. Excess of such proteins would interfere and can be cytotoxic, and therefore components of proteostasis network ensures the proper protein function by regulating its quality and quantity in cytoplasm and cellular organelles. These proteins reside in the cellular compartment together with the misfolded, unfolded, properly folded, and intermediate-folded state protein species. To prevent the cell from such an unamiable environment proteostasis network system has machinery like molecular chaperons, ubiquitin, proteasome, and autophagy (Fig. 9.1). These components of proteostasis help in either refolding or clearance of the accumulated nonfunctional/misfolded proteins. However, with age, the ability to maintain and preserve proteostasis network declines [116, 241], which might lead to the generation of proteinopathies eventually linking it with the neurodegenerative diseases. Proteinopathy refers to the accumulation of unstructured proteins that aggregate into higher order toxic insoluble structures. Proteostasis networks have mechanisms to prevent proteinopathies. It involves a stress response generated by the generation of unfolded proteins. For example, endoplasmic reticulum (ER) provides suitable environment to the newly synthesized protein to attain biologically active confirmation with the help of molecular chaperones. The folding process is acutely sensitive to change in surrounding signals, e.g., aggregation of misfolded/unfolded proteins generate an unfolded protein response (UPR) [91]. These stresses are sensed by the receptors of

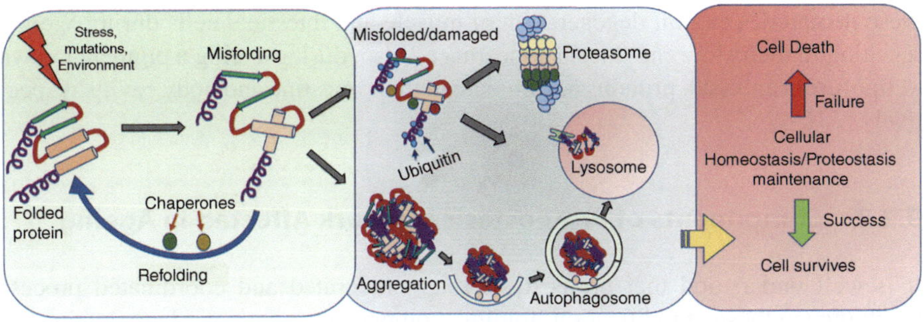

Fig. 9.1 Schematic representation of proteostasis network system. It has machineries like molecular chaperons, ubiquitin, proteasome, and autophagy. Failure of proteostasis leads to cell death

endoplasmic reticulum and further the UPR signaling through pro-survival pathways helping in tuning the protein folding status in the reticulum. In mammalian cell, UPR is an intricate mechanism and operates through three main reticulum transmembrane receptors: inositol-requiring enzyme 1 alpha, the eukaryotic initiation factor 2 alpha kinase, and activating transcription factor. The UPR seeks to manage the build-up of unfolded proteins by reducing the transcription of secretory proteins and enhancing the elimination of misfolded proteins via endoplasmic reticulum-associated or lysosomal degradation pathways. Moreover, the production of ER-resident chaperones and foldases are elevated to enhance the folding capacity of the endoplasmic reticulum and to relieve the load of unfolded proteins. Further if UPR is unable to resolve the issue, cellular apoptosis through the activation of JNK protein kinase and caspases takes place [76, 174]. Now we will discuss how proteostasis components are affected during ageing in the following section.

9.2.1 Ageing and Proteostasis

In humans, ageing can be defined as progressive accumulation of diverse physiological and behavioral changes with time. It escalates the likelihood to become susceptible to environmental stressors including cellular stress, oxidative damage, and pathogenic infections [66]. Most of these stresses lead to health anomalies and diseases, which may increase the chances of death. Although ageing is a universal process, its progression rate may be variable between individuals. Some individuals at younger age appear older than others of the same age; however, some are energetic at old age. Ageing is a reflection of what is going inside the cell or how healthy or stressful the cells are. Failure of proteostasis results in aggregation and loss of solubility of proteins and is one of the hallmark features of cellular ageing [164]. The ability to maintain the proteostasis network decline with age and thus considered as one of the major factors of age-induced cellular anomalies and causes of neurodegenerative diseases [77, 116]. *Caenorhabditis elegans*, due to its short lifespan, is widely used as model in the field of ageing [236]. It has been shown that in these nematodes exhibit degeneration of muscle and intestinal cells during ageing, resulting in their slow and flaccid appearance. Also, during ageing a pigment known as lipofuscin and yolk proteins accumulate in their intestine and body cavity, respectively [24].

9.2.2 Components of Proteostasis Network Affected in Ageing

It is well understood that proteostasis is an integrated and coordinated process involving synthesis, modification, folding, compartmentalization, degradation, and clearance of polypeptides (Table 9.1). Under normal cellular conditions, stress on proteostasis network is sensed by heat shock proteins (hsp) and UPR molecules. These proteins help in rectifying and rebalancing the proteostasis. Recent studies on *C. elegans* reveals that the expression of several genes is required in maintaining proteostasis, the loss of function of which could lead to ageing [189]. A quantitative analysis shows accumulation of insoluble proteins were much higher in old *C. elegans* compared to young adults. Half of such insoluble proteins were related to translation machinery, cellular energy production, and proteostasis. Cohort of such transcript when silenced using RNA interference results in increased lifespan. The increase in longevity was not due to a delay in the development process as they attain adulthood similar to the same aged controls. The studies reveal that during ageing along with increased protein load, composition and stoichiometry of proteins dramatically change, which affects the proteostasis [175]. Now we will look into the proteostasis network and its components in context of ageing in subsequent sections.

9.2.2.1 Protein Synthesis

In an organism, an uninterrupted supply of protein must be available for its proper growth and development. Protein synthesis is a necessary function of a cell and it plays an important role in development, immunity, metabolism, growth, and repair. We should note that across species (yeast, nematodes, fly, rodents, and humans) age-dependent specific gene expression helps in maintaining important sets of proteins across their life spans [94, 134]. A significant increase in protein requirement during early developmental stages is fulfilled by activation of a large number of genes which are downregulated when the organism advances toward old age has been reported in all species including humans. Gene expression studies in yeast, *C. elegans*, and *Drosophila* provide a deep understanding on suppression of protein expression of certain transcript mainly insulin/Insulin like growth factors-1 receptor (IGF-1R), TOR (target of rapamycin) kinase, and their homologs/orthologs (DAF/FOXO/*chico*, PHA-4/FOX A) which are the key components regulating longevity [5, 24, 84, 144, 239]. An overall reduction in protein synthesis has been observed during ageing. Studies have shown that ageing is associated with either a reduction in protein synthesis or an increase in the transcription of a subset of genes [95, 116, 236]. In a separate study, when protein translation was inhibited using cycloheximide (CHX) in early, late, and entire adulthood stage of *Drosophila*, it increased

Table 9.1 Components of proteostasis, their functions, and failure outcomes

Stage	Components	Description	Function	Effect of ageing	Failure outcome
Translation	Translational factors and ribosomes	Cellular structures composed of rRNA and proteins	Synthesize proteins from mRNA	Reduced efficiency of translation	Decreased protein synthesis, impaired cellular function
Folding	Chaperones	HSP70, HSP90, BiP, etc.	Assist in proper protein folding and prevent aggregation	Decreased expression and activity	Increased aggregation, cellular stress, apoptosis
Protein modification	Posttrans-lational modifiers	Kinases, phosphatases, ubiquitin ligases, etc.	Modify proteins (e.g., glycosylation, phosphory-lation)	Altered modification patterns	Increased protein misfolding, altered interactions
Transport	Vesicular trafficking proteins	Clathrin, COPI, COPII	Mediate transport of proteins and lipids	Disrupted transport leading to mislocalized proteins	Impaired secretion, loss of membrane integrity
Membrane fusion	SNARE proteins	Syntaxin, SNAP-25, VAMP	Facilitate membrane fusion during vesicle transport	Impaired fusion processes	Secretory defects, accumulation of vesicles
Quality Control	Proteasomes	26S Proteasome complex	Degrade damaged or misfolded proteins	Reduced efficiency and activity	Accumulation of damaged proteins, proteotoxic stress
Autophagy	Autophagy machinery	Macroautophagy, chaperone-mediated autophagy	Degrade and recycle cellular components	Impaired autophagic flux	Build-up of dysfunctional organelles, cellular ageing
Cellular repair	Cellular Stress response proteins	NRF2, p53	Manage oxidative stress and cellular repair	Compromised response leading to damage	Increased oxidative damage, genomic instability
Degradation	Ubiquitin-proteasome system (UPS)	Ubiquitin, E3 ligases (e.g., MDM2)	Tags proteins for degradation	Impaired tagging and degradation processes	Cellular senescence, loss of homeostasis

lifespan of early and entire adulthood phenotypes only. CHX treatment does not affect gut health, cognition, and age-dependent spontaneous activity in the flies. However, the CHX treatment delayed fertility and egg development during the early stage of adulthood [98]. In a similar study, overexpression of protein S6K, loss of which is associated with age-dependent reduction in translation, decreased lifespan of the flies. Similar changes in transcriptome and proteome have been conserved from rodents to humans [158, 199]. Controversial results on translational frequency have been reported in many studies. However, in an experiment with *Drosophila* chico variant, a long-lived homozygous strain that maintains a constant level of protein translation, some interesting results have been found. When a transient increase in protein synthesis was induced in these strains, it led to accumulation of old-age related insoluble ubiquitined proteins (IUPs). This suggests build-up of IUPs aggregation could lead to proteostasis dysfunction during old age. Till now, it has become evident that reduction in protein synthesis accelerates during ageing in order to reduce the proteostatic burden. Along with reduction, the accumulation of ribonucleoprotein complex in old *C. elegans* has also been reported [175]. Such accumulation has been observed due to ribosome pausing during ageing in *C. elegans* and *Saccharomyces cerevisiae*. Thousands of such pause sites with notably higher ribosome occupancy where ribosome translation kinetics slowdown gradually increase as the organism ages. These ribosomal pause sites were enriched for codons Arg, Glu, and Pro. Extended pausing causes ribosome collisions and results in di- and trisomes, which are harmful and require to be resolved by the ribosome quality control mechanism. Such incidents lead to protein aggregation followed by load, which may lead to disturbed proteostasis [214].

The process of ageing can be accelerated by external environmental factors such as lifestyle and nutrition/diet [132]. A dysregulated nutrient sensing can also lead to altered metabolism and further disturbance to proteostasis. In *Drosophila* model of dietary restriction, overexpression of d4E-BP, a translational factor has been found to extend longevity [258]. Microarray analysis of transcripts showed age-regulated differential expression of proteins such as translational initiation factor, proteasomal components, mitochondrial proteins, Hsp, allergenic, actins, and akt-2 kinase in wild strain and long-lived mutants of *C. elegans* [61]. It has been shown that reduction in expression of eEF1A and eEF2, protein translation elongation factors, is associated with decreased protein expression in old *Drosophila melanogaster* and aged rats, respectively [5, 144]. In one of the studies, using mass spectroscopy, it has been found that the cellular proteomic analysis from mitochondria, endoplasmic reticulum, and cytosol of young and aged rats' brain tissue showed pronounced variation in the expression profile of various protein groups. This study demonstrated that the brain of the old rats have depleted intracellular signaling mediators such as G-protein coupled receptors, protein kinases, and multiple ion-channels. Also, altered permeability of the nuclear pore complex due to deterioration in its scaffold components has been observed in post-mitotic cells of aged rats [158]. Recently, gene expression and proteome studies on young and old Baker's yeast have shown fold-changes in protein expression and its transcript. The study reveals that changes in the level of proteins such as cytoskeletal, cell wall, and sterol biosynthesis were accompanied by the changes in the level of their transcript. However, changes in protein levels of translation regulation, metabolism, and endoplasmic reticulum were not coupled with their transcript levels. The study reveals that during ageing, protein levels are largely uncoupled with the level of their transcripts [84].

Researchers have also studied the accuracy of protein translation, known as translational fidelity, in cell culture and cell-free translation using ribosomes isolated from animals of different age groups. The rate of error of translation in an organism and codons is estimated between 10^{-3} and 10^{-6}, which is higher than mutations. Such error may be due to the transfer of mismatched tRNA at the A-site of the larger subunit of ribosome and tRNAs misacylation. The error in protein translation was found to be associated with generation of increased heat shock response (HSR) in *C. elegans*. Such silent mutations during ageing results in production of protein "phenotypes" after substitution of amino acids [103, 128]. The stoichiometry of these proteins may also be perturbed leading to their reduced activity during ageing and also affecting precision of DNA repair. L. E. Orgel in 1963 proposed that accumulation of mutation might be associated with reduced accuracy of protein synthesis in non-dividing or slow dividing cells during ageing [157].

Streptomycin is a well-known agent affecting fidelity of protein synthesis through misreading the transcripts. Treatment of *Escherichia coli* with streptomycin leads to increased protein carbonylation, a modification which is marker of irreversible oxidative damage. It was increased several folds similar to that reported in aged *E. coli*. Through this experiment, it has been demonstrated that a decrease in fidelity of translation produces misfolded proteins, which makes them susceptible to oxidative damage such as carbonylation [43, 103]. Oxidative damage to proteins of transcription machinery induces spontaneous mutation in DNA, thereby leading to generation of a less accurate polypeptide. Such incidences are known where less accurate polypeptides accumulate or aggregate, leading to neurodegenerative disorder (e.g., synuclein accumulation/aggregation). In a recent finding by Martinez-Miguel et al. [128], it is demonstrated that enhancing translation fidelity in *Schizosaccharomyces pombe*, *C. elegans*, and flies by changing a single amino acid in the ribosome's decoding core is sufficient to increase health and lifespan. Working on RPS23, a key protein in the ribosomal decoding center, revealed a conserved lysine residue across all life forms. A few hyperthermophilic archaea have arginine at this conserved site. When they introduced this mutation into eukaryotic RPS23 homologs, it improved translation accuracy, elevated heat shock resistance, and extended lifespan in yeast, worms, and flies. These findings suggest error in translation machinery is one of the consequence that creates a burden on the proteostasis network failure of which results in ageing and associated diseases.

9.2.2.2 Protein Folding and Modification

During the process of translation, a linear polypeptide chain is generated. In order to perform biochemical, mechanical, and signal transduction activities, this linear polypeptide must be folded properly into a 3D conformation or a higher order confirmation known as secondary, tertiary, or quaternary structures. The newly synthesised polypeptide possesses all the information required to attain a biologically active conformation. Endoplasmic reticulum is the major site for these processes along with the posttranslational modifications (PTM) [174]. During the thermodynamic process of protein folding, a polypeptide undergoes various intermediate conformations and can end in a non-native but thermodynamically stable stage, which requires lesser energy. Such misfolded or partially folded proteins tend to aggregate and can form an amorphous structure or an insoluble aggregate, also known as amyloid. These amyloids interact and interfere with other normal cellular proteins and affect their functions, as observed in many neurodegenerative diseases. Almost 90% of cellular proteins (larger than >100 amino acids) are assisted by several chaperones classified as ubiquitous "Hsp system," high molecular weight oligomeric "chaperonins," and companion proteins during folding. The chaperons bind to exposed hydrophobic amino acids of nascent polypeptides and assist in correct folding either co-translationally, known as Chaperones linked to protein synthesis (CLIPS), or after the completion of translation. Whereas chaperonins perform refolding by trapping them in its multiunit complex cage-like structure. Historically, the chaperones were identified after their induction followed by stress to refold and prevent aggregation of the misfolded proteins [25]. Apart from folding, chaperones are responsible for translocation, assembly in the organelles/cytosol, and targeted degradation of the damaged or misfolded proteins. The main or major chaperone components are Hsp and these are monomeric and classified on the basis of their molecular weight like Hsp70, Hsp60, Hsp40, etc. Whereas chaperonins are high molecular weight oligomeric cage-like complex machines. Both perform ATP-dependent functions [2, 40]. Now, we will discuss how ageing affects molecular chaperones and vice versa.

Where from the misfolded proteins arise? The source of misfolded proteins is within the cytosol and the cellular organelles. Along with misfolded newly synthesized proteins, defective mRNAs and translational machinery, external stressors like elevated heat, radiation, hypoxia, heavy metals, and chemical factors are also responsible for a subset of proteins to misfold/unfold which further increases their chances of aggregation [77, 251]. These stress-induced unfolded proteins in cytosol activate an evolutionary conserved and ubiquitous component known as heat shock factor-1 (HSF-1). Activation of HSF-1 leads to the transcription of chaperones and associated factors responsible for protein refolding [40], which is otherwise kept inactivated by associated Hsp70 and Hsp90 [257, 260]. After the successful restoration of the cellular function, the Hsp70/Hsp90 reassociate and inactivate the HSF1 [62]. Morley and Morimoto [143] demonstrated that silencing of HSF-1 using RNAi reduced longevity in long-lived mutants (daf-16, daf-2, and age-1) of *C. elegans*. These mutants otherwise have an increased expression of HSF-1 which is a key factor responsible for their longevity. They explained that HSF-1 acts downstream to the insulin like signalling (ILS) pathway which influences ageing. A number of studies on animal models delivers the message that overexpression (Hsp16, Hsp22) and deficit (ubiquitin ligase) in chaperone expression increases longevity and accelerate ageing related protein phenotypes, respectively [122, 143, 254].

In a recent study by Paukštytė et al. [164], using limited proteolysis and mass spectroscopy (Lip-MS), it has been found that conformational changes at peptide level alters the structure of the age-related proteins between aged and young yeast strains. In aged yeast, these age-altered proteins were present in the nucleus, mitochondria, and cytosol. Lip-MS analysis of Sssp1, an Hsp70 isoform, revealed a structural alteration in its substrate binding domain, which promotes harboring of the protein aggregates [164]. They have demonstrated that Sssp1 was associated with increased protein aggregation in aged yeast. Further, they have shown that the age-altered proteins belonged to the family of major chaperones protein Hsp70, Hsp40, Hsp90, Hsp100, Hsp110, and TCP-1 ring complex of chaperonin family, along with other protein regulators involved in translation, metabolic regulators, and oxidative stress. These aggregates are nothing but the misfolded, carbonylated, and oxidative damaged proteins generated in response to environmental stress [47, 182]. The studies demonstrated that ageing in organisms might be associated with compartmentalization of Hsp70s with increased occupancy of aggregates, which otherwise is responsible for protein folding of nascent polypeptides, could further contribute to collapse of proteostasis in late stage of ageing. However, these aggregates were not transferred to newly divided daughter cells, signifying its asymmetric age-associated distribution, suggesting the role of Hsp42 and Hsp104/Hsp70. Deletion of Hsp42 and Hsp104 results in the transfer of these aggregates to daughter cells, extending the life span of the mother yeast cell. These results showed Hsp104/Hsp70 activity extends the lifespan by countering the aggregation, while Hsp42 helps in forming the protein deposits that promote ageing in the mother yeast cell [182].

Chaperone is a key factor in mediating cell cycle progression by allowing nuclear accumulation of cyclins. Moreno et al. [142] demonstrated that multifunctional chaperone systems are crucial for nuclear accumulation of the Cdc28–Cln3 complex, a cell-cycle regulator required during G1-S phase transition in yeast cells. Additionally, in aged cells, increased protein aggregation depletes free chaperone pools, limiting Cln3's nuclear accumulation by reallocating chaperones to protein synthesis and trafficking [142]. As we age, the accumulation of misfolded and aggregated proteins surpasses the capacity of available chaperones, leading to a depletion of chaperones and thereby reducing its availability for other essential tasks. This diverts them from performing their crucial functions, such as ensuring the proper folding of newly synthesized proteins [107]. Studies in *C. elegans* show that chaperone expression declines with age, exacerbating the strain on the proteome. This impaired chaperone function contributes to the accumulation of protein damage over time and it provides a causal link between cellular senescence and damage during ageing [254]. These studies reveal chaperones as a key mediator of maintaining proteostasis and progression of ageing.

9.2.2.3 Protein Modification

In an organism, newly synthesized polypeptides undergo posttranslational modifications (PTMs). These modifications occur co-translationally, along with folding. PTMs are required for regulating protein activity, stability, localization, and interactions. Alterations in PTMs like phosphorylation, acetylation, SUMOylation and ubiquitination, and various other chemical modifications can accumulate, affecting cellular homeostasis and contributing to ageing and age-related diseases [45, 188, 254].

Hutchinson–Gilford progeria syndrome (HGPS) shows a distinct feature of a rare preterm ageing, which is caused by mutation in the nuclear laminin "LMNA" gene. Lamin proteins are intermediate filaments, and these are the principle constituents of the nuclear lamina, involved in maintaining the nuclear shape, chromatin organization, and helps in smooth nuclear function of replication, transcription, and repair. In healthy humans, lamin-A protein undergoes farnesylation followed by cleavage and again methylation at c-terminal of the protein, which is further cleaved to produce a mature lamin protein. However, due to lack of Zmpste24 site in case of HGPS, the farnesylated tail remains attached and cannot be cleaved. Therefore, the terminal remains farnesylated, which results in lamin-A accumulation and abnormal nuclear morphology [45]. This shows that faulty protein modification could also be a causative factor for ageing.

In *C. elegans*, phosphoproteomic study of ILS downstream mediators such as GTP binding protein-1 at 20 and 25 °C temperatures has been shown to increase longevity and resistance to oxidative and heat stress [80]. Other studies have shown that the kinase ataxia-telangiectasia mutated (ATM, a serine/threonine kinase) phosphorylates the actin-binding protein drebrin, which helps in regulating lifespan and stress tolerance by making drebrin more stable and improving actin remodeling [102]. In *C. elegans*, removing the germ cells results in an extended lifespan. Likewise, disrupting the activity of the ILS pathway also leads to a longer lifespan in these animals. Moll et al. [140] demonstrated that decreasing the activity of the ILS pathway not only slowed ageing in these worms but also lowered the levels of SUMOylation on specific proteins, including CAR-1, which is responsible for producing germ cells. This study reveals that inhibiting the SUMOylation of CAR-1 prolonged the lifespan. When they replaced the protein with a mutated form of CAR-1, which cannot undergo SUMOylation, it resulted in longer life. Such studies reveal the importance of the PTMs in cellular ageing [140]. Ageing also leads to a widespread loss of ubiquitination, caused by increased activity of deubiquitinases. Such observable changes in the ubiquitinated proteins appear as organisms age, which are worsened by approaches that promote longevity, like dietary restrictions and reduced insulin signaling [101, 225]. By examining worms with a damaged proteasome, it was discovered that certain proteins accumulate with age due to a lack of proper ubiquitination. Reducing the accumulation of these proteins extends longevity, whereas preventing their breakdown leads to a shorter lifespan [101]. Baldensperger et al. [8] quantified age-induced enzymatic and nonenzymatic PTMs in young and aged mouse hepatocytes by utilizing HPLC-MS/MS approach. They found accumulation of nonenzymatic protein glycation, acylation, formylation, citrullination, and acetylation in subcellular compartments. However, enzymatic acetylation of histone proteins was reduced with age in mitochondrial and cytoplasmic proteins [8].

9.2.2.4 Protein Degradation

Lysosome-mediated autophagy and ubiquitin-mediated proteasomal degradation, also known as ubiquitin-proteasome system (UPS), are two distinct proteolytic systems responsible for the breakdown and clearance of proteins that have completed their lifespan and proteins that are irreversibly damaged. These mechanisms are required to clear the unnecessary burden of misfolded and aggregated proteins in the cellular compartments along with recycling of the amino acids to ensure the normal cellular functions, which otherwise leads to cytotoxic protein accumulation and stress resulting in neurodegenerative diseases such as Alzheimer's disease, Parkinson's disease, and Huntington disease, which will be discussed later in this chapter. In subsequent paragraphs, we will discuss how a faulty or altered degradation can lead to proteostasis imbalance.

In a study on wild young *C. elegans*, an overall reduction in mitochondrial and cytoplasmic ribosomal proteins was observed during the initial 12 days of normal development. However, age-matched nematodes showed a significant decrease only in cytoplasmic ribosomal subunits. In contrary, an age-dependent increase in 20s and 19s proteasomal subunits were observed with enhanced proteasomal activity in aged nematodes. Also, E3 ubiquitin ligase, a component of ubiquitin-mediated proteasomal degradation, was moderately increased [77, 236]. This indicates that dysregulated ribosome could lead to proteostasis burden. An increase in proteasomal response is generated as an adaptive mechanism to eliminate the faulty proteins. Transgenic (Tg) mouse model with reduced proteasomal chymotrypsin-like activity was produced to study the role of proteosome during ageing. These Tg mice had a shorter lifespan and showed signs of ageing. Also, they showed polyubiquitinated and oxidized protein accumulation. When these Tg mice were given high-fat diet they developed obesity and hepatic stenosis. Further, these models showed increased expression of adipose-related proteins, a metabolic malfunctional condition which is associated with aged mice. It shows that reduced proteasomal chymotrypsin-like activity impacts lifespan and worsens age-related metabolic conditions [226]. Hamer et al. [69] developed and expressed photoconversion-based proteasomal substrate in *C. elegans*. Through this approach, they revealed that by fifth day of adulthood, the turnover of the ubiquitin-proteasomal proteins in dorsorectal neurons occurred much more slowly compared to day 2, indicating a significant decrease in proteasome activity in these cells [69]. However, no such reduction in muscles of the body wall was observed, signifying that the age-related changes in proteasomal degradation might be tissue-specific and temporally regulated [107].

Stem cells are more resistant to the age-related decline in the proteostasis network than differentiated cells. Studies have shown that human embryonic stem cells have higher proteasome activity, which helps them in eliminating the misfolded proteins. Imbalance in proteostasis due to the overproduction of misfolded proteins and restricted proteasome capacity is linked to many age-related diseases. A decrease in proteasome function has been shown in different tissues and cell types of model organisms [77, 232].

One of the key changes seen with ageing is in autophagy, which has been recognized as an important factor in ageing across different species. It utilizes lysosome to degrade and clear damaged components like deteriorated organelles and misfolded proteins of the cells [66]. Sufficient evidence shows that autophagy activity reduces with age in various tissues across species. Ageing reduces the hydrolysis of proteins in lysosomes, which diminish autophagy, increases cellular damage, and contributes to the development of age-related disorders [66, 73, 92]. Sarkis et al. [192] described age-dependent decrease in cathepsin proteins of lysosomal protease and considered this as the reason for reduced turnover of proteins in old animals. Autophagy-related protein 8a (Atg-8a), a key protein for autophagosome formation was suppressed in the neural tissue and Atg8a mutant *Drosophila*, and this reduction was associated with increased level of ubiquitinated proteins, which are insoluble in nature [192]. However, an enhanced neural expression of this Atg8a results in extended longevity in these flies [205]. Moreover, in human nerve cells, the expression of autophagy-related genes BECN1, ATG-5, and ATG-7 decreases with age. On the other hand, in *C. elegans* increased expression of similar autophagy genes (ATG-1, ATG-7, and ATG-18) is associated with higher autophagy and extended longevity. Lapierre et al. [109] showed that HLH-30, a helix-loop-helix transcription factor, is involved in the transcription regulation of several autophagy-related genes in *C. elegans*. These genes have functions related to the formation of autophagosomes, synthesis of proteolytic enzymes, maintaining lysosomal acidic environment, and synthesis of proteins responsible for fusion with lysosomes. Overexpressing this transcription factor in transgenic *C. elegans* markedly extends its lifespan, supporting the role of autophagy in ageing [109]. Similarly, selective autophagy has also been shown to be associated with age-related pathology. These include degradation of defective mitochondria (mitophagy), lysosomes (lysophagy), endoplasmic reticulum (ER-phagy), aggrephagy, and lipophagy for protein aggregates and lipid droplets, respectively [73, 92]. Accumulation of iron has been shown in senescent cells. It may be due to the buildup of iron because of impaired autophagic breakdown of ferritin within lysosomes [252]. Moreover, mitochondrial ferritin (FTMT) gathers on the outer membrane of malfunctioning mitochondria, subsequently enhancing mitophagy through a targeted interaction with the autophagic cargo receptor associated with phagophore [66]. Additionally, higher amounts of copper (Cu) in embryonic fibroblasts of older mice and astrocytes have been observed alongside increased levels of high-affinity Cu transport protein 1 and decreased levels of Cu-transporting ATPase 1 (Atp7a). Additional studies have revealed that the absence of functional Atp7a prevents the autophagic-lysosomal breakdown of Cu in older MEFs, which contributes to age-related degenerative diseases [131, 170].

Endoplasmic reticulum and mitochondria are the cellular compartments that are involved in proteostasis maintenance through in-built stress response mechanisms. These organelles contain molecular machinery that quickly detects and responds to proteotoxic stress through conserved signaling pathways, triggering specific transcriptional responses to maintain proteostasis and cellular health. Together, these complex networks of stress detection, signalling, and resolution are known as UPR, as discussed earlier. These responses help in protecting and sustaining the function of both the organelles and the cell. The well-known components responsible for UPR in reticulum are BiP [202], IRE1α/XBP1 signaling [27], PERK activity [15],

ATF-6 [202], and in mitochondria are Hsp60, Lonp1, ATF-5, SIRT-3 along with electron transport chain complex [241]. Overall, studies from postmortem human tissues and other animal species indicate that UPR in both the organelles undergo significant changes with ageing. As age progresses, the adaptive signaling capacity of reticulum and mitochondrial responses declines, potentially resulting in a harmful situation where the UPR can no longer resolve the stresses caused by age-related cellular dysfunction [241].

So far, we have discussed how imbalance in the proteostasis due to reduced translational fidelity, defect in maintaining the correct protein conformation, and disturbed protein degradation could lead to the production of aggregation-prone proteins. If this load reaches to critical level, these aggregates would retain an insoluble state, which would negatively impact normal cellular function and cause cytotoxicity (Fig. 9.2).

Now we will discuss neurodegenerative diseases due to dysfunctional proteostasis which are often related to age. Therefore, most of the neurodegenerative diseases exhibiting late onset are associated with ageing related decline in protein homeostasis. All the neurodegenerative disease are due to proteostasis network dysfunction. Therefore, neurodegenerative disorders are linked to proteinopathy. Ageing is a significant risk factor for neurodegenerative diseases and is associated with the accumulation of insoluble proteins across various tissues in different species. It remains unclear whether the age-related accumulation causes cellular dysfunction or it results from diminished proteostasis. Research using *C. elegans* as a model has revealed that hundreds of proteins become misfolded and insoluble with age. Importantly, small molecules targeting specific misfolded proteins have been shown to restore protein homeostasis in vivo and extend lifespan, highlighting the connection between protein aggregation and ageing [181]. Now, we will discuss in detail about the proteostasis disturbances that lead to neurodegenerative disorders in subsequent section and subsections.

9.3 Neurodegenerative Disorders and Proteostasis

Under normal physiological conditions, proteins are structured and fully organized, having active domains in their secondary or quaternary structures. On the other hand, proteins such as α-synuclein, amyloid, tau, and polyglutamine-expanded proteins, which are linked to neurodegenerative diseases, are intrinsically disordered and tend to be unstable under normal physiological conditions (Fig. 9.3).

These disordered and misfolded proteins have exposed beta sheets that are susceptible to abnormal interactions with other key proteins and mediators, disrupting various intracellular processes, including protein synthesis, cytoskeletal integrity, vesicular trafficking, protein misfolding, impaired protein degradation, and cellular metabolism. Among these proteins, misfolding and aggregation are common, causing various neurodegeneration diseases. Additionally, these beta sheet aggregates are highly resistant to proteolytic degradation and ultimately contribute to the formation of inclusion bodies or extracellular plaques. Each protein associated with a

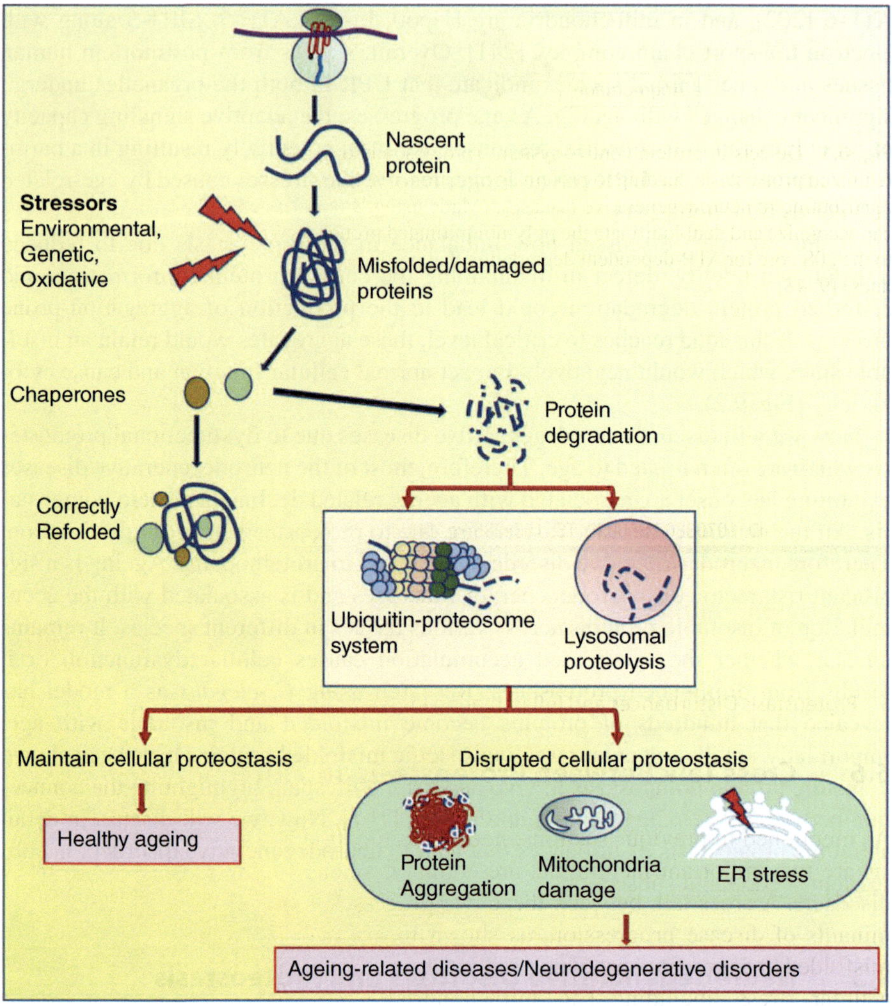

Fig. 9.2 Imbalance in proteostasis network system associated with ageing. An imbalance in proteostasis resulting from diminished translational fidelity, deficiencies in maintaining proper protein conformation, and disrupted protein degradation may result in protein aggregation, mitochondrial damage, and ER stress causes damage to cells or cell death, which subsequently leads to age-related diseases, including neurodegenerative disorders

specific neurodegenerative disease disrupts a distinct cellular signaling network, which in turn contributes to proteostasis dysfunction and selective neuronal vulnerability. Although genetic and environmental factors are the main cause of the neurodegenerative disorders, ageing is one of the important risk factors compromising the proteostasis, which we have already discussed. We will discuss the altered proteostasis in neurodegenerative disorder in detail. Before discussing about proteostasis imbalance in different neurodegenerative disorders in detail, we should have a basic idea about the proteostasis network/components affected in these diseases.

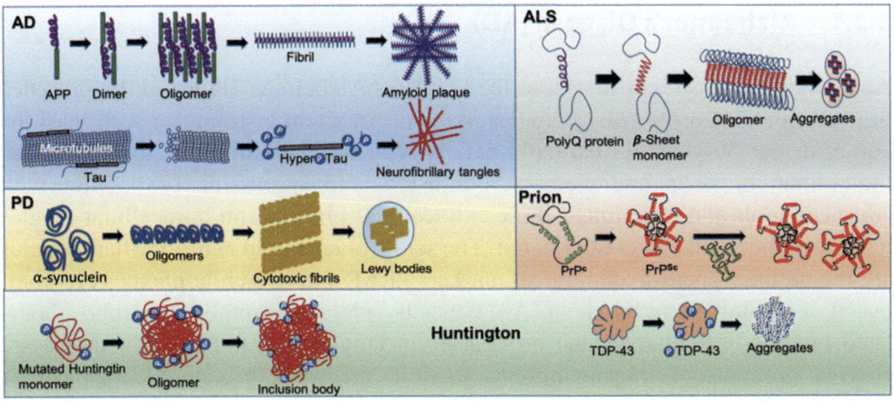

Fig. 9.3 Schematic representation of protein misfolding and associated neurodegenerative disorders. It should be noted that these disorders result from the failure of proteostasis system which ultimately leads to formation of protein aggregates responsible for triggering neurodegeneration. Abbreviations: AD = Alzheimer's disease, PD = Parkinson's disease, ALS = amyotrophic lateral sclerosis

(A) Impaired translation eIF2alpha, along with $_{met}$tRNA and 40s ribosome, forms a ternary complex for translation initiation. In response to various cellular stresses, eIF2α kinases can phosphorylate eIF2α, reducing overall translation by affecting the formation of ternary complex [17]. However, protein aggregates in neurodegenerative disorders and impaired mTOR signaling disrupt the regulated level of protein synthesis [227]. Malfunction in eukaryotic elongation and termination has been found in various neurodegenerative disorders [10, 83].

(B) Impaired posttranslation protein modifications Impaired and aberrant posttranslational modification of cellular protein leads to an altered conformation, which has shown to be associated with pathophysiological symptoms in neurodegenerative disorders. It includes increased phosphorylation, acetylation, glycosylation, palmitoylation of proteins such as tau, α-synuclein, huntingtin, and various other cellular proteins disturbing their native structure and functions [115].

(C) Impaired protein degradation system To preserve proteostasis, aggregated faulty proteins need to be cleared by chaperones, lysosomes, or the UPS. These machineries for degradation and refolding are impaired by aggregated proteins, which eventually leads to cellular toxicity. A decreased proteasomal activity, defective or accumulation of autophagosome, and abnormal autophagy have been shown in various neurodegenerative disorders [38, 129, 201, 256]. Accumulation of ubiquitin has also been shown in yeast model of α-synuclein-mediated pathology [162]. Now, we will discuss in detail how the impaired proteostasis leads to the progression of different neurodegenerative disorders.

9.3.1 Alzheimer's Disease (AD)

Alzheimer's disease was first described by Lois Alzheimer. He noted the presence of plaques and neurofibrillary tangles in the postmortem histological section of the cortex in her 50-year old paranoia patient who was admitted due to progressive impairment of sleep and memory, confusion, and aggression [78]. Now, with advanced tools and techniques, these extracellular plaques and intracellular tangles have been unravelled in greater detail. The senile plaques and neurofibrillary tangles are the hallmark neuropathological markers of AD [194]. Senile plaques are composed of insoluble amyloid beta (Aβ) deposits, while abnormally hyperphosphorylated intermediate filament tau proteins form tangles [197]. The Aβ peptides are cleaved products of transmembrane amyloid precursor protein (APP). APP is cleaved by α-secretase to generate sAPPα, which helps in neuronal development, synaptogenesis, and has neuroprotective role. However, in an alternate cleavage mediated by γ-secretase truncated Aβ40 and Aβ42 peptides are generated, which are otherwise detrimental to the neurons [195, 197]. These peptides are made throughout life but in AD these peptides accelerate their self-production and form a toxic aggregate. They build up due to either more production or less breakdown and removal [126]. Once these amyloids are established in the cellular milieu, they are extremely resilient to breakdown. Proteasomes can only break down single-chain polypeptides and also necessitate that the proteins be partially or entirely unfolded. Furthermore, the amyloid state is thermodynamically very stable due to the extensive interactions among the protein chains of their polymer. The thermodynamic resilience of amyloid aggregates also aids in their capacity to transform native proteins into amyloid structures [221] (discussed later under section on prion disease).

In a study conducted by Shankar et al. [200], it was demonstrated that when Aβ from AD patients is microinjected in rats, it impairs learning and memory [200]. Moreover, extracellular soluble Aβ dimers impair synaptic plasticity by disturbing NMDA receptor signaling responsible for the generation of LTP (long-term potentiation). Aβ also induces hyperphosphorylation of tau through activation of GSK-3β and MAPK signaling [126]. Hyperphosphorylated tau aggregates and develops into filamentous structure, resulting in the disruption of the assembly of microtubules, eventually affecting the vesicular transport [253]. Although intracellular Aβ peptides have regions that are intrinsically unstructured, they undergo anomalous interaction with newly synthesized and preexisting cellular proteins mediating cytoskeletal integrity, chromatin stability, nucleocytoplasmic transport, vesicular transport, transcription-translational regulator, and proteins involved in quality control. Expression of these peptides also causes reduced cellular stress response, limiting the cellular defense [156, 221, 242].

The APP gene can cause AD either by overexpression of extra copies (as in Down's syndrome) or microduplication of the gene, and also due to an increase in the production of Aβ peptides through rare mutation in BACE or PSEN 1 and 2 genes responsible for β- or γ-secretase enzymes. β-secretase 1, β-secretase 2, and cathepsin B are involved independently in the processing of APP to Aβ peptides in an amyloidogenic pathway [159, 195]. Also, it may be due to defect in components of cholesterol metabolism such as apolipoprotein E (ApoE). Ageing is one of the major non-genetic risk factors for sporadic occurrence of AD. Amyloidogenic term describes a protein or peptide that interacts with similar molecules, forming insoluble fibres that link together to create a beta sheet-like stable structure. The structure of these amyloid sheets makes them highly resistant to breakdown by proteolytic enzymes [20]. The failure in Aβ clearance seems to be a major cause of late onset of AD [23]. Since the causative factors and proteins involved in the progression of AD is well established (Table 9.2), what prevents these Aβ plaques and tangles to be cleared from the cellular milieu will be important to deal with. In the earlier sections, we have discussed that efficiency of proteosome and autophagy declines as the organism ages. Inability to restore the balance between protein synthesis and degradation leads to proteotoxic stress which eventually results in cellular death.

Table 9.2 Components of proteostasis system impaired in Alzheimer's disease

Gene	Protein	Associated with	Pathological hallmark	Reference
APP	Amyloid precursor protein	Precursor to amyloid-β; involved in synaptic function	Amyloid plaques	Hampel et al. [70]; O'Brien and Wong [153]
PSEN1	Presenilin 1	Component of the gamma-secretase complex; cleaves APP	Amyloid plaques	Kelleher and Shen [93]; Xiao et al. [246]
PSEN2	Presenilin 2	Involved in gamma-secretase activity	Amyloid plaques	Xiao et al. [246]
APOE	Apolipoprotein E	Involved in lipid metabolism and amyloid-beta clearance	Amyloid deposition	Liu et al. [118]; Raulin et al. [172]
LAMP2	Lysosome-associated membrane protein 2	Involved in autophagy and lysosomal function	Impaired lysosomal degradation	Armstrong et al. [6]; Nixon [149]
ATG5 and 7	Autophagy-related protein5 and 7	Involved in autophagosome formation and maturation	Impaired autophagy	Uchiyama et al. [229]; Uddin et al. [230]
mTOR	Mechanistic target of rapamycin	Regulates cell growth, proliferation, and protein synthesis	Altered protein synthesis, neuroinflammation	Perluigi et al. [165]; Yates et al. [248]
eIF2α	Eukaryotic initiation factor 2 alpha	Controls translation initiation; responds to stress	Impaired translation, stress responses	Ding et al. [41]; Oliveira and Klann [155]

(continued)

Table 9.2 (continued)

Gene	Protein	Associated with	Pathological hallmark	Reference
GSK-3β	Glycogen synthase kinase 3 beta	Phosphorylates tau; involved in cell signaling	Tau hyperphosphorylation, neurofibrillary tangles	Llorens-Martín et al. [119]; Takashima [222]
MAPT	Tau protein	Stabilize microtubules, neuronal transport	Neurofibrillary tangles	Rawat et al. [173]; Strang et al. [217]
P62/ SQSTM1	Sequestosome 1	Adaptor protein linking ubiquitinated proteins to autophagy	Accumulation of protein aggregates	Mary et al. [130]; Salminen et al. [186]
SIRT1	Sirtuin 1	NAD+-dependent deacetylase; regulates protein acetylation and stress responses	Impaired proteostasis, neuroinflammation	D. Kim et al. [97]; Manjula et al. [127]

9.3.1.1 Failure of Proteostasis in AD

First, we are going to discuss how autophagy gets derailed in AD. In one of the earlier studies [211], using APP chimeric protein expression in CHO cells, it was demonstrated that the APP peptides are generated in all the three intracellular compartments, namely, ER, Golgi, and trans-Golgi network. However, amyloidogenic peptide $A\beta_{1-42}$ primarily originates from the endocytes and ER and to lesser extent from the intracellular compartments-derived secretory pathway [198, 240]. In a similar study it has been shown that PSEN1 and APP interact in the ER and Golgi, with PSEN1 helping to process APP correctly. Mutations in PSEN1 lead to higher $A\beta_{1-42}$ levels, mainly in Golgi-like vesicles [112, 244]. In the sporadic model of AD, morphological volume of early endosomes in neuron was increased to threefold compared with normal neurons. These enlarged endosomes have prolonged expression of cathepsin B and D in their early and late forms, confirming that endosomal pathway abnormalities can lead to both production and clearance of Aβ peptides in ageing, which coexist with ApoE and Presenilin in human patients [29]. PSEN1 also plays a key function in lysosome acidification, which is crucial for activating lysosomal proteases during autophagy [112].

Fibroblasts from patients with familial AD caused by PSEN1 mutations show significant defects in lysosome acidification and autolysosome maturation [112]. The findings from studies pertaining to both human and other animal models suggest that lysosomal dysfunction results in inefficient clearance of proteins and peptides causes exacerbation of autophagic-lysosomal, amyloid, and tau pathologies, and also facilitates neuronal death [30]. The progressive failure of autophagy leads to a large-scale accumulation of partially digested substrates in deteriorating axons and dendrites [150], and affected neurons show compromised maturation and transport of autophagic vacuoles [151]. Wolfe et al. [243], using pH probes, showed that elevated lysosomal pH and disruption of lysosomal proteolysis are major contributors to the failure of autophagy and its pathological consequences in Alzheimer's disease [243]. Sirtuin 1 and 3 play important roles in neuroprotection by mediating autophagy and mitophagy to clear damaged mitochondria and protein aggregates [48].

Dysregulation of UPS has been implicated in the pathogenesis of various neurodegenerative diseases including Alzheimer's disease, Parkinson's disease, Huntington's disease, and amyotrophic lateral sclerosis (ALS) all of which have been described elsewhere in this chapter [230]. The dysfunction of the UPS may be associated with the inability to clear misfolded proteins, resulting in intracellular protein buildup, cytotoxicity, and neuronal death.

Impairment in Chaperon network such as enzymes for ubiquitination (E1 and E2), accumulation of mutated ubiquitin, oxidation of deubiquitinating enzymes, alteration in proteosome subunits 20s and 26s, and reduced proteasomal activity has been observed in AD patients and transgenic animal AD models resulting in lower α-secretase activity, and promoting the production of Aβ [1, 57, 82, 121]. In a study by Almeida et al. [3], it was shown that the accumulation of Aβ reduces the ubiquitination and disrupts the sorting of endocytic trafficking of the membrane receptors in APP transgenic mice. Similarly, hyperphosphorylated tau and its synaptic accumulation have been associated with impaired proteasome activity. The oligomers of Aβ have been shown to block proteasome activity by keeping the 20S core particle in a closed state [3].

In a study by Ukmar-Godec et al. [231], using NMR and MS, it was found that phosphorylation of tau by the Ca2+/calmodulin-dependent protein kinase-II prevents its degradation by the 20S proteasome. On the other hand, phosphorylation of tau by GSK3, a key proline-directed tau kinase, affects tau degradation rates in a residue-specific way [231]. Earlier studies have shown that ubiquitinated proteins accumulate around amyloid plaques and NFTs. A frameshift mutation in the ubiquitin B peptide has been associated with resistance to deubiquitination and disruption of proteasome activity in AD [1, 37]. Posttranslational modification in neurodegenerative diseases includes extensive phosphorylation at sites and locations of amyloid-prone proteins and their aggregation [49]. Also, 214 dysregulated phosphorylation, both hypo and hyper, in human AD patients has been shown in frontal and entorhinal cortex using mass spectroscopy [52].

9.3.2 Parkinson's Disease (PD)

The initial clinical description of Parkinson's disease was provided by James Parkinson in 1817. It includes a tremor that occurs when the patient is at rest, rather than during movement, along with slowness of movement (Bradykinesia), muscle stiffness, and difficulty in maintaining balance [60]. These are due to the loss of midbrain dopaminergic neurons [59, 168]. Most cases of PD are sporadic, with about 5–10% being familial. Over 20 genes associated with Parkinson's disease have been identified, including synuclein alpha (SNCA), Parkin, PINK1, leucine-rich repeat kinase 2 (LRRK2), DJ-1, glucocerebrosidase (GBA), etc. (Table 9.3) [71, 179, 245].

Table 9.3 Components of proteostasis system impaired in Parkinson's disease

Gene	Protein	Associated with	Pathological hallmark	Reference
SNCA	α-Synuclein	Regulation of synaptic function	Lewy body	Spillantini et al. [212]
PRKN	Parkin	E3 ubiquitin ligase; regulates protein degradation and mitochondrial quality control	Dysfunctional mitophagy	Doherty and Hardy [42]; Farrer et al. [50]
PINK1	PINK1	Protects mitochondria; initiates mitophagy	Dysfunctional mitophagy	Ge et al. [58]; Samaranch et al. [187]
GBA	Glucocerebrosidase	Breaks down glucocerebroside; involved in lysosomal function	Lysosomal dysfunction	Blauwendraat et al. [18]; Tsuang et al. [228]
UCHL1	Ubiquitin C-terminal hydrolase L1	Ubiquitin-proteasome system; protein degradation	Protein aggregation, impaired degradation	Barrachina et al. [11]; Kumar et al. [104]
FBXO7	FBXO7	Ubiquitin-proteasome pathway regulation	Protein aggregation	Zhao et al. [255]
mTOR	Mechanistic target of rapamycin	Regulates cell growth, proliferation, and protein synthesis	Altered protein synthesis, neuroinflammation	Lipton and Sahin [117]
eIF2α	Eukaryotic initiation factor 2 alpha	Controls translation initiation; responds to stress	Impaired translation, stress responses	Hrabos et al. [79]
ATF4	Activating transcription factor 4	Regulates stress response and translation during stress	Altered protein homeostasis, stress responses	Sun et al. [220]
LAMP2	Lysosome-associated membrane protein 2	Involved in autophagy and lysosomal function	Impaired lysosomal degradation	Murphy et al. [145]
P62/ SQSTM1	Sequestosome 1	Adaptor protein linking ubiquitinated proteins to autophagy	Accumulation of protein aggregates	Narendra et al. [147]
SIRT1	Sirtuin 1	NAD+-dependent deacetylase; regulates protein acetylation and stress responses	Impaired proteostasis, neuroinflammation	Chao et al. [33]

9.3.2.1 α-Synuclein Misfolding and Aggregation

The main pathological characteristic of PD is intraneural aggregation of misfolded α-synuclein proteins. These aggregates are known as Lewy bodies (LBs). Genetic cause includes mutation in synuclein alpha (SCNA) coding gene resulting in A53T, A30P, E46K amino acid change, duplication and triplication in SCNA locus [213, 245, 250]. The primary role of α-synuclein seems to be regulating neurotransmitter release by influencing the SNARE complex.

In a recent experiment by Mahul-Mellier et al. [125], using neuronal seeding model, formation of LBs has been demonstrated which was not known clearly. By employing an integrative approach combining omics, biochemical analysis, and imaging techniques, they analyzed the molecular processes involved in the various stages of LBs formation and their impact on neuronal dysfunction and degeneration. By seeding preformed fibrils in mice after 21 day LB-like inclusions were observed along with other morphological structures. These inclusions were transitioned from filamentous aggregates, which were predominately present at day 14 post-seeding. They have shown that LBs are spherically organized, made up of α-synuclein fibrils as well as other proteins, lipids, and membranous organelles, such as mitochondria and lysosomal structures. Formation of LBs is driven by a complex interaction between α-synuclein fibrillization, posttranslational modifications, and the interaction of α-synuclein aggregates with membranous organelles [125].

9.3.2.2 Failure of Proteostasis in PD

It is becoming more widely acknowledged that the disruption of cellular proteostasis is a common characteristic in laboratory models of PD, with signs of impairment in chaperone proteins, ubiquitin-proteasome system, and autophagy, which have been observed in the later stages of the disease.

Proteasome studies from postmortem PD brain tissues found a notable reduction in chymotrypsin-like and trypsin-like proteasome activity in the substantia nigra (SN) of PD patients compared to age-matched controls. However, no signs of impaired proteasome activity were detected in other brain regions, where increased activity was actually observed in the unaffected areas [4, 56]. In the substantia nigra (SN) of PD patients, several genes encoding proteasome subunits were downregulated, leading to reduced levels of the 20S proteasome core, α-subunit, and 19S regulatory caps [26, 137, 213]. The suggested mechanisms by which α-synuclein inhibits proteasome activity include binding to the S6′ or Rpt5 subunit of the 19S proteasome or to the β5 subunit of the 20S proteasome. In order to study the impact on ubiquitin-proteasome activity, in a cell culture study when wild-type or mutated α-synuclein proteins were overexpressed, it reduced activity of the proteasome [208, 223] in dopaminergic neurons of substantia nigra [136].

Mutation in genes related to PD as mentioned above along with genes such as LRKK2, VPS 35, GAB1, PARKIN 2,6,7, and 15 are also linked to autophagy. For example, Parkin, an enzyme that attaches ubiquitin to proteins, is activated during UPR and prevent cell death. When parkin function is lost, PAEL-R, a substrate of parkin, misfolds and accumulates in the ER of substantia nigra neurons, causing ER stress and ultimately leading to cell death [21, 71]. Deregulation of several autophagy-mediating proteins such as LAMP1, P-type lysosomal ATPase, Hsp73, and

CGpase are associated with the aggregation of α-synuclein. Protein levels of AMBRA1, Beclin1, and ULK1 were found to increase in the peripheral blood cells of PD patients [113].

Apart from cellular stress and mutation, molecular-level analysis revealed that posttranslational modifications of α-synuclein, including phosphorylation, glycosylation, SUMOylation, nitrosylation, ubiquitination, and nitration, play a role in the misfolding and aggregation of α-synuclein. It also binds with tyrosine hydroxylase, rate-limiting enzyme in the synthesis of dopamine, and dopamine transporter and modulates their activity [113, 168, 213]. In a recent work, El Manaa et al. [46] have observed a correlation between the interaction of XBP1s and PINK1 and the activation of the apoptotic response in human substantia nigra neurons, primary cell culture, and transgenic mice. They suggest that the protective effects of the PINK1–XBP1s interaction are outweighed by the prolonged PINK1-mediated mitophagy response in the later stages of PD [46].

9.3.3 Huntington's Disease (HD)

George Huntington, a physician, observed abnormal involuntary movements, impaired motor coordination, psychiatric disturbance, and cognitive decline, known as "Chorea," which runs within family. In 1872, he provided a detailed clinical characterization and proposed the symptoms that follow autosomal dominant inheritance and hence also known as Huntington's Chorea [16, 85, 87].

HD is caused by mutations in the huntingtin (HTT) gene, which encodes a widely expressed 350 kDa neuronal protein. HTT gene contains a polyglutamine (polyQ) region, which is encoded by uninterrupted CAG trinucleotide repeats in the first exon of the HTT gene. The length of this repeat varies in the general population, and when it exceeds around 35 CAG repeats, it can lead to HD [67, 87]. It is characterized by degeneration in basal ganglia, more of striatum GABA-containing medium spiny neurons [67, 85]. Huntingtin is largely an α-helical cytosolic protein with a polyQ sequence at its N-terminal, proline rich domain, and several consensus HEAT sequence required for protein–protein interaction with proteins involved in cellular processes such as intracellular trafficking, synapse formation, endocytosis, cell adhesion, transcription regulator, RNA processing, mitochondria function, and protein turnover [65, 74, 85, 87].

9.3.3.1 Proteostasis Failure in HD

The aggregation of mutated Htt proteins in HD is due to its long N-terminal glutamine repeats which acquires excessively high affinity for each other and partners proteins like ubiquitin, chaperones, and proteasomal subunits resulting in agglomeration and formation of amyloid like beta-sheet structures leading to toxicity in neurons [34, 166]. These aggregates are also known as inclusion bodies [209]. Also, the proteolytic cleavage of Htt through various proteases, such as calpains and caspases, generates N-terminal fragments that are pathogenic [87, 213]. These pathogenic fragments and aggregates disrupt ribosomal biogenesis, intracellular trafficking, membrane structure, and cytoskeleton [209, 234] (Table 9.4).

Table 9.4 Components of proteostasis system impaired in Huntington's disease

Gene	Protein	Associated with	Pathological hallmark	Reference
HTT	Huntingtin	Involved in intracellular transport and cell signaling	Aggregation of mutant huntingtin	Saudou and Humbert [193]; Schulte and Littleton [196]
ATXN3	Ataxin-3	Involved in ubiquitin-proteasome system	Impair cellular proteostasis	Yue et al. [249]
ATG5 and 7	Autophagy-related protein5 and 7	Involved in autophagosome formation and maturation	Impaired autophagy	Jimenez-Sanchez et al. [86]; Sarkar and Rubinsztein [190]
Caspase 2 and 3	Caspase 2 and 3	Executing apoptosis and regulating cellular stress responses	Apoptosis, neuronal cell death	Hermel et al. [75]; Y. J. Kim et al. [96]
HSPs	Heat shock proteins	Assist in protein folding and prevent aggregation	Impaired chaperone function, protein aggregation	Chai et al. [32]; Stenoien et al. [215]
GSK-3β	Glycogen synthase kinase 3 beta	Phosphorylates tau; involved in cell signaling	Neurofibrillary tangles	Carmichael et al. [28]; Rippin et al. [178]
Ub	Ubiquitin	Tags proteins for degradation via the proteasome	Impaired protein degradation, accumulation of damaged proteins	Bennett et al. [13]; Ortega and Lucas [160]
P62/SQSTM1	Sequestosome 1	Adaptor protein linking ubiquitinated proteins to autophagy	Accumulation of protein aggregates	Huang et al. [81]; Kurosawa et al. [105]
SIRT1	Sirtuin 1	NAD+-dependent deacetylase; regulates protein acetylation and stress responses	Impaired proteostasis, neuroinflammation	Naia and Rego [146]; Salamon et al. [185]
HDAC6	Histone deacetylase 6	Involved in protein deacetylation and aggresome formation	Impaired proteostasis, neurodegeneration	Sadri-Vakili and Cha [183]; Suelves et al. [218]

It has been suggested that diffuse mutant Htt aggregates may indirectly block or disassemble the proteasome by overloading the protein folding machinery, causing an accumulation of misfolded and ubiquitinated proteins, ultimately impairing its function [74, 209]. In yeast expressing mutant Htt and brain tissues of HD patients, it has been observed that Hsp40 and Hsp70 chaperones, component of proteasomal system, were trapped in the inclusion bodies [209].

In HD, the impaired ability of autophagosomes to recognize cargo, as well as the subsequent processes of axonal transport and material degradation, have been observed. Mutated Htt reduces autophagy by sequestering and inhibiting Rhes GTPase, which is required to activate Beclin-1 and promote autophagy. It also negatively modulates mTOR-dependent autophagy [74, 209]. In a recent study, a monomeric form of mutated Htt proteins has been shown that lack polyubiquitylation site required for its recognition by the autophagic receptors and further its clearance [54]. Mitochondria also play an important role in preserving the homeostasis through its quality control mechanism and dynamics. Abnormalities in mitochondrial function lead to metabolic impairment reported in neurodegenerative disorders [259]. Functionally, mutated Htt is believed to impact mitochondria and energy metabolism indirectly through transcriptional deregulation and directly by interacting with the organelle and its associated proteins. Mitochondrial fragmentation is elevated in several HD models and in the patients, with a link to the dysregulation of fission and fusion proteins [64, 154]. Mutant Htt either directly binds to the fission-promoting protein Drp1 or induces its posttranslational modifications, modulating its activity [35]. Furthermore, aggregated Htt disrupts mitochondrial trafficking, mitophagy and physically hinders the movement of intracellular organelles. Additionally, diffuse mHtt may interfere with the function of trafficking adaptors and regulatory proteins, impacting both mitochondrial and autophagosomal processes. In addition, similar disruption in cellular redox balance causes proteotoxic stress in ageing as well as in expanded polyQ protein expressing *C. elegans* model [100].

9.3.4 Amyotrophic Lateral Sclerosis (ALS)

ALS is a neurodegenerative disorder affecting motor neurons in central and peripheral nervous system. The pathological symptoms include loss of myelination, axonal degeneration, and atrophy [224]. ALS is categorized into familial (disease runs in family through genetic mutation) and sporadic (occurs at any stage of life, no family history of disease). Either gain-of-function or loss-of-function mutation in genes such as C9orf72 (chromosome 9 open reading frame 72), TDP-43 (TAR DNA-binding protein 43), SOD1 (superoxide dismutase1), and FUS (fused in sarcoma) lead to development of ALS [120, 237]. Patients having ALS show clinical features like paralysis of muscles responsible for speech (dysarthria), difficulty in swallowing (dysphagia), weakness of upper and/or lower limb muscles along with behavioral and cognitive impairment.

9.3.4.1 Failure of Proteostasis in ALS

It is a major area of concern in ALS. However, the mechanism of proteostasis failure in ALS is still not fully uncovered. Several proteins encoded by ALS-associated genes have a major role in maintaining the cellular proteostasis. Mutations which lead to dysfunction of the proteostasis components are being investigated. Postmortem studies from ALS patients show SOD-1, FUS, TDP-43, and ubiquitin-positive cytoplasmic or intranuclear inclusions. These inclusions are formed due to the supersaturation of the mutated proteins, which have a tendency to aggregate and impair proteostasis [36]. These inclusions have monomeric/oligomeric peptide precursors, misfolded proteins, and aggregates with toxic properties (Table 9.5).

These inclusions lead to impaired RNA metabolism, defects in the cytoskeleton, impaired intracellular trafficking, disruptions in proteostasis and autophagy, and mitochondrial abnormalities [51]. One of the prevalent gene C9ORF72, which is associated with ALS, is a hexa GGGGCC nucleotide repeat, translation of which results in generation of dipeptide repeat proteins which are toxic for the motor neurons. These toxic repeats are poly-glycine-arginine (Poly-GR), poly-glycine-alanine (Poly-GA), poly-proline-alanine (Poly-PA), poly-glycine-alanine (Poly-GA), and poly-proline-arginine (Poly-PR) [53, 237]. In a recent study using animal model, to induce C9ORF72-associated ALS, functional loss of non-canonical initiation factor eIF2D and DENR (DNA encoded RNA) in *C. elegans* and *Drosophila*, respectively, leads to a decrease in dipeptide repeats, resulting in improvements in both lifespan

Table 9.5 Components of proteostasis system impaired in amyotrophic lateral sclerosis

Gene	Protein	Associated with	Pathological hallmark	Reference
SOD1	Superoxide dismutase 1	Antioxidant enzyme that detoxifies superoxide radicals	Mutant SOD1 aggregates, oxidative stress	Berdyński et al. [14]; Rosen et al. [180]
UBQLN2	Ubiquilin-2	Involved in protein degradation and quality control	Impaired proteostasis, aggregation	Deng et al. [39]; Renaud et al. [177]
C9orf72	C9ORF72	Role in RNA processing and transport	Hexanucleotide repeat expansion leads to TDP-43 pathology	Balendra and Isaacs [9]; Smeyers et al. [206]
TARDBP	TDP-43	RNA binding and regulation of gene expression	Aggregation of TDP-43 in cytoplasmic inclusions	Mackenzie and Rademakers [124]; Suk and Rousseaux [219]
FUS	Fused in sarcoma	RNA processing and transport	Aggregation of FUS in cytoplasmic inclusions	Kamelgarn et al. [90]; Nolan et al. [152]
P4HB	Protein disulfide isomerase	Catalyzes the formation and rearrangement of disulfide bonds in proteins	Impaired protein folding, contributing to aggregation	Kwok et al. [106]; Walker [235]

and movement abnormalities [63, 210]. The presence of Poly-R in nucleoli forms RNA foci, hindering the function of RNA binding proteins and thus reduces rRNA, which impairs nucleolar functions and dynamics [53]. Cytosolic poly-GR protein is found alongside translation initiation factors and ribosome proteins in both animal models and brain tissues from C9ORF72-ALS/FTD patients. This suggests that poly-GR may influence protein translation through other mechanisms, in addition to affecting ribosome assembly in the nucleolus [237].

Deng et al. [39] showed that mutation in ubiquillin 2 (UBQLN2), a ubiquitin-like protein that acts as a shuttle mediating proteasomal degradation, is linked to familial ALS and frontoparietal dementia [39]. Proteins associated with ALS, such as optineurin, endoplasmic ATPase/p97, VAMP-associated protein B, and cyclin F, are involved in protein degradation processes and encode key components of the ubiquitin-proteasome or autophagy mutation which leads to dysregulated proteostasis [108]. The E3 ubiquitin ligase ubiquitinates TDP-43, promoting its accumulation in the cytoplasm as inclusions without degradation of the protein. The ubiquitin-conjugating enzyme and the ubiquitin-isopeptidase Y have been found to interact with TDP-43, promoting its ubiquitination and the formation of insoluble, high-molecular-weight aggregates [72].

9.3.5 Prion Diseases

Prion diseases are a group of lethal diseases of central nervous system affecting both humans and animals [169]. It was first discovered in sheep and goats as early as 1853 as "Scrapie" [216] and is transmissible. It has been now established that neurodegenerative disorders namely spongiform encephalopathy (SE), Creutzfeldt-Jakob Disease (CJD), Kuru, and Scrapie possess similar pathological symptoms and are caused by a proteinaceous agent known as "Prion" [68, 88, 169].

Prion is a membrane-bound glycoprotein predominately present on the cell membrane of neurons of the brain and spinal cord. It is encoded by PRNP gene. These prions recycle between GPI-anchored membrane and Golgi vesicle in soluble PrP^c state. The function of membrane-bound and soluble prion proteins has been extensively studied and found to be involved in neurodevelopment, formation of synapse, myelination, cellular differentiation, and circadian rhythm [203]. Structural studies using Fourier transform infrared spectroscopy and circular dichroism revealed prion proteins are rich in alpha-helix [163], while after transformation into beta sheet structure which is resistant to proteases it attains a neurotoxic scrapie form "PrP^{sc}" [31, 163, 184]. PrP^{sc}-induced neurodegenerative disorders can be genetic, sporadic, or transmitted through infection. Although not well established, the transmitted pathogenic PrP^{sc} has been postulated to have intrinsic ability to convert the soluble PrP^c into pathogenic state [7, 12, 163]. The transformation leads to loss of function along with tau hyperphosphorylation [253] and amyloid-like assembly in the neurons (Table 9.6).

Table 9.6 Components of proteostasis system impaired in prion diseases

Gene	Protein	Associated with	Pathological hallmark	Reference
PRNP	Prion protein (PrP^{c})	Cellular function related to cell signaling and protection against oxidative stress	Misfolding into pathogenic isoform (PrP^{SC}), aggregation, and neurodegeneration	Brundin et al. [22]; Polymenidou and Cleveland [167]
CWD	Chronic wasting disease-related proteins	Involved in prion propagation in affected species	Accumulation of misfolded proteins in neural tissues	Duque Velásquez et al. [44]
NGB	Neuroglobin	Oxygen transport and neuroprotection	Aggregation	Lechauve et al. [110]
TUBB	Tubulin	RNA binding and regulation of gene expression	Disruption of microtubule dynamics may be implicated in prion pathology	Nieznanski et al. [148]; Osiecka et al. [161]
APP	Amyloid precursor protein	Precursor to amyloid-β; involved in synaptic function	Interaction with prion proteins may influence pathological aggregation	Kwok et al. [106]; Miranzadeh Mahabadi and Taghibiglou [138]
SNCA	Alpha-synuclein	Involved in synaptic function and vesicle transport	Potential interaction with prions, although primarily linked to other neurodegenerative diseases	Kwok et al. [106]
MAPT	Tau	Microtubule stabilization and neuronal transport	Leads to reduction in association with tubulin protein	X.-F. Wang et al. [238]; J. Zhang and Dong [253]
BCL2	Bcl-2	Regulates apoptosis and cell survival	Aggregation, apoptosis	Rambold et al. [171]

9.3.5.1 Failure of Proteostasis in Prion Diseases

Mckinnon et al. [135], in transgenic ubiquitin-proteasomal system reporter mice model infected with prion, demonstrated various behavioral and cognitive changes such as early impairment in burrowing behavior, ataxia, impaired rightening reflex, and altered breathing. These symptom were linked with pathophysiology associated with enhanced PrP^{sc} accumulation, reactive astrocytosis, and dysfunction of proteasomal system, which lead to a reduction in half of the thalamic neurons. They have shown impaired chymotrypsin-like catalytic activity system leading to loss of substrate clearance and eventually to neural apoptosis [135]. Postmortem study of brain tissue from prion disease patients such as CJD, SE, fatal familial insomnia, and Gerstmann-Sträussler-Scheinker disease showed the presence of autophagic vacuoles in the synapse, which contributed to the synaptic loss [204]. Similar autophagic vacuoles have been shown in animal models infected with prion proteins, which was postulated to impair the cellular protein metabolism [19]. Xu et al. [247] studied

PrP^{sc} infected animals and proposed that dysfunction of autophagy may be due to oligomeric PrP^{sc} aggregates which are inaccessible to small barrel of proteasome or inhibition of the 26s component of proteasome which is responsible for the proteolytic clearance of misfolded proteins [247]. In order to study the effect of overexpression of PrP in mice model on translation initiation, Moreno et al. [141] used transgenic mice expressing 3x level of PrP than wild type and shown that accumulation of prion proteins generates prolonged UPR which in term phosphorylates and inhibits eIF2α, which further results in synaptic and neuronal loss [141].

9.4 Proteostasis Components as Therapeutic Targets in Ageing and Neurodegenerative Disorders

So far, we have discussed about the proteostasis components affected during ageing and in neurodegenerative disorders. As we know that ageing is an unavoidable pathophysiological process driven by numerous factors causing a gradual decline in the capacity to withstand stress and contribute to a range of age-related diseases. Therefore, in this section, we will discuss about some components of proteostasis as potential targets for therapeutic intervention to attain a healthy ageing and reduce pathological symptoms of neurodegenerative disorders.

9.4.1 Therapeutic Targets of Proteostasis Components in Ageing

As mentioned earlier, the components of proteostasis derailed during ageing can be potential target to lead a healthy ageing. First of all, there is protein synthesis which sharply decreases with age, remaining at low baseline levels throughout middle to late ages across various animal groups. Reducing the protein translation by inhibiting the factors (like eIF1, 2, 3, eIF4A, E, eIF5A) pharmacologically, using RNA interference or by genetic manipulation in animal models have been shown to increase lifespan and slow down the process of ageing [95].

For instance, reducing the rate of protein translation by inhibiting ribosomal biogenesis improved the physiological ageing of human mesenchymal stem cells and in mice, it helped in counteracting the onset of osteoarthritis and frailty [176]. Additionally, anti-ageing drugs like rapamycin, Torin1, and trametinib reduce translation errors, with rapamycin further extending lifespan in RPS23 hyperaccuracy mutants. [128]. Further, the protein clearance machinery, i.e., proteasome and autophagy, are also a potential target of intervention for healthy ageing. Elevated proteasome activity has been observed in exceptionally long-lived creatures like the 100-year-old humans known as centenarians, giant clams, and naked mole rats. Enhancing the proteosome activity either by overexpression or inhibiting the repressor of proteosome components (such as19s, 26s/30s) or autophagy (Atg1, 8, LAMP-2A, sestrin) has been shown to reduce oxidative stress and extends the longevity. Rapamycin, spermidine, resveratrol, and metformin are recognized pharmaceutical agents that effectively stimulate the autophagy system [139]. Also, regulating the cellular activity through diet restriction and insulin/IGF-1R signaling has been shown to be associated with enhancing the proteostasis stability and extend the lifespan in animal models and humans [233].

Numerous studies over the years have highlighted the importance of miRNAs in modulating elements of the proteostasis network, which operate at different stages, including translation, folding, and degradation. Notably, the role of miRNAs in regulating protein translation is well-documented [133]. The emerging discovery that the pioneer miRNAs lin-4 and lin-14 in *C. elegans* influence lifespan and its implication in the regulation of ageing processes has only recently been recognized. So far, many miRNAs have been found to be significantly modulated with age; several of these for, e.g., miR-71 and miR-17–92 have been identified as key regulators of ageing at the tissue or organism level, as well as in cellular senescence in *C. elegans* and in mammals, respectively [99, 207]. These findings have enhanced our understanding of specific factors that modify ageing signaling pathways across various species, from *C. elegans* to humans.

9.4.2 Therapeutic Targets of Proteostasis Components in Neurodegenerative Disorders

Many of the neurodegenerative disorders are linked with impaired correction of misfolded proteins and protein aggregate clearance machinery. Targeting these components of proteostasis network pharmacologically might help to overcome the neurotoxic effect of such aggregates. Enhancing the expression of heat shock factor-1 and molecular chaperones through heat shock response in separate studies have been shown to provide neuroprotective effect in cellular models and suppress the protein aggregation and repress the toxicity in cellular and animal model of polyQ aggregation [107, 221], respectively. Furthermore, distinct chaperones have been identified to provide neuroprotection when overexpressed in cellular or animal models of various neurodegenerative disorders. In numerous instances, specific chaperones seem to offer protection against multiple disease-associated proteins. Moreover, repressing the activity of deubiquitinating enzymes has been shown to overcome the toxicity of protein aggregates by elevating the proteasome activity [111]. In a study by Sarkar et al. [191], a novel small molecule was identified which activates the process of autophagy similar to rapamycin showed enhanced clearance of synuclein and huntingtin aggregates in cellular and animal models [191]. Nonetheless, mTOR is a versatile protein that governs numerous cellular processes beyond protein degradation, and so far, mTOR inhibitors in clinical trials have resulted in unpredictable and unwanted side effects. Removal of extracellular misfolded proteins through monoclonal antibodies, which help in clearance, could be an effective yet challenging approach [55].

miRNAs (miR-7/miR-153, miR 106a, 520c, miR-29a, miR-9, miR-196a, miR-206) are promising biomolecules that can regulate gene expression by binding to mRNA, either inhibiting its translation or causing its degradation. Additionally, they can interfere with protein synthesis, promote the degradation of specific mRNA targets, or affect apoptosis signaling in various neurodegenerative disorders [89, 114, 123].

9.5 Conclusion

Proteostasis is important for healthy living. Homeostatic balance among fundamental processes like protein synthesis, protein modifications, protein folding, and protein clearance is crucial for sustenance. This balance is disturbed during ageing and neurodegenerative disorders (Fig. 9.4).

Age-related decline in proteostasis is inevitable but therapeutic interventions targeting components of proteostasis can enhance life quality. Proteostasis dysregulation is a major cause of neurodegenerative disorders which are often related to ageing. This imbalance culminates with an accumulation of toxic protein aggregates inside cells, which ultimately leads to cellular damage and death. This process

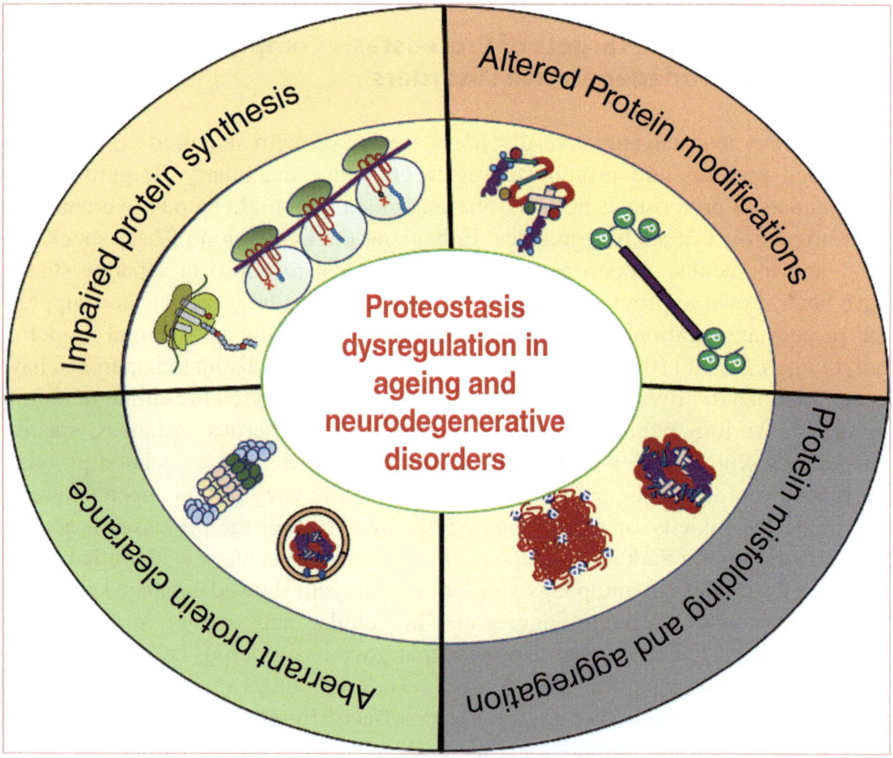

Fig. 9.4 Proteostasis dysregulation in ageing and neurodegenerative disorders. Four fundamental processes, namely, protein synthesis, protein modifications, protein folding, and protein clearance, are affected due to proteostasis dysregulation leading to occurrence of age-related pathologies and neurodegenerative disorders

involves complex pathways or signaling cascades involving a large number of signaling components that form an important part of the proteostasis network. Therefore, proteostasis components are also important therapeutic targets for the treatment of neurodegenerative disorders. This chapter stresses on the importance of proteostasis for normal functioning of living system. It highlights the exquisite and complex regulation of living systems at the cellular level.

References

1. Al Mamun A, Uddin MS, Kabir MT, Khanum S, Sarwar MS, Mathew B, Rauf A, Ahmed M, Ashraf GM. Exploring the promise of targeting ubiquitin-proteasome system to combat Alzheimer's disease. Neurotox Res. 2020;38(1):8–17. https://doi.org/10.1007/s12640-020-00185-1.
2. Albanèse V, Yam AY-W, Baughman J, Parnot C, Frydman J. Systems analyses reveal two chaperone networks with distinct functions in eukaryotic cells. Cell. 2006;124(1):75–88. https://doi.org/10.1016/j.cell.2005.11.039.
3. Almeida CG, Takahashi RH, Gouras GK. β-Amyloid accumulation impairs multivesicular body sorting by inhibiting the ubiquitin-proteasome system. J Neurosci. 2006;26(16):4277–88. https://doi.org/10.1523/JNEUROSCI.5078-05.2006.
4. Anglade P, Vyas S, Javoy-Agid F, Herrero MT, Michel PP, Marquez J, Mouatt-Prigent A, Ruberg M, Hirsch EC, Agid Y. Apoptosis and autophagy in nigral neurons of patients with Parkinson's disease. Histol Histopathol. 1997;12(1):25–31.
5. Anisimova AS, Meerson MB, Gerashchenko MV, Kulakovskiy IV, Dmitriev SE, Gladyshev VN. Multifaceted deregulation of gene expression and protein synthesis with age. Proc Natl Acad Sci. 2020;117(27):15581–90. https://doi.org/10.1073/pnas.2001788117.
6. Armstrong A, Mattsson N, Appelqvist H, Janefjord C, Sandin L, Agholme L, Olsson B, Svensson S, Blennow K, Zetterberg H, Kågedal K. Lysosomal network proteins as potential novel CSF biomarkers for Alzheimer's disease. NeuroMolecular Med. 2014;16(1):150–60. https://doi.org/10.1007/s12017-013-8269-3.
7. Atkinson CJ, Zhang K, Munn AL, Wiegmans A, Wei MQ. Prion protein scrapie and the normal cellular prion protein. Prion. 2016;10(1):63–82. https://doi.org/10.1080/19336896.2015.1110293.
8. Baldensperger T, Eggen M, Kappen J, Winterhalter PR, Pfirrmann T, Glomb MA. Comprehensive analysis of posttranslational protein modifications in aging of subcellular compartments. Sci Rep. 2020;10(1):7596. https://doi.org/10.1038/s41598-020-64265-0.
9. Balendra R, Isaacs AM. C9orf72-mediated ALS and FTD: multiple pathways to disease. Nat Rev Neurol. 2018;14(9):544–58. https://doi.org/10.1038/s41582-018-0047-2.
10. Barmada SJ, Ju S, Arjun A, Batarse A, Archbold HC, Peisach D, Li X, Zhang Y, Tank EMH, Qiu H, Huang EJ, Ringe D, Petsko GA, Finkbeiner S. Amelioration of toxicity in neuronal models of amyotrophic lateral sclerosis by hUPF1. Proc Natl Acad Sci USA. 2015;112(25):7821–6. https://doi.org/10.1073/pnas.1509744112.
11. Barrachina M, Castaño E, Dalfó E, Maes T, Buesa C, Ferrer I. Reduced ubiquitin C-terminal hydrolase-1 expression levels in dementia with Lewy bodies. Neurobiol Dis. 2006;22(2):265–73. https://doi.org/10.1016/j.nbd.2005.11.005.
12. Benetti F, Legname G. De novo mammalian prion synthesis. Prion. 2009;3(4):213–9. https://doi.org/10.4161/pri.3.4.10181.
13. Bennett EJ, Shaler TA, Woodman B, Ryu K-Y, Zaitseva TS, Becker CH, Bates GP, Schulman H, Kopito RR. Global changes to the ubiquitin system in Huntington's disease. Nature. 2007;448(7154):704–8. https://doi.org/10.1038/nature06022.

14. Berdyński M, Miszta P, Safranow K, Andersen PM, Morita M, Filipek S, Żekanowski C, Kuźma-Kozakiewicz M. SOD1 mutations associated with amyotrophic lateral sclerosis analysis of variant severity. Sci Rep. 2022;12(1):103. https://doi.org/10.1038/s41598-021-03891-8.
15. Bertolotti A, Zhang Y, Hendershot LM, Harding HP, Ron D. Dynamic interaction of BiP and ER stress transducers in the unfolded-protein response. Nat Cell Biol. 2000;2(6):326–32. https://doi.org/10.1038/35014014.
16. Bhattacharyya K. The story of George Huntington and his disease. Ann Indian Acad Neurol. 2016;19(1):25. https://doi.org/10.4103/0972-2327.175425.
17. Bhattarai KR, Chaudhary M, Kim H-R, Chae H-J. Endoplasmic reticulum (ER) stress response failure in diseases. Trends Cell Biol. 2020;30(9):672–5. https://doi.org/10.1016/j.tcb.2020.05.004.
18. Blauwendraat C, Reed X, Krohn L, Heilbron K, Bandres-Ciga S, Tan M, Gibbs JR, Hernandez DG, Kumaran R, Langston R, Bonet-Ponce L, Alcalay RN, Hassin-Baer S, Greenbaum L, Iwaki H, Leonard HL, Grenn FP, Ruskey JA, Sabir M, et al. Genetic modifiers of risk and age at onset in GBA associated Parkinson's disease and Lewy body dementia. Brain. 2020;143(1):234–48. https://doi.org/10.1093/brain/awz350.
19. Boellaard JW, Kao M, Schlote W, Diringer H. Neuronal autophagy in experimental scrapie. Acta Neuropathol. 1991;82(3):225–8. https://doi.org/10.1007/BF00294449.
20. Brothers HM, Gosztyla ML, Robinson SR. The physiological roles of amyloid-β peptide hint at new ways to treat Alzheimer's disease. Front Aging Neurosci. 2018;10:118. https://doi.org/10.3389/fnagi.2018.00118.
21. Brown MK, Naidoo N. The endoplasmic reticulum stress response in aging and age-related diseases. Front Physiol. 2012;3:263. https://doi.org/10.3389/fphys.2012.00263.
22. Brundin P, Melki R, Kopito R. Prion-like transmission of protein aggregates in neurodegenerative diseases. Nat Rev Mol Cell Biol. 2010;11(4):301–7. https://doi.org/10.1038/nrm2873.
23. Bu G. Apolipoprotein E and its receptors in Alzheimer's disease: pathways, pathogenesis and therapy. Nat Rev Neurosci. 2009;10(5):333–44. https://doi.org/10.1038/nrn2620.
24. Budovskaya YV, Wu K, Southworth LK, Jiang M, Tedesco P, Johnson TE, Kim SK. An elt-3/elt-5/elt-6 GATA transcription circuit guides aging in C. elegans. Cell. 2008;134(2):291–303. https://doi.org/10.1016/j.cell.2008.05.044.
25. Bukau B, Horwich AL. The Hsp70 and Hsp60 chaperone machines. Cell. 1998;92(3):351–66. https://doi.org/10.1016/s0092-8674(00)80928-9.
26. Bukhatwa S, Zeng B-Y, Rose S, Jenner P. A comparison of changes in proteasomal subunit expression in the substantia nigra in Parkinson's disease, multiple system atrophy and progressive supranuclear palsy. Brain Res. 2010;1326:174–83. https://doi.org/10.1016/j.brainres.2010.02.045.
27. Calfon M, Zeng H, Urano F, Till JH, Hubbard SR, Harding HP, Clark SG, Ron D. IRE1 couples endoplasmic reticulum load to secretory capacity by processing the XBP-1 mRNA. Nature. 2002;415(6867):92–6. https://doi.org/10.1038/415092a.
28. Carmichael J, Sugars KL, Bao YP, Rubinsztein DC. Glycogen synthase kinase-3β inhibitors prevent cellular polyglutamine toxicity caused by the Huntington's disease mutation. J Biol Chem. 2002;277(37):33791–8. https://doi.org/10.1074/jbc.M204861200.
29. Cataldo AM, Peterhoff CM, Troncoso JC, Gomez-Isla T, Hyman BT, Nixon RA. Endocytic pathway abnormalities precede amyloid beta deposition in sporadic Alzheimer's disease and Down syndrome: differential effects of APOE genotype and presenilin mutations. Am J Pathol. 2000;157(1):277–86. https://doi.org/10.1016/s0002-9440(10)64538-5.
30. Cataldo AM, Peterhoff CM, Schmidt SD, Terio NB, Duff K, Beard M, Mathews PM, Nixon RA. Presenilin mutations in familial Alzheimer disease and transgenic mouse models accelerate neuronal lysosomal pathology. J Neuropathol Exp Neurol. 2004;63(8):821–30. https://doi.org/10.1093/jnen/63.8.821.
31. Caughey BW, Dong A, Bhat KS, Ernst D, Hayes SF, Caughey WS. Secondary structure analysis of the scrapie-associated protein PrP 27-30 in water by infrared spectroscopy. Biochemistry. 1991;30(31):7672–80. https://doi.org/10.1021/bi00245a003.

32. Chai Y, Koppenhafer SL, Bonini NM, Paulson HL. Analysis of the role of heat shock protein (Hsp) molecular chaperones in polyglutamine disease. J Neurosci. 1999;19(23):10338–47. https://doi.org/10.1523/JNEUROSCI.19-23-10338.1999.
33. Chao C-C, Huang C-L, Cheng J-J, Chiou C-T, Lee I-J, Yang Y-C, Hsu T-H, Yei C-E, Lin P-Y, Chen J-J, Huang N-K. SRT1720 as an SIRT1 activator for alleviating paraquat-induced models of Parkinson's disease. Redox Biol. 2022;58:102534. https://doi.org/10.1016/j.redox.2022.102534.
34. Chen S, Berthelier V, Hamilton JB, O'Nuallai B, Wetzel R. Amyloid-like features of polyglutamine aggregates and their assembly kinetics. Biochemistry. 2002;41(23):7391–9. https://doi.org/10.1021/bi011772q.
35. Cherubini M, Ginés S. Mitochondrial fragmentation in neuronal degeneration: toward an understanding of HD striatal susceptibility. Biochem Biophys Res Commun. 2017;483(4):1063–8. https://doi.org/10.1016/j.bbrc.2016.08.042.
36. Chisholm CG, Lum JS, Farrawell NE, Yerbury JJ. Ubiquitin homeostasis disruption, a common cause of proteostasis collapse in amyotrophic lateral sclerosis? Neural Regen Res. 2022;17(10):2218–20. https://doi.org/10.4103/1673-5374.335786.
37. Cozachenco D, Ribeiro FC, Ferreira ST. Defective proteostasis in Alzheimer's disease. Ageing Res Rev. 2023;85:101862. https://doi.org/10.1016/j.arr.2023.101862.
38. Dehay B, Bové J, Rodríguez-Muela N, Perier C, Recasens A, Boya P, Vila M. Pathogenic lysosomal depletion in Parkinson's disease. J Neurosci. 2010;30(37):12535–44. https://doi.org/10.1523/JNEUROSCI.1920-10.2010.
39. Deng H-X, Chen W, Hong S-T, Boycott KM, Gorrie GH, Siddique N, Yang Y, Fecto F, Shi Y, Zhai H, Jiang H, Hirano M, Rampersaud E, Jansen GH, Donkervoort S, Bigio EH, Brooks BR, Ajroud K, Sufit RL, et al. Mutations in UBQLN2 cause dominant X-linked juvenile and adult-onset ALS and ALS/dementia. Nature. 2011;477(7363):211–5. https://doi.org/10.1038/nature10353.
40. Díaz-Villanueva JF, Díaz-Molina R, García-González V. Protein folding and mechanisms of proteostasis. Int J Mol Sci. 2015;16(8):17193–230. https://doi.org/10.3390/ijms160817193.
41. Ding Q, Markesbery WR, Chen Q, Li F, Keller JN. Ribosome dysfunction is an early event in Alzheimer's disease. J Neurosci. 2005;25(40):9171–5. https://doi.org/10.1523/JNEUROSCI.3040-05.2005.
42. Doherty KM, Hardy J. Parkin disease and the Lewy body conundrum. Mov Disord. 2013;28(6):702–4. https://doi.org/10.1002/mds.25486.
43. Dukan S, Farewell A, Ballesteros M, Taddei F, Radman M, Nyström T. Protein oxidation in response to increased transcriptional or translational errors. Proc Natl Acad Sci USA. 2000;97(11):5746–9. https://doi.org/10.1073/pnas.100422497.
44. Duque Velásquez C, Kim C, Herbst A, Daude N, Garza MC, Wille H, Aiken J, McKenzie D. Deer prion proteins modulate the emergence and adaptation of chronic wasting disease strains. J Virol. 2015;89(24):12362–73. https://doi.org/10.1128/JVI.02010-15.
45. Ebert T, Tran N, Schurgers L, Stenvinkel P, Shiels PG. Ageing – oxidative stress, PTMs and disease. Mol Asp Med. 2022;86:101099. https://doi.org/10.1016/j.mam.2022.101099.
46. El Manaa W, Duplan E, Goiran T, Lauritzen I, Vaillant Beuchot L, Lacas-Gervais S, Morais VA, You H, Qi L, Salazar M, Ozcan U, Chami M, Checler F, Alves da Costa C. Transcription- and phosphorylation-dependent control of a functional interplay between XBP1s and PINK1 governs mitophagy and potentially impacts Parkinson disease pathophysiology. Autophagy. 2021;17(12):4363–85. https://doi.org/10.1080/15548627.2021.1917129.
47. Erjavec N, Larsson L, Grantham J, Nyström T. Accelerated aging and failure to segregate damaged proteins in Sir2 mutants can be suppressed by overproducing the protein aggregation-remodeling factor Hsp104p. Genes Dev. 2007;21(19):2410–21. https://doi.org/10.1101/gad.439307.
48. Eshraghi M, Adlimoghaddam A, Mahmoodzadeh A, Sharifzad F, Yasavoli-Sharahi H, Lorzadeh S, Albensi BC, Ghavami S. Alzheimer's disease pathogenesis: role of autophagy and mitophagy focusing in microglia. Int J Mol Sci. 2021;22(7):3330. https://doi.org/10.3390/ijms22073330.

49. Espargaró A, Sabate R. Phosphorylation-driven aggregative proteins in neurodegenerative diseases: implications and therapeutics. Neural Regen Res. 2024;19(5):966–8. https://doi.org/10.4103/1673-5374.382250.
50. Farrer M, Chan P, Chen R, Tan L, Lincoln S, Hernandez D, Forno L, Gwinn-Hardy K, Petrucelli L, Hussey J, Singleton A, Tanner C, Hardy J, Langston JW. Lewy bodies and parkinsonism in families with parkin mutations. Ann Neurol. 2001;50(3):293–300. https://doi.org/10.1002/ana.1132.
51. Feldman EL, Goutman SA, Petri S, Mazzini L, Savelieff MG, Shaw PJ, Sobue G. Amyotrophic lateral sclerosis. Lancet. 2022;400(10360):1363–80. https://doi.org/10.1016/S0140-6736(22)01272-7.
52. Ferrer I, Andrés-Benito P, Ausín K, Pamplona R, del Rio JA, Fernández-Irigoyen J, Santamaría E. Dysregulated protein phosphorylation: a determining condition in the continuum of brain aging and Alzheimer's disease. Brain Pathol. 2021;31(6):e12996. https://doi.org/10.1111/bpa.12996.
53. Freibaum BD, Taylor JP. The role of dipeptide repeats in C9ORF72-related ALS-FTD. Front Mol Neurosci. 2017;10:35. https://doi.org/10.3389/fnmol.2017.00035.
54. Fu Y, Wu P, Pan Y, Sun X, Yang H, Difiglia M, Lu B. A toxic mutant huntingtin species is resistant to selective autophagy. Nat Chem Biol. 2017;13(11):1152–4. https://doi.org/10.1038/nchembio.2461.
55. Fuller JP, Stavenhagen JB, Christensen S, Kartberg F, Glennie MJ, Teeling JL. Comparing the efficacy and neuroinflammatory potential of three anti-abeta antibodies. Acta Neuropathol. 2015;130(5):699–711. https://doi.org/10.1007/s00401-015-1484-2.
56. Furukawa Y, Vigouroux S, Wong H, Guttman M, Rajput AH, Ang L, Briand M, Kish SJ, Briand Y. Brain proteasomal function in sporadic Parkinson's disease and related disorders. Ann Neurol. 2002;51(6):779–82. https://doi.org/10.1002/ana.10207.
57. Gadhave K, Bolshette N, Ahire A, Pardeshi R, Thakur K, Trandafir C, Istrate A, Ahmed S, Lahkar M, Muresanu DF, Balea M. The ubiquitin proteasomal system: a potential target for the management of Alzheimer's disease. J Cell Mol Med. 2016;20(7):1392–407. https://doi.org/10.1111/jcmm.12817.
58. Ge P, Dawson VL, Dawson TM. PINK1 and Parkin mitochondrial quality control: a source of regional vulnerability in Parkinson's disease. Mol Neurodegener. 2020;15(1):20. https://doi.org/10.1186/s13024-020-00367-7.
59. German DC, Manaye K, Smith WK, Woodward DJ, Saper CB. Midbrain dopaminergic cell loss in Parkinson's disease: computer visualization. Ann Neurol. 1989;26(4):507–14. https://doi.org/10.1002/ana.410260403.
60. Goetz CG. The history of Parkinson's disease: early clinical descriptions and neurological therapies. Cold Spring Harb Perspect Med. 2011;1(1):a008862. https://doi.org/10.1101/cshperspect.a008862.
61. Golden TR, Melov S. Microarray analysis of gene expression with age in individual nematodes. Aging Cell. 2004;3(3):111–24. https://doi.org/10.1111/j.1474-9728.2004.00095.x.
62. Gomez-Pastor R, Burchfiel ET, Thiele DJ. Regulation of heat shock transcription factors and their roles in physiology and disease. Nat Rev Mol Cell Biol. 2018;19(1):4–19. https://doi.org/10.1038/nrm.2017.73.
63. Green KM, Miller SL, Malik I, Todd PK. Non-canonical initiation factors modulate repeat-associated non-AUG translation. Hum Mol Genet. 2022;31(15):2521–34. https://doi.org/10.1093/hmg/ddac021.
64. Guedes-Dias P, Pinho BR, Soares TR, de Proença J, Duchen MR, Oliveira JMA. Mitochondrial dynamics and quality control in Huntington's disease. Neurobiol Dis. 2016;90:51–7. https://doi.org/10.1016/j.nbd.2015.09.008.
65. Guo Q, Huang B, Cheng J, Seefelder M, Engler T, Pfeifer G, Oeckl P, Otto M, Moser F, Maurer M, Pautsch A, Baumeister W, Fernández-Busnadiego R, Kochanek S. The cryo-electron microscopy structure of huntingtin. Nature. 2018;555(7694):117–20. https://doi.org/10.1038/nature25502.

66. Guo J, Huang X, Dou L, Yan M, Shen T, Tang W, Li J. Aging and aging-related diseases: from molecular mechanisms to interventions and treatments. Signal Transduct Target Ther. 2022;7(1):391. https://doi.org/10.1038/s41392-022-01251-0.
67. Gusella JF, Lee J-M, MacDonald ME. Huntington's disease: nearly four decades of human molecular genetics. Hum Mol Genet. 2021;30(R2):R254–63. https://doi.org/10.1093/hmg/ddab170.
68. Hadlow WJ, Prusiner SB, Kennedy RC, Race RE. Brain tissue from persons dying of Creutzfeldt-Jakob disease causes scrapie-like encephalopathy in goats. Ann Neurol. 1980;8(6):628–32. https://doi.org/10.1002/ana.410080615.
69. Hamer G, Matilainen O, Holmberg CI. A photoconvertible reporter of the ubiquitin-proteasome system in vivo. Nat Methods. 2010;7(6):473–8. https://doi.org/10.1038/nmeth.1460.
70. Hampel H, Hardy J, Blennow K, Chen C, Perry G, Kim SH, Villemagne VL, Aisen P, Vendruscolo M, Iwatsubo T, Masters CL, Cho M, Lannfelt L, Cummings JL, Vergallo A. The amyloid-β pathway in Alzheimer's disease. Mol Psychiatry. 2021;26(10):5481–503. https://doi.org/10.1038/s41380-021-01249-0.
71. Han D, Zheng W, Wang X, Chen Z. Proteostasis of α-synuclein and its role in the pathogenesis of Parkinson's disease. Front Cell Neurosci. 2020;14:45. https://doi.org/10.3389/fncel.2020.00045.
72. Hans F, Fiesel FC, Strong JC, Jäckel S, Rasse TM, Geisler S, Springer W, Schulz JB, Voigt A, Kahle PJ. UBE2E ubiquitin-conjugating enzymes and ubiquitin isopeptidase Y regulate TDP-43 protein ubiquitination. J Biol Chem. 2014;289(27):19164–79. https://doi.org/10.1074/jbc.M114.561704.
73. Hansen M, Rubinsztein DC, Walker DW. Autophagy as a promoter of longevity: insights from model organisms. Nat Rev Mol Cell Biol. 2018;19(9):579–93. https://doi.org/10.1038/s41580-018-0033-y.
74. Harding RJ, Tong Y. Proteostasis in Huntington's disease: disease mechanisms and therapeutic opportunities. Acta Pharmacol Sin. 2018;39(5):754–69. https://doi.org/10.1038/aps.2018.11.
75. Hermel E, Gafni J, Propp SS, Leavitt BR, Wellington CL, Young JE, Hackam AS, Logvinova AV, Peel AL, Chen SF, Hook V, Singaraja R, Krajewski S, Goldsmith PC, Ellerby HM, Hayden MR, Bredesen DE, Ellerby LM. Specific caspase interactions and amplification are involved in selective neuronal vulnerability in Huntington's disease. Cell Death Differ. 2004;11(4):424–38. https://doi.org/10.1038/sj.cdd.4401358.
76. Hetz C. The unfolded protein response: controlling cell fate decisions under ER stress and beyond. Nat Rev Mol Cell Biol. 2012;13(2):89–102. https://doi.org/10.1038/nrm3270.
77. Hipp MS, Kasturi P, Hartl FU. The proteostasis network and its decline in ageing. Nat Rev Mol Cell Biol. 2019;20(7):421–35. https://doi.org/10.1038/s41580-019-0101-y.
78. Hippius H, Neundörfer G. The discovery of Alzheimer's disease. Dialogues Clin Neurosci. 2003;5(1):101–8. https://doi.org/10.31887/DCNS.2003.5.1/hhippius.
79. Hrabos D, Poggiolini I, Civitelli L, Galli E, Esapa C, Saarma M, Lindholm P, Parkkinen L. Unfolded protein response markers Grp78 and eIF2alpha are upregulated with increasing alpha-synuclein levels in Lewy body disease. Neuropathol Appl Neurobiol. 2024;50(4):e12999. https://doi.org/10.1111/nan.12999.
80. Huang N, Erie C, Lu ML, Wei J. Aberrant subcellular localization of SQSTM1/p62 contributes to increased vulnerability to proteotoxic stress recovery in Huntington's disease. Mol Cell Neurosci. 2018a;88:43–52. https://doi.org/10.1016/j.mcn.2017.12.005.
81. Huang J, Wu Z, Wang J, Zhang X. Quantitative phosphoproteomics reveals GTBP-1 regulating C. elegans lifespan at different environmental temperatures. Biochem Biophys Res Commun. 2018b;503(3):1962–7. https://doi.org/10.1016/j.bbrc.2018.07.142.
82. Inda MC, Joshi S, Wang T, Bolaender A, Gandu S, Koren Iii J, Che AY, Taldone T, Yan P, Sun W, Uddin M, Panchal P, Riolo M, Shah S, Barlas A, Xu K, Chan LYL, Gruzinova A, Kishinevsky S, et al. The epichaperome is a mediator of toxic hippocampal stress and

leads to protein connectivity-based dysfunction. Nat Commun. 2020;11(1):319. https://doi.org/10.1038/s41467-019-14082-5.
83. Jan A, Jansonius B, Delaidelli A, Bhanshali F, An YA, Ferreira N, Smits LM, Negri GL, Schwamborn JC, Jensen PH, Mackenzie IR, Taubert S, Sorensen PH. Activity of translation regulator eukaryotic elongation factor-2 kinase is increased in Parkinson disease brain and its inhibition reduces alpha synuclein toxicity. Acta Neuropathol Commun. 2018;6(1):54. https://doi.org/10.1186/s40478-018-0554-9.
84. Janssens GE, Meinema AC, González J, Wolters JC, Schmidt A, Guryev V, Bischoff R, Wit EC, Veenhoff LM, Heinemann M. Protein biogenesis machinery is a driver of replicative aging in yeast. elife. 2015;4:e08527. https://doi.org/10.7554/eLife.08527.
85. Jiang A, Handley RR, Lehnert K, Snell RG. From pathogenesis to therapeutics: a review of 150 years of Huntington's disease research. Int J Mol Sci. 2023;24(16):13021. https://doi.org/10.3390/ijms241613021.
86. Jimenez-Sanchez M, Thomson F, Zavodszky E, Rubinsztein DC. Autophagy and polyglutamine diseases. Prog Neurobiol. 2012;97(2):67–82. https://doi.org/10.1016/j.pneurobio.2011.08.013.
87. Jimenez-Sanchez M, Licitra F, Underwood BR, Rubinsztein DC. Huntington's disease: mechanisms of pathogenesis and therapeutic strategies. Cold Spring Harb Perspect Med. 2017;7(7):a024240. https://doi.org/10.1101/cshperspect.a024240.
88. Joshi T, Ahuja N. The prion basis of progressive neurodegenerative disorders. Interdiscip Perspect Infect Dis. 2023;2023:6687264. https://doi.org/10.1155/2023/6687264.
89. Junn E, Mouradian MM. MicroRNAs in neurodegenerative diseases and their therapeutic potential. Pharmacol Ther. 2012;133(2):142–50. https://doi.org/10.1016/j.pharmthera.2011.10.002.
90. Kamelgarn M, Chen J, Kuang L, Jin H, Kasarskis EJ, Zhu H. ALS mutations of FUS suppress protein translation and disrupt the regulation of nonsense-mediated decay. Proc Natl Acad Sci USA. 2018;115(51):E11904–13. https://doi.org/10.1073/pnas.1810413115.
91. Kaufman RJ. Stress signaling from the lumen of the endoplasmic reticulum: coordination of gene transcriptional and translational controls. Genes Dev. 1999;13(10):1211–33. https://doi.org/10.1101/gad.13.10.1211.
92. Kaushik S, Cuervo AM. Proteostasis and aging. Nat Med. 2015;21(12):1406–15. https://doi.org/10.1038/nm.4001.
93. Kelleher RJ, Shen J. Presenilin-1 mutations and Alzheimer's disease. Proc Natl Acad Sci USA. 2017;114(4):629–31. https://doi.org/10.1073/pnas.1619574114.
94. Kenyon C. The plasticity of aging: insights from long-lived mutants. Cell. 2005;120(4):449–60. https://doi.org/10.1016/j.cell.2005.02.002.
95. Kim HS, Pickering AM. Protein translation paradox: implications in translational regulation of aging. Front Cell Dev Biol. 2023;11:1129281. https://doi.org/10.3389/fcell.2023.1129281.
96. Kim YJ, Yi Y, Sapp E, Wang Y, Cuiffo B, Kegel KB, Qin ZH, Aronin N, DiFiglia M. Caspase 3-cleaved N-terminal fragments of wild-type and mutant huntingtin are present in normal and Huntington's disease brains, associate with membranes, and undergo calpain-dependent proteolysis. Proc Natl Acad Sci USA. 2001;98(22):12784–9. https://doi.org/10.1073/pnas.221451398.
97. Kim D, Nguyen MD, Dobbin MM, Fischer A, Sananbenesi F, Rodgers JT, Delalle I, Baur JA, Sui G, Armour SM, Puigserver P, Sinclair DA, Tsai L-H. SIRT1 deacetylase protects against neurodegeneration in models for Alzheimer's disease and amyotrophic lateral sclerosis. EMBO J. 2007;26(13):3169–79. https://doi.org/10.1038/sj.emboj.7601758.
98. Kim HS, Parker DJ, Hardiman MM, Munkácsy E, Jiang N, Rogers AN, Bai Y, Brent C, Mobley JA, Austad SN, Pickering AM. Early-adulthood spike in protein translation drives aging via juvenile hormone/germline signaling. Nat Commun. 2023;14(1):5021. https://doi.org/10.1038/s41467-023-40618-x.
99. Kinser HE, Pincus Z. MicroRNAs as modulators of longevity and the aging process. Hum Genet. 2020;139(3):291–308. https://doi.org/10.1007/s00439-019-02046-0.

100. Kirstein J, Morito D, Kakihana T, Sugihara M, Minnen A, Hipp MS, Nussbaum-Krammer C, Kasturi P, Hartl FU, Nagata K, Morimoto RI. Proteotoxic stress and ageing triggers the loss of redox homeostasis across cellular compartments. EMBO J. 2015;34(18):2334–49. https://doi.org/10.15252/embj.201591711.
101. Koyuncu S, Loureiro R, Lee HJ, Wagle P, Krueger M, Vilchez D. Rewiring of the ubiquitinated proteome determines ageing in C. elegans. Nature. 2021;596(7871):285–90. https://doi.org/10.1038/s41586-021-03781-z.
102. Kreis P, Gallrein C, Rojas-Puente E, Mack TGA, Kroon C, Dinkel V, Willmes C, Murk K, tom-Dieck S, Schuman EM, Kirstein J, Eickholt BJ. ATM phosphorylation of the actin-binding protein drebrin controls oxidation stress-resistance in mammalian neurons and C. elegans. Nat Commun. 2019;10(1):486. https://doi.org/10.1038/s41467-019-08420-w.
103. Krisko A, Radman M. Phenotypic and genetic consequences of protein damage. PLoS Genet. 2013;9(9):e1003810. https://doi.org/10.1371/journal.pgen.1003810.
104. Kumar R, Jangir DK, Verma G, Shekhar S, Hanpude P, Kumar S, Kumari R, Singh N, Sarovar Bhavesh N, Ranjan Jana N, Kanti Maiti T. S-nitrosylation of UCHL1 induces its structural instability and promotes α-synuclein aggregation. Sci Rep. 2017;7(1):44558. https://doi.org/10.1038/srep44558.
105. Kurosawa M, Matsumoto G, Kino Y, Okuno M, Kurosawa-Yamada M, Washizu C, Taniguchi H, Nakaso K, Yanagawa T, Warabi E, Shimogori T, Sakurai T, Hattori N, Nukina N. Depletion of p62 reduces nuclear inclusions and paradoxically ameliorates disease phenotypes in Huntington's model mice. Hum Mol Genet. 2015;24(4):1092–105. https://doi.org/10.1093/hmg/ddu522.
106. Kwok CT, Morris AG, Frampton J, Smith B, Shaw CE, de Belleroche J. Association studies indicate that protein disulfide isomerase is a risk factor in amyotrophic lateral sclerosis. Free Radic Biol Med. 2013;58:81–6. https://doi.org/10.1016/j.freeradbiomed.2013.01.001.
107. Labbadia J, Morimoto RI. The biology of proteostasis in aging and disease. Annu Rev Biochem. 2015;84(1):435–64. https://doi.org/10.1146/annurev-biochem-060614-033955.
108. Lambert-Smith IA, Shephard VK, McAlary L, Yerbury JJ, Saunders DN. High-content analysis of proteostasis capacity in cellular models of amyotrophic lateral sclerosis (ALS). Sci Rep. 2024;14(1):13844. https://doi.org/10.1038/s41598-024-64366-0.
109. Lapierre LR, De Magalhaes Filho CD, McQuary PR, Chu C-C, Visvikis O, Chang JT, Gelino S, Ong B, Davis AE, Irazoqui JE, Dillin A, Hansen M. The TFEB orthologue HLH-30 regulates autophagy and modulates longevity in Caenorhabditis elegans. Nat Commun. 2013;4:2267. https://doi.org/10.1038/ncomms3267.
110. Lechauve C, Rezaei H, Celier C, Kiger L, Corral-Debrinski M, Noinville S, Chauvierre C, Hamdane D, Pato C, Marden MC. Neuroglobin and prion cellular localization: investigation of a potential interaction. J Mol Biol. 2009;388(5):968–77. https://doi.org/10.1016/j.jmb.2009.03.047.
111. Lee B-H, Lee MJ, Park S, Oh D-C, Elsasser S, Chen P-C, Gartner C, Dimova N, Hanna J, Gygi SP, Wilson SM, King RW, Finley D. Enhancement of proteasome activity by a small-molecule inhibitor of USP14. Nature. 2010a;467(7312):179–84. https://doi.org/10.1038/nature09299.
112. Lee J-H, Yu WH, Kumar A, Lee S, Mohan PS, Peterhoff CM, Wolfe DM, Martinez-Vicente M, Massey AC, Sovak G, Uchiyama Y, Westaway D, Cuervo AM, Nixon RA. Lysosomal proteolysis and autophagy require presenilin 1 and are disrupted by Alzheimer-related PS1 mutations. Cell. 2010b;141(7):1146–58. https://doi.org/10.1016/j.cell.2010.05.008.
113. Lehtonen Š, Sonninen T-M, Wojciechowski S, Goldsteins G, Koistinaho J. Dysfunction of cellular proteostasis in Parkinson's disease. Front Neurosci. 2019;13:457. https://doi.org/10.3389/fnins.2019.00457.
114. Li S, Lei Z, Sun T. The role of microRNAs in neurodegenerative diseases: a review. Cell Biol Toxicol. 2023;39(1):53–83. https://doi.org/10.1007/s10565-022-09761-x.
115. Li W, Li H-L, Wang J-Z, Liu R, Wang X. Abnormal protein post-translational modifications induces aggregation and abnormal deposition of protein, mediating neurodegenerative diseases. Cell Biosci. 2024;14(1):22. https://doi.org/10.1186/s13578-023-01189-y.

116. Liang V, Ullrich M, Lam H, Chew YL, Banister S, Song X, Zaw T, Kassiou M, Götz J, Nicholas HR. Altered proteostasis in aging and heat shock response in C. elegans revealed by analysis of the global and de novo synthesized proteome. Cell Mol Life Sci. 2014;71(17):3339–61. https://doi.org/10.1007/s00018-014-1558-7.
117. Lipton JO, Sahin M. The neurology of mTOR. Neuron. 2014;84(2):275–91. https://doi.org/10.1016/j.neuron.2014.09.034.
118. Liu C-C, Liu C-C, Kanekiyo T, Xu H, Bu G. Apolipoprotein E and Alzheimer disease: risk, mechanisms and therapy. Nat Rev Neurol. 2013;9(2):106–18. https://doi.org/10.1038/nrneurol.2012.263.
119. Llorens-Martín M, Jurado J, Hernández F, Avila J. GSK-3β, a pivotal kinase in Alzheimer disease. Front Mol Neurosci. 2014;7:46. https://doi.org/10.3389/fnmol.2014.00046.
120. Logan R, Dubel-Haag J, Schcolnicov N, Miller SJ. Novel genetic signatures associated with sporadic amyotrophic lateral sclerosis. Front Genet. 2022;13:851496. https://doi.org/10.3389/fgene.2022.851496.
121. López Salon M, Morelli L, Castaño EM, Soto EF, Pasquini JM. Defective ubiquitination of cerebral proteins in Alzheimer's disease. J Neurosci Res. 2000;62(2):302–10. https://doi.org/10.1002/1097-4547(20001015)62:2<302::AID-JNR15>3.0.CO;2-L.
122. López-Otín C, Blasco MA, Partridge L, Serrano M, Kroemer G. The hallmarks of aging. Cell. 2013;153(6):1194–217. https://doi.org/10.1016/j.cell.2013.05.039.
123. Maciotta S, Meregalli M, Torrente Y. The involvement of microRNAs in neurodegenerative diseases. Front Cell Neurosci. 2013;7:265. https://doi.org/10.3389/fncel.2013.00265.
124. Mackenzie IRA, Rademakers R. The role of transactive response DNA-binding protein-43 in amyotrophic lateral sclerosis and frontotemporal dementia. Curr Opin Neurol. 2008;21(6):693–700. https://doi.org/10.1097/WCO.0b013e3283168d1d.
125. Mahul-Mellier A-L, Burtscher J, Maharjan N, Weerens L, Croisier M, Kuttler F, Leleu M, Knott GW, Lashuel HA. The process of Lewy body formation, rather than simply α-synuclein fibrillization, is one of the major drivers of neurodegeneration. Proc Natl Acad Sci USA. 2020;117(9):4971–82. https://doi.org/10.1073/pnas.1913904117.
126. Majd S, Power JH, Grantham HJM. Neuronal response in Alzheimer's and Parkinson's disease: the effect of toxic proteins on intracellular pathways. BMC Neurosci. 2015;16(1):69. https://doi.org/10.1186/s12868-015-0211-1.
127. Manjula R, Anuja K, Alcain FJ. SIRT1 and SIRT2 activity control in neurodegenerative diseases. Front Pharmacol. 2021;11:585821. https://doi.org/10.3389/fphar.2020.585821.
128. Martinez-Miguel VE, Lujan C, Espie-Caullet T, Martinez-Martinez D, Moore S, Backes C, Gonzalez S, Galimov ER, Brown AEX, Halic M, Tomita K, Rallis C, von der Haar T, Cabreiro F, Bjedov I. Increased fidelity of protein synthesis extends lifespan. Cell Metab. 2021;33(11):2288–2300.e12. https://doi.org/10.1016/j.cmet.2021.08.017.
129. Martinez-Vicente M, Talloczy Z, Wong E, Tang G, Koga H, Kaushik S, de Vries R, Arias E, Harris S, Sulzer D, Cuervo AM. Cargo recognition failure is responsible for inefficient autophagy in Huntington's disease. Nat Neurosci. 2010;13(5):567–76. https://doi.org/10.1038/nn.2528.
130. Mary A, Eysert F, Checler F, Chami M. Mitophagy in Alzheimer's disease: molecular defects and therapeutic approaches. Mol Psychiatry. 2023;28(1):202–16. https://doi.org/10.1038/s41380-022-01631-6.
131. Masaldan S, Clatworthy SAS, Gamell C, Smith ZM, Francis PS, Denoyer D, Meggyesy PM, La Fontaine S, Cater MA. Copper accumulation in senescent cells: interplay between copper transporters and impaired autophagy. Redox Biol. 2018;16:322–31. https://doi.org/10.1016/j.redox.2018.03.007.
132. Matai L, Sarkar GC, Chamoli M, Malik Y, Kumar SS, Rautela U, Jana NR, Chakraborty K, Mukhopadhyay A. Dietary restriction improves proteostasis and increases life span through endoplasmic reticulum hormesis. Proc Natl Acad Sci. 2019;116(35):17383–92.
133. Matai L, Slack FJ. MicroRNAs in age-related proteostasis and stress responses. Noncoding RNA. 2023;9(2):26. https://doi.org/10.3390/ncrna9020026.

134. McCormick MA, Kennedy BK. Genome-scale studies of aging: challenges and opportunities. Curr Genomics. 2012;13(7):500–7. https://doi.org/10.2174/138920212803251454.
135. McKinnon C, Goold R, Andre R, Devoy A, Ortega Z, Moonga J, Linehan JM, Brandner S, Lucas JJ, Collinge J, Tabrizi SJ. Prion-mediated neurodegeneration is associated with early impairment of the ubiquitin-proteasome system. Acta Neuropathol. 2016;131(3):411–25. https://doi.org/10.1007/s00401-015-1508-y.
136. McKinnon C, De Snoo ML, Gondard E, Neudorfer C, Chau H, Ngana SG, O'Hara DM, Brotchie JM, Koprich JB, Lozano AM, Kalia LV, Kalia SK. Early-onset impairment of the ubiquitin-proteasome system in dopaminergic neurons caused by α-synuclein. Acta Neuropathol Commun. 2020;8(1):17. https://doi.org/10.1186/s40478-020-0894-0.
137. McNaught KSP, Belizaire R, Isacson O, Jenner P, Olanow CW. Altered proteasomal function in sporadic Parkinson's disease. Exp Neurol. 2003;179(1):38–46. https://doi.org/10.1006/exnr.2002.8050.
138. Miranzadeh Mahabadi H, Taghibiglou C. Cellular prion protein (PrPc): putative interacting partners and consequences of the interaction. Int J Mol Sci. 2020;21(19):7058. https://doi.org/10.3390/ijms21197058.
139. Mishra SK, Balendra V, Esposto J, Obaid AA, Maccioni RB, Jha NK, Perry G, Moustafa M, Al-Shehri M, Singh MP, Khan AA, Vamanu E, Singh SK. Therapeutic antiaging strategies. Biomedicines. 2022;10(10):2515. https://doi.org/10.3390/biomedicines10102515.
140. Moll L, Roitenberg N, Bejerano-Sagie M, Boocholez H, Carvalhal Marques F, Volovik Y, Elami T, Siddiqui AA, Grushko D, Biram A, Lampert B, Achache H, Ravid T, Tzur YB, Cohen E. The insulin/IGF signaling cascade modulates SUMOylation to regulate aging and proteostasis in Caenorhabditis elegans. elife. 2018;7:e38635. https://doi.org/10.7554/eLife.38635.
141. Moreno JA, Radford H, Peretti D, Steinert JR, Verity N, Martin MG, Halliday M, Morgan J, Dinsdale D, Ortori CA, Barrett DA, Tsaytler P, Bertolotti A, Willis AE, Bushell M, Mallucci GR. Sustained translational repression by eIF2α-P mediates prion neurodegeneration. Nature. 2012;485(7399):507–11. https://doi.org/10.1038/nature11058.
142. Moreno DF, Jenkins K, Morlot S, Charvin G, Csikasz-Nagy A, Aldea M. Proteostasis collapse, a hallmark of aging, hinders the chaperone-Start network and arrests cells in G1. elife. 2019;8:e48240. https://doi.org/10.7554/eLife.48240.
143. Morley JF, Morimoto RI. Regulation of longevity in Caenorhabditis elegans by heat shock factor and molecular chaperones. Mol Biol Cell. 2004;15(2):657–64. https://doi.org/10.1091/mbc.e03-07-0532.
144. Muñoz MF, Argüelles S, Cano M, Marotta F, Ayala A. Aging and oxidative stress decrease pineal elongation factor 2: in vivo protective effect of melatonin in young rats treated with cumene hydroperoxide. J Cell Biochem. 2017;118(1):182–90. https://doi.org/10.1002/jcb.25624.
145. Murphy KE, Gysbers AM, Abbott SK, Spiro AS, Furuta A, Cooper A, Garner B, Kabuta T, Halliday GM. Lysosomal-associated membrane protein 2 isoforms are differentially affected in early Parkinson's disease. Mov Disord. 2015;30(12):1639–47. https://doi.org/10.1002/mds.26141.
146. Naia L, Rego AC. Sirtuins: double players in Huntington's disease. Biochim Biophys Acta. 2015;1852(10):2183–94. https://doi.org/10.1016/j.bbadis.2015.07.003.
147. Narendra D, Kane LA, Hauser DN, Fearnley IM, Youle RJ. p62/SQSTM1 is required for Parkin-induced mitochondrial clustering but not mitophagy; VDAC1 is dispensable for both. Autophagy. 2010;6(8):1090–106. https://doi.org/10.4161/auto.6.8.13426.
148. Nieznanski K, Podlubnaya ZA, Nieznanska H. Prion protein inhibits microtubule assembly by inducing tubulin oligomerization. Biochem Biophys Res Commun. 2006;349(1):391–9. https://doi.org/10.1016/j.bbrc.2006.08.051.
149. Nixon RA. Amyloid precursor protein and endosomal-lysosomal dysfunction in Alzheimer's disease: inseparable partners in a multifactorial disease. FASEB J. 2017;31(7):2729–43. https://doi.org/10.1096/fj.201700359.

150. Nixon RA, Yang D-S. Autophagy failure in Alzheimer's disease--locating the primary defect. Neurobiol Dis. 2011;43(1):38–45. https://doi.org/10.1016/j.nbd.2011.01.021.
151. Nixon RA, Wegiel J, Kumar A, Yu WH, Peterhoff C, Cataldo A, Cuervo AM. Extensive involvement of autophagy in Alzheimer disease: an immuno-electron microscopy study. J Neuropathol Exp Neurol. 2005;64(2):113–22. https://doi.org/10.1093/jnen/64.2.113.
152. Nolan M, Talbot K, Ansorge O. Pathogenesis of FUS-associated ALS and FTD: insights from rodent models. Acta Neuropathol Commun. 2016;4(1):99. https://doi.org/10.1186/s40478-016-0358-8.
153. O'Brien RJ, Wong PC. Amyloid precursor protein processing and Alzheimer's disease. Annu Rev Neurosci. 2011;34:185–204. https://doi.org/10.1146/annurev-neuro-061010-113613.
154. Oliveira JMA. Nature and cause of mitochondrial dysfunction in Huntington's disease: focusing on huntingtin and the striatum. J Neurochem. 2010;114(1):1–12. https://doi.org/10.1111/j.1471-4159.2010.06741.x.
155. Oliveira MM, Klann E. eIF2-dependent translation initiation: memory consolidation and disruption in Alzheimer's disease. Semin Cell Dev Biol. 2022;125:101–9. https://doi.org/10.1016/j.semcdb.2021.07.009.
156. Olzscha H, Schermann SM, Woerner AC, Pinkert S, Hecht MH, Tartaglia GG, Vendruscolo M, Hayer-Hartl M, Hartl FU, Vabulas RM. Amyloid-like aggregates sequester numerous metastable proteins with essential cellular functions. Cell. 2011;144(1):67–78. https://doi.org/10.1016/j.cell.2010.11.050.
157. Orgel LE. The maintenance of the accuracy of protein synthesis and its relevance to ageing. Proc Natl Acad Sci USA. 1963;49(4):517–21. https://doi.org/10.1073/pnas.49.4.517.
158. Ori A, Toyama BH, Harris MS, Bock T, Iskar M, Bork P, Ingolia NT, Hetzer MW, Beck M. Integrated transcriptome and proteome analyses reveal organ-specific proteome deterioration in old rats. Cell Syst. 2015;1(3):224–37. https://doi.org/10.1016/j.cels.2015.08.012.
159. Orr ME, Oddo S. Autophagic/lysosomal dysfunction in Alzheimer's disease. Alzheimers Res Ther. 2013;5(5):53. https://doi.org/10.1186/alzrt217.
160. Ortega Z, Lucas JJ. Ubiquitin–proteasome system involvement in Huntington's disease. Front Mol Neurosci. 2014;7:77. https://doi.org/10.3389/fnmol.2014.00077.
161. Osiecka KM, Nieznanska H, Skowronek KJ, Karolczak J, Schneider G, Nieznanski K. Prion protein region 23-32 interacts with tubulin and inhibits microtubule assembly. Proteins. 2009;77(2):279–96. https://doi.org/10.1002/prot.22435.
162. Outeiro TF, Lindquist S. Yeast cells provide insight into alpha-synuclein biology and pathobiology. Science. 2003;302(5651):1772–5.
163. Pan KM, Baldwin M, Nguyen J, Gasset M, Serban A, Groth D, Mehlhorn I, Huang Z, Fletterick RJ, Cohen FE. Conversion of alpha-helices into beta-sheets features in the formation of the scrapie prion proteins. Proc Natl Acad Sci USA. 1993;90(23):10962–6. https://doi.org/10.1073/pnas.90.23.10962.
164. Paukštytė J, López Cabezas RM, Feng Y, Tong K, Schnyder D, Elomaa E, Gregorova P, Doudin M, Särkkä M, Sarameri J, Lippi A, Vihinen H, Juutila J, Nieminen A, Törönen P, Holm L, Jokitalo E, Krisko A, Huiskonen J, et al. Global analysis of aging-related protein structural changes uncovers enzyme-polymerization-based control of longevity. Mol Cell. 2023;83(18):3360–3376.e11. https://doi.org/10.1016/j.molcel.2023.08.015.
165. Perluigi M, Di Domenico F, Barone E, Butterfield DA. mTOR in Alzheimer disease and its earlier stages: links to oxidative damage in the progression of this dementing disorder. Free Radic Biol Med. 2021;169:382–96. https://doi.org/10.1016/j.freeradbiomed.2021.04.025.
166. Perutz MF, Johnson T, Suzuki M, Finch JT. Glutamine repeats as polar zippers: their possible role in inherited neurodegenerative diseases. Proc Natl Acad Sci USA. 1994;91(12):5355–8. https://doi.org/10.1073/pnas.91.12.5355.
167. Polymenidou M, Cleveland DW. Prion-like spread of protein aggregates in neurodegeneration. J Exp Med. 2012;209(5):889–93. https://doi.org/10.1084/jem.20120741.
168. Popova B, Kleinknecht A, Braus GH. Posttranslational modifications and clearing of α-synuclein aggregates in yeast. Biomol Ther. 2015;5(2):617–34. https://doi.org/10.3390/biom5020617.

169. Prusiner SB. Novel proteinaceous infectious particles cause scrapie. Science. 1982;216(4542):136–44. https://doi.org/10.1126/science.6801762.
170. Pushkar Y, Robison G, Sullivan B, Fu SX, Kohne M, Jiang W, Rohr S, Lai B, Marcus MA, Zakharova T, Zheng W. Aging results in copper accumulations in glial fibrillary acidic protein-positive cells in the subventricular zone. Aging Cell. 2013;12(5):823–32. https://doi.org/10.1111/acel.12112.
171. Rambold AS, Miesbauer M, Rapaport D, Bartke T, Baier M, Winklhofer KF, Tatzelt J. Association of Bcl-2 with misfolded prion protein is linked to the toxic potential of cytosolic PrP. Mol Biol Cell. 2006;17(8):3356–68. https://doi.org/10.1091/mbc.e06-01-0083.
172. Raulin A-C, Doss SV, Trottier ZA, Ikezu TC, Bu G, Liu C-C. ApoE in Alzheimer's disease: pathophysiology and therapeutic strategies. Mol Neurodegener. 2022;17(1):72. https://doi.org/10.1186/s13024-022-00574-4.
173. Rawat P, Sehar U, Bisht J, Selman A, Culberson J, Reddy PH. Phosphorylated tau in Alzheimer's disease and other tauopathies. Int J Mol Sci. 2022;23(21):12841. https://doi.org/10.3390/ijms232112841.
174. Read A, Schröder M. The unfolded protein response: an overview. Biology. 2021;10(5):384. https://doi.org/10.3390/biology10050384.
175. Reis-Rodrigues P, Czerwieniec G, Peters TW, Evani US, Alavez S, Gaman EA, Vantipalli M, Mooney SD, Gibson BW, Lithgow GJ, Hughes RE. Proteomic analysis of age-dependent changes in protein solubility identifies genes that modulate lifespan. Aging Cell. 2012;11(1):120–7. https://doi.org/10.1111/j.1474-9726.2011.00765.x.
176. Ren X, Hu B, Song M, Ding Z, Dang Y, Liu Z, Zhang W, Ji Q, Ren R, Ding J, Chan P, Jiang C, Ye K, Qu J, Tang F, Liu G-H. Maintenance of nucleolar homeostasis by CBX4 alleviates senescence and osteoarthritis. Cell Rep. 2019;26(13):3643–3656.e7. https://doi.org/10.1016/j.celrep.2019.02.088.
177. Renaud L, Picher-Martel V, Codron P, Julien J-P. Key role of UBQLN2 in pathogenesis of amyotrophic lateral sclerosis and frontotemporal dementia. Acta Neuropathol Commun. 2019;7(1):103. https://doi.org/10.1186/s40478-019-0758-7.
178. Rippin I, Bonder K, Joseph S, Sarsor A, Vaks L, Eldar-Finkelman H. Inhibition of GSK-3 ameliorates the pathogenesis of Huntington's disease. Neurobiol Dis. 2021;154:105336. https://doi.org/10.1016/j.nbd.2021.105336.
179. Rocha EM, De Miranda B, Sanders LH. Alpha-synuclein: pathology, mitochondrial dysfunction and neuroinflammation in Parkinson's disease. Neurobiol Dis. 2018;109:249–57. https://doi.org/10.1016/j.nbd.2017.04.004.
180. Rosen DR, Siddique T, Patterson D, Figlewicz DA, Sapp P, Hentati A, Donaldson D, Goto J, O'Regan JP, Deng HX. Mutations in Cu/Zn superoxide dismutase gene are associated with familial amyotrophic lateral sclerosis. Nature. 1993;362(6415):59–62. https://doi.org/10.1038/362059a0.
181. Ruegsegger C, Saxena S. Proteostasis impairment in ALS. Brain Res. 2016;1648:571–9. https://doi.org/10.1016/j.brainres.2016.03.032.
182. Saarikangas J, Barral Y. Protein aggregates are associated with replicative aging without compromising protein quality control. elife. 2015;4:e06197. https://doi.org/10.7554/eLife.06197.
183. Sadri-Vakili G, Cha J-HJ. Mechanisms of disease: histone modifications in Huntington's disease. Nat Clin Pract Neurol. 2006;2(6):330–8. https://doi.org/10.1038/ncpneuro0199.
184. Safar J, Roller PP, Gajdusek DC, Gibbs CJ. Conformational transitions, dissociation, and unfolding of scrapie amyloid (prion) protein. J Biol Chem. 1993;268(27):20276–84.
185. Salamon A, Maszlag-Török R, Veres G, Boros FA, Vágvölgyi-Sümegi E, Somogyi A, Vécsei L, Klivényi P, Zádori D. Cerebellar predominant increase in mRNA expression levels of Sirt1 and Sirt3 isoforms in a transgenic mouse model of Huntington's disease. Neurochem Res. 2020;45(9):2072–81. https://doi.org/10.1007/s11064-020-03069-0.
186. Salminen A, Kaarniranta K, Haapasalo A, Hiltunen M, Soininen H, Alafuzoff I. Emerging role of p62/sequestosome-1 in the pathogenesis of Alzheimer's disease. Prog Neurobiol. 2012;96(1):87–95. https://doi.org/10.1016/j.pneurobio.2011.11.005.

187. Samaranch L, Lorenzo-Betancor O, Arbelo JM, Ferrer I, Lorenzo E, Irigoyen J, Pastor MA, Marrero C, Isla C, Herrera-Henriquez J, Pastor P. PINK1-linked parkinsonism is associated with Lewy body pathology. Brain. 2010;133(Pt 4):1128–42. https://doi.org/10.1093/brain/awq051.
188. Santos AL, Lindner AB. Protein posttranslational modifications: roles in aging and age-related disease. Oxidative Med Cell Longev. 2017;2017:5716409. https://doi.org/10.1155/2017/5716409.
189. Santra M, Dill KA, de Graff AMR. Proteostasis collapse is a driver of cell aging and death. Proc Natl Acad Sci USA. 2019;116(44):22173–8. https://doi.org/10.1073/pnas.1906592116.
190. Sarkar S, Rubinsztein DC. Huntington's disease: degradation of mutant huntingtin by autophagy. FEBS J. 2008;275(17):4263–70. https://doi.org/10.1111/j.1742-4658.2008.06562.x.
191. Sarkar S, Perlstein EO, Imarisio S, Pineau S, Cordenier A, Maglathlin RL, Webster JA, Lewis TA, O'Kane CJ, Schreiber SL, Rubinsztein DC. Small molecules enhance autophagy and reduce toxicity in Huntington's disease models. Nat Chem Biol. 2007;3(6):331–8. https://doi.org/10.1038/nchembio883.
192. Sarkis G, Ashcom J, Hawdon J, Jacobson L. Decline in protease activities with age in the nematode Caenorhabditis elegans. Mech Ageing Dev. 1988;45(3):191–201. https://doi.org/10.1016/0047-6374(88)90001-2.
193. Saudou F, Humbert S. The biology of huntingtin. Neuron. 2016;89(5):910–26. https://doi.org/10.1016/j.neuron.2016.02.003.
194. Schachter AS, Davis KL. Alzheimer's disease. Dialogues Clin Neurosci. 2000;2(2):91–100. https://doi.org/10.31887/DCNS.2000.2.2/asschachter.
195. Schreiner TG, Schreiner OD, Adam M, Popescu BO. The roles of the amyloid beta monomers in physiological and pathological conditions. Biomedicines. 2023;11(5):1411. https://doi.org/10.3390/biomedicines11051411.
196. Schulte J, Littleton JT. The biological function of the Huntingtin protein and its relevance to Huntington's disease pathology. Curr Trends Neurol. 2011;5:65–78.
197. Selkoe DJ. Alzheimer's disease. Cold Spring Harb Perspect Biol. 2011;3(7):a004457. https://doi.org/10.1101/cshperspect.a004457.
198. Selkoe DJ, Podlisny MB, Joachim CL, Vickers EA, Lee G, Fritz LC, Oltersdorf T. Beta-amyloid precursor protein of Alzheimer disease occurs as 110- to 135-kilodalton membrane-associated proteins in neural and nonneural tissues. Proc Natl Acad Sci USA. 1988;85(19):7341–5. https://doi.org/10.1073/pnas.85.19.7341.
199. Selman C, Tullet JMA, Wieser D, Irvine E, Lingard SJ, Choudhury AI, Claret M, Al-Qassab H, Carmignac D, Ramadani F, Woods A, Robinson ICA, Schuster E, Batterham RL, Kozma SC, Thomas G, Carling D, Okkenhaug K, Thornton JM, et al. Ribosomal protein S6 kinase 1 signaling regulates mammalian life span. Science. 2009;326(5949):140–4. https://doi.org/10.1126/science.1177221.
200. Shankar GM, Li S, Mehta TH, Garcia-Munoz A, Shepardson NE, Smith I, Brett FM, Farrell MA, Rowan MJ, Lemere CA, Regan CM, Walsh DM, Sabatini BL, Selkoe DJ. Amyloid-beta protein dimers isolated directly from Alzheimer's brains impair synaptic plasticity and memory. Nat Med. 2008;14(8):837–42. https://doi.org/10.1038/nm1782.
201. Sharoar MG, Hu X, Ma X-M, Zhu X, Yan R. Sequential formation of different layers of dystrophic neurites in Alzheimer's brains. Mol Psychiatry. 2019;24(9):1369–82. https://doi.org/10.1038/s41380-019-0396-2.
202. Shen J, Snapp EL, Lippincott-Schwartz J, Prywes R. Stable binding of ATF6 to BiP in the endoplasmic reticulum stress response. Mol Cell Biol. 2005;25(3):921–32. https://doi.org/10.1128/MCB.25.3.921-932.2005.
203. Sigurdson CJ, Bartz JC, Glatzel M. Cellular and molecular mechanisms of prion disease. Annu Rev Pathol. 2019;14:497–516. https://doi.org/10.1146/annurev-pathmechdis-012418-013109.
204. Sikorska B, Liberski PP, Giraud P, Kopp N, Brown P. Autophagy is a part of ultrastructural synaptic pathology in Creutzfeldt-Jakob disease: a brain biopsy study. Int J Biochem Cell Biol. 2004;36(12):2563–73. https://doi.org/10.1016/j.biocel.2004.04.014.

205. Simonsen A, Cumming RC, Brech A, Isakson P, Schubert DR, Finley KD. Promoting basal levels of autophagy in the nervous system enhances longevity and oxidant resistance in adult Drosophila. Autophagy. 2008;4(2):176–84. https://doi.org/10.4161/auto.5269.
206. Smeyers J, Banchi E-G, Latouche M. C9ORF72: what it is, what it does, and why it matters. Front Cell Neurosci. 2021;15:661447. https://doi.org/10.3389/fncel.2021.661447.
207. Smith-Vikos T, Slack FJ. MicroRNAs and their roles in aging. J Cell Sci. 2012;125(Pt 1):7–17. https://doi.org/10.1242/jcs.099200.
208. Snyder H, Mensah K, Theisler C, Lee J, Matouschek A, Wolozin B. Aggregated and monomeric alpha-synuclein bind to the S6′ proteasomal protein and inhibit proteasomal function. J Biol Chem. 2003;278(14):11753–9. https://doi.org/10.1074/jbc.M208641200.
209. Soares TR, Reis SD, Pinho BR, Duchen MR, Oliveira JMA. Targeting the proteostasis network in Huntington's disease. Ageing Res Rev. 2019;49:92–103. https://doi.org/10.1016/j.arr.2018.11.006.
210. Sonobe Y, Aburas J, Krishnan G, Fleming AC, Ghadge G, Islam P, Warren EC, Gu Y, Kankel MW, Brown AEX, Kiskinis E, Gendron TF, Gao F-B, Roos RP, Kratsios P. A C. elegans model of C9orf72-associated ALS/FTD uncovers a conserved role for eIF2D in RAN translation. Nat Commun. 2021;12(1):6025. https://doi.org/10.1038/s41467-021-26303-x.
211. Soriano S, Chyung AS, Chen X, Stokin GB, Lee VM, Koo EH. Expression of beta-amyloid precursor protein-CD3gamma chimeras to demonstrate the selective generation of amyloid beta(1-40) and amyloid beta(1-42) peptides within secretory and endocytic compartments. J Biol Chem. 1999;274(45):32295–300. https://doi.org/10.1074/jbc.274.45.32295.
212. Spillantini MG, Schmidt ML, Lee VM-Y, Trojanowski JQ, Jakes R, Goedert M. α-Synuclein in Lewy bodies. Nature. 1997;388(6645):839–40. https://doi.org/10.1038/42166.
213. Stefanis L. α-Synuclein in Parkinson's disease. Cold Spring Harb Perspect Med. 2012;2(2):a009399. https://doi.org/10.1101/cshperspect.a009399.
214. Stein KC, Morales-Polanco F, van der Lienden J, Rainbolt TK, Frydman J. Ageing exacerbates ribosome pausing to disrupt cotranslational proteostasis. Nature. 2022;601(7894):637–42. https://doi.org/10.1038/s41586-021-04295-4.
215. Stenoien DL, Cummings CJ, Adams HP, Mancini MG, Patel K, DeMartino GN, Marcelli M, Weigel NL, Mancini MA. Polyglutamine-expanded androgen receptors form aggregates that sequester heat shock proteins, proteasome components and SRC-1, and are suppressed by the HDJ-2 chaperone. Hum Mol Genet. 1999;8(5):731–41. https://doi.org/10.1093/hmg/8.5.731.
216. Stockman S. Scrapie: an obscure disease of sheep. J Comp Pathol Ther. 1913;26:317–27. https://doi.org/10.1016/S0368-1742(13)80060-4.
217. Strang KH, Golde TE, Giasson BI. MAPT mutations, tauopathy, and mechanisms of neurodegeneration. Lab Invest. 2019;99(7):912–28. https://doi.org/10.1038/s41374-019-0197-x.
218. Suelves N, Kirkham-McCarthy L, Lahue RS, Ginés S. A selective inhibitor of histone deacetylase 3 prevents cognitive deficits and suppresses striatal CAG repeat expansions in Huntington's disease mice. Sci Rep. 2017;7(1):6082. https://doi.org/10.1038/s41598-017-05125-2.
219. Suk TR, Rousseaux MWC. The role of TDP-43 mislocalization in amyotrophic lateral sclerosis. Mol Neurodegener. 2020;15(1):45. https://doi.org/10.1186/s13024-020-00397-1.
220. Sun X, Liu J, Crary JF, Malagelada C, Sulzer D, Greene LA, Levy OA. ATF4 protects against neuronal death in cellular Parkinson's disease models by maintaining levels of parkin. J Neurosci. 2013;33(6):2398–407. https://doi.org/10.1523/JNEUROSCI.2292-12.2013.
221. Sweeney P, Park H, Baumann M, Dunlop J, Frydman J, Kopito R, McCampbell A, Leblanc G, Venkateswaran A, Nurmi A, Hodgson R. Protein misfolding in neurodegenerative diseases: implications and strategies. Transl Neurodegener. 2017;6(1):6. https://doi.org/10.1186/s40035-017-0077-5.
222. Takashima A. GSK-3 is essential in the pathogenesis of Alzheimer's disease. J Alzheimers Dis. 2006;9(3 Suppl):309–17. https://doi.org/10.3233/jad-2006-9s335.
223. Tanaka Y. Inducible expression of mutant alpha-synuclein decreases proteasome activity and increases sensitivity to mitochondria-dependent apoptosis. Hum Mol Genet. 2001;10(9):919–26. https://doi.org/10.1093/hmg/10.9.919.

224. Tandan R, Bradley WG. Amyotrophic lateral sclerosis: Part 1. Clinical features, pathology, and ethical issues in management. Ann Neurol. 1985;18(3):271–80. https://doi.org/10.1002/ana.410180302.
225. Tawo R, Pokrzywa W, Kevei É, Akyuz ME, Balaji V, Adrian S, Höhfeld J, Hoppe T. The ubiquitin ligase CHIP integrates proteostasis and aging by regulation of insulin receptor turnover. Cell. 2017;169(3):470–482.e13. https://doi.org/10.1016/j.cell.2017.04.003.
226. Tomaru U, Takahashi S, Ishizu A, Miyatake Y, Gohda A, Suzuki S, Ono A, Ohara J, Baba T, Murata S, Tanaka K, Kasahara M. Decreased proteasomal activity causes age-related phenotypes and promotes the development of metabolic abnormalities. Am J Pathol. 2012;180(3):963–72. https://doi.org/10.1016/j.ajpath.2011.11.012.
227. Tseng C-S, Chao Y-W, Liu Y-H, Huang Y-S, Chao H-W. Dysregulated proteostasis network in neuronal diseases. Front Cell Dev Biol. 2023;11:1075215. https://doi.org/10.3389/fcell.2023.1075215.
228. Tsuang D, Leverenz JB, Lopez OL, Hamilton RL, Bennett DA, Schneider JA, Buchman AS, Larson EB, Crane PK, Kaye JA, Kramer P, Woltjer R, Kukull W, Nelson PT, Jicha GA, Neltner JH, Galasko D, Masliah E, Trojanowski JQ, et al. GBA mutations increase risk for Lewy body disease with and without Alzheimer disease pathology. Neurology. 2012;79(19):1944–50. https://doi.org/10.1212/WNL.0b013e3182735e9a.
229. Uchiyama Y, Shibata M, Koike M, Yoshimura K, Sasaki M. Autophagy–physiology and pathophysiology. Histochem Cell Biol. 2008;129(4):407–20. https://doi.org/10.1007/s00418-008-0406-y.
230. Uddin MS, Stachowiak A, Al Mamun A, Tzvetkov NT, Takeda S, Atanasov AG, Bergantin LB, Abdel-Daim MM, Stankiewicz AM. Autophagy and Alzheimer's disease: from molecular mechanisms to therapeutic implications. Front Aging Neurosci. 2018;10:04. https://doi.org/10.3389/fnagi.2018.00004.
231. Ukmar-Godec T, Fang P, Ibáñez de Opakua A, Henneberg F, Godec A, Pan K-T, Cima-Omori M-S, Chari A, Mandelkow E, Urlaub H, Zweckstetter M. Proteasomal degradation of the intrinsically disordered protein tau at single-residue resolution. Sci Adv. 2020;6(30):eaba3916. https://doi.org/10.1126/sciadv.aba3916.
232. Vilchez D, Boyer L, Morantte I, Lutz M, Merkwirth C, Joyce D, Spencer B, Page L, Masliah E, Berggren WT, Gage FH, Dillin A. Increased proteasome activity in human embryonic stem cells is regulated by PSMD11. Nature. 2012;489(7415):304–8. https://doi.org/10.1038/nature11468.
233. Vilchez D, Saez I, Dillin A. The role of protein clearance mechanisms in organismal ageing and age-related diseases. Nat Commun. 2014;5:5659. https://doi.org/10.1038/ncomms6659.
234. Wagner M, Zhu G, Khalid F, Phan T, Maity P, Lupu L, Agyeman-Duah E, Wiese S, Lindenberg KS, Schön M, Landwehrmeyer GB, Penzo M, Kochanek S, Scharffetter-Kochanek K, Mulaw M, Iben S. General loss of proteostasis links Huntington disease to Cockayne syndrome. Neurobiol Dis. 2024;201:106668. https://doi.org/10.1016/j.nbd.2024.106668.
235. Walker AK. Protein disulfide isomerase and the endoplasmic reticulum in amyotrophic lateral sclerosis. J Neurosci. 2010;30(11):3865–7. https://doi.org/10.1523/JNEUROSCI.0408-10.2010.
236. Walther DM, Kasturi P, Zheng M, Pinkert S, Vecchi G, Ciryam P, Morimoto RI, Dobson CM, Vendruscolo M, Mann M, Hartl FU. Widespread proteome remodeling and aggregation in aging C. elegans. Cell. 2015;161(4):919–32. https://doi.org/10.1016/j.cell.2015.03.032.
237. Wang S, Sun S. Translation dysregulation in neurodegenerative diseases: a focus on ALS. Mol Neurodegener. 2023;18(1):58. https://doi.org/10.1186/s13024-023-00642-3.
238. Wang X-F, Dong C-F, Zhang J, Wan Y-Z, Li F, Huang Y-X, Han L, Shan B, Gao C, Han J, Dong X-P. Human tau protein forms complex with PrP and some GSS- and fCJD-related PrP mutants possess stronger binding activities with tau in vitro. Mol Cell Biochem. 2008;310(1–2):49–55. https://doi.org/10.1007/s11010-007-9664-6.
239. Webster GC, Webster SL. Decline in synthesis of elongation factor one (EF-1) precedes the decreased synthesis of total protein in aging Drosophila melanogaster. Mech Ageing Dev. 1983;22(2):121–8. https://doi.org/10.1016/0047-6374(83)90105-7.

240. Weidemann A, König G, Bunke D, Fischer P, Salbaum JM, Masters CL, Beyreuther K. Identification, biogenesis, and localization of precursors of Alzheimer's disease A4 amyloid protein. Cell. 1989;57(1):115–26. https://doi.org/10.1016/0092-8674(89)90177-3.
241. Wodrich APK, Scott AW, Shukla AK, Harris BT, Giniger E. The unfolded protein responses in health, aging, and neurodegeneration: recent advances and future considerations. Front Mol Neurosci. 2022;15:831116. https://doi.org/10.3389/fnmol.2022.831116.
242. Woerner AC, Frottin F, Hornburg D, Feng LR, Meissner F, Patra M, Tatzelt J, Mann M, Winklhofer KF, Hartl FU, Hipp MS. Cytoplasmic protein aggregates interfere with nucleocytoplasmic transport of protein and RNA. Science. 2016;351(6269):173–6. https://doi.org/10.1126/science.aad2033.
243. Wolfe DM, Lee J-H, Kumar A, Lee S, Orenstein SJ, Nixon RA. Autophagy failure in Alzheimer's disease and the role of defective lysosomal acidification. Eur J Neurosci. 2013;37(12):1949–61. https://doi.org/10.1111/ejn.12169.
244. Xia W, Zhang J, Ostaszewski BL, Kimberly WT, Seubert P, Koo EH, Shen J, Selkoe DJ. Presenilin 1 regulates the processing of beta-amyloid precursor protein C-terminal fragments and the generation of amyloid beta-protein in endoplasmic reticulum and Golgi. Biochemistry. 1998;37(47):16465–71. https://doi.org/10.1021/bi9816195.
245. Xia N, Cabin DE, Fang F, Reijo Pera RA. Parkinson's disease: overview of transcription factor regulation, genetics, and cellular and animal models. Front Neurosci. 2022;16:894620. https://doi.org/10.3389/fnins.2022.894620.
246. Xiao X, Liu H, Liu X, Zhang W, Zhang S, Jiao B. APP, PSEN1, and PSEN2 variants in Alzheimer's disease: systematic re-evaluation according to ACMG guidelines. Front Aging Neurosci. 2021;13:695808. https://doi.org/10.3389/fnagi.2021.695808.
247. Xu Y, Tian C, Wang S-B, Xie W-L, Guo Y, Zhang J, Shi Q, Chen C, Dong X-P. Activation of the macroautophagic system in scrapie-infected experimental animals and human genetic prion diseases. Autophagy. 2012;8(11):1604–20. https://doi.org/10.4161/auto.21482.
248. Yates SC, Zafar A, Hubbard P, Nagy S, Durant S, Bicknell R, Wilcock G, Christie S, Esiri MM, Smith AD, Nagy Z. Dysfunction of the mTOR pathway is a risk factor for Alzheimer's disease. Acta Neuropathol Commun. 2013;1(1):3. https://doi.org/10.1186/2051-5960-1-3.
249. Yue H-W, Hong J-Y, Zhang S-X, Jiang L-L, Hu H-Y. PolyQ-expanded proteins impair cellular proteostasis of ataxin-3 through sequestering the co-chaperone HSJ1 into aggregates. Sci Rep. 2021;11(1):7815. https://doi.org/10.1038/s41598-021-87382-w.
250. Zarranz JJ, Alegre J, Gómez-Esteban JC, Lezcano E, Ros R, Ampuero I, Vidal L, Hoenicka J, Rodriguez O, Atarés B, Llorens V, Tortosa EG, del Ser T, Muñoz DG, de Yebenes JG. The new mutation, E46K, of α-synuclein causes Parkinson and Lewy body dementia. Ann Neurol. 2004;55(2):164–73. https://doi.org/10.1002/ana.10795.
251. Zečić A, Braeckman BP. DAF-16/FoxO in Caenorhabditis elegans and its role in metabolic remodeling. Cells. 2020;9(1):109. https://doi.org/10.3390/cells9010109.
252. Zeidan RS, Han SM, Leeuwenburgh C, Xiao R. Iron homeostasis and organismal aging. Ageing Res Rev. 2021;72:101510. https://doi.org/10.1016/j.arr.2021.101510.
253. Zhang J, Dong X-P. Dysfunction of microtubule-associated proteins of MAP2/tau family in prion disease. Prion. 2012;6(4):334–8. https://doi.org/10.4161/pri.20677.
254. Zhang WH, Koyuncu S, Vilchez D. Insights into the links between proteostasis and aging from C. elegans. Front Aging. 2022;3:854157. https://doi.org/10.3389/fragi.2022.854157.
255. Zhao T, Severijnen L-A, van der Weiden M, Zheng PP, Oostra BA, Hukema RK, Willemsen R, Kros JM, Bonifati V. FBXO7 immunoreactivity in α-synuclein-containing inclusions in Parkinson disease and multiple system atrophy. J Neuropathol Exp Neurol. 2013;72(6):482–8. https://doi.org/10.1097/NEN.0b013e318293c586.
256. Zheng S, Clabough EBD, Sarkar S, Futter M, Rubinsztein DC, Zeitlin SO. Deletion of the huntingtin polyglutamine stretch enhances neuronal autophagy and longevity in mice. PLoS Genet. 2010;6(2):e1000838. https://doi.org/10.1371/journal.pgen.1000838.
257. Zheng X, Krakowiak J, Patel N, Beyzavi A, Ezike J, Khalil AS, Pincus D. Dynamic control of Hsf1 during heat shock by a chaperone switch and phosphorylation. elife. 2016;5:e18638. https://doi.org/10.7554/eLife.18638.

258. Zid BM, Rogers AN, Katewa SD, Vargas MA, Kolipinski MC, Lu TA, Benzer S, Kapahi P. 4E-BP extends lifespan upon dietary restriction by enhancing mitochondrial activity in Drosophila. Cell. 2009;139(1):149–60. https://doi.org/10.1016/j.cell.2009.07.034.
259. Zong Y, Li H, Liao P, Chen L, Pan Y, Zheng Y, Zhang C, Liu D, Zheng M, Gao J. Mitochondrial dysfunction: mechanisms and advances in therapy. Signal Transduct Target Ther. 2024;9(1):124. https://doi.org/10.1038/s41392-024-01839-8.
260. Zou J, Guo Y, Guettouche T, Smith DF, Voellmy R. Repression of heat shock transcription factor HSF1 activation by HSP90 (HSP90 complex) that forms a stress-sensitive complex with HSF1. Cell. 1998;94(4):471–80. https://doi.org/10.1016/s0092-8674(00)81588-3.

10 The Role of Chaperones and Co-Chaperones in Synaptic Proteostasis and Neurodegenerative Disease

Neha Sharma, Ragini Singh, Sharmistha Banerjee, and Shuchi Kaushik

10.1 Introduction

Neurons are characterized by their high metabolic demand and complex morphology, making them particularly susceptible to protein misfolding and aggregation. Synapses, the communication points between neurons, require precise protein composition for effective signaling. Chaperones and co-chaperones are critical in maintaining synaptic proteostasis by ensuring proteins reach their correct conformations thus safeguarding synaptic integrity and function [1]. Proteins must fold into unique three-dimensional structures to fulfill their biological functions [17]. Proteins carry numerous cellular functions that are vital for proper cellular functioning and the health of an organism. Consequently, the cell has evolved a complex molecular system of integrated processes that balance the conformation, concentration, and subcellular localization of proteins for providing protein homeostasis (proteostasis). It is essential for the dynamic changes required for a cell to respond to a given stimulus [66].

Molecular chaperones and co-chaperones are a class of proteins that play an essential role in maintaining cellular proteostasis and interact with non-native proteins to promote protein compartmentalization, folding, and degradation of their substrates [17]. Protein degradation is a fundamental part of the cellular machinery [66] (Fig. 10.1).

N. Sharma (✉) · R. Singh
Amity Institute of Biotechnology, Amity University, Gwalior, Madhya Pradesh, India
e-mail: nsharma2@gwa.amity.edu

S. Banerjee
Chhattisgarh Swami Vivekanand Technical University, Bhilai, Chhattisgarh, India

S. Kaushik
State Forensic Science Laboratory, Sagar, Madhya Pradesh, India

N. Puranik (ed.), *Proteostasis*, https://doi.org/10.1007/978-981-96-6202-9_10

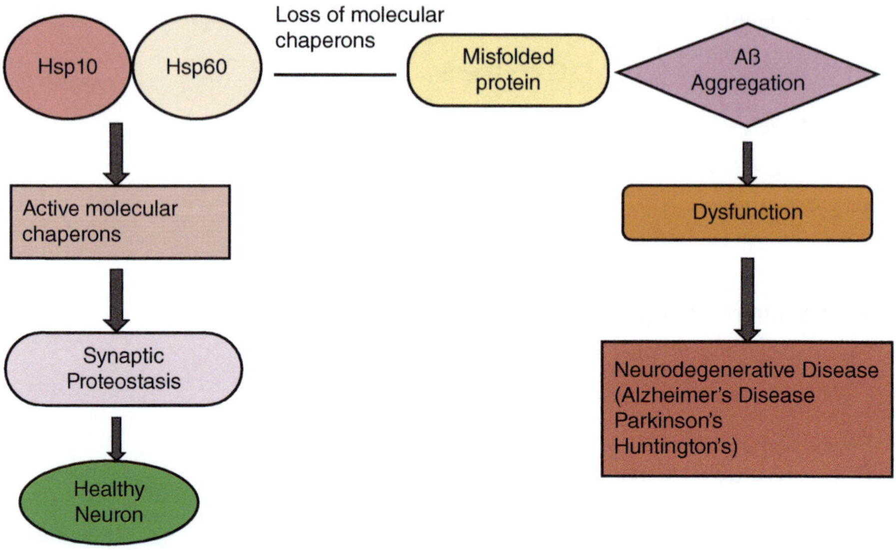

Fig. 10.1 Role of chaperons and co-chaperons in neurogenerative disease

They are highly ubiquitous and help the newly synthesized proteins to fold and partially folded proteins to refold to their native structures [22, 23, 62]. Chaperones are often referred to as stress proteins or heat shock proteins (Hsps) because they are upregulated during conformational stress [17]. The major chaperone families are classified by molecular weight (Hsp40, Hsp60, Hsp70, Hsp90, Hsp100) and the small Hsp [49]. They are localized in the nucleus, cytoplasm, and cellular organelles like endoplasmic reticulum (ER) and mitochondria [49]. Most importantly, HSPs are tightly regulated, requiring a precise balance, while overproduction of certain HSPs can lead to diseases, including cancer [3]. However, among all HSPs, low molecular weight HSPs are ATP-independent for their chaperone activity, whereas high molecular weight HSPs are ATP-dependent [11, 27, 47].

10.2 Role of Chaperones

Chaperones are various positions of evolutionarily conserved proteins that are involved in protein congregation, folding, and stability in order to achieve proteostasis. Chaperones comprise up to 10% of the mass of human cells and unfold, dissociate, refold, or degrade the other 90% of cellular proteins as required [13].

All of the chaperone members, except tiny Hsps, catalyze ATP-dependent processes. Chaperones play an important role in synaptic proteostasis and thus help in maintaining protein homeostasis and avert protein misfolding and aggregation. In neurodegenerative diseases (NDDs), chaperones are frequently overwhelmed, leading to protein aggregation and synaptic dysfunction. For example, Hsp70s work in conjunction with Hsp40 co-chaperones (also known as DNAJ or J proteins) [45].

Stress-inducible Hsp70s and their constitutive relative heat shock cognate 70 (Hsc70) are the first line of respondents to cellular stress and accumulation of misfolded proteins. In the nervous system, induction of Hsp70 by stress is weak, and therefore Hsc70 is the main Hsp70-class chaperone [4, 31, 41, 46].

Hsp90 is an essential protein in all eukaryotes and a major regulator of protein folding in the cell. Hsp90s contribute to conformational maturation and maintenance in addition to aiding in protein folding. Hsp90 coordinates with the Hsp70 system through the Hsp70-Hsp90 Organizing Protein (HOP) co-chaperone [9]. Eukaryotic Hsp60 class chaperone TriC is a nano-cage that aids in the folding of cytoskeletal proteins such as actin and tubulin [15]. Misfolded and aggregated proteins during proteomic stress induce the upregulation of disaggregates, including the Hsp110 family member. Small Hsps exhibit chaperone-like activity in preventing aggregation of target proteins, maintaining them in a folding–competent state and refolding them independently or in concert with other ATP-dependent chaperones such as Hsp70s [42].

Within the cell, chaperone families have been separated into several subcellular divisions that each have unique members that efficiently fold proteins. The precise position of ubiquitous cytosolic chaperone proteins, Hsp70s and Hsp90s, is dictated by their partner Hsp40 co-chaperones. Mitochondria-specific chaperones include mHsp70 and Hsp60, which are involved in importing mitochondrial proteins into the matrix and folding them [28, 38]. The endoplasmic reticulum (ER), the region where proteins are synthesized, folded, separated, and secreted, contains a different complement of chaperones. ER chaperones, such as ER Hsp70 and BiP, function there to stop stress brought on by excessive unfolded protein levels. Recent papers have highlighted the importance of the chaperones involved in the unfolded protein response at the ER in neuronal proteostasis and neurodegenerative disease [25]. Finally, Hsp100s and small Hsps are nearly ubiquitous in terms of cellular localization [24, 68].

Protein chaperones can prevent the aggregation of unfolded polypeptide chains, promote their de novo folding or refolding [23], cooperate with proteases to facilitate protein degradation, and disaggregate protein aggregates [26, 64, 65].

In addition to these well-established functions, molecular chaperones have also recently been found to suppress the cytotoxic effects of protein aggregates after their formation [5, 8, 39, 44]. At present, there is no successful remedy that tackles and resolves the underlying causes of multiple debilitating neurological disorders, which include selective neuronal vulnerability and protein misfolding events. Therapeutic agents that reverse deleterious protein misfolding and aggregation in a manner that restores previously misfolded proteins to native form and function could be highly beneficial as they would simultaneously eliminate any deleterious loss of function or toxic gain of function caused by misfolded conformers [29].

Neurons are particularly susceptible to proteostasis network dysfunction compared to other cell types [35]. An inability for the neuron to maintain its proteome leads to the accumulation of misfolded/aggregating proteins, subsequently causing proteome collapse and the formation of inclusions [33]. An accumulation of misfolded/aggregated proteins and inclusions can be observed within distinct neuronal cell types and anatomical regions in several neurodegenerative disorders including Alzheimer's (AD), Parkinson's (PD), Huntington's disease (HD) [66]. Similarly, amyotrophic lateral sclerosis (ALS), which causes the progressive loss of upper and lower motor neurons, also exhibits the accumulation of misfolded proteins and inclusions within motor neurons [37, 59].

These inclusions are essentially created up of misfolded, aggregated, and at times hazardous kinds of one or more particular proteins. As an example, α-synuclein protein deposition into Lewy bodies (which is usually in substantia nigra) shows Parkinson's disease (PD); amyloid-β plaques and phosphorylated tau tangles point to Alzheimer's disease (AD); and aggregates of huntingtin with expanded glutamine repeats in the cytoplasm and nucleus indicate Huntington's disease. In contrast, ALS pathology is characterized by a number of different types of inclusions, including Bunina bodies, and ubiquitinated, hyaline, and skein-like inclusions [10]. Thus, the presence of inclusions, visible products of protein aggregation in a number of neurodegenerative diseases, indicates that a significant disruption to proteostasis has occurred [2].

Synapses, referring to sites located between two neuronal junctions, are essential for connecting neurons to create circuits. Synapses are necessary for an organism's continuing existence in order to sustain the brain's normal activity and functionality. Synaptic functions require a large complement of specialized proteins, such as neurotransmitter receptors, synaptic vesicles, endo- and exocytosis proteins, and organizers of the active zone and postsynaptic density, all of which are essential for neurotransmission [17]. When they undergo methods designed for precise changes in shape and high-energy protein–protein interactions, synaptic proteins remain constantly under proteostatic stress. It has also been noted that synapses are enriched in proteins susceptible to misfolding and aggregation, such as a-synuclein, tau, and amyloid-b, all of which are linked to neurodegenerative diseases [14, 19, 34, 54, 55].

Neurodegenerative diseases, such as Parkinson's disease and Alzheimer's disease, have common features like the deposition and aggregation of misfolded proteins inside and outside of brain neurons, as well as the selective dying of neurons in the central nervous system [30].

Some notable chaperones involved in synaptic proteostasis include the following (Table 10.1):

1. Small HSPs [30]
2. Hsp 40 (Heat Shock Protein 40) [30]
3. Hsp 60 (Heat Shock Protein 60) [30]
4. Hsp70 (Heat Shock Protein 70) [7, 48]
5. Hsp90 (Heat Shock Protein 90) [67]
6. Hsp110 (Heat Shock Protein 110)
7. Hsp84 (Heat Shock Protein 84) [7]
8. Hspa5 [67]
9. HspA4/A4L [67]
10. Hsp8 (Heat Shock Protein 8) [7, 48]
11. Hsp70-4 (Heat Shock Protein 70-4) [67]

12. FKBP (FK506-binding protein) [30]
13. TRiC (TCP-1 ring complex) [30]
14. Hsc70 (Heat Shock Cognate 70) [7, 48]
15. Hsc71 (Heat Shock Cognate 71) [7, 48]
16. DNAJC6 (Auxilin) [7]
17. DNAJC5 (CSPα) [58, 67]
18. DNAJ homologus [48]
19. Hsp40-3 cognate [48]
20. DNAJA [67]
21. Hsp40 (DNAJB1) [57]
22. Chaperonin TriC [67]
23. Hsp105/110 [67]

Table 10.1 List of chaperones with their functions

S. no.	Name of chaperons	Location	Molecular weight (kDa)	Functions	References
1	Small HSPs	Cytosol, endoplasmic reticulum, and nucleus	12–43	Attenuate protein aggregation Involved in thermotolerance and cellular development and differentiation Ubiquitin proteasome degradation of misfolded or denatured proteins	Powers and Workman [49], Sun and MacRae [56], Sarto et al. [52]
2	HSP 40 (DnaJ)	Cytosol	40	Co-chaperones for HSP70 Stimulate ATPase activity of HSP70	Han and Christen [20]
3	HSP60	Mitochondria	60	Transportation and transmission of mitochondrial protein Key role in apoptosis	Tutar and Tutar [61], Ranford et al. [50]
4	HSP 70	Cytosol and endoplasmic reticulum	70	Binds to misfolded peptides and increases the ATP hydrolysis Neuroprotective role in protein misfolding diseases including all neurodegenerative disorders	Fan et al. [12], Mapa et al. [40], Turturici et al. [60]
5	HSP90 (dimeric protein)	Cytosol, endoplasmic reticulum, and nucleus	90	Stabilize certain protein complexes and assist in protein degradation	Gallo et al. [16]
6	HSP104/110	Cytosol and endoplasmic reticulum	104/110	Protein degradation and transcription factor	Schirmer et al. [53], Gottesman et al. [18]

10.3 Role of Co-Chaperones

Co-chaperones play a crucial role in synaptic proteostasis and neurodegenerative disease by regulating the activity of chaperones, modifying their substrate specificity, and influencing protein degradation pathways. Co-chaperones can perform the following:

1. Modulate chaperone activity: Co-chaperones can activate or inhibit chaperone function, fine-tuning protein folding and degradation.
2. Alter substrate specificity: Co-chaperones can change the binding preferences of chaperones, targeting specific proteins for folding or degradation.
3. Regulate protein degradation: Co-chaperones can direct proteins to specific degradation pathways, such as the proteasome or lysosome.
4. Influence synaptic plasticity: Co-chaperones can impact synaptic strength and learning by regulating the dynamics of synaptic proteins [17].

In neurodegenerative diseases, co-chaperones may perform the following:

1. Disrupt chaperone function: Co-chaperone dysregulation can impair chaperone activity, leading to protein misfolding and aggregation.
2. Alter protein degradation: Co-chaperone changes can redirect proteins to harmful degradation pathways, contributing to neurodegeneration.
3. Exacerbate synaptic loss: Co-chaperone dysfunction can disrupt synaptic protein homeostasis, accelerating synaptic loss.

Some notable co-chaperones involved in synaptic proteostasis include the following [36] (Table 10.2):

- Hop (Hsp70/Hsp90 organizing protein)
- Chip (Carboxyl terminus of Hsc70-interacting protein)
- BAG (Bcl-2-associated athanogene) family
- SGT (Small glutamine-rich tetratricopeptide repeat-containing protein)
- PPIases (Peptidyl-prolyl cis-trans isomerases)
- HSF binding protein1 [48]
- Calnexin [7]
- CCT family [7]
- FLJ10737 [48]
- HIP [67]

Dysregulation of co-chaperones has been implicated in various neurodegenerative diseases, including Alzheimer's, Parkinson's, and Huntington's. Understanding co-chaperone roles may reveal new therapeutic targets for neurodegenerative diseases.

Table 10.2 List of co-chaperones with their functions

S. no.	Name of co-chaperons	Functions	References
1	CDC37 (cell division cycle 37)	Prevents closure of the "lid" in HSP90 Specific for maturation of kinases	Wang et al. [63], Bohush et al. [6]
2	Aha1/AHSA1 (activator of Hsp90 ATPase activity 1)	Stimulates ATPase activity of HSP90	
3	p23/PTGES3 Co-chaperone p23 (Sba1 in yeast)	Stabilizes the HSP90 and inhibits Hsp90 ATPase activity	
4	Hop/Sti1 (stress-inducible phosphoprotein 1/Hop)	Transfers clients from Hsp70 to Hsp90 and inhibits Hsp90 ATPase activity	
5	PP5/Ppt1 in yeast (protein phosphatase 5)	Dephosphorylates Hsp90 and dephosphorylates Cdc37	
6	CHIP/STUB1 (carboxyl terminus of Hsp70-interacting protein)	E3 ubiquitin ligase	
7	FKBP51 (FK506 binding protein 5)	Peptidyl-prolyl isomerase activity; participates in Hsp90-steroid receptor complex; generally regulates Hsp90 conformational cycle	
8	CacyBP/SIP/CACYBP (calcyclin binding protein/Siah-1 interacting protein)	Calcyclin binding protein/Siah-1 interacting protein	
9	USP19/USP19 (cytoplasmic ubiquitin-specific protease 19)	Ubiquitin-specific protease activity	
10	FKBP52/FKBP4 (FK506 binding protein 4)	Peptidyl-prolyl isomerase activity; participates in Hsp90-steroid receptor complex; generally, regulates Hsp90 conformational cycle	

Neurodegenerative illnesses come out as consequence of the four synaptic chaperones' role in sustaining the nerve terminal and its genetic makeup. Three of these are Hsp40 co-chaperones (DNAJs):

1. Cysteine String Protein alpha (CSPα; DNAJC5)
2. Auxilin (DNAJC6)
3. Receptor-Mediated Endocytosis 8 (RME-8; DNAJC13)

These co-chaperones contain a conserved J domain that allows them to interact with heat shock cognate 70 (Hsc70) to create a complex that improves chaperones' ATPase activity. A protein that exists in the synaptic vesicles CSPα chaperones the endocytic GTPase dynamin-1 and the t-SNARE SNAP-25, which in turn regulates the endocytosis and exocytosis of the vesicles. Auxilin is a clathrin cage binder that facilitates the uncoating of clathrin-coated vesicles through its interaction with Hsc70. This process is essential for the regeneration of synaptic vesicles [17].

RME-8 is a co-chaperone on endosomes and may be active in the clathrin-coated vesicle's endocytosis on this specific organelle. These three co-chaperones do not disintegrate preexisting aggregates; rather, they preserve assembly and folding to prevent client aggregation hence sustaining client function. Heat shock protein 110 (Hsp110), which interacts with Hsc70, DNAJAs, and DNAJBs to form a disaggregase enzyme, is the fourth synaptic chaperone. At the synapse, hsp110-related disaggregase activity is known to shield protein aggregations like α-synuclein. These co-chaperones are vital to the neurological system [17].

10.4 Therapeutic Implications

Given the critical roles of chaperones and co-chaperones in synaptic proteostasis, therapeutic strategies targeting these proteins show promise [21, 32, 43, 51].

- *Enhancing Chaperone Activity*: Small molecules that enhance the expression or activity of chaperones may help alleviate protein aggregation.
- *Gene Therapy*: Increasing the levels of protective chaperones through gene delivery systems could restore proteostasis in vulnerable neurons.
- *Cochaperone Modulators*: Drugs that modulate the activity of co-chaperones may improve the specificity and efficiency of the chaperone machinery.

10.5 Conclusion

Chaperones and co-chaperones are indispensable for maintaining synaptic proteostasis and protecting against neurodegenerative diseases. Here is a simplified outline of the role of chaperones in synaptic proteostasis and neurodegenerative disease:

1. Protein Synthesis: Chaperones help fold newly synthesized proteins into their correct conformation.
2. Protein Quality Control: Chaperones recognize and refold misfolded proteins, preventing aggregation.
3. Protein Degradation: Chaperones facilitate the degradation of misfolded proteins through proteasomal and lysosomal pathways.
4. Synaptic Plasticity: Chaperones regulate synaptic protein dynamics, influencing synaptic strength and learning.
5. Neurodegenerative Disease: Chaperone overload or dysfunction leads to protein aggregation, synaptic loss, and neurodegeneration.

Dysregulation of chaperone networks contributes to various neurodegenerative disorders like Parkinson's, ALS, Alzheimer's, and Huntington's disease. By understanding the role of chaperones in synaptic proteostasis, researchers may uncover new therapeutic strategies for neurodegenerative diseases. Future research should focus on elucidating the specific interactions and regulatory mechanisms of these proteins, paving the way for innovative therapeutic approaches aimed at restoring proteostasis in neurodegenerative conditions. When co-chaperones are impaired, it disrupts the stability and maintenance of synaptic proteins, leading to the loss of synapses. This synaptic degeneration is a hallmark of many neurodegenerative diseases, including Alzheimer's, Parkinson's, and Huntington's diseases. As synapses deteriorate, neuronal communication weakens, resulting in cognitive decline and loss of motor function.

Consequently, chaperones and co-chaperones play an important role in preventing synaptic loss and maintaining neural communication. Continued research into their mechanisms and interactions with other cellular components may lead to new therapeutic targets for neurodegenerative disorders. Restoring or enhancing the function of chaperones and co-chaperones could potentially prevent or slow down the progression of these diseases, making them a key area of study in neuroscience.

References

1. Alecki C, Rizwan J, Le P, Jacob-Tomas S, Xu S, Minotti S, et al. Localized synthesis of molecular chaperones sustains neuronal proteostasis. bioRxiv. 2023. https://doi.org/10.1101/2023.10.03.560761.
2. Altinok S, Hodge RS. With or without you: co-chaperones mediate health and disease by modifying chaperone function and protein triage. Cells. 2021;2021(10):3121.
3. Bagatell R, Whitesell L. Altered Hsp90 function in cancer: a unique therapeutic opportunity. Mol Cancer Ther. 2004;3(8):1021–30.
4. Batulan Z, Shinder GA, Minotti S, He BP, Doroudchi MM, Nalbantoglu J, et al. High threshold for induction of the stress response in motor neurons is associated with failure to activate HSF1. J Neurosci. 2003;23:5789–98.
5. Binger KJ, Ecroyd H, Yang S, Carver JA, Howlett GJ, Griffin MD. Avoiding the oligomeric state: αB-crystallin inhibits fragmentation and induces dissociation of apolipoprotein C-II amyloid fibrils. FASEB J. 2013;27:1214–22.
6. Bohush A, Bieganowski P, Filipek A. Hsp90 and its co-chaperones in neurodegenerative diseases. Int J Mol Sci. 2019;20(20):4976.
7. Burré J, Beckhaus T, Corvey C, Karas M, Zimmermann H, Volknandt W. Synaptic vesicle proteins under conditions of rest and activation: analysis by 2-D difference gel electrophoresis. Electrophoresis. 2006;27:3488–96. https://doi.org/10.1002/elps.200500864.
8. Cascella R, Conti S, Mannini B, Li X, Buxbaum JN, Tiribilli B, Chiti F, Cecchi C. Transthyretin suppresses the toxicity of oligomers formed by misfolded proteins in vitro. Biochim Biophys Acta. 2013;1832:2302–14.
9. Chen S, Smith DF. Hop as an adaptor in the heat shock protein 70 (Hsp70) and hsp90 chaperone machinery. J Biol Chem. 1998;273(52):35194–200.
10. Ciechanover A. Intracellular protein degradation: from a vague idea, through the lysosome and the ubiquitin–proteasome system, and onto human diseases and drug targeting (Nobel lecture). Angew Chem Int Ed. 2005;44(37):5944–67.
11. Ellis RJ. Molecular chaperones: assisting assembly in addition to folding. Trends Biochem Sci. 2006;31(7):395–401.
12. Fan CY, Lee S, Cyr DM. Mechanisms for regulation of Hsp70 function by Hsp40. Cell Stress Chaperones. 2003;8(4):309.
13. Finka A, Mattoo RU, Goloubinoff P. Experimental milestones in the discovery of molecular chaperones as polypeptide unfolding enzymes. Annu Rev Biochem. 2016;85:715–42.
14. Fortin DL, Troyer MD, Nakamura K, Kubo S-I, Anthony MD, Edwards RH. Lipid rafts mediate the synaptic localization of α-synuclein. J Neurosci. 2004;24:6715–23.

15. Frydman J, Nimmesgern E, Erdjument-Bromage H, Wall JS, Tempst P, Hartl FU. Function in protein folding of TRiC, a cytosolic ring complex containing TCP-1 and structurally related subunits. EMBO J. 1992;11(13):4767–78.
16. Gallo LI, Lagadari M, Piwien-Pilipuk G, Galigniana MD. The 90-kDa heat-shock protein (Hsp90)-binding immunophilin FKBP51 is a mitochondrial protein that translocates to the nucleus to protect cells against oxidative stress. J Biol Chem. 2011;286(34):30152–60.
17. Gorenberg EL, Chandra SS. The role of co-chaperones in synaptic proteostasis and neurodegenerative disease. Front Neurosci. 2017;11:248. https://doi.org/10.3389/fnins.2017.00248. PMID: 28579939; PMCID: PMC5437171.
18. Gottesman S, Squires C, Pichersky E, Carrington M, Hobbs M, Mattick JS, et al. Conservation of the regulatory subunit for the Clp ATP-dependent protease in prokaryotes and eukaryotes. Proc Natl Acad Sci. 1990;87(9):3513–7.
19. Greten-Harrison B, Polydoro M, Morimoto-Tomita M, Diao L, Williams AM, Nie EH, et al. αβγ-Synuclein triple knockout mice reveal age-dependent neuronal dysfunction. Proc Natl Acad Sci. 2010;107:19573–8. https://doi.org/10.1073/pnas.1005005107.
20. Han W, Christen P. (2004). Cis-Effect of DnaJ on DnaK in ternary complexes with chimeric DnaK/DnaJ-binding peptides. FEBS letters. 2004;563(1–3):146–150.
21. Hartl FU, Hayer-Hartl M. Molecular chaperones in the cytosol: from nascent chain to folded protein. Nat Rev Mol Cell Biol. 2002;3(10):783–96.
22. Hartl FU, Hayer-Hartl M. Converging concepts of protein folding in vitro and in vivo. Nat Struct Mol Biol. 2009;16:574–81. https://doi.org/10.1038/nsmb.1591.
23. Hartl FU, Bracher A, Hayer-Hartl M. Molecular chaperones in protein folding and proteostasis. Nature. 2011;475:324–32. https://doi.org/10.1038/nature10317.
24. Haslbeck M, Vierling E. A first line of stress defense: small heat shock proteins and their function in protein homeostasis. J Mol Biol. 2015;427(7):1537–48.
25. Hetz C, Chevet E, Oakes SA. Proteostasis control by the unfolded protein response. Nat Cell Biol. 2015;17(7):829–38.
26. Hodson S, Marshall JJ, Burston SG. Mapping the road to recovery: the ClpB/Hsp104 molecular chaperone. J Struct Biol. 2012;179:161–71. https://doi.org/10.1016/j.jsb.2012.05.015.
27. Hoffmann JH, Linke K, Graf PC, Lilie H, Jakob U. Identification of a redox-regulated chaperone network. EMBO J. 2004;23(1):160–8.
28. Itoh H, Komatsuda A, Ohtani H, et al. Mammalian HSP60 is quickly sorted into the mitochondria under conditions of dehydration. Eur J Biochem. 2002;269(23):5931–8.
29. Jackrel ME, Shorter J. Engineering enhanced protein disaggregases for neurodegenerative disease. Prion. 2015;9:90–109.
30. Jee H. Size dependent classification of heat shock proteins: a mini-review. J Exerc Rehabil. 2016;12(4):255.
31. Kaarniranta K, Oksala N, Karjalainen HM, Suuronen T, Sistonen L, Helminen HJ, et al. Neuronal cells show regulatory differences in the hsp70 gene response. Mol Brain Res. 2002;101:136–40. https://doi.org/10.1016/S0169-328X(02)00179-1.
32. Klaips CL, Jayaraj GG, Hartl FU. Pathways of cellular proteostasis and their implications for neurodegenerative disease. Nat Rev Neurosci. 2018;19(5):253–68.
33. Knowles TP, Vendruscolo M, Dobson CM. The amyloid state and its association with protein misfolding diseases. Nat Rev Mol Cell Biol. 2014;15(6):384–96.
34. Kramer ML, Schulz-Schaeffer WJ. Presynaptic alpha-synuclein aggregates, not Lewy bodies, cause neurodegeneration in dementia with Lewy bodies. J Neurosci. 2007;27:1405–10. https://doi.org/10.1523/JNEUROSCI.4564-06.2007.
35. Kundra R, Dobson CM, Vendruscolo M. A cell-and tissue-specific weakness of the protein homeostasis system underlies brain vulnerability to protein aggregation. iScience. 2020;23(3):100934.
36. Lackie RE, Maciejewski A, Ostapchenko VG, Marques-Lopes J, Choy WY, Duennwald ML, et al. The Hsp70/Hsp90 chaperone machinery in neurodegenerative diseases. Front Neurosci. 2017;11:254.

37. Leigh PN, Whitwell H, Garofalo O, Buller J, Swash M, Martin JE, et al. Ubiquitin-immunoreactive intraneuronal inclusions in amyotrophic lateral sclerosis: morphology, distribution, and specificity. Brain. 1991;114(2):775–88.
38. Manning-Krieg UC, Scherer PE, Schatz G. Sequential action of mitochondrial chaperones in protein import into the matrix. EMBO J. 1991;10:3273–80.
39. Mannini B, Mulvihill E, Sgromo C, Cascella R, Khodarahmi R, Ramazzotti M, Dobson CM, Cecchi C, Chiti F. Toxicity of protein oligomers is rationalized by a function combining size and surface hydrophobicity. ACS Chem Biol. 2014;9:2309–17.
40. Mapa K, Sikor M, Kudryavtsev V, Waegemann K, Kalinin S, Seidel CA, et al. The conformational dynamics of the mitochondrial Hsp70 chaperone. Mol Cell. 2010;38(1):89–100.
41. Marcuccilli CJ, Mathur SK, Morimoto RI, Miller RJ. Regulatory differences in the stress response of hippocampal neurons and glial cells after heat shock. J Neurosci. 1996;16:478–85.
42. McGreal RS, Kantorow WL, Chauss DC, Wei J, Brennan LA, Kantorow M. αB-crystallin/sHSP protects cytochrome c and mitochondrial function against oxidative stress in lens and retinal cells. Biochim Biophys Acta. 2012;1820(7):921–30.
43. Muchowski PJ, Wacker JL. Modulation of neurodegeneration by molecular chaperones. Nat Rev Neurosci. 2005;6(11):859–67.
44. Ojha J, Masilamoni G, Dunlap D, Udoff RA, Cashikar AG. Sequestration of toxic oligomers by HspB1 as a cytoprotective mechanism. Mol Cell Biol. 2011;31:3146–57.
45. Otto H, Conz C, Maier P, Wölfle T, Suzuki CK, Jenö P, et al. The chaperones MPP11 and Hsp70L1 form the mammalian ribosome-associated complex. Proc Natl Acad Sci. 2005;102(29):10064–9.
46. Pardue S, Groshan K, Raese JD, Morrison-Bogorad M. Hsp70 mRNA induction is reduced in neurons of aged rat hippocampus after thermal stress. Neurobiol Aging. 1992;13:661–72. https://doi.org/10.1016/0197-4580(92)90088-F.
47. Pauwels K, Van Molle I, Tommassen J, Van Gelder P. Chaperoning Anfinsen: the steric foldases. Mol Microbiol. 2007;64(4):917–22.
48. Phillips GR, Florens L, Tanaka H, Khaing ZZ, Fidler L, Yates JR, et al. Proteomic comparison of two fractions derived from the transsynaptic scaffold. J Neurosci Res. 2005;81:762–75. https://doi.org/10.1002/jnr.20614.
49. Powers MV, Workman P. Inhibitors of the heat shock response: biology and pharmacology. FEBS Lett. 2007;581:3758–69.
50. Ranford JC, Coates AR, Henderson B. Chaperonins are cell-signalling proteins: the unfolding biology of molecular chaperones. Expert Rev Mol Med. 2000;2(8):1–17.
51. Robinson DR, et al. Chaperone-mediated autophagy in the degradation of aggregated proteins: implications for neurodegeneration. Trends Biochem Sci. 2011;36(10):546–54.
52. Sarto C, Binz PA, Mocarelli P. Heat shock proteins in human cancer. Electrophoresis. 2000;21(6):1218–26.
53. Schirmer EC, Glover JR, Singer MA, Lindquist S. HSP100/Clp proteins: a common mechanism explains diverse functions. Trends Biochem Sci. 1996;21(8):289–96.
54. Schubert W, Prior R, Weidemann A, Dircksen H, Multhaup G, Masters CL, et al. Localization of Alzheimer beta A4 amyloid precursor protein at central and peripheral synaptic sites. Brain Res. 1991;563:184–94. https://doi.org/10.1016/0006-8993(91)91532-6.
55. Schulz-Schaeffer WJ. The synaptic pathology of α-synuclein aggregation in dementia with Lewy bodies, Parkinson's disease and Parkinson's disease dementia. Acta Neuropathol. 2010;120:131–43. https://doi.org/10.1007/s00401-010-0711-0.
56. Sun Y, MacRae TH. The small heat shock proteins and their role in human disease. FEBS J. 2005;272(11):2613–27.
57. Suzuki T, Usuda N, Murata S, Nakazawa A, Ohtsuka K, Takagi H. Presence of molecular chaperones, heat shock cognate (Hsc) 70 and heat shock proteins (Hsp) 40, in the post-synaptic structures of rat brain. Brain Res. 1999;816:99–110. https://doi.org/10.1016/S0006-8993(98)01083-X.

58. Takamori S, Holt M, Stenius K, Lemke EA, Grønborg M, Riedel D, et al. Molecular anatomy of a trafficking organelle. Cell. 2006;127:831–46. https://doi.org/10.1016/j.cell.2006.10.030.
59. Ticozzi N, Ratti A, Silani V. Protein aggregation and defective RNA metabolism as mechanisms for motor neuron damage. CNS Neurol Disord Drug Targets. 2010;9(3):285–96.
60. Turturici G, Sconzo G, Geraci F. Hsp70 and its molecular role in nervous system diseases. Biochem Res Int. 2011;2011(1):618127.
61. Tutar L, Tutar Y. Heat shock proteins; an overview. Curr Pharm Biotechnol. 2010;11(2):216–22.
62. Tyedmers J, Mogk A, Bukau B. Cellular strategies for controlling protein aggregation. Nat Rev Mol Cell Biol. 2010;11:777–88. https://doi.org/10.1038/nrm2993.
63. Wang L, Bergkvist L, Kumar R, Winblad B, Pavlov PF. Targeting chaperone/co-chaperone interactions with small molecules: a novel approach to tackle neurodegenerative diseases. Cells. 2021;10:2596. https://doi.org/10.3390/cells10102596.
64. Weibezahn J, Schlieker C, Tessarz P, Mogk A, Bukau B. Novel insights into the mechanism of chaperone-assisted protein disaggregation. Biol Chem. 2005;386:739–44.
65. Winkler J, Tyedmers J, Bukau B, Mogk A. Chaperone networks in protein disaggregation and prion propagation. J Struct Biol. 2012;179:152–60.
66. Yerbury JJ, Ooi L, Dillin A, Saunders DN, Hatters DM, Beart PM, Cashman NR, Wilson MR, Ecroyd H. Walking the tightrope: proteostasis and neurodegenerative disease. J Neurochem. 2016;137(4):489–505.
67. Zhang Y-Q, Henderson MX, Colangelo CM, Ginsberg SD, Bruce C, Wu T, et al. Identification of CSPα clients reveals a role in dynamin 1 regulation. Neuron. 2012;74:136–50. https://doi.org/10.1016/j.neuron.2012.01.029.
68. Zuo D, Subjeck J, Wang XY. Unfolding the role of large heat shock proteins: new insights and therapeutic implications. Front Immunol. 2016;7:180703.

Proteasomal Degradation and Autophagy in Neurodegenerative Diseases

11

Neha Sharma, Sharmistha Banerjee, and Shuchi Kaushik

11.1 Introduction

All proteins in the cell are continuously synthesized and altered by translation and post-translation mechanisms, respectively, for proper cellular growth and differentiation. Proteins inside the cell occur in a dynamic state to regulate its amount and functions. Two main automated cell machineries continuously work on protein regulation, i.e., proteasomal degradation and autophagy. Both ensure proteolysis and are involved in homeostatic control of protein in eukaryotic cells. The proteasome ubiquitin system targets the misfolded or abnormal proteins or short-lived regulatory proteins for degradation. It is involved in the homeostatic control of proteins within the cell and regulates the accumulation of surplus unusual proteins, which is usually harmful to the cells [50]. On the other hand, the lysosomal proteolysis was assumed to largely degrade bulk proteins and engulf cytoplasmic components non-selectively [30]. It seems likely that proteasomal destruction and autophagy are interrelated, with ubiquitin serving as a shared signal for misfolded proteins. Proteolysis is also extremely important to the elaborate quality-control steady-state levels of proteins (proteostasis), and it is becoming evident that failure in proteostasis causes a number of neurodegenerative disorders [10, 32].

Neurodegenerative diseases such as Alzheimer's, Huntington's, Parkinson's, and amyotrophic lateral sclerosis (ALS) include inadequate protein quality control mechanisms in neuronal cells, which are briefly described in Table 11.1.

N. Sharma (✉)
Amity Institute of Biotechnology, Amity University, Gwalior, Madhya Pradesh, India
e-mail: nsharma2@gwa.amity.edu

S. Banerjee
Chhattisgarh Swami Vivekanand Technical University, Bhilai, India

S. Kaushik
State Forensic Science Laboratory, Sagar, Madhya Pradesh, India

N. Puranik (ed.), *Proteostasis*, https://doi.org/10.1007/978-981-96-6202-9_11

Table 11.1 Overview of neurodegenerative diseases

S.No	Name of disease	Symptoms	Effects	Regulation	References
1	Alzheimer's disease	Dementia (memory decline)	Extracellular accumulation of Aβ Formation of intracellular neuro-fibrillary tangles Hyper-phosphorylation tau proteins	Proteolytic degradation balances Aβ synthesis and transports it from the brain to the peripheral bloodstream via passive and active mechanisms	[7, 40]
2	Parkinson's disease	Motor system	Harm of dopaminergic (DA) neurons Progressive accumulation of α-SYN Affect the roles of endoplasmic reticulum and Golgi bodies Inhibit the activity of proteasomal and lysosomal proteins	Aggregates degrades by macro-autophagy and chaperone-mediated autophagy (CMA)	[24, 26, 28]
3	Amyotrophic lateral sclerosis	Progressive destruction to motor neurons Loss of muscular control	Deposition of ubiquitinated TAR DNA-binding protein 43 (TDP-43) in cytoplasm, which is insoluble and toxic	Polyubiquitin insoluble chains of TDP-43 are degraded by ubiquitin proteasomal complex and autophagy	[1, 34, 55]
4	Huntington's disease	CAG trinucleotide repeat expansion in exon 1 Less muscle coordination and psychosis	40 repeats of CAG trinucleotide cause disease conditions	Aggregates degraded by autophagy	[8, 16, 27, 45]

It was reported that protein misfolding, abnormal aggregation and deposition of proteins in cytoplasm, are the common factors of neurodegenerative disorders. The ineffective clearance of misfolded proteins by the proteasome and autophagy may be one of the causes of neurodegenerative illness. It causes the accumulation of harmful misfolded protein in the form of oligomers and aggregates in neurons, disrupting cellular activities and eventually leading to the death of brain cells, indicating neurodegenerative disease [13, 49]. The major roles of proteostasis and

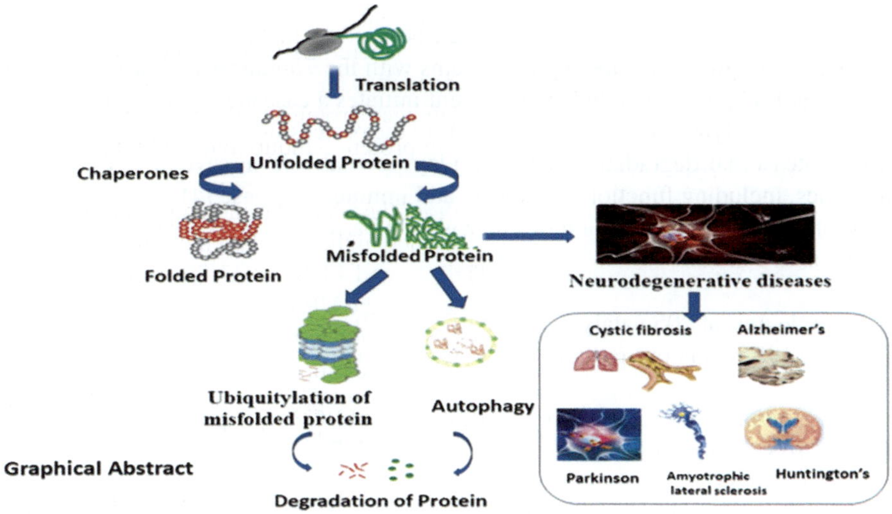

Fig. 11.1 Process of folding, misfolding protein, and degradation, along with the connection to neurodegenerative diseases

autophagy are maintaining cellular health and preventing the formation of aggregation of misfolded protein [36, 48].

There are two main protein degradation systems: the ubiquitin–proteasome system (UPS), which is responsible for the degradation of both functional and dysfunctional short-lived proteins associated with ubiquitin molecules [11, 39], and the second one is autophagy–lysosomal system which is responsible for degradation of long-lived proteins, accumulated misfolded proteins, nonfunctional organelles and cellular components [38].

So, it is an important event for the proper maintenance of the cellular observation system, including the degradation of misfolded or aggregated proteins through degradative pathways, i.e., proteostasis or proteasomal degradation and autophagy (Fig. 11.1).

11.2 Proteostasis or Proteasomal Degradation

Proteostasis, also known as protein homeostasis, is the process of ensuring that intracellular and extracellular proteins function properly. Proteostasis ensures that newly synthesized proteins fold properly, as well as that misfolded proteins unfold, refold, and degrade [33]. With aging, the control of proteostasis in postmitotic neurons becomes exclusively critical, as the selection of proteostasis machinery is limited, resulting in the aggregation of damaged or nonfunctional proteins and cell organelles [21].

Proteostasis is maintained by the proteasome complex. Proteasomes are huge multisubunit protease complexes that degrade misfolded, unnecessary, or damaged proteins through proteolysis [51]. Damaged proteins are degraded by proteasomes, which produce amino acids that can then be used to make new proteins. Protein breakdown requires the labeling of proteins with the 76-amino acid protein known as "ubiquitin" [56]. A single labeling event initiates a cascade, resulting in the production of a polyubiquitin chain that binds to the proteasome for proteolysis [4, 25]. The proteasomal degradation pathway plays a vital role in several physiological activities, including function, cell cycle, and immune response [4].

Proteosome is a 2.5-MDa molecular weight barrel-shaped regulatory protein. It contains two units:

(i) Single unit of 20S catalytic core particle (CP)
(ii) Two 19S regulatory particles (RPs) terminal units

RP units are connected by both sides of central CP in opposite directions and formed 26S holo-proteasome functional unit [2, 5, 9].

The CP consists of four heptameric rings, two outer α-rings, and two inner β-rings. These two rings contain seven structurally related α and β subunits. The CP contains three types of β subunits:

(i) β1, which is a caspase-like subunit,
(ii) β2, which is trypsin-like subunit,
(iii) β5, which is chymotrypsin-like activities.

These subunits have catalytically active threonine residues at their N termini, as well as an N-terminal nucleophile (Ntn) hydrolase that degrades peptide bonds at post-acidic, basic, and hydrophobic amino acid residues. The crystal structures of the CP show that the core of the α-ring is nearly completely closed, preventing substrate proteins from reaching the inner cavity of the β-ring, which includes proteolytically active sites [12, 53]. The N-terminal α subunits act as a physical barrier, preventing substrate proteins from entering active regions in the β-ring. This shows that CP is inactive in cells. To access active sites, a substrate protein must pass through a narrow opening in the closed gate at the middle of the α-ring. The RP consists of a number of distinct integral components divided into two groups: The regulatory particle comprises of six triple-ATPase (Rpt) subunits, followed by 13 non-ATPase (Rpn) subunits. The proteasome detects and degrades ubiquitinated proteins, resulting in peptides. The RP subunit opens the α-ring within the CP's closed gate. Thus, the RP contributes significantly to the target's annihilation. Rpn10 and Rpn13 have been discovered as two ubiquitin receptors located at the apex of the RP complex, toward the periphery [3, 6, 22, 23, 43]. Rpn11, a deubiquitinylating enzyme (DUB) involved in the deubiquitinylation of captured substrates, is another important RP receptor. Metalloisopeptidase aids in the recycling of ubiquitin in this process as well. Importantly, Rpn11 is located near the substrate entrance point in the ATPase ring, which is composed of six homologous AAA–ATPase subunits (Rpt1-Rpt6) arrayed in a crown-like configuration. The ATPase ring serves three functions [47]:

1. Facilitates substrate unfolding after the Rpt ring drags it into the central pore.
2. Opens a channel in the closed α-ring.
3. Transports substrates to the CP, ensuring their destruction.

The endoplasmic reticulum (ER) is a significant contributor to proteostasis, and the unfolded protein response (UPR) regulates it. The UPR and ER-associated degradation (ERAD) work along with the UPS and the autophagy–lysosomal system to prevent protein misfolding and its effects. Proteostasis deficits are widely thought to be linked to a number of neurodegenerative diseases, including Alzheimer's, Parkinson's, and ALS, which are distinguished by neuronal death in various regions of the central nervous system. Despite their disparate clinical consequences, all of these illnesses show a buildup of protease-resistant misfolded and aggregated pathogenic proteins [48].

11.3 Autophagy

Autophagy is a degradative recycling system that acts alongside the lysosome. Lysosomes serve as an intracellular digesting system, removing and destroying unnecessary or useless cellular components. It promotes cellular health by recycling metabolic components, mending cell membranes, signaling, and supplying nutrients during nutrient deficiency. Mutations in lysosomal protein synthesis genes have been linked to around 40 human genetic disorders, including lysosomal storage diseases [17, 35, 37, 46].

The primary autophagic pathway is known as macroautophagy or simply autophagy; however, there are other forms of autophagy, such as microautophagy and chaperone-mediated autophagy, whose physiological roles are mostly unknown [14] (Table 11.2).

On the other hand, starvation-induced adaptive autophagy and basal or constitutive autophagies have functionally different roles, but they share the same molecular genes, i. e., Atg (autophagy-related gene) proteins [29]. Furthermore, autophagy is responsible for the global turnover of long-lived but not short-lived proteins, as well as the MHC class II presentation of endogenous proteins (adaptive immunity) [52].

The earlier type's primary function is to deliver nutrients to cells during an energy crisis, and it is significantly activated in response to a variety of stimuli, including starvation (nutritional step-down) and humoral/trophic hormones (e.g., glucagon). Indeed, in response to nutritional stress, an isolation membrane emerges in the cytosol and eventually elongates to sequester vast amounts of cytoplasmic components. To adjust to starvation, autophagosomes randomly secure cytoplasmic components such as organelles. Subsequently, the membrane's edges fuse together to form autophagosomes, which are double membrane structures. An autophagosome quickly fuses with a lysosome, and its contents, along with the inner membrane, are destroyed by a variety of lysosomal digestive hydrolases. Interestingly, autophagy-deficient mice manufacture fewer amino acids throughout the neonatal starving period and die within 1 day after birth [20]. Similarly, autophagy is required

Table 11.2 Types of autophagy

S.No.	Types	Mechanism	References
Major autophagy			
1	Macro-autophagy	Cytoplasmic substances are bundled into vesicular structures called autophagosomes, which subsequently interact with lysosomes to assist in the breakdown and recycling of the contents	[44, 54]
2	Micro-autophagy	The cytoplasm is directly engulfed into lysosomes by invaginating the lysosomal membrane	[19, 44]
3	Chaperone-associated autophagy	Chaperone-mediated autophagy is a process in which particular proteins called "chaperones" recognize, direct, and move substrate proteins across the lysosomal membrane into the lumen for breakdown	[19]
Minor autophagy			
4	Adaptive autophagy	Induced by cell starvation and a nonselective mechanism	[29]
5	Basal or constitutive autophagies	Even under nutrient-rich conditions, it is constitutively active at low levels, and it plays an important role in the overall turnover of cellular components and quality control of cytoplasmic materials, including organelles. It is activated by cell deprivation and is not a selective process	[14, 29]
5 A.	Mitophagy	Excess and damaged organelles such as mitochondria, peroxisomes, and endoplasmic reticulum are selectively disposed of	[52]
5 B.	Pexophagy		
5 C.	Reticulophagy		
5 D.	Ribophagy	Selective disposal of ribosomes	
5 E.	Macrolipophagy	Selective removal of lipid droplets	
5 F	Xenophagy	Selective removal of pathogens	
5 G	Aggrephagy	Selective clearance of aggregate-prone proteins	

for the preimplantation development of rat embryos under energy-deprivation conditions. Adaptive autophagy is typically described as a nonselective process (Fig. 11.2).

11.4 Importance of Ubiquitin in Neurogenerative Disease

Ubiquitin directs the protein for destruction via the proteosome complex or autophagy. Ubiquitin is ligated via a post-translation mechanism that involves an isopeptide bond between its terminal carboxyl residue. Ubiquitin primarily targets the lysine residue on the target protein. It has seven lysin residues, and K63 (lysin at the 63rd position) coupled polyubiquitinylation serves as an autophagy marker. This action is catalyzed by three distinct enzymes:

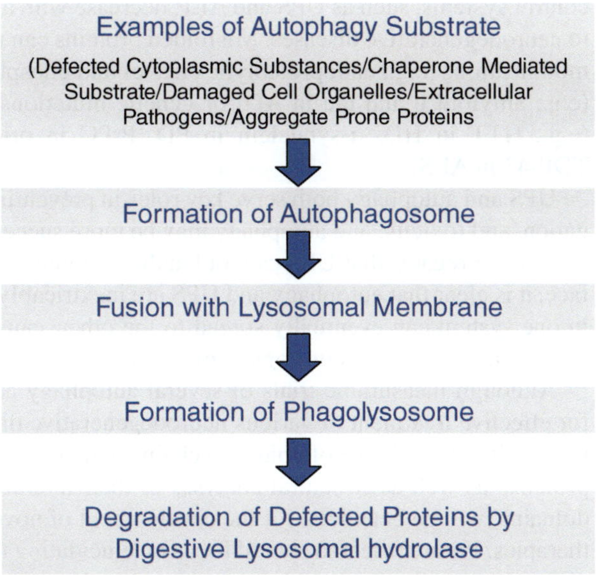

Fig. 11.2 Simplified process of autophagy

(i) Ubiquitin activating enzyme.
(ii) Ubiquitin conjugative enzyme.
(iii) Ubiquitin-ligating enzyme [15, 18, 41].

Ubiquitin has the potential to target the next ubiquitin molecule and construct a polyubiquitin chain. Ubiquitin is produced by mono- and polyubiquitin genes. Under normal conditions, all ubiquitin genes operate normally, but under stress, their function rapidly increases. Excess ubiquitin levels are detrimental; hence, ubiquitin homeostasis must be strictly maintained. Ubiquitin has been linked to various neurodegenerative diseases, including Alzheimer's disease, Parkinson's and Lewy body dementia, Huntington's disease, frontotemporal lobar degeneration, and amyotrophic lateral sclerosis. In these neurodegenerative illnesses, it was thought that both the ubiquitin–proteasome system (UPS) and autophagy would work due to a buildup of polyubiquitinylated proteins [5, 31, 42].

11.5 Conclusion

Proteasomal degradation and autophagy are vital processes in maintaining neuronal health, and their dysfunction is a hallmark of neurodegenerative diseases. Understanding the mechanisms that govern these pathways and their interactions may reveal novel therapeutic targets. Proteasome degradation and autophagy play critical roles in cellular responses. Ubiquitin signals provide a wide range of roles, both proteolytic and nonproteolytic. The ubiquitin–proteasome system and the autophagy–lysosome pathway both recognize ubiquitin signals. Protein quality control systems, such as UPS and ALP, decrease with age, contributing significantly to neurodegenerative diseases. Misfolded proteins can result from post-translational modifications (e.g., endoproteolytic cleaves and phosphorylation) of native proteins (e.g., amyloid β and tau in AD) or genetic mutations of non-pathogenic proteins (e.g., HTT in HD, α-synuclein in PD, PrPC in prion disease, and SOD1 and TDP-43 in ALS).

UPS and autophagy both serve key roles in preventing protein misfolding, aggregation, and toxicity, and autophagy may be more successful in dealing with massive protein aggregates that UPS cannot handle. Although we are only scraping the surface, it is clear that autophagy and UPS are inextricably linked, and early alterations in one system can eventually spread to the other, causing protein aggregation and toxicity in a range of neurological disorders.

Although measurable trials of several autophagy activators and UPS inhibition for effective treatment of various neurodegenerative diseases have been conducted, but detailed knowledge of ubiquitin chains in neurodegenerative disease-associated proteins, as well as structural analyses of their interactions with ubiquitin-binding domains, will be beneficial for the development of novel neurodegenerative disease therapies. Future research should focus on elucidating the precise molecular mechanisms involved and exploring combination therapies that could restore cellular homeostasis in neurodegenerative contexts.

References

1. Arai T, Hasegawa M, Akiyama H, Ikeda K, Nonaka T, Mori H, et al. TDP-43 is a component of ubiquitin-positive tau-negative inclusions in frontotemporal lobar degeneration and amyotrophic lateral sclerosis. Biochem Biophys Res Commun. 2006;351(3):602–11.
2. Baumeister W, Walz J, Zühl F, Seemüller E. The proteasome: paradigm of a self-compartmentalizing protease. Cell. 1998;92(3):367–80.
3. Beck F, Unverdorben P, Bohn S, Schweitzer A, Pfeifer G, Sakata E, et al. Near-atomic resolution structural model of the yeast 26S proteasome. Proc Natl Acad Sci. 2012;109(37):14870–5.
4. Ciechanover A, Schwartz AL. The ubiquitin-proteasome pathway: the complexity and myriad functions of proteins death. Proc Natl Acad Sci. 1998;95(6):2727–30. https://doi.org/10.1073/pnas.95.6.2727.
5. Coux O, Tanaka K, Goldberg AL. Structure and functions of the 20S and 26S proteasomes. Annu Rev Biochem. 1996;65(1):801–47.
6. Da Fonseca PC, He J, Morris EP. Molecular model of the human 26S proteasome. Mol Cell. 2012;46(1):54–66.
7. Eckman EA, Reed DK, Eckman CB. Degradation of the Alzheimer's amyloid β peptide by endothelin-converting enzyme. J Biol Chem. 2001;276(27):24540–8.
8. Finkbeiner S. Huntington's disease. Cold Spring Harb Perspect Biol. 2011;3(6):a007476.
9. Finley D. Recognition and processing of ubiquitin-protein conjugates by the proteasome. Annu Rev Biochem. 2009;78:477–513.
10. Gestwicki JE, Garza D. Protein quality control in neurodegenerative disease. Prog Mol Biol Transl Sci. 2012;107:327–53.
11. Glickman MH, Ciechanover A. The ubiquitin-proteasome proteolytic pathway: destruction for the sake of construction. Physiol Rev. 2002;82:373.
12. Groll M, Ditzel L, Löwe J, Stock D, Bochtler M, Bartunik HD, Huber R. Structure of 20S proteasome from yeast at 2.4 Å resolution. Nature. 1997;386(6624):463–71.

13. Harding RJ, Tong YF. Proteostasis in Huntington's disease: disease mechanisms and therapeutic opportunities. Acta Pharmacol Sin. 2018;39(5):754–69.
14. He C, Klionsky DJ. Regulation mechanisms and signaling pathways of autophagy. Annu Rev Genet. 2009;43:67–93.
15. Hershko A, Ciechanover A. The ubiquitin system for protein degradation. Annu Rev Biochem. 1992;61(1):761–807.
16. Holmberg CI, Staniszewski KE, Mensah KN, Matouschek A, Morimoto RI. Inefficient degradation of truncated polyglutamine proteins by the proteasome. EMBO J. 2004;23(21):4307–18.
17. Hubert V, Peschel A, Langer B, Gröger M, Rees A, Kain R. LAMP-2 is required for incorporating syntaxin-17 into autophagosomes and for their fusion with lysosomes. Biol Open. 2016;5(10):1516–29.
18. Iwai K. Diverse ubiquitin signaling in NF-κB activation. Trends Cell Biol. 2012;22(7):355–64.
19. Kaushik S, Cuervo AM. The coming of age of chaperone-mediated autophagy. Nat Rev Mol Cell Biol. 2018;19(6):365–81.
20. Kuma A, Hatano M, Matsui M, Yamamoto A, Nakaya H, Yoshimori T, et al. The role of autophagy during the early neonatal starvation period. Nature. 2004;432(7020):1032–6.
21. Labbadia J, Morimoto RI. The biology of proteostasis in aging and disease. Annu Rev Biochem. 2015;84:435–64.
22. Lander GC, Estrin E, Matyskiela ME, Bashore C, Nogales E, Martin A. Complete subunit architecture of the proteasome regulatory particle. Nature. 2012;482(7384):186–91.
23. Lasker K, Förster F, Bohn S, Walzthoeni T, Villa E, Unverdorben P, et al. Molecular architecture of the 26S proteasome holocomplex determined by an integrative approach. Proc Natl Acad Sci. 2012;109(5):1380–7.
24. Lehtonen Š, Sonninen TM, Wojciechowski S, Goldsteins G, Koistinaho J. Dysfunction of cellular proteostasis in Parkinson's disease. Front Neurosci. 2019;13:457790.
25. Li W, Ye Y. Polyubiquitin chains: functions, structures, and mechanisms. Cell Mol Life Sci. 2008;65:2397–406.
26. Lindersson E, Beedholm R, Højrup P, Moos T, Gai W, Hendil KB, Jensen PH. Proteasomal inhibition by α-synuclein filaments and oligomers. J Biol Chem. 2004;279(13):12924–34.
27. MacDonald ME, Ambrose CM, Duyao MP, Myers RH, Lin C, Srinidhi L, et al. A novel gene containing a trinucleotide repeat that is expanded and unstable on Huntington's disease chromosomes. Cell. 1993;72(6):971–83.
28. Malkus KA, Ischiropoulos H. Regional deficiencies in chaperone-mediated autophagy underlie α-synuclein aggregation and neurodegeneration. Neurobiol Dis. 2012;46(3):732–44.
29. Mizushima N, Komatsu M. Autophagy: renovation of cells and tissues. Cell. 2011;147(4):728–41.
30. Mizushima N, Yoshimori T, Ohsumi Y. The role of Atg proteins in autophagosome formation. Annu Rev Cell Dev Biol. 2011;27:107–32.
31. Mori H, Kondo J, Ihara Y. Ubiquitin is a component of paired helical filaments in Alzheimer's disease. Science. 1987;235(4796):1641–4.
32. Murrow L, Debnath J. Autophagy as a stress-response and quality-control mechanism: implications for cell injury and human disease. Annu Rev Pathol. 2013;8:105–37.
33. Mymrikov EV, Daake M, Richter B, Haslbeck M, Buchner J. The chaperone activity and substrate spectrum of human small heat shock proteins. J Biol Chem. 2017;292(2):672–84.
34. Neumann M, Sampathu DM, Kwong LK, Truax AC, Micsenyi MC, Chou TT, et al. Ubiquitinated TDP-43 in frontotemporal lobar degeneration and amyotrophic lateral sclerosis. Science. 2006;314(5796):130–3.
35. Nieto-Torres JL, Hansen M. Macroautophagy and aging: the impact of cellular recycling on health and longevity. Mol Asp Med. 2021;82:101020.
36. Panwar S, Uniyal P, Kukreti N, Hashmi A, Verma S, Arya A, Joshi G. Role of autophagy and proteostasis in neurodegenerative diseases: exploring the therapeutic interventions. Chem Biol Drug Des. 2024;103(4):e14515.
37. Parenti G, Pignata C, Vajro P, Salerno M. New strategies for the treatment of lysosomal storage diseases. Int J Mol Med. 2013;31(1):11–20.

38. Parzych KR, Klionsky DJ. An overview of autophagy: morphology, mechanism, and regulation. Antioxid Redox Signal. 2014;20(3):460–73.
39. Pickart CM. Mechanisms underlying ubiquitination. Annu Rev Biochem. 2001;70(1):503–33.
40. Pivovarova O, Höhn A, Grune T, Pfeiffer AF, Rudovich N. Insulin-degrading enzyme: new therapeutic target for diabetes and Alzheimer's disease? Ann Med. 2016;48(8):614–24.
41. Ravid T, Hochstrasser M. Diversity of degradation signals in the ubiquitin–proteasome system. Nat Rev Mol Cell Biol. 2008;9(9):679–89.
42. Ryu KY, Maehr R, Gilchrist CA, Long MA, Bouley DM, Mueller B, et al. The mouse polyubiquitin gene UbC is essential for fetal liver development, cell-cycle progression and stress tolerance. EMBO J. 2007;26(11):2693–706.
43. Sakata E, Bohn S, Mihalache O, Kiss P, Beck F, Nagy I, et al. Localization of the proteasomal ubiquitin receptors Rpn10 and Rpn13 by electron cryomicroscopy. Proc Natl Acad Sci. 2012;109(5):1479–84.
44. Schuck S. Microautophagy–distinct molecular mechanisms handle cargoes of many sizes. J Cell Sci. 2020;133(17):jcs246322.
45. Semaka A, Kay C, Doty C, Collins JA, Bijlsma EK, Richards F, et al. CAG size-specific risk estimates for intermediate allele repeat instability in Huntington disease. J Med Genet. 2013;50(10):696–703.
46. Settembre C, Fraldi A, Medina DL, Ballabio A. Signals from the lysosome: a control Centre for cellular clearance and energy metabolism. Nat Rev Mol Cell Biol. 2013;14(5):283–96.
47. Smith DM, Fraga H, Reis C, Kafri G, Goldberg AL. ATP binds to proteasomal ATPases in pairs with distinct functional effects, implying an ordered reaction cycle. Cell. 2011;144(4):526–38.
48. Sonninen TM, Goldsteins G, Laham-Karam N, Koistinaho J, Lehtonen Š. Proteostasis disturbances and inflammation in neurodegenerative diseases. Cells. 2020;9(10):2183.
49. Tanaka K, Matsuda N. Proteostasis and neurodegeneration: the roles of proteasomal degradation and autophagy. Biochimica et Biophysica Acta (BBA)-molecular. Cell Res. 2014;1843(1):197–204.
50. Tanaka K, Mizushima T, Saeki Y. The proteasome: molecular machinery and pathophysiological roles. Biol Chem. 2012;393(4):217–34.
51. Tanaka K, Suzuki T, Hattori N, Mizuno Y. Ubiquitin, proteasome and parkin. Biochimica et Biophysica Acta (BBA)-molecular. Cell Res. 2004;1695(1–3):235–47.
52. Tsukamoto S, Kuma A, Murakami M, Kishi C, Yamamoto A, Mizushima N. Autophagy is essential for preimplantation development of mouse embryos. Science. 2008;321(5885):117–20.
53. Unno M, Mizushima T, Morimoto Y, Tomisugi Y, Tanaka K, Yasuoka N, Tsukihara T. The structure of the mammalian 20S proteasome at 2.75 Å resolution. Structure. 2002;10(5):609–18.
54. Wang L, Klionsky DJ, Shen HM. The emerging mechanisms and functions of microautophagy. Nat Rev Mol Cell Biol. 2023;24(3):186–203.
55. Wang XJ, Yu J, Wong SH, Cheng AS, Chan FK, Ng SS, et al. A novel crosstalk between two major protein degradation systems: regulation of proteasomal activity by autophagy. Autophagy. 2013;9(10):1500–8.
56. Weissman AM. Themes and variations on ubiquitylation. Nat Rev Mol Cell Biol. 2001;2(3):169–78.

The Insight into the Role of Protein Misfolding and Tau Oligomers in Alzheimer's Disease

12

Pooja Pandey, Chandan Kumar, and Rekha Khandia

12.1 Introduction

Proteins are considered to be building blocks of a living system, which also can play several roles as biochemical catalysts, enzymes, hormones, etc. They are assemblies of amino acids that are connected with each other via peptide linkages. These peptides make multiple polypeptide chain that folded in a way to form complex proteins that perform various biological functions [62]. In standard conditions, protein folding is a formation of soluble, highly organised and thermodynamically stable molecules which serve precursors of several other biologically relevant molecules. [81]. But proteins also get misfolded when these standard conditions are not provided or if there is a change in environmental conditions. This may lead to improper folding of the whole protein or some of the parts of it [61].

Protein misfolding is one of the phenomena where the soluble proteins assemble themselves to make certain polymers of proteins (identical units) eventually called amyloid fibrils [91]. These amyloids are made up of β-amyloid (42-residual unit of peptide) which produced through large amyloid precursor protein via enzymatic processing [93]. Misfolding of protein may cause cellular toxicity which contributes to cellular proteostatic collapse [105]. Aggregation of proteins gives rise to different neurodegenerative disorders such as Alzheimer's and Parkinson's diseases ([22]). During misfolding, some other monomeric proteins tau (Alzheimer's disease) and a-syn (Parkinson's disease) also change their conformation from native to highly ordered b-sheet-rich amyloid fibrils [102].

Alzheimer's disease is a type of dementia mostly found in elderly humans in which a person experiences behavioural disorders such as depression, aggressiveness and apathy together with loss in cognitive function and episodic memory [97]. Alzheimer's disease typically considered to be one of the devastating brain

P. Pandey (✉) · C. Kumar · R. Khandia
Department of Biochemistry and Genetics, Barkatullah University, Bhopal, India

N. Puranik (ed.), *Proteostasis*, https://doi.org/10.1007/978-981-96-6202-9_12

disorders that manifest after the age of 65 [75]. However, early-onset forms of Alzheimer's disease, that is, before the age of 65 are less common, and about ~2–5% is reported among all Alzheimer's disease cases [88].

In recent estimations, Alzheimer's disease became a major public health problem as 57.4 million people are hugely affected from it. According to data reported by Centres for Disease Control and Prevention (CDC), the seventh-ranked disease was Alzheimer itself in the United States in the year 2022 before the COVID-19 pandemic [59]. Whereas, last decade witnessed India being the fourth largest contributor of this disease and also expected to become the second largest (First is China) in 2050 probably going to surpass Japan and the United States in global dementia cases (*Early Detection & Management of Alzheimer's Disease & Dementia in India: A Policy Perspective*, 1 C.E.)

Tau oligomer aggregation is the major component that is involved in the pathophysiology of Alzheimer's disease. Soluble (protofibrils) and insoluble (fibrils) are two major types of tau aggregates in the brain cells [125]. Soluble aggregates considered highly toxic, which gave rise to neurodegeneration and the diseases they cause known as tauopathies. The functions of Tau protein are microtubule stabilisers, in monitoring axonal transport of organelles, and in regulation of neurite growth [74].

The stabilising role of tau was affected because of this oligomer aggregation, which spread in brain cells and transferred to neighbouring cells [77]. Post-translational modifications such as hyperphosphorylation, conformational changes, ubiquitination and truncations considered to be chief reason for tau accumulation whereas RNA, heparin or arachidonic acid also been reported to induce tau aggregation [125].

The disease is characterised by accumulation of the amyloid plaques made up of the amyloid β extracellularly and microtubule-associated tau protein in the form of neurofibrillary tangles (NFT) which created intracellularly [3]. These NFTs are made up of either 4 repeat (4R)-tau or 3 repeat (3R)-tau considered to be neuropathological lesion of Alzheimer's disease [111]. NFTs includes three levels that is pretangles, mature tangles and ghost tangles [80]. In 2022, Centre for Cognitive Health reported that phosphorylated tau protein (p-Tau) 217 considered to be a potent mediator between accumulation of amyloid plaque and NFTs. It also stated that factors like environment, genes and various strands of Tau protein also plays key role in NFT propagation. In 2018, Brici et al. revealed the three different morphological stages involved in NFT formation: pre-NFTs, extraneuronal NFTs (eNFTs) and intraneuronal NFTs (iNFTs). They explained pre-NFTs have normal looking neurons with central nuclei, iNFTs have deteriorated neurons with displaced nucleus whereas eNFTs comprise dead neurons with the absence of neuronal nuclei.

In this chapter, we are going to discuss about role of protein misfolding, dimensions of toxic tau protein and role of synergistic behaviour of β amyloid and Tau in the development of Alzheimer's disease. This chapter also reviews the occurrence of tau-mediated inflammatory and neuroimmune changes, in addition to that neurochemical imbalance during Alzheimer's disease. Quantification techniques, biophysical characterisation and consequences of their dysregulated presence of these oligomers are also being discussed. Moreover, use of biomarkers in the diagnosis of Alzheimer's disease and recent developments in different drugs against tau oligomers are going to be concluded.

12.2 Concept of Protein Folding

12.2.1 Protein Structure, Conformation and Folding

Proteins are considered to be molecular machines that change their structure to perform different functions. There are four different levels in which the proteins exist: primary, secondary, tertiary and quaternary. The proteins made up of amino acids connected with polypeptide chains considered to be primary structures. Whereas Secondary structures made up of folded primary assemblies in the shape of alpha-helices and beta-sheets, which are stabilised by hydrogen bonds from the polypeptide backbone. Moreover, the tertiary structure (three-dimensional) of proteins is shaped because of side-chain interactions between 2 or more polypeptides and quaternary also formed as same as tertiary structures [94]. However, quaternary structures are shown by some of the proteins like haemoglobin which are made up of several polypeptide chains [36]. So, the process of protein folding is for the formation of highly organised molecular structures that have highly stable conformations thermodynamically [4]. These molecular structures govern the ability to bind with other molecules, which basically helps in catalysing different metabolic reactions. Eventually, protein folding helps in performing different biological functions and also take major part in cell signalling. The changes occur in the proteins can lead to denaturation of proteins which is a condition of molecules that loses its structure and function [38]. The changes occur in these protein structures due to any of the reason like environmental factors, post-translational modification, or mutations can alter the functioning of the proteins and ultimately lead to diseases.

Usually, the conversion of alpha helix into beta-sheet occurs in various functional native proteins such as immunoglobulins. But protein toxicity arise due to the extensive conformational change in proteins can make them toxic which leads to the aggregation of proteins [89]. Aggregation of proteins precisely occurs due to clumping of proteins where the hydrophobic interaction residues exposed on the protein surface. The toxicity can be transferred to native proteins through the interactions between them. The multiplication rate of the toxicity can weaken the cell functions or ultimately lead to cell death. Moreover, in normal conditions, cell have a preventive mechanism to get rid of the protein misfolding that we will discuss further in this chapter. Nevertheless, the misfolded protein can also be detected by cell's quality-control mechanisms that tag them to be sent to the cytoplasm, where they will be degraded.

12.2.2 Major Causes and Their Analysis in Protein Misfolding

Protein misfolding depends upon several factors including post-translational modifications (PTMs), conformational changes, environmental stress, genetic mutations or imbalances in proteostasis. PTMs such as acetylation, oxidation, glycosylation, ubiquitination and phosphorylation tend to lead protein folding. If an error occurs in these PTMs process, they impact the folding patterns, which ultimately results in protein misfolding. As discussed before, senile plaques and tau tangles, considered to be two most trademarks of Alzheimer's disease, which was the outcome of these PTMs [52]. Mass Spectrometry used as a primary analytical tool used for the analysis of these PTMs involved in protein misfolding [37]. This technique works on mass spectra shift because of PTMs, excess od modified peptide in the sample, modification effect on ionising efficiency of peptides and also the stability of modified peptide in the sample [86]. Conformational change also encourages the protein misfolding by gain of toxic activity and loss of biological functions of the native folded protein [101]. Several in vivo assays and in vitro cell culture techniques are used to test the structural activity of native and toxic peptides [103]. Use of synthetic β-sheet breaker peptides can be used for the conversion the β-sheet-rich abnormal conformation into its normal conformation [100]. Environmental factors also contribute to the development and progression of protein misfolding. For example, all types of pollution as chronic exposure of heavy metals or transition metals (Mn, Fe, Cu, Zn) can cause conformational changes in protein leads to misfolding [114]. Some other metals like As, Pb and MeHg are considered to be highly neurotoxic, which cause cognitive dysfunction by disrupting electron transport chain, the ubiquitin–proteasome system, mRNA splicing and oxidative stress [51]. Faecal microbiota transplantation (FMT) used for restoration of the gut microbiota as prevention and treatment strategies in any neuronal impairment occur due to environmental stress [114]. Proteostatis is a healthy state of cell in which protein folding, trafficking, synthesis and clearance are controlled to reduce the aggregation of toxic misfolded proteins [9]. If the imbalances occur in proteostatis, lead to neuroinflammation in brain and disrupt the protein folding environment [54]. The amount of proteostasis balance turn to be limited via ageing because of the protein accumulation [55]. It is important to maintain the proteostatis network for sustainable protein synthesis, unfolding, refolding and damaging the misfolded proteins. Development of novel therapeutic and pharmacological strategies can be fruitful in treating the imbalance in proteostatis at the cellular level [99].

12.3 Factors Affecting Protein Folding

Protein folding is considered to nonlinear, non-equilibrium, dissipative process which defines the various active characteristics of proteins [33]. Several factors like chemicals, pH, temperature, magnetic fields, electric fields, molecular crowding and space limitations influence the protein to fold into their correct functional forms.

These factors affect the stability of proteins but sometimes lead to denaturation too. Other factors like biochemical environment, amino acid sequences, geometric interactions and constraints also influence the structures, functions and stability of the proteins [121]. Some heat shock proteins may also support the unfolding of proteins and refolding in the correct manner. The forces that affect the protein structure are electrostatic forces, hydrogen bonding, hydrophobic forces and disulphide bonds. According to the studies, an abnormal self-metabolite produced in some individuals tends to modify the peptides and cause misfolding of proteins [124].

12.4 Mechanisms of Misfolded Protein Toxicity

Protein toxicity is a process of changes occur in pathological events arises from aggregation, mislocalisation, and/or multimerisation of disease-causing peptides or proteins [24]. The toxicity generates through protein misfolding in all neurodegenerative diseases can lead to loss in brain ability, synaptic dysfunction and eventually neuronal cell death [105]. The most devastating disorders in neurodegenerative diseases are alteration in emotional feelings, abstract thinking, skilled movements, memory and cognition [68]. Genetic mutations are one of the major mechanisms through which protein toxicity would occur. Hunter, in 2018, illustrated 5 different types of point mutations in the genes encoding synuclein alpha (SNCA; G51D, H50Q, A30P, E46K and A53T) and 52 mutations in amyloid precursor protein (APP) have so far been recognised to be linked with Parkinson's disease and Alzheimer's disease, respectively. Protein toxicity is based on subcellular localisation of toxic peptides by which toxicity move from one subcellular compartment to another such as nucleus, cytoplasm and mitochondria. Twenty different neurodegenerative diseases were observed due to protein toxicity in the nucleus [118]. "PolyQ diseases" is the terminology given to those neurodegenerative diseases that are generated through nucleus protein toxicity. There are 9 polyQ diseases reported, majorly spinocerebellar ataxias (SCAs) caused by nuclear inclusions of ataxin proteins and spinal and bulbar muscular atrophy (SBMA) by mutant androgen receptor aggregated in the cytoplasm [113]. Protein quality control (PQC) system resides in cytoplasm as it contains extravagant cytoskeleton in comparison to nucleus. Moreover, the aggregation of misfolded proteins certainly affects the PQC system, which includes macroautophagy, and ER-associated degradation (ERAD), chaperone-mediated autophagy (CMA) and macroautophagy and ubiquitin–proteasome system (UPS) [25]. Protein toxicity also occur in mitochondria and mitochondrial dysfunction can be harmful for the cell to survive. A mitochondrion is the cellular organelle basically responsible for ATP production. Neurites need local supply of ATP to maintain its growth and function but protein toxicity in the mitochondrion will trouble the elongation process of neurites [1]. Apart from the brain, Valastyan & Lindquist in 2014 reported about, dual toxicity by α1-antitrypsin (protease inhibitor), which lead to emphysema (lung disease) and liver disease also [112].

12.5 Therapeutic Targets in Protein Misfolding

Chaperones are the class of therapeutic molecules, used in unfolding and refolding the protein (if misfolded) or for the removal of aggregated proteins. They are required to ensure protein homeostasis (proteostasis), and they support the ubiquitin-proteasome system (UPS) and autophagy system [44]. They help in promoting stability and cell viability and also participate in several processes like degradation, translocation and folding of proteins Cellular molecular chaperones are fundamentally used in reducing the severity of numerous neurodegenerative diseases and several other protein-misfolding disorders [21]. It has been reported that chaperones and co-chaperons are expressed in different sets inside the cells [20]. There are two different mechanisms of action to neutralise protein oligomer toxicity: removing toxicity by binding chaperones to the hydrophobic sites on the protein oligomer directly and other is conversion of toxic oligomers into harmless species through chaperones resulting reduction in the surface–volume ratio [72]. HsP70, HsP90 and Hsp100 are some examples of molecular chaperones which assist a large variety of protein folding processes. The folding of protein occurs through chaperonins which are large and cylindrical shape complexes that provide a central section for a protein chain to fold unaffected by aggregation [56].

Chaperones are generally classified into 3 major groups, molecular, pharmacological and chemical. Molecular chaperones not considered to have functional protein structure but it associated with first line of defence in protein misfolding and aggregation of protein [82]. Pharmacological chaperones involved in restoring function, structure stabilisation by inducing protein refolding [12]. Chemical chaperones usually not considered as therapeutic agents like pharmacological one as they are low molecular weight compounds and not able to bind to the protein directly [26]. They are further classified into hydrophobic compounds and osmolytes. Hydrophobic segments of the unfolded protein are exposed to hydrophobic regions of the chaperone, the interaction protects the protein from accumulation. Whereas Osmolytes work against stress conditions for example, amino acids protect against high salt concentrating extracellular environments, polyols protect cells against dehydration and extreme temperature, methylamines (found in urea-rich cells) provide protection against urea's effect on structure of protein [26] (Fig. 12.1).

12.6 Introduction to Tau Oligomers

Alzheimer's disease is determined by two key features like accumulation of amyloid-β (Aβ) peptides in the extracellular space and intracellular accumulation of hyperphosphorylated tau protein. Amyloid-beta peptides undergo extracellular fibrillisation, which yields senile plaques. Intracellularly, hyperphosphorylated tau protein undergoes conformational changes, which produce neurofibrillary tangles (NFTs) characterised by paired helical filaments (PHFs) and straight filaments [23]. The debate surrounding amyloid-β 1–42 deposits has been a defining feature of Alzheimer's disease research. Nevertheless, soluble Tau protein oligomers are

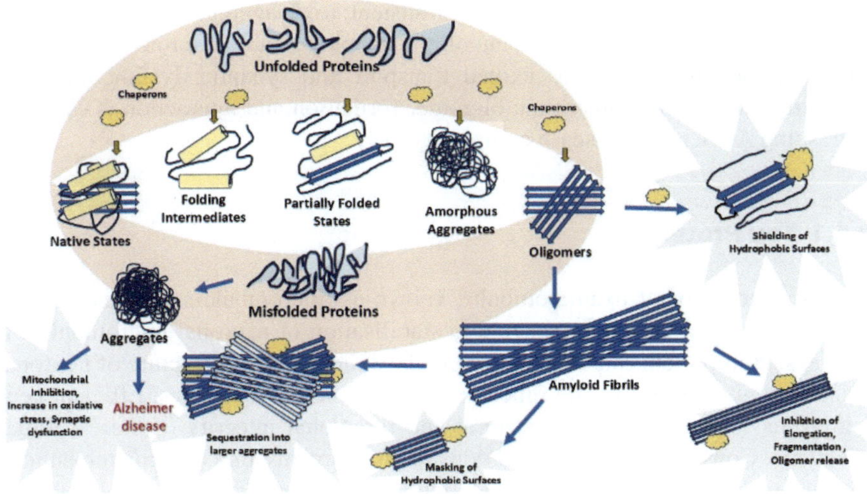

Fig. 12.1 Chaperons reaction in protein aggregation and folding

suspected to contribute to the disease's initial development [81]. The field of protein misfolding diseases has undergone a significant shift in perspective, from implicating fibrillar amyloidogenic proteins as primary toxic entities to exploring alternative mechanisms. Evidence indicates that misfolded protein oligomers, smaller and more dynamic aggregates, are often more toxic than mature fibrils (4,5,6–12). Amyloid fibrils exert cytotoxicity in various protein-misfolding diseases through two primary mechanisms, first by functional proteins sequestration and second by direct cellular and tissue damage. Fibrils increases the rate of formation of new oligomers. So, fibrillar aggregates remain essential pathological biomarkers, meanwhile oligomers are identified as critical targets for drug development and screening initiatives [91]. Therefore, Tau protein is an important aspect in Alzheimer's disease progression, significantly impacting its pathological trajectory. A groundbreaking study revealed that injecting tau oligomers into the brains of healthy mice triggered a cascade of neurodegenerative events, including memory loss, synaptic disruption and mitochondrial devastation. For decades, neurofibrillary tangles were considered the representative of tau-mediated neurotoxicity. Challenging long-held assumptions, cutting-edge research has illuminated a multifaceted dynamic between neurofibrillary tangle formation, cognitive function and neurodegeneration in the early stages of Alzheimer's disease. Groundbreaking discoveries in Alzheimer's disease research challenge traditional notions, suggesting that filamentous tau aggregates, once believed to be harmful, may instead play a protective role, potentially altering our understanding of disease pathogenesis. Detecting oligomers outside a living organism for diagnostic purposes is challenging due to their diverse structure, fleeting existence and low levels [73]. The presence of tau oligomers in CSF has been confirmed at remarkably low levels, underscoring the sensitivity of current detection methods [78]. Unravelling Alzheimer's pathogenesis has led researchers to a critical pivot, prioritising tau oligomers—transient, toxic aggregates forming before filamentous tau. Evidence suggests tau oligomers play a causative role in early neurodegeneration, manifesting as axonal transport and synaptic dysfunction [41]. Elucidating the mechanisms of tau oligomer formation and interaction is essential for grasping Alzheimer's disease complexity [19].

12.6.1 Tau Protein

Tau is a protein bound to microtubules known as microtubule-associated protein (MAP) which plays an important role in stabilisation of neuronal microtubules. It also plays an important role in maintaining the delicate micro-structure of neuronal cells, ensuring the stability and functionality of axons and dendrites. Efficient neural transmission relies on the structural and functional integrity of these cellular projections. Consequently, any alterations in tau's organisation or arrangement can have far-reaching functional and pathological consequences. Microtubules (MTs) are dynamic structures essential for various cellular processes, including the development of cell processes such as axons and dendrites, establishment of cellular polarity and facilitation of signal transduction pathways that regulate cellular communication. Disruption in this signalling pathways lead to kinase/phosphatase imbalance, resulting in tau hyperphosphorylation, microtubules instability and disassembly. Tau protein is mainly axonal. However, during pathological states, tau delocalises and accumulates in the cell soma. Consequently, hyperphosphorylated tau accumulates as Paired Helical Filaments (PHFs), disrupting intracellular transport and causing neuronal degradation [5, 84]. The intercellular spread of tau protein is facilitated by tau oligomers, which can penetrate cells through various pathways, which include compromising cell membrane integrity, creating pore-like structures and enabling passive diffusion or by endocytic processes, such as macropinocytosis or clathrin-mediated endocytosis [119]. Tau protein secretion involves multiple cellular pathways which include exosome-mediated release in which Tau is packaged within exosomes and secreted through synaptic vesicle exocytosis or through non-vesicular secretion in which Tau is directly released from cells without an enclosing membrane, potentially facilitated by interaction with Fyn receptor (Src kinase) [39]. The presence of Fyn–tau interactions within exosomes, as demonstrated by colocalisation studies, underscores the potential importance of Fyn kinase in regulating tau protein transport, potentially influencing tau spreading and neurodegeneration [67].

12.6.2 Orientation of Tau Protein

Tau protein is a heat-stable molecule. Its molecular weight is approximately 60 kDa made up of 352–441 amino acids. The Tau gene found on 17q21, contains 16 exons, whose complex transcriptional regulation yields a multitude of protein variants. The alternative splicing of tau's exons E2, E3 and E10 forms six distinct isoforms of tau.

These are 3R0N, 3R1N, 3R2N, 4R0N, 4R1N and 4R2N. The six tau isoforms exhibit diversity due to variations in tubulin-binding repeat (TBR) count, resulting in 3R and 4R classifications based on the presence of three or four TBRs. The adult brain maintains a delicate balance between 3R and 4R tau isoforms, with an equal ratio. Alteration in this ratio causes neurodegenerative frontotemporal dementias (FTDs) and tauopathies [10, 40]. During early developmental stages, findings show that tau proteins lacking exon 10 are primarily transcribed. Alternative splicing shapes tau isoform expression across developmental processes. Exons 2, 3 and 10 alternative splicing gives rise to tau isoforms which are exclusively expressed in the adult brain, which sheds light on complex tau protein regulation [8]. In adult humans, tau protein isoforms show heterogeneity in their microtubule binding sites, with alternative splicing of exons 2 and 3. This variability gives rise to two main isoform categories: 3R tau, comprising 0N3R, 1N3R and 2N3R, and 4R tau, consisting of 0N4R, 1N4R and 2N4R. Tau protein isoforms exhibit distinct structural variations. The 2N4R isoform, predominantly expressed in the brain, contains 441 amino acids. In contrast, the fetal isoform, 0N3R, comprises 352 amino acids, lacking two amino-terminal inserts and possessing only three microtubule-binding repeats [27]. Tau proteins in the brain have a two-domain structure. It consists of a carboxy-terminal and amino-terminal domain. The amino terminal projection domain accounts for approximately two-thirds of the molecule's length and plays a crucial role in tau's interaction with other proteins and cellular components. The carboxy-terminal microtubule-binding domain, encompassing approximately one-third of the molecule, is responsible for tau's association with microtubules and regulation of microtubule dynamics. The projection domain of tau protein encompasses two sub-regions, an amino-terminal region and a proline-rich region. An amino-terminal region, rich in acidic amino acid residues, potentially involved in protein–protein interactions, while a proline-rich region, characterised by a high density of proline residues, may facilitate protein–ligand interactions and conformational flexibility. The projection domain of tau protein is a critical region for mediating diverse molecular interactions. Acidic residues facilitate binding to other molecules and cations. Specific motifs, such as KKXK, helps in heparin binding. Proline-rich regions contain PPXXP and PXXP motifs, which interact with SH3 domain-containing proteins, regulating signal transduction and cytoskeletal dynamics [7, 108].

12.6.3 Features of Typical Tau

Neurons, the fundamental units of the nervous system rely on microtubules as a structural framework. Microtubules, stabilised by tau, facilitate axon and dendrite formation [31]. Tau's dominant role in stabilising microtubules governs MT dynamics, modulating cellular architecture and regulatory processes. The microtubule network facilitates axonal transport systems, enabling precise trafficking of vital cellular components, including organelles and signalling molecules. Therefore, tau is indispensable for microtubule network-mediated axonal transport systems and

MT stabilisation. Moreover, tau plays a crucial role in maintaining neuronal architecture, rendering neurons the most asymmetrical cells in the human body [13].

12.6.4 Tau-Mediated Microtubule Dynamics—Stabilisation and Destabilisation

Tau protein plays a crucial role in maintaining microtubule stability, thereby facilitating efficient axonal transport. Tau protein exhibits a distinct structural organisation, featuring a carboxy-terminal (C-terminal) domain on one half and two distinct regions on the other half: a positively charged proline-rich region and an acidic amino-terminal (N-terminal) region. Tau's tubulin-binding repeats specifically interact with distinct pockets on the inner surface of microtubules, while its positively charged proline-rich regions tightly bind to the negatively charged MT surfaces, conferring stability [13, 31, 66]. Any aberrations or pathological alterations in tau protein function can precipitate severe tau dysfunction, culminating in microtubule depolymerisation and destabilisation. In Alzheimer's disease, tau protein undergoes pathological phosphorylation at Ser262 and Ser214 sites, leading to its detachment from microtubules. While tau phosphorylation is crucial for establishing cell polarity, excessive activity of microtubule affinity regulating kinase (MARK) leads to aberrant phosphorylation and ultimately, cell death [92, 110].

12.7 Pathological Effects of Tau and Its Regulation

12.7.1 Tauopathies

The scientific community has shifted attention to tau protein's conformational changes and multimeric aggregates as a possible triggering event in Alzheimer's disease pathology. This reevaluation is fuelled by evidence suggesting amyloid-β deposition in plaques occurs relatively late and is nonspecific, challenging the long-held amyloid hypothesis [42, 83]. The amyloid hypothesis, a longstanding framework for understanding Alzheimer's disease, suggested that the gradual self-polymerisation of amyloid-β into plaques leads to brain damage. However, recent studies have shifted focus toward oligomeric Aβ and/or tau species as primary drivers of neurodegeneration, particularly synaptic dysfunction, which occurs prior to the deposition of fibrillar Aβ (PHFs) or tau (NFTs) [35, 43, 106]. Pathological modifications in tau, including prion-like propagation, trigger aberrant conformational changes, aggregation and fibrillisation, collectively known as Tauopathies. This elevated level of tau modifications induces self-aggregation and hyperphosphorylation, yielding pathological tau (p-tau), a hallmark of various tauopathies [47, 48, 53, 116]. Hyperphosphorylation of tau changes its configuration and polarity, revealing the microtubule-binding domain, leading to self-aggregation and oligomerisation, ultimately giving neurofibrillary tangles (NFTs). Prior to neurofibrillary tangle (NFT) formation, hyperphosphorylation at tau's C-terminus initiates self-assembly, yielding paired helical filaments (PHFs). This conformational change disrupts axonal transport and destabilises microtubules (19). Compromised Tau–MT functioning disrupts synaptic plasticity and axonal transport, culminating in cognitive deficiencies. Hyperphosphorylated tau (p-tau) facilitates the activation of microtubule-severing proteins, such as katanin, exacerbating microtubule disassembly [120]. Notably, pretangle tau aggregates, characterised as oligomers, have been identified in brains during the early stages of Alzheimer's disease, supporting this paradigm shift [69]. In the early stages of neuropathology, tau protein oligomers or multimers have been implicated in synaptic toxicity, leading to disruptions in synaptic function and contributing to neurodegenerative disease progression [30]. Under pathological conditions, tau oligomers exhibit enhanced stability, evading proteolytic degradation and facilitating their transport via cytoplasmic flow to nerve terminals where they disrupt synaptic function and are released into the extracellular environment, where they can be taken up by adjacent neurons, initiating a cascade of damage [19]. The squid axoplasm assay, employed by Binder et al., provides evidence that aggregated oligomeric tau, rather than monomeric soluble tau, disrupts anterograde fast axonal transport (FAT), supporting the hypothesis that prefibrillar tau oligomers are neurotoxic species contributing to early-stage neurodegeneration [63, 115]. Studies by Lasagna-Reeves et al. provide compelling evidence that tau oligomers extracted from human Alzheimer's disease brains or generated from recombinant full-length human tau protein significantly disrupt synaptic plasticity and memory in wild-type mice. Specifically, injections near the hippocampus induced acute impairment of anterograde memory storage [65]. Irregular tau phosphorylation is an important event escalating tau pathology, which is more associated to Alzheimer's disease progression than Aβ accumulation [49, 50] (Fig. 12.2).

12.7.2 Regulation of Tau by Post-translational Modification

Tau protein's activity and stability are regulated by diverse post-translational modifications. Research has demonstrated that tau post-translational modification is regulated by the activities of protein kinases and phosphatases [78]. Protein kinases (PKs) phosphorylating tau are classified into Proline-Directed Protein Kinases (PDPKs), Non-Proline Directed Protein Kinases (Non-PDPKs) and Tyrosine Protein Kinases (TPKs). Proline-Directed Protein Kinases (PDPKs) are pivotal in tau phosphorylation at proline sites and have been implicated in neurodegenerative disorders characterised by early tau deposits and tangles in brain tissues [84]. In contrast to PDPKs, non-proline-directed protein kinases (non-PDPKs) phosphorylate tau at non-proline residues. The complex process of tau polymerisation is orchestrated by a concerted action of PDPKs and non-PDPKs, which synergistically regulate tau's aggregation dynamics under both normal and disease-related conditions.

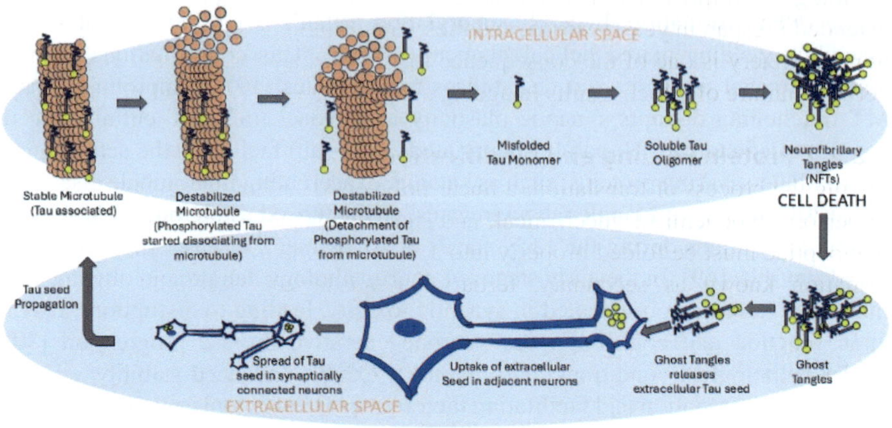

Fig. 12.2 Tau neurofibrillary tangles (NFTs) development in Alzheimer's disease

12.7.3 Protein Interactions Exacerbating Tau Pathology

Tau pathology is intensified by various proteins such as Amyloid-β, Pin1, Fyn Kinase, Heat Shock Proteins, FKBP51/52 Immunophilins, α-Synuclein and PACSIN1, which significantly impact tau's phosphorylation and dephosphorylation. Amyloid beta promotes excessive phosphorylation of tau and NFT formation. Impairment in Pin1 causes neuronal death. Fyn kinase leads to the phosphorylation of tyrosine. Heat Shock Protein 90 regulates tau's configuration, enabling GSK-3β mediated phosphorylation. This causes filamentous structure formation. Conversely, FKBP51 protects tau from ubiquitination, whereas FKBP52 reduces tau's microtubule-stabilising ability, inducing instability. Furthermore, α-Synuclein enhances tau phosphorylation via Protein Kinase A (PKA), and tau's interaction with PACSIN1 decreases its microtubule affinity [84]. The complex regulation of tau protein is orchestrated by two critical factors: post-translational modifications and dynamic protein interactions, which provide opportunities for therapeutical interventions [78].

12.7.4 Tau–Aβ Synergy in Alzheimer's Disease

Tau's involvement is prerequisite for Amyloid beta associated cognitive incapacity, indicating that the connection between Aβ and tau is pivotal in progression of dementia in Alzheimer's disease. Understanding the link between Aβ and tau pathology can reveal how Alzheimer's disease progresses. A comprehensive array of in vitro experiments, spanning neuronal cell lines to primary hippocampal cultures, unequivocally revealed Amyloid-β's detrimental impact on tau, manifesting as augmented phosphorylation and spatial dysregulation [14, 104]. Amyloid-β significantly impacts tau phosphorylation by inducing the activation of GSK-3β, leading to enhanced tau phosphorylation.

12.8 Therapeutic Approach

Pharmacological approaches to inhibiting tau aggregation exploit the distinct structural features of tau–tau and tau–tubulin binding. This selective targeting is crucial, as disrupting the normal interaction between tau and tubulin could have unintended consequences. By leveraging these structural differences, tau aggregation inhibitors can effectively prevent pathological tau–tau binding while preserving essential tau–tubulin interactions. The use of chemically derived small molecules presents several benefits, including enhanced delivery to target tissues due to their compact size and lower manufacturing costs [2]. Kinases as a Therapeutic Target such as GSK3 Inhibitors, Cdk 5 Inhibitors, MARK/PAR1 as a Target for Inhibition considered to be primary targets in the discovery of drugs [16]. Targeting the dysregulation of tau protein holds significant therapeutic promise, especially when treatments are applied in a timely and localised manner. While there is potential for various treatment options, further advancements are needed to enhance their specificity, efficacy and safety. Overall, these developments are generating excitement within the tau research community, marking a pivotal moment for future therapies.

12.9 Conclusion

This chapter reviews about the role of protein misfolding and various Tau species in synaptic dysfunction and memory loss in Alzheimer's disease. In the past decade, research on tau protein has progressive impact, revealing its alternative functions in both normal and disease states. The recent identification of toxic soluble tau oligomers is reforming our understanding of tau's role in disease pathogenesis. While the amyloid beta cascade has dominated descriptions of AD pathology for 25 years, this theory is now seen as insufficient to account for all aspects of the disease. Recent research has focused on oligomeric forms of Aβ as potentially toxic, but this assessment still fails to explain all pathological features. Given that Aβ plaques and neurofibrillary tangles appear together in Alzheimer's disease with distinct patterns, attention has shifted to Tau protein, the main component of these aggregates, as a key player in the disease's development. Oligomers are now viewed as the primary toxic form of tau pathology in tauopathies, while fibrillary tau aggregates, such as neurofibrillary tangles (NFTs), are considered a late-stage event that may serve protective functions rather than directly causing early synaptic dysfunction. These oligomers or multimers are thought to be synaptotoxic, leading to memory deficits in the early stages of the disease. Additionally, evidence suggests that tau pathology spreads intercellularly, starting in specific brain areas and progressively affecting previously unaffected regions. There is an urgent need for simple, reliable and sensitive biomarkers for Alzheimer's disease that can facilitate early diagnosis and treatment monitoring. Blood-based biomarkers are preferable due to the ease of collection compared to lumbar punctures. Tau oligomers in plasma or serum may be promising candidates, but further research is necessary before they can be used clinically. Validating Tau oligomers as diagnostic tools will require robust, standardised assays with high sensitivity, followed by confirmation in large, multicentre patient associates. Further research is essential to understand the initial mechanisms that lead to the formation of tau oligomers. This knowledge is crucial for developing effective therapeutic interventions aimed at blocking or potentially preventing the progression of neurodegeneration.

12.10 Future Prospects

Targeting tau is a promising strategy for treating Alzheimer's disease, but several challenges hinder drug development. These include the blood–brain barrier limiting drug entry, lack of selectivity in target proteins, differences between animal models and humans and the timing of treatment. Advancements in diagnostic methods allow for earlier interventions at the mild cognitive impairment stage, potentially before symptoms arise. Active immunotherapy may prevent early disease but can have side effects due to lack of specificity. As the disease progresses, a single treatment is insufficient; a combination of passive immunotherapy and methods to inhibit tau aggregation and expression may enhance treatment efficacy. Despite many tau-targeting candidates being developed, few are clinically effective. Continued research into the underlying mechanisms of Alzheimer's disease is essential for identifying new intervention targets. The perspective of tau protein-targeting therapy for Alzheimer's disease remains unclear due to insufficient evidence and conflicting observations. The inadequate number of studies and inconsistent measures hinder the ability to draw general conclusions, indicating that the field is still in its early stages. Despite the challenges posed by Alzheimer's, the pharmaceutical industry has seen little progress in treatment options, with existing drugs showing low effectiveness and recent phase 3 clinical studies yielding negative results over the past two decades [87]. There is an urgent need for simple, reliable and sensitive biomarkers for Alzheimer's disease to facilitate early diagnosis and treatment monitoring. Blood-based biomarkers are favoured for their ease of collection. Tau oligomers in plasma or serum show potential as candidates, but more research is needed before they can be clinically implemented. Developing validated tests for Tau oligomers will require robust, standardised assays with high sensitivity, confirmed in large, multi-centre studies (Tables 12.1 and 12.2).

Table 12.1 Molecular Chaperones and their inhibitory effects in neurodegenerative diseases

Molecular chaperones	Gene name	Origin	Targeted species of A beta	Disease link	Inhibited microscopic nucleation event	References
Clustrin	CLU	Extracellular	Fibril end	Alzheimer's disease Abeta 40 plaques	Elongation ↓↓	[34, 46, 95]
Hsp10	Hsp10	Intracellular and extracellular	Fibril end	Alzheimer's disease-associated Aβ1–42	Primary nucleation or elongation	[64]
Progranulin	GRN	Extracellular	Heterodimer	Alzheimer's disease Abeta plaques and DNA-binding protein 43 encephalopathy	–	[76, 90]
Hsp1	Hsp1	Intracellular and extracellular	Fibril end	Causative mutations	Nucleation and elongation	[60, 96]
Nucleobindin 1	NUCB1	Intracellular and extracellular	Pre-fibrillar species	Aβ42 associated with Alzheimer's disease	Primary and secondary nucleation	[15]
αB-crystallin	αB-cry or HspB5	Intracellular	Fibril surface and fibril-ends	Aβ1–40 peptide associated with Alzheimer's disease	Secondary nucleation ↓↓, elongation ↓	[6, 11, 32]
HspB6	HspB6	Extracellular	Fibril-ends	Aβ1–40 (80–90%), Aβ42 (5–10%), increased protein level	Secondary nucleation ↓↓, elongation ↓	[58, 85]
HspB8	HspB8	Extracellular	Fibrillar structures	Abeta (1–42), Abeta (1–40), causative mutations	Dominated by secondary nucleation	[58, 98, 117]

Table 12.2 Characterisation of major tau isoforms and their neurotoxicological profiles

Tau species	Molecular wt. and dimension	Features	Toxicity	References
Tau monomer	60 kDa weight made up of 352–441 amino acids	Produces diverse forms of tau strains	Do not cause toxicity	[45, 81]
Non-PHF tau with altered phosphorylation	67–70 kDa	3–4 times more phosphorylated and phosphorylation occur at ser 199/202	Leads to disassembly of P-tau in axons and also causes synapse loss	[29, 122, 123]
Tau dimer/trimer	120–180 kDa weight and 22–25 nm length	Dimer found in cerebrospinal fluid of AD patient and trimer is the fundamental unit that causes aggregation of tau	Even small quantity causes neuron toxicity	[57, 79, 107]
Small soluble tau oligomers	300–500 kDa	It contains 6–8 tau molecules	May derived from soluble dimeric form and also soluble and causes toxicity	[18, 28]
Granular tau oligomers	1800 kDa weight and 20–50 nm diameter	It contains 36–40 molecules that are derived from brain of patient suffering from AD	Leads to formation of paired helical filaments and straight filaments and causes toxicity and neuronal cell death	[28, 70, 71]
Straight filaments of tau	>50 nm length and 10 nm width	Tau hyperphosphorylation leads to its assembly into NFTs	Not always show toxic effect	[28, 123]
Paired helical filaments	>220 nm length and 10–20 nm width	Key component of NFTs and highly insoluble	Depending on the context, it can be toxic or non-toxic	[28, 109]
Neurofibrillary tangles	Variable, not assessed	Formed by SFs and PHFs	Probably not toxic Primary biomarker of AD	[17, 109]
Ghost tangles	Smaller than NFTs	These are NFTs that are subjected to proteolytic cleavage or degradation	Can alter cellular calcium level that leads to neuronal dysfunction	[28, 126]

References

1. Abramov AY, Berezhnov AV, Fedotova EI, Zinchenko VP, Dolgacheva LP. Interaction of misfolded proteins and mitochondria in neurodegenerative disorders. Biochem Soc Trans. 2017;45(4):1025–33. https://doi.org/10.1042/BST20170024.
2. Akoury E, Gajda M, Pickhardt M, Biernat J, Soraya P, Griesinger C, Mandelkow E, Zweckstetter M. Inhibition of tau filament formation by conformational modulation. J Am Chem Soc. 2013;135(7):2853–62. https://doi.org/10.1021/ja312471h.
3. Al Mamun A, Uddin MS, Mathew B, Ashraf GM. Toxic tau: structural origins of tau aggregation in Alzheimer's disease. Neural Regen Res. 2020;15(8):1417–20. https://doi.org/10.4103/1673-5374.274329.
4. Alberts B, Johnson A, Lewis J, Walter P, Raff M, Roberts K. Molecular biology of the cell, International Student Edition. 4th ed. Routledge; 2002. https://books.google.co.in/books?id=ozigkQEACAAJ
5. Amir Mishan M, Rezaei Kanavi M, Shahpasand K, Ahmadieh H. Pathogenic tau protein species: promising therapeutic targets for ocular neurodegenerative diseases. J Ophthalmic Vis Res. 2019;14(4):491–505. https://doi.org/10.18502/jovr.v14i4.5459.
6. Arosio P, Michaels TCT, Linse S, Månsson C, Emanuelsson C, Presto J, Johansson J, Vendruscolo M, Dobson CM, Knowles TPJ. Kinetic analysis reveals the diversity of microscopic mechanisms through which molecular chaperones suppress amyloid formation. Nat Commun. 2016;7:10948. https://doi.org/10.1038/ncomms10948.
7. Arrasate M, Pérez M, Avila J. Tau dephosphorylation at tau-1 site correlates with its association to cell membrane. Neurochem Res. 2000;25(1):43–50. https://doi.org/10.1023/a:1007583214722.
8. Avila J, Lucas JJ, Perez M, Hernandez F. Role of tau protein in both physiological and pathological conditions. Physiol Rev. 2004;84(2):361–84. https://doi.org/10.1152/physrev.00024.2003.
9. Balch WE, Morimoto RI, Dillin A, Kelly JW. Adapting proteostasis for disease intervention. Science New York, N.Y. 2008;319(5865):916–9. https://doi.org/10.1126/science.1141448.
10. Ballatore C, Lee VM-Y, Trojanowski JQ. Tau-mediated neurodegeneration in Alzheimer's disease and related disorders. Nat Rev Neurosci. 2007;8(9):663–72. https://doi.org/10.1038/nrn2194.
11. Barati A, Rezaei Somee L, Shahsavani MB, Ghasemi A, Hoshino M, Hong J, Saboury AA, Moosavi-Movahedi AA, Agnetti G, Yousefi R. Insights into the dual nature of αB-crystallin chaperone activity from the p.P39L mutant at the N-terminal region. Sci Rep. 2024;14(1):7353. https://doi.org/10.1038/s41598-024-57651-5.
12. Bernier V, Lagacé M, Bichet DG, Bouvier M. Pharmacological chaperones: potential treatment for conformational diseases. Trends Endocrinol Metabol. 2004;15(5):222–8. https://doi.org/10.1016/j.tem.2004.05.003.
13. Binder LI, Frankfurter A, Rebhun LI. The distribution of tau in the mammalian central nervous system. J Cell Biol. 1985;101(4):1371–8. https://doi.org/10.1083/jcb.101.4.1371.
14. Blurton-Jones M, Laferla FM. Pathways by which Abeta facilitates tau pathology. Curr Alzheimer Res. 2006;3(5):437–48. https://doi.org/10.2174/156720506779025242.
15. Bonito-Oliva A, Barbash S, Sakmar TP, Graham WV. Nucleobindin 1 binds to multiple types of pre-fibrillar amyloid and inhibits fibrillization. Sci Rep. 2017;7(1):42880. https://doi.org/10.1038/srep42880.
16. Boutajangout A, Sigurdsson EM, Krishnamurthy PK. Tau as a therapeutic target for Alzheimer's disease. Curr Alzheimer Res. 2011;8(6):666–77. https://doi.org/10.2174/156720511796717195.
17. Brion JP, Anderton BH, Authelet M, Dayanandan R, Leroy K, Lovestone S, Octave JN, Pradier L, Touchet N, Tremp G. Neurofibrillary tangles and tau phosphorylation. Biochem Soc Symp. 2001;67:81–8. https://doi.org/10.1042/bss0670081.

18. Brunden KR, Trojanowski JQ, Lee VM-Y. Evidence that non-fibrillar tau causes pathology linked to neurodegeneration and behavioral impairments. J Alzheimer's Dis: JAD. 2008;14(4):393–9. https://doi.org/10.3233/jad-2008-14406.
19. Cárdenas-Aguayo Mdel C, Gómez-Virgilio L, DeRosa S, Meraz-Ríos MA. The role of tau oligomers in the onset of Alzheimer's disease neuropathology. ACS Chem Neurosci. 2014;5(12):1178–91. https://doi.org/10.1021/cn500148z. Epub 2014. PMID: 25268947.
20. Carnemolla A, Lazell H, Moussaoui S, Bates GP. In vivo profiling reveals a competent heat shock response in adult neurons: implications for neurodegenerative disorders. PLoS One. 2015;10(7):1–11. https://doi.org/10.1371/journal.pone.0131985.
21. Chaudhuri TK, Paul S. Protein-misfolding diseases and chaperone-based therapeutic approaches. FEBS J. 2006;273(7):1331–49. https://doi.org/10.1111/j.1742-4658.2006.05181.x.
22. Choi ML, Gandhi S. Crucial role of protein oligomerization in the pathogenesis of Alzheimer's and Parkinson's diseases. FEBS J. 2018;285(19):3631–44. https://doi.org/10.1111/febs.14587.
23. Choi H, Kim HJ, Yang J, Chae S, Lee W, Chung S, Kim J, Choi H, Song H, Lee CK, Jun JH, Lee YJ, Lee K, Kim S, Sim H-R, Choi YI, Ryu KH, Park J-C, Lee D, et al. Acetylation changes tau interactome to degrade tau in Alzheimer's disease animal and organoid models. Aging Cell. 2020;19(1):e13081. https://doi.org/10.1111/acel.13081.
24. Chung CG, Lee H, Lee SB. Mechanisms of protein toxicity in neurodegenerative diseases. Cell Mol Life Sci. 2018;75(17):3159–80. https://doi.org/10.1007/s00018-018-2854-4.
25. Ciechanover A, Kwon YT. Degradation of misfolded proteins in neurodegenerative diseases: therapeutic targets and strategies. Exp Mol Med. 2015;47(3):e147. https://doi.org/10.1038/emm.2014.117.
26. Cortez L, Sim V. The therapeutic potential of chemical chaperones in protein folding diseases. Prion. 2014;8(2):197–202. https://doi.org/10.4161/pri.28938.
27. Couchie D, Mavilia C, Georgieff IS, Liem RK, Shelanski ML, Nunez J. Primary structure of high molecular weight tau present in the peripheral nervous system. Proc Natl Acad Sci USA. 1992;89(10):4378–81. https://doi.org/10.1073/pnas.89.10.4378.
28. Cowan CM, Mudher A. Are tau aggregates toxic or protective in tauopathies? Front Neurol. 2013;4:114. https://doi.org/10.3389/fneur.2013.00114.
29. Cowan CM, Bossing T, Page A, Shepherd D, Mudher A. Soluble hyper-phosphorylated tau causes microtubule breakdown and functionally compromises normal tau in vivo. Acta Neuropathol. 2010;120(5):593–604. https://doi.org/10.1007/s00401-010-0716-8.
30. Crespo-Biel N, Theunis C, Van Leuven F. Protein tau: prime cause of synaptic and neuronal degeneration in Alzheimer's disease. Int J Alzheimers Dis. 2012;2012:251426. https://doi.org/10.1155/2012/251426.
31. Das BC, Pradhan S, Ojha DP, Das A, Hosmane NS, Das S. The role of Tau protein in diseases. Ann Adv Chem. 2018; https://www.advancechemjournal.com/apdf/aac-aid1010.pdf
32. Dehle FC, Ecroyd H, Musgrave IF, Carver JA. αB-Crystallin inhibits the cell toxicity associated with amyloid fibril formation by κ-casein and the amyloid-β peptide. Cell Stress Chaperones. 2010;15(6):1013–26. https://doi.org/10.1007/s12192-010-0212-z.
33. Dutta P, Roy P, Sengupta N. Effects of external perturbations on protein systems: a microscopic view. ACS Omega. 2022;7(49):44556–72. https://doi.org/10.1021/acsomega.2c06199.
34. Elias AK, Wilson MR, Carver JA, Musgrave IF. The extracellular molecular chaperone Clusterin inhibits amyloid fibril formation and suppresses cytotoxicity associated with semen-derived enhancer of virus infection (SEVI). Cells. 2022;11(20):3259. https://doi.org/10.3390/cells11203259.
35. Ferreira ST, Vieira MNN, De Felice FG. Soluble protein oligomers as emerging toxins in Alzheimer's and other amyloid diseases. IUBMB Life. 2007;59(4–5):332–45. https://doi.org/10.1080/15216540701283882.
36. Fersht A. Enzyme structure and mechanism (2nd ed.) (vol. 186, Issue 1). W.H. Freeman. 1985. https://core.ac.uk/download/pdf/82589184.pdf

37. García-Blanco A, Baquero M, Vento M, Gil E, Bataller L, Cháfer-Pericás C. Potential oxidative stress biomarkers of mild cognitive impairment due to Alzheimer disease. J Neurol Sci. 2017;373:295–302. https://doi.org/10.1016/j.jns.2017.01.020.
38. Garrett RH, Grisham CM. Biochemistry. 4th ed. Brooks/Cole, Cengage Learning; 2010. https://worldcat.org/title/297392560
39. Gerson JE, Kayed R. Formation and propagation of tau oligomeric seeds. Front Neurol. 2013;4:93. https://doi.org/10.3389/fneur.2013.00093.
40. Götz J, Halliday G, Nisbet RM. Molecular pathogenesis of the Tauopathies. Annu Rev Pathol. 2019;14:239–61. https://doi.org/10.1146/annurev-pathmechdis-012418-012936.
41. Haass C, Selkoe DJ. Soluble protein oligomers in neurodegeneration: lessons from the Alzheimer's amyloid beta-peptide. Nat Rev Mol Cell Biol. 2007;8(2):101–12. https://doi.org/10.1038/nrm2101.
42. Hardy JA, Higgins GA. Alzheimer's disease: the amyloid cascade hypothesis. Science (New York, N.Y.). 1992;256(5054):184–5. https://doi.org/10.1126/science.1566067.
43. Hardy J, Selkoe DJ. The amyloid hypothesis of Alzheimer's disease: progress and problems on the road to therapeutics. Science (New York, N.Y.). 2002;297(5580):353–6. https://doi.org/10.1126/science.1072994.
44. Hartl FU, Bracher A, Hayer-Hartl M. Molecular chaperones in protein folding and proteostasis. Nature. 2011;475(7356):324–32. https://doi.org/10.1038/nature10317.
45. Helbok R, Beer R. Chapter 14 – cerebrospinal fluid and brain extracellular fluid in severe brain trauma. In: Deisenhammer F, Teunissen CE, Tumani H, editors. Cerebrospinal fluid in neurologic disorders, vol. 146. Elsevier; 2018. p. 237–58. https://doi.org/10.1016/B978-0-12-804279-3.00014-9.
46. Howlett DR, Hortobágyi T, Francis PT. Clusterin associates specifically with Aβ40 in Alzheimer's disease brain tissue. Brain Pathol (Zurich, Switzerland). 2013;23(6):623–32. https://doi.org/10.1111/bpa.12057.
47. Iqbal K, Grundke-Iqbal I. Ubiquitination and abnormal phosphorylation of paired helical filaments in Alzheimer's disease. Mol Neurobiol. 1991;5(2–4):399–410. https://doi.org/10.1007/BF02935561.
48. Iqbal K, Liu F, Gong C-X. Tau and neurodegenerative disease: the story so far. Nat Rev Neurol. 2016;12(1):15–27. https://doi.org/10.1038/nrneurol.2015.225.
49. Johnson GVW, Stoothoff WH. Tau phosphorylation in neuronal cell function and dysfunction. J Cell Sci. 2004;117(Pt 24):5721–9. https://doi.org/10.1242/jcs.01558.
50. Kametani F, Hasegawa M. Reconsideration of amyloid hypothesis and TAU hypothesis in Alzheimer's disease. Front Neurosci. 2018;12:25. https://doi.org/10.3389/fnins.2018.00025.
51. Karri V, Ramos D, Martinez JB, Odena A, Oliveira E, Coort SL, Evelo CT, Mariman ECM, Schuhmacher M, Kumar V. Differential protein expression of hippocampal cells associated with heavy metals (Pb, As, and MeHg) neurotoxicity: deepening into the molecular mechanism of neurodegenerative diseases. J Proteome. 2018;187:106–25. https://doi.org/10.1016/j.jprot.2018.06.020.
52. Kelley AR, Bach SBH, Perry G. Analysis of post-translational modifications in Alzheimer's disease by mass spectrometry. Biochim Biophys Acta Mol basis Dis. 2019;1865(8):2040–7. https://doi.org/10.1016/j.bbadis.2018.11.002.
53. Khatoon S, Grundke-Iqbal I, Iqbal K. Levels of normal and abnormally phosphorylated tau in different cellular and regional compartments of Alzheimer disease and control brains. FEBS Lett. 1994;351(1):80–4. https://doi.org/10.1016/0014-5793(94)00829-9.
54. Kikis EA. Nature versus nurture: does proteostasis imbalance underlie the genetic, environmental, and age-related risk factors for Alzheimer's disease? Healthcare (Basel, Switzerland). 2017;5(3):46. https://doi.org/10.3390/healthcare5030046.
55. Kikis EA. The proteostatic effects of traffic-derived air pollution on Alzheimer's disease risk. Open Biol. 2020;10(8):200146. https://doi.org/10.1098/rsob.200146.
56. Kim YE, Hipp MS, Bracher A, Hayer-Hartl M, Hartl FU. Molecular chaperone functions in protein folding and proteostasis. Annu Rev Biochem. 2013;82:323–55. https://doi.org/10.1146/annurev-biochem-060208-092442.

57. Kim D, Lim S, Haque MM, Ryoo N, Hong HS, Rhim H, Lee D-E, Chang Y-T, Lee J-S, Cheong E, Kim DJ, Kim YK. Identification of disulfide cross-linked tau dimer responsible for tau propagation. Sci Rep. 2015;5:15231. https://doi.org/10.1038/srep15231.
58. Koopman MB, Rüdiger SGD. Alzheimer cells on their way to derailment show selective changes in protein quality control network. Front Mol Biosci. 2020;7:214. https://doi.org/10.3389/fmolb.2020.00214.
59. Kumar A, Sidhu J, Lui F, Tsao JW. Alzheimer disease. 2024. https://www.ncbi.nlm.nih.gov/books/NBK499922/.
60. Kundel F, De S, Flagmeier P, Horrocks MH, Kjaergaard M, Shammas SL, Jackson SE, Dobson CM, Klenerman D. Hsp70 inhibits the nucleation and elongation of tau and sequesters tau aggregates with high affinity. ACS Chem Biol. 2018;13(3):636–46. https://doi.org/10.1021/acschembio.7b01039.
61. Kundu D, Prerna K, Chaurasia R, Bharty MK, Dubey VK. Advances in protein misfolding, amyloidosis and its correlation with human diseases. 3 Biotech. 2020;10(5):193. https://doi.org/10.1007/s13205-020-2166-x.
62. LaPelusa A, Kaushik R. Physiology, Proteins. 2022 Nov 14. In: StatPearls [Internet]. Treasure Island (FL): StatPearls Publishing. 2025. PMID: 32310450.
63. LaPointe NE, Morfini G, Pigino G, Gaisina IN, Kozikowski AP, Binder LI, Brady ST. The amino terminus of tau inhibits kinesin-dependent axonal transport: implications for filament toxicity. J Neurosci Res. 2009;87(2):440–51. https://doi.org/10.1002/jnr.21850.
64. Larsson JNK, Nyström S, Hammarström P. HSP10 as a chaperone for neurodegenerative amyloid fibrils. Front Neurosci. 2022;16:902600. https://doi.org/10.3389/fnins.2022.902600.
65. Lasagna-Reeves CA, Castillo-Carranza DL, Sengupta U, Guerrero-Munoz MJ, Kiritoshi T, Neugebauer V, Jackson GR, Kayed R. Alzheimer brain-derived tau oligomers propagate pathology from endogenous tau. Sci Rep. 2012;2:700. https://doi.org/10.1038/srep00700.
66. Lee G, Neve RL, Kosik KS. The microtubule binding domain of tau protein. Neuron. 1989;2(6):1615–24. https://doi.org/10.1016/0896-6273(89)90050-0.
67. Lee S, Kim W, Li Z, Hall GF. Accumulation of vesicle-associated human tau in distal dendrites drives degeneration and tau secretion in an in situ cellular tauopathy model. Int J Alzheimers Dis. 2012;2012:172837. https://doi.org/10.1155/2012/172837.
68. Levenson RW, Sturm VE, Haase CM. Emotional and behavioral symptoms in neurodegenerative disease: a model for studying the neural bases of psychopathology. Annu Rev Clin Psychol. 2014;10:581–606. https://doi.org/10.1146/annurev-clinpsy-032813-153653.
69. Luna-Muñoz J, Chávez-Macías L, García-Sierra F, Mena R. Earliest stages of tau conformational changes are related to the appearance of a sequence of specific phospho-dependent tau epitopes in Alzheimer's disease. J Alzheimer's Dis. 2007;12(4):365–75. https://doi.org/10.3233/jad-2007-12410.
70. Maeda S, Sahara N, Saito Y, Murayama S, Ikai A, Takashima A. Increased levels of granular tau oligomers: an early sign of brain aging and Alzheimer's disease. Neurosci Res. 2006;54(3):197–201. https://doi.org/10.1016/j.neures.2005.11.009.
71. Maeda S, Sahara N, Saito Y, Murayama M, Yoshiike Y, Kim H, Miyasaka T, Murayama S, Ikai A, Takashima A. Granular tau oligomers as intermediates of tau filaments. Biochemistry. 2007;46(12):3856–61. https://doi.org/10.1021/bi061359o.
72. Mannini B, Chiti F. Chaperones as suppressors of protein misfolded oligomer toxicity. Front Mol Neurosci 2017;10:98. https://doi.org/10.3389/fnmol.2017.00098.
73. Martin L, Latypova X, Wilson CM, Magnaudeix A, Perrin M-L, Yardin C, Terro F. Tau protein kinases: involvement in Alzheimer's disease. Ageing Res Rev. 2013;12(1):289–309. https://doi.org/10.1016/j.arr.2012.06.003.
74. Medina M, Avila J. The role of extracellular tau in the spreading of neurofibrillary pathology. Front Cell Neurosci. 2014;8:113. https://doi.org/10.3389/fncel.2014.00113.
75. Mendez MF. Early-onset Alzheimer disease. Neurol Clin. 2017;35(2):263–81. https://doi.org/10.1016/j.ncl.2017.01.005.
76. Mendsaikhan A, Tooyama I, Bellier J-P, Serrano GE, Sue LI, Lue L-F, Beach TG, Walker DG. Characterization of lysosomal proteins Progranulin and Prosaposin and their interac-

tions in Alzheimer's disease and aged brains: increased levels correlate with neuropathology. Acta Neuropathol Commun. 2019;7(1):215. https://doi.org/10.1186/s40478-019-0862-8.
77. Meraz-Ríos MA, Lira-De León KI, Campos-Peña V, De Anda-Hernández MA, Mena-López R. Tau oligomers and aggregation in Alzheimer's disease. J Neurochem. 2010;112(6):1353–67. https://doi.org/10.1111/j.1471-4159.2009.06511.x.
78. Mietelska-Porowska A, Wasik U, Goras M, Filipek A, Niewiadomska G. Tau protein modifications and interactions: their role in function and dysfunction. Int J Mol Sci. 2014;15(3):4671–713. https://doi.org/10.3390/ijms15034671.
79. Mirbaha H, Holmes BB, Sanders DW, Bieschke J, Diamond MI. Tau trimers are the minimal propagation unit spontaneously internalized to seed intracellular aggregation. J Biol Chem. 2015;290(24):14893–903. https://doi.org/10.1074/jbc.M115.652693.
80. Moloney CM, Lowe VJ, Murray ME. Visualization of neurofibrillary tangle maturity in Alzheimer's disease: a clinicopathologic perspective for biomarker research. Alzheimer's Dementia: J Alzheimer's Assoc. 2021;17(9):1554–74. https://doi.org/10.1002/alz.12321.
81. Mroczko B, Groblewska M, Litman-Zawadzka A. The role of protein Misfolding and tau oligomers (TauOs) in Alzheimer's disease (AD). Int J Mol Sci. 2019;20(19):4661. https://doi.org/10.3390/ijms20194661.
82. Muchowski PJ, Wacker JL. Modulation of neurodegeneration by molecular chaperones. Nat Rev Neurosci. 2005;6(1):11–22. https://doi.org/10.1038/nrn1587.
83. Mullane K, Williams M. Alzheimer's therapeutics: continued clinical failures question the validity of the amyloid hypothesis-but what lies beyond? Biochem Pharmacol. 2013;85(3):289–305. https://doi.org/10.1016/j.bcp.2012.11.014.
84. Muralidar S, Ambi SV, Sekaran S, Thirumalai D, Palaniappan B. Role of tau protein in Alzheimer's disease: the prime pathological player. Int J Biol Macromol. 2020;163:1599–617. https://doi.org/10.1016/j.ijbiomac.2020.07.327.
85. Murphy MP, LeVine H 3rd. Alzheimer's disease and the amyloid-beta peptide. J Alzheimer's Dis. 2010;19(1):311–23. https://doi.org/10.3233/JAD-2010-1221.
86. Parker CE, Mocanu V, Mocanu M, Dicheva N, Warren MR. Mass spectrometry for post-translational modifications. In: Alzate O, editor. Neuroproteomics. Boca Raton (FL): CRC Press/Taylor & Francis; 2010.
87. Pluta R, Ulamek-Koziol M. TAU protein-targeted therapies in Alzheimer's disease: current state and future perspectives. Exon Publications eBooks; 2020. https://doi.org/10.36255/exonpublications.alzheimersdisease.2020.ch4.
88. Qiu C, Kivipelto M, von Strauss E. Epidemiology of Alzheimer's disease: occurrence, determinants, and strategies toward intervention. Dialogues Clin Neurosci. 2009;11(2):111–28. https://doi.org/10.31887/DCNS.2009.11.2/cqiu.
89. Reynaud E. Protein misfolding and degenerative diseases. Nat. Educ. 2010;3(9):28.
90. Rhinn H, Tatton N, McCaughey S, Kurnellas M, Rosenthal A. Progranulin as a therapeutic target in neurodegenerative diseases. Trends Pharmacol Sci. 2022;43(8):641–52. https://doi.org/10.1016/j.tips.2021.11.015.
91. Rinauro DJ, Chiti F, Vendruscolo M, Limbocker R. Misfolded protein oligomers: mechanisms of formation, cytotoxic effects, and pharmacological approaches against protein misfolding diseases. Mol Neurodegener. 2024;19(1):20. https://doi.org/10.1186/s13024-023-00651-2.
92. Roy S, Zhang B, Lee VM-Y, Trojanowski JQ. Axonal transport defects: a common theme in neurodegenerative diseases. Acta Neuropathol. 2005;109(1):5–13. https://doi.org/10.1007/s00401-004-0952-x.
93. Salvadores N, Shahnawaz M, Scarpini E, Tagliavini F, Soto C. Detection of misfolded Aβ oligomers for sensitive biochemical diagnosis of Alzheimer's disease. Cell Rep. 2014;7(1):261–8. https://doi.org/10.1016/j.celrep.2014.02.031.
94. Sanvictores T, Farci F. Biochemistry, primary protein structure. 2022. https://www.ncbi.nlm.nih.gov/books/NBK564343/
95. Scheidt T, Łapińska U, Kumita JR, Whiten DR, Klenerman D, Wilson MR, Cohen SIA, Linse S, Vendruscolo M, Dobson CM, Knowles TPJ, Arosio P. Secondary nucleation and elongation

occur at different sites on Alzheimer's amyloid-β aggregates. Sci Adv. 2019;5(4):eaau3112. https://doi.org/10.1126/sciadv.aau3112.
96. Scieglinska D, Krawczyk Z, Sojka DR, Gogler-Pigłowska A. Heat shock proteins in the physiology and pathophysiology of epidermal keratinocytes. Cell Stress Chaperones. 2019;24(6):1027–44. https://doi.org/10.1007/s12192-019-01044-5.
97. Silva MVF, de Mello Gomide Loures C, Alves LCV, de Souza LC, Borges KBG, das Graças Carvalho M. Alzheimer's disease: risk factors and potentially protective measures. J Biomed Sci. 2019;26(1):33. https://doi.org/10.1186/s12929-019-0524-y.
98. Sinnige T. Molecular mechanisms of amyloid formation in living systems. Chem Sci. 2022;13(24):7080–97. https://doi.org/10.1039/D2SC01278B.
99. Sonninen T-M, Goldsteins G, Laham-Karam N, Koistinaho J, Lehtonen Š. Proteostasis disturbances and inflammation in neurodegenerative diseases. Cells. 2020;9(10):2183. https://doi.org/10.3390/cells9102183.
100. Soto C. Alzheimer's and prion disease as disorders of protein conformation: implications for the design of novel therapeutic approaches. J Mol Med (Berl). 1999;77(5):412–8. https://doi.org/10.1007/s001090050371.
101. Soto C. Protein misfolding and disease; protein refolding and therapy. FEBS Lett. 2001;498(2–3):204–7. https://doi.org/10.1016/s0014-5793(01)02486-3.
102. Soto C. Unfolding the role of protein misfolding in neurodegenerative diseases. Nat Rev Neurosci. 2003;4(1):49–60. https://doi.org/10.1038/nrn1007.
103. Soto C, Saborio GP, Permanne B. Inhibiting the conversion of soluble amyloid-beta peptide into abnormally folded amyloidogenic intermediates: relevance for Alzheimer's disease therapy. Acta Neurol Scand Suppl. 2000;176:90–5. https://doi.org/10.1034/j.1600-0404.2000.00313.x.
104. Stancu I-C, Vasconcelos B, Terwel D, Dewachter I. Models of β-amyloid induced tau-pathology: the long and "folded" road to understand the mechanism. Mol Neurodegener. 2014;9:51. https://doi.org/10.1186/1750-1326-9-51.
105. Sweeney P, Park H, Baumann M, Dunlop J, Frydman J, Kopito R, McCampbell A, Leblanc G, Venkateswaran A, Nurmi A, Hodgson R. Protein misfolding in neurodegenerative diseases: implications and strategies. Transl Neurodegener. 2017;6(1):6. https://doi.org/10.1186/s40035-017-0077-5.
106. Tabaton M, Mandybur TI, Perry G, Onorato M, Autilio-Gambetti L, Gambetti P. The widespread alteration of neurites in Alzheimer's disease may be unrelated to amyloid deposition. Ann Neurol. 1989;26(6):771–8. https://doi.org/10.1002/ana.410260614.
107. Tian H, Davidowitz E, Lopez P, Emadi S, Moe J, Sierks M. Trimeric tau is toxic to human neuronal cells at low nanomolar concentrations. Int J Cell Biol. 2013;2013:260787. https://doi.org/10.1155/2013/260787.
108. Trinczek B, Biernat J, Baumann K, Mandelkow EM, Mandelkow E. Domains of tau protein, differential phosphorylation, and dynamic instability of microtubules. Mol Biol Cell. 1995;6(12):1887–902. https://doi.org/10.1091/mbc.6.12.1887.
109. Trojanowski JQ, Schmidt ML, Shin RW, Bramblett GT, Rao D, Lee VM. Altered tau and neurofilament proteins in neuro-degenerative diseases: diagnostic implications for Alzheimer's disease and Lewy body dementias. Brain Pathol (Zurich, Switzerland). 1993;3(1):45–54. https://doi.org/10.1111/j.1750-3639.1993.tb00725.x.
110. Trojanowski JQ, Smith AB, Huryn D, Lee VM-Y. Microtubule-stabilising drugs for therapy of Alzheimer's disease and other neurodegenerative disorders with axonal transport impairments. Expert Opin Pharmacother. 2005;6(5):683–6. https://doi.org/10.1517/14656566.6.5.683.
111. Uematsu M, Nakamura A, Ebashi M, Hirokawa K, Takahashi R, Uchihara T. Brainstem tau pathology in Alzheimer's disease is characterized by increase of three repeat tau and independent of amyloid β. Acta Neuropathol Commun. 2018;6(1):1. https://doi.org/10.1186/s40478-017-0501-1.
112. Valastyan JS, Lindquist S. Mechanisms of protein-folding diseases at a glance. Dis Model Mech. 2014;7(1):9–14. https://doi.org/10.1242/dmm.013474.

113. Anna von Mikecz. Protein aggregation in the cell nucleus: structure, function and topology. Open Biol J (vol. 2). Bentham Science. 2009. https://doi.org/10.2174/1874196700902010193
114. Wang H, Yang F, Zhang S, Xin R, Sun Y. Genetic and environmental factors in Alzheimer's and Parkinson's diseases and promising therapeutic intervention via fecal microbiota transplantation. Npj Parkinson's Dis. 2021;7(1):70. https://doi.org/10.1038/s41531-021-00213-7.
115. Ward SM, Himmelstein DS, Lancia JK, Binder LI. Tau oligomers and tau toxicity in neurodegenerative disease. Biochem Soc Trans. 2012;40(4):667–71. https://doi.org/10.1042/BST20120134.
116. Weingarten MD, Lockwood AH, Hwo SY, Kirschner MW. A protein factor essential for microtubule assembly. Proc Natl Acad Sci USA. 1975;72(5):1858–62. https://doi.org/10.1073/pnas.72.5.1858.
117. Wilhelmus MMM, Boelens WC, Otte-Höller I, Kamps B, Kusters B, Maat-Schieman MLC, de Waal RMW, Verbeek MM. Small heat shock protein HspB8: its distribution in Alzheimer's disease brains and its inhibition of amyloid-beta protein aggregation and cerebrovascular amyloid-beta toxicity. Acta Neuropathol. 2006;111(2):139–49. https://doi.org/10.1007/s00401-005-0030-z.
118. Woulfe J. Nuclear bodies in neurodegenerative disease. Biochim Biophys Acta. 2008;1783(11):2195–206. https://doi.org/10.1016/j.bbamcr.2008.05.005.
119. Wu JW, Herman M, Liu L, Simoes S, Acker CM, Figueroa H, Steinberg JI, Margittai M, Kayed R, Zurzolo C, Di Paolo G, Duff KE. Small misfolded tau species are internalized via bulk endocytosis and anterogradely and retrogradely transported in neurons. J Biol Chem. 2013;288(3):1856–70. https://doi.org/10.1074/jbc.M112.394528.
120. Xia Y, Sorrentino ZA, Kim JD, Strang KH, Riffe CJ, Giasson BI. Impaired tau-microtubule interactions are prevalent among pathogenic tau variants arising from missense mutations. J Biol Chem. 2019;294(48):18488–503. https://doi.org/10.1074/jbc.RA119.010178.
121. Xu J, Han M, Ren Y, Li JH. Exploring the mechanisms of protein folding. 2013.
122. Zempel H, Thies E, Mandelkow E, Mandelkow E-M. Abeta oligomers cause localized ca(2+) elevation, missorting of endogenous tau into dendrites, tau phosphorylation, and destruction of microtubules and spines. J Neurosci: Off J Soc Neurosci. 2010;30(36):11938–50. https://doi.org/10.1523/JNEUROSCI.2357-10.2010.
123. Zeng Y, Yang J, Zhang B, Gao M, Su Z, Huang Y. The structure and phase of tau: from monomer to amyloid filament. Cell Mol Life Sci: CMLS. 2021;78(5):1873–86. https://doi.org/10.1007/s00018-020-03681-x.
124. Zhang Q, Powers ET, Nieva J, Huff ME, Dendle MA, Bieschke J, Glabe CG, Eschenmoser A, Wentworth PJ, Lerner RA, Kelly JW. Metabolite-initiated protein misfolding may trigger Alzheimer's disease. Proc Natl Acad Sci USA. 2004;101(14):4752–7. https://doi.org/10.1073/pnas.0400924101.
125. Zheng H, Sun H, Cai Q, Tai H-C. The enigma of tau protein aggregation: mechanistic insights and future challenges. Int J Mol Sci. 2024;25(9):4969. https://doi.org/10.3390/ijms25094969.
126. Zwang TJ, Woost B, Bailey J, Hoglund Z, Richardson DS, Bennett RE, Hyman BT. Spatial characterization of tangle-bearing neurons and ghost tangles in the human inferior temporal gyrus with three-dimensional imaging. Brain Commun. 2023;5(3):fcad130. https://doi.org/10.1093/braincomms/fcad130.

Therapeutic Strategies Targeting Amyloid-β in Alzheimer's Disease

13

Neha Tyagi, Shilpi Pal, and Divya Bisht

13.1 Introduction

13.1.1 Overview of Alzheimer's Disease

Alzheimer's disease is the leading cause of dementia and is rapidly becoming one of the most costly, deadly, and burdensome diseases of this century [27]. Alzheimer's disease (AD) is a progressive, relentless neurodegenerative disorder that impacts extensive regions of the cerebral cortex and hippocampus [2]. The abnormalities are typically first observed in brain tissue within the frontal and temporal lobes, gradually spreading to other areas of the neocortex at varying rates among individuals. The disease is characterized by the accumulation of insoluble amyloid-β (Aβ) in plaques found in extracellular spaces and in the walls of blood vessels, as well as the aggregation of the microtubule protein tau into neurofibrillary tangles within neurons. Aβ results from the proteolytic cleavage of amyloid precursor protein (APP) by a complex array of enzymes, including presenilin 1 (PS1; encoded by PSEN1) and presenilin 2 (PS2; encoded by PSEN2), which are part of the γ-secretase and β-secretase families [13].

N. Tyagi (✉)
Department of Pharmaceutics, School of Pharmaceutical Education & Research, Jamia Hamdard, New Delhi, India

S. Pal
Hi-Tech Dental College and Hospital, Utkal University, Bhubaneswar, Odisha, India

D. Bisht
Biochemistry Discipline, School of Sciences, Indira Gandhi National Open University, New Delhi, India

N. Puranik (ed.), *Proteostasis*, https://doi.org/10.1007/978-981-96-6202-9_13

13.2 Amyloid Cascade Hypothesis

Changes in Aβ metabolism result in increased production of Aβ, an elevated Aβ42/Aβ40 ratio, and reduced degradation or clearance of Aβ. This cascade contributes to the oligomerization and deposition of Aβ42. Soluble Aβ42 oligomers subtly disrupt synaptic function, triggering inflammatory responses and the formation of amyloid plaques, which ultimately leads to neuronal injury. The disruption of ionic homeostasis, oxidative stress, abnormal oligomerization, and hyperphosphorylation of tau contributes to widespread neuronal dysfunction and cell death, accompanied by neurotransmitter deficits. This process is linked to dementia characterized by both plaque and tangle pathology (Fig. 13.1).

13.3 Amyloid-β Pathophysiology

The "amyloid cascade hypothesis" proposes that the proteolysis of the approximately 770-amino acid, polytopic transmembrane beta-amyloid precursor protein (βAPP; found at chr 21q21.1; GenBank: BAA22264.1) involves interactions with associated membrane proteins such as nicastrin [34]. This process, facilitated by the actions of beta (β)- and gamma (γ)-secretases, results in the production of various amyloid-beta (Aβ) peptide monomers. Importantly, γ-secretase is a unique aspartic protease that cleaves bAPP within its hydrophobic transmembrane domain, and its catalytic activity is considered the rate-limiting step in Aβ peptide production. The cleavage requires a water molecule, indicating that factors such as membrane

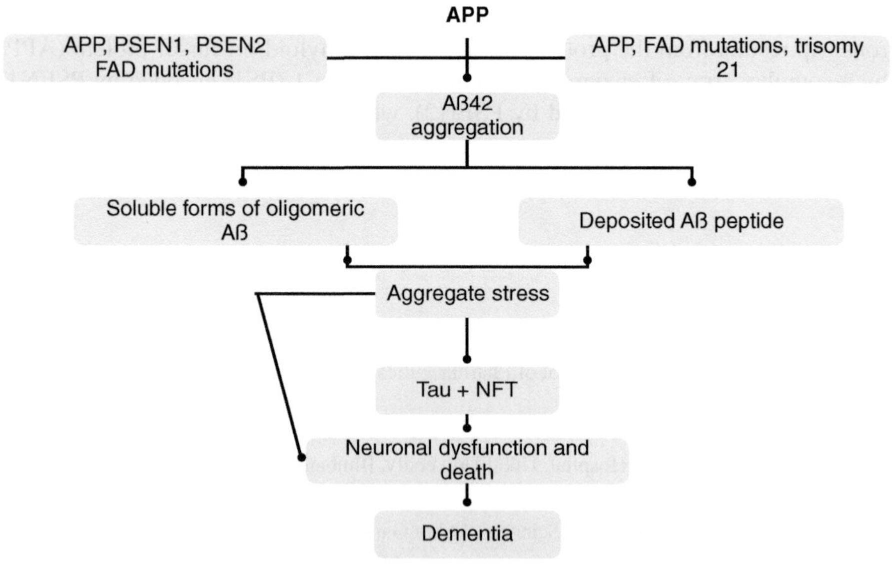

Fig. 13.1 Amyloid cascade hypothesis

disruption or the dynamic movement of bAPP within the membrane may enhance this hydrolysis [43].

The γ-secretase complex consists of several integral membrane proteins, including presenilin 1 (PS1; PSEN1), nicastrin, Aph-1, and PSEN-2 [48]. Although all four components are essential for γ-secretase activity, the exact mechanism by which this complex recognizes and cleaves APP remains inadequately understood, representing a crucial area for developing drugs that specifically modulate β- and γ-secretase functions.

The generated Aβ peptides primarily include two key isoforms: Aβ40, a 40-amino acid peptide that associates with the cerebral vasculature, and the more neurotoxic, though less abundant, Aβ42, which comprises 42 amino acids and is prone to self-aggregation, forming the core of senile plaques. The additional two amino acids in Aβ42 are believed to confer significant toxic properties on this slightly larger molecule. The brain's microglial surveillance system recognizes these end-stage Aβ peptides, but microglia often struggle to manage the toxic, insoluble aggregates, contributing to the inflammatory and oxidative stress characteristic of AD [9].

A primary therapeutic strategy for AD has focused on targeting the secretases responsible for the production of Aβ40 and Aβ42. An informative website offers animated explanations of the βAPP cleavage mechanism and the potential for modulating Aβ42 levels to reduce AD-related deposits.

In contrast to the toxic Aβ peptides, a non-amyloidogenic form known as soluble amyloid precursor protein alpha (sAPPα), generated through alternative α-secretase cleavage of βAPP, possesses various neuroprotective and growth-promoting qualities. APP has been shown to enhance neuronal excitability and synaptic plasticity, and it aids in brain cell regeneration after injuries such as stroke. While the neurotoxic effects of Aβ peptides lead to oxidative stress, inflammation, neuronal dysfunction, and cell death, sAPPα production diverts the formation of harmful Aβ peptides while promoting neurotrophic benefits.

The single-transmembrane lipoprotein receptor SorLA/LR11 plays a crucial role in regulating bAPP trafficking from the cell surface. Its abundance influences whether neurotoxic or neurotrophic pathways are favored, with genetic variants linked to SorLA being associated with late-onset forms of AD [29].

Several alternative hypotheses to the amyloid cascade theory exist. These include the notion that amyloid deposition may serve as a neuroprotective response to AD, that Aβ42 increases could be a reaction to infections, and the idea that AD may be an autoimmune disorder triggered by pathogens like herpes simplex virus (HSV-1) in a gene-dependent manner [25]. Interestingly, antiviral medications such as acyclovir have been found to reduce HSV-1-related increases in pro-inflammatory micro-RNA (miRNA-146a) in human brain cells, diminish Aβ peptide accumulation in AD model systems, and potentially provide therapeutic benefits for the condition.

13.4 Current Therapeutic Approaches

13.4.1 Amyloid-Based Therapeutic Approaches

Significant efforts in clinical research have focused on developing new therapies for Alzheimer's disease (AD). While all currently approved medications provide symptomatic relief, they do not halt disease progression. These drugs can often lead to serious side effects and are typically recommended only for severe cases. As a result, ongoing research into amyloid-based therapies aims to identify more effective treatments for this devastating condition. The amyloid hypothesis, bolstered by genetic studies over the past two decades, posits that the accumulation of Aβ peptide in the brain is a key factor in the development of AD [50].

Current research is centered on developing treatment strategies that either reduce Aβ production or enhance its clearance. Various therapeutic approaches are being explored to counteract the effects of the Aβ peptide, including the inhibition of β- and γ-secretases and the activation of α-secretase. These strategies offer promising avenues for advancing therapies for Alzheimer's disease.

13.4.1.1 Inhibition of β-Secretase

β-Secretase, also referred to as BACE-1, Asp2, or memapsin2, is a transmembrane aspartate protease that cleaves amyloid precursor protein (APP) at the β-secretase site, resulting in the production of Aβ peptide in the brain. In sporadic Alzheimer's disease (AD), there is a documented increase in β-secretase activity, which plays a crucial role in the production of Aβ peptide. Interestingly, studies involving BACE knockout mice—known for being normal, healthy, and fertile—reveal that these mice produce significantly lower levels of Aβ from amyloid precursor protein (APP) compared to their wild-type counterparts. Remarkably, despite this substantial reduction in Aβ production, BACE knockout mice do not display any notable phenotypic differences, suggesting that the absence of β-secretase does not adversely affect their overall health or development. This raises important questions about the role of β-secretase in AD and the potential implications for therapeutic strategies targeting this enzyme [51]. Consequently, β-secretase is a key therapeutic target for developing inhibitor drugs.

Since the discovery of β-secretase, more than 400 publications and patents have been released in the past 8 years, all centered on β-secretase inhibitors. The first potent inhibitor, OM99-2, was developed using a substrate model that features a phenylnorstatine moiety as the transition-state isostere [3]. Subsequent advancements included the development of hydroxyethylene-based inhibitors, hydroxyethylamine-based inhibitors, carbinamine-based inhibitors, macrocyclic inhibitors, and non-peptidomimetic β-secretase inhibitors, as outlined by Ghosh and colleagues in 2012 [20].

Despite these advancements, only a limited number of β-secretase inhibitors have progressed to phase I clinical trials. The first publicly announced phase I trial for a β-secretase inhibitor, CTS-21166, was carried out by CoMentis [35]. Although β-secretase inhibitors have shown promise in reducing Aβ concentrations in both animal models and cellular systems, their ability to effectively cross the blood-brain barrier remains a significant limitation. This barrier poses a critical challenge for many therapeutic agents, as it restricts the passage of compounds into the central nervous system. As a result, even with demonstrated efficacy in lowering Aβ levels in peripheral systems, the limited penetration of these inhibitors into the brain may hinder their overall effectiveness in treating Alzheimer's disease. Addressing this challenge is crucial for the development of more effective β-secretase inhibitors that can reach therapeutic levels in the brain [10]. Furthermore, β-secretase inhibitors are not fully specific to β-secretase; they can also inhibit other aspartate proteases [23]. Therefore, there is an urgent need to develop effective and safe β-secretase inhibitors using structure-based design to realize their potential as valuable therapeutic targets for Alzheimer's disease.

13.4.1.2 Inhibition of γ-Secretase

γ-Secretase is a membrane-embedded aspartate protease complex that consists of four essential subunits: Aph1, PEN2, glycosylated nicastrin, and endoproteolyzed presenilin, which acts as the catalytic core of the complex [5]. Presenilin undergoes endoproteolysis to produce an N-terminal fragment (NTF) and a C-terminal fragment (CTF) that remain associated. Two conserved aspartates within adjacent transmembrane domains are crucial for both presenilin endoproteolysis and γ-secretase activity.

Presenilins are expressed in the brain and various peripheral tissues and are involved in Notch signaling, cell adhesion, and cell signaling [32]. Approximately 185 mutations in the PSEN1 gene and 13 in the PSEN2 gene have been linked to familial Alzheimer's disease (FAD). These mutations can alter the γ-secretase active site, potentially shifting the cleavage site of substrates. This altered APP processing leads to increased production of the more pathogenic Aβ1–42 peptide [16]. Thus, inhibiting γ-secretase could be a viable therapeutic strategy for Alzheimer's disease.

Sulfonamide-based inhibitors selectively target PSEN1 and PSEN2, while other inhibitors like DAPT and L685458 show less selectivity [44]. γ-Secretase inhibitors such as BMS-299897 and MRK-560 are notably effective in reducing Aβ levels. Another inhibitor, LY45139, was tested in a randomized controlled trial involving patients with mild to moderate Alzheimer's disease and demonstrated reduced Aβ concentrations [22]. However, γ-secretase is involved in multiple physiological processes, and inhibitors affecting this enzyme can disrupt important pathways, particularly Notch signaling [54].

Therefore, there is a need for γ-secretase modulators that can adjust the ratios of Aβ1–38, Aβ1–40, and Aβ1–42 production while preserving Notch activity. In vivo studies have shown that while γ-secretase inhibitors can prevent the formation of new Aβ plaques and significantly suppress Aβ production, they do not enhance the clearance of existing plaques [1]. Some nonsteroidal anti-inflammatory drugs (NSAIDs), such as ibuprofen, indomethacin, and sulindac sulfide, can decrease Aβ1–40 production but increase Aβ1–38 levels [53]. Currently, a detailed 3D atomic-level structure of the γ-secretase complex has not been fully resolved.

Additionally, concerns remain about the potential negative effects on the Notch signaling pathway due to γ-secretase inhibition and modulation [46].

13.4.1.3 Activation of α-Secretase

α-Secretases are responsible for cleaving the amyloid precursor protein (APP) within its transmembrane region[38]. Key proteases involved in this process include members of the ADAM family (A Disintegrin and Metalloprotease), such as ADAM17, ADAM10, and ADAM9, as well as tumor necrosis factor-α convertase (TACE) [7].

The cleavage of APP by α-secretase occurs within the Aβ domain, resulting in the release of soluble APP, which has neuroprotective and memory-enhancing effects [4]. Therefore, enhancing α-secretase activity is considered a promising therapeutic approach for Alzheimer's disease as it prevents the formation of Aβ. Studies have shown that overexpression of ADAM10 in transgenic mice results in reduced amyloid deposition in the hippocampus and improved neurological function [56]. The PKC activator TPPB [(2S, 5S)-(E, E)-8-(5-(4-(trifluoromethyl) phenyl)-2,4-pentadienoylamino)benzolactam] has been found to increase α-secretase activity and decrease Aβ production. Additionally, Sirtuin-1, which upregulates ADAM10 expression, promotes α-secretase cleavage and lowers Aβ levels [42]. M1 muscarinic agonists such as AF267B have also been shown to enhance ADAM17 activation and increase α-secretase cleavage of APP in AD mouse models. Various methods are being explored to regulate α-secretase activity at transcriptional, translational, and posttranslational levels.

However, ADAM10 is involved in numerous physiological and pathological processes, including embryonic development, cell adhesion, signal transduction, cancer, and AD [24]. Despite these advances, the precise cellular mechanisms of α-secretase cleavage are not fully understood, indicating that further research is needed to assess the therapeutic potential of increasing α-secretase activity in AD [31].

13.4.1.4 Inhibitors of Aβ Oligomerization or Fibrillization

Following γ-secretase action on APP, the released Aβ peptide monomers initially adopt an α-helical conformation but rapidly transition to a β-sheet structure, leading to self-aggregation into oligomers, protofibrils, and fibrils, ultimately forming Aβ plaques. Considerable research is underway to develop compounds that interfere with Aβ polymerization and fibrillogenesis to prevent or treat AD [14]. Several molecules have been identified that either prevent Aβ aggregation and fibrillogenesis or inhibit Aβ toxicity. For instance, the Aβ fibril-binding dye Congo red reduces fibril formation and neurotoxicity. Additionally, sulfonated dyes like benzofuran-based compounds, sulfonated anions, and amphiphilic surfactants (di-C6-PC and di-C7-PC) have been used as Aβ aggregation inhibitors [5]. Other compounds such as melatonin, nicotine, estrogen, and anthracycline 49-iodo-49-deoxydoxorubicin also show promise in inhibiting Aβ aggregation or reducing Aβ neurotoxicity.

The development of specific peptide ligands that can bind to Aβ peptides, such as N-methylated (NMe) derivatives of Aβ (25–35), Aβ (16–20), and lysine hexamer

(KLVFFKKKKKK), has been shown to alter fibril morphology and reduce Aβ cellular toxicity [21]. The recently developed drug iAβ5p, a β-sheet breaker, decreases Aβ plaque deposits and improves spatial memory [36]. However, small molecules as aggregation inhibitors face challenges such as lack of specificity, toxicity, and unknown mechanisms of action, while peptide derivatives encounter issues related to administration, delivery, and rapid degradation.

In TgCRND8 mice, which carry human APP mutations K670N/M671L and V717F, tramiprosate administered subcutaneously reduced soluble plasma Aβ1–40 and Aβ1–42 levels in a dose-dependent manner [19]. Nevertheless, a double-blind, placebo-controlled trial with tramiprosate did not show a reduction in plasma Aβ1–40 levels in patients [15]. Bapineuzumab, a humanized IgG1 version of the 3D6 mouse monoclonal antibody that binds to the N-terminus of Aβ peptide, failed to significantly reduce Aβ plaque levels in a Phase 2 trial involving mild-to-moderate AD patients. Similarly, Solanezumab, a humanized IgG1 antibody derived from the mouse monoclonal antibody m266, showed binding to Aβ peptides in Phase 3 trials but did not improve cognition or functional ability in patients [33]. Future efforts should focus on identifying safe antifibrilizing agents while addressing the limitations of earlier drug candidates.

13.4.2 Aβ Degradation and Clearance

In Alzheimer's disease (AD), the imbalance between the production and removal of Aβ peptide leads to complex pathological events. Elevated levels of Aβ peptides in the brain are due to either overproduction or impaired removal of the peptide. Reduced clearance of Aβ can result from several factors, including increased aggregation, defective degradation, disrupted transport across the blood-brain barrier, or inefficient peripheral removal [55]. Consequently, pharmacological interventions have focused on these mechanisms, although many compounds have shown limited clinical efficacy in enhancing Aβ clearance.

A promising new therapeutic approach involves targeting Aβ peptide degradation and clearance. Recent research has highlighted the importance of proteases involved in Aβ degradation as potential therapeutic targets. Key enzymes identified include Neprilysin (NEP), Insulin Degrading Enzyme (IDE), Endothelin Converting Enzyme (ECE-1), plasmin, Angiotensin Converting Enzyme (ACE), cathepsin B, aminopeptidase, gelatinase A, matrix metalloendopeptidase-9, coagulation factor Xia, antibody light chains c23.5 and hk14, and the Upa-/tPA-plasmin system, among others. These enzymes play a crucial role in degrading Aβ peptides, and increasing their expression could be an alternative strategy to lower Aβ peptide concentrations.

These Aβ-degrading enzymes can cleave Aβ peptides at single or multiple sites, resulting in products that are less neurotoxic and less prone to aggregation compared to the Aβ peptide itself. They are essential in maintaining Aβ homeostasis in neurons [12]. Enzymes such as NEP, IDE, ACE, and ECE are well established as effective Aβ-degrading agents. However, further research is needed to fully understand the relative contributions of these enzymes and their roles in the amyloid cascade, which will be crucial for developing more effective therapies for AD.

13.4.2.1 Neprilysin (NEP)

Neutral Endopeptidase (NEP), also known as enkephalinase, neutrophil cluster-differentiation antigen 10, or common acute lymphoblastic leukemia antigen, belongs to the M13 zinc metallopeptidase subfamily of type II integral membrane proteins. NEP is located on the cell surface and is composed of three domains: a short N-terminal cytoplasmic tail, a membrane-spanning domain, and a large C-terminal extracellular catalytic domain containing the characteristic HExxH zinc-binding motif. As an ectoenzyme, NEP catalyzes peptide hydrolysis outside the cell. It is expressed in various normal tissues, including the brush border of intestinal and kidney epithelial cells, neutrophils, thymocytes, lung, prostate, testis, and brain.

NEP is a zinc-dependent endopeptidase that cleaves peptides with hydrophobic residues. It hydrolyzes and inactivates several peptides, such as enkephalins, atrial natriuretic peptides, endothelin, and substance P. NEP can also degrade Aβ peptides in vivo. For instance, injecting NEP-expressing lentivirus into the hippocampus of transgenic mice expressing human amyloid precursor protein (hAPP) resulted in a reduction of Aβ plaque load. Conversely, NEP-deficient mice exhibited increased levels of endogenous Aβ1–40 and Aβ1–42. Reduced NEP expression and activity have been observed in the brains of AD patients. Understanding and manipulating NEP's regulatory mechanisms could be valuable for preventing AD progression.

13.4.2.2 Insulin Degrading Enzyme (IDE)

Insulin Degrading Enzyme (IDE), also known as insulysin or insulinase, is a zinc metalloprotease (EC 3.4.24.56) that is highly expressed in the liver, testis, muscle, and brain. IDE degrades various peptides, including insulin, glucagon, amylin, and insulin-like growth factors I and II. IDE has been shown to degrade Aβ both in vivo and in vitro. Mice with homozygous deletions of the IDE gene exhibit increased cerebral Aβ accumulation, hyperinsulinemia, and glucose intolerance. Conversely, upregulation of IDE in transgenic mice has resulted in decreased Aβ peptide levels, reduced amyloid plaque formation, and improved associated pathology [30]. Purified IDE from rat liver and brain effectively degrades synthetic Aβ peptides in vitro. Additionally, both extracellular and cell surface IDE have been shown to degrade Aβ peptides in cultured neurons [52].

Given its role, upregulating IDE and enhancing its activity could represent a promising therapeutic approach for AD. Insulin and its receptors are abundantly present in the brain, and both have been implicated in AD pathology. Moreover, hyperinsulinemia, insulin resistance, type 2 diabetes mellitus, and variations in the IDE gene are associated with increased AD risk. While current evidence supports the potential of IDE upregulation as a therapeutic strategy for AD, further research is needed to evaluate the physiological consequences of increased proteolytic activity.

13.4.2.3 Angiotensin-Converting Enzyme (ACE)

Angiotensin-converting enzyme (ACE), also known as dipeptidyl carboxypeptidase, is a membrane-bound zinc metalloprotease (EC 3.4.15.1) widely expressed in blood vessels throughout the body. ACE plays a crucial role in regulating blood pressure by converting angiotensin I into the potent vasoconstrictor angiotensin II. Mammalian ACE exists in two distinct isoforms: somatic (sACE) and testis (tACE). Somatic ACE comprises two identical catalytic domains (C- and N-domains) that are proteolytically active, while testis ACE consists of a single domain corresponding to the C-terminal domain of sACE. The highest concentrations of ACE are found in the circumventricular organs of the brain, such as the subfornical organ, area postrema, and median eminence [6].

Genetic studies suggest a potential link between ACE and Alzheimer's disease (AD). ACE has been shown to significantly inhibit the aggregation, deposition, and cytotoxicity of Aβ peptides in vitro. Recent research indicates that ACE can convert Aβ1–42 to Aβ1–40 in both mouse and human brain homogenates. Purified ACE degrades Aβ peptides and decreases the Aβ1–42/Aβ1–40 ratio [57]. Inhibition of ACE has been associated with increased Aβ levels in cells transfected with APP and ACE. However, some studies have found that ACE gene inactivation or ACE inhibition in mice does not affect Aβ peptide levels, making the role of ACE inhibitors in AD still a matter of debate. Consequently, more research is needed to clarify their effects. Upregulating ACE could be considered a therapeutic strategy for AD patients who do not have hypertension.

13.4.2.4 Cathepsin B

Cathepsin B is a member of the cysteine protease family and is involved in various physiological and pathological processes. It degrades proteins entering the endolysosomal system and can truncate Aβ1–42 at the C-terminal, significantly reducing Aβ peptide deposition. Enhanced activity of endogenous cathepsin B may lower Aβ1–42 levels in AD. Research by Sun and colleagues in 2008 showed that increasing cathepsin B levels reduced amyloid-beta plaque load in transgenic mice overexpressing human APP [47]. Additionally, increasing cathepsin B activity by deleting its endogenous inhibitor, cystatin C (CysC), significantly reduced levels of total Aβ1–40 and Aβ1–42 in mice expressing wild-type hAPP. Molecular modeling studies also demonstrated that Aβ peptides fit well into the active site of cathepsin B, leading to cleavage near Lys16 of the Aβ peptide.

13.4.2.5 Human Endothelin-Converting Enzymes (ECEs)

Human Endothelin-Converting Enzymes, hECE-1 and hECE-2 (EC 3.4.24.71), are type II integral membrane zinc metalloendopeptidases belonging to the M13 subfamily. They are expressed in endothelial cells of various tissues, including the aorta, lungs, ovary, and testis. Unlike neprilysin, which primarily degrades smaller peptides, hECE-1 can hydrolyze larger biologically active peptides such as big endothelin-1, neurotensin, substance P, bradykinin, and the oxidized insulin B chain. Specifically, hECE-1 cleaves big endothelin-1 at the Trp21-Val/Ile22 site to form endothelin-1 (ET-1), a potent vasoconstrictor. Immunolabeling studies have shown that ECE-1 is present in neurons, particularly in pyramidal neurons of the hippocampus and neocortex in both rats and cows. In the human brain, hECE-1 expression has been detected in various regions, including the hippocampus and pyramidal neurons of layers III and V of the neocortex. Several studies have demonstrated that ECE-1 and ECE-2 mRNAs are localized in neurons of the rat brain.

Experimental studies have shown that treating ECE-expressing cell lines with the metalloprotease inhibitor phosphoramidon leads to a two- to three fold increase in extracellular Aβ concentrations [18]. Additionally, overexpression of ECE-1 results in a significant reduction of Aβ1–40 by 75–90% and Aβ1–42 by 45–60%, indicating that hECE-1 can hydrolyze Aβ peptides both in vitro and in vivo. Increased Aβ levels have been observed in the brains of ECE-deficient mice compared to age-matched controls. Eckman and colleagues found that oral administration and intracerebroventricular (ICV) injection of selective ECE inhibitors, along with a dual NEP–ECE inhibitor, caused a rapid increase in plasma Aβ concentrations in mice. This was followed by a small but significant rise in brain Aβ levels. The hECE-2 mRNA level was found to be increased in the temporal cortex of AD brains and upregulated in SH-SY5Y cells after 24 h of treatment with both monomeric and oligomeric Aβ1–42. hECE-1 mRNA levels were also elevated following treatment with HNE and Aβ peptide. Double fluorescence staining and confocal analysis of oxidatively stressed SH-SY5Y cells showed hECE-1 localized in both the cytosol and membrane [5]. Overexpression of protein kinase C, which induces ECE-1 activity, significantly reduced plaque and Aβ deposit areas. ECE-1 participates in regulating Aβ levels intracellularly via the endosomal-autophagic-lysosomal pathways. Intracranial administration of a recombinant adeno-associated viral vector (rAAV) carrying the ECE1 gene into the anterior cortex and hippocampus of 6-month-old transgenic mice (APP+PSEN1) led to decreased Aβ levels in these regions, as detected by immunohistochemical and Congo red staining analyses.

A genetic association study identified a functional genetic variant (hECE-1b C-338A) in a regulatory region of the ECE1 gene. Increased ECE-1 mRNA expression in the human neocortex was linked to a reduced risk of AD, especially in individuals with homozygous A allele carriers. The polymorphism in the CpG-CA repeat within the hECE-1c promoter also contributes to AD pathogenesis. These findings suggest that hECE-1 plays a crucial role in Aβ degradation in AD, although there is limited information on the genetic variation of hECE-2 in AD development. Only two studies have been conducted on hECE-2, yielding conflicting results: one study found downregulation of hECE-2 gene and protein levels in AD patients, while another study reported increased hECE-2 mRNA levels in the AD cortex. Further research is needed to clarify the role of ECE-2 in AD. Overall, in vitro, in vivo, and animal model studies suggest that enhancing hECE-1 activity could help prevent or delay AD progression.

13.4.2.6 Aβ-Degrading Microbial Proteases

Proteases play a critical role in cellular metabolic processes and have garnered significant attention. While animal and plant sources of proteases face limitations due to climatic and ethical concerns, microbes offer an advantageous alternative due to their rapid growth, minimal space requirements, and ease of genetic manipulation to produce proteases with desirable properties. The increasing demand for proteases has highlighted the importance of microbial sources.

Microbial proteases are crucial hydrolytic enzymes and have been extensively studied for their potential therapeutic applications. Several microbial proteases have been shown to degrade Aβ peptides. For example, Hsu et al. [26] reported the amyloid-degrading properties of nattokinase, a fibrinolytic subtilisin-like serine protease from *Bacillus subtili* [26]. Similarly, aminopeptidase from *Streptomyces griseus* KK565 (SKAP) was found to cleave N-terminal amino acids of Aβ1–42 monomers, inhibit amyloid fibril formation, and protect against Aβ1–42-induced neurotoxicity. Structural studies by revealed that cathepsin B from Xanthomonas campestrisplays a role in Aβ degradation. These findings suggest that microbial proteases could be valuable in developing new strategies for controlling Aβ levels [17].

13.4.2.7 Structural Studies of Aβ-Degrading Enzymes

Several metalloproteases in humans have been identified, characterized by a short N-terminal cytoplasmic region, a membrane-spanning section, and a large C-terminal extracellular catalytic domain. The extracellular catalytic domain contains the conserved HExxH zinc-binding motif crucial for zinc ligation and catalysis. Structural studies have provided significant insights into substrate recognition and catalysis of these enzymes in Alzheimer's disease.

To date, experimental structures are available for the extracellular domains of NEP, ACE, ECE, and IDE complexed with substrates and inhibitors [11]. The crystal structure of the soluble extracellular domain of human NEP (residues 52–749) in complex with phosphoramidon shows the inhibitor nestled in the enzyme's cavity, binding near the conserved consensus sequences 583HExxH687 and 646ExxxD650, and the 542NAFY545 motif. The S1′ subsite of NEP forms a large hydrophobic pocket that interacts with aromatic or large hydrophobic moieties, while the S2 subsite accommodates bulky side chains.

The crystal structures of human IDE with four substrates reveal that its N- and C-terminal domains form an enclosed cage around the substrate, which undergoes conformational changes to form β-sheets with IDE for degradation. The structure of the tACE-lisinopril complex includes the HExxH zinc-binding motif in the enzyme's active site, with chloride ions playing a role in substrate-dependent activation. The three-dimensional structure of monomeric hECE-1(C428S) ectodomain complexed with phosphoramidon shows two large α-helical domains connected by an interdomain linker. The C-terminal domain contains substrate recognition subsites (S1, S1′, and S2′) and a zinc-binding site, with conserved motifs 607HExxH611, 667ExxD671, 542NAYY545, and 565VNA567 present in its catalytic domain [45].

Recent molecular modeling studies of hECE-1, ACE, cathepsin B, and aminopeptidase with Aβ peptides and zinc ions have explored the binding modes and cleavage mechanisms of these enzymes. These studies provide fundamental insights into substrate recognition and Aβ peptide degradation at the atomic level. Further structural studies are needed to understand the active site specificities and substrate recognition of these enzymes, which could aid in developing pharmacological strategies to manipulate their activity in disease states [28].

13.4.3 Other Therapeutic Approaches

Aβ peptide deposits and neurofibrillary tangles in Alzheimer's disease trigger inflammatory responses. These include activation of the classical and alternative complement pathways, cytokine and chemokine pathways, acute phase reactants, and other inflammatory mediators, exacerbating AD pathogenesis. Targeting the inflammatory cascade to reduce harmful effects and promote abnormal protein clearance presents a promising therapeutic option.

The endosomal–lysosomal pathway plays a crucial role in recycling cellular components essential for cell health. Dysfunction in this pathway is an early pathological event in AD. Nixon and colleagues reported that lysosomal abnormalities lead to the accumulation and development of fibrillar Aβ into senile plaques. Expression of a dominant-negative dynamin mutant that inhibits endocytosis in transfected cell lines results in decreased Aβ1–42/Aβ1–40 ratios [40]. Lysosomes are essential for autophagy, and loss of lysosomal membrane impermeability leads to Aβ accumulation. Reduced autophagy efficiency results in Aβ and α-synuclein oligomer accumulation in the mitochondrial membrane, cytochrome c release, activation of the caspase cascade, and ultimately, cell death and neurodegeneration [39]. Detailed studies on endosomal-lysosomal dysfunction in AD are needed to explore these mechanisms further.

13.5 Emerging and Experimental Therapies

An intuitive intervention strategy has been proposed to use immunization against Aβ peptides as a means to effectively reduce excessive Aβ accumulation in the brains of Alzheimer's disease (AD) patients, thereby potentially mitigating Aβ-triggered pathological changes. Early studies in transgenic mice that overexpress βAPP demonstrated significant reductions in AD-like neuropathology. However, initial human trials of immunization against Aβ42 were halted due to serious inflammatory reactions, including meningoencephalitis, which posed life-threatening risks [49].

Despite extensive preclinical validation in mouse models, the development of effective immunization strategies that are free of significant side effects in humans remains a challenge. AAB-001 (Bapineuzumab), a humanized monoclonal anti-Aβ antibody administered intravenously for mild to moderate AD, received fast-track designation from the US FDA [37]. Unfortunately, one unexpected side effect of AAB-001 was vasogenic cerebral edema, leading to the removal of patients on the highest doses from the trials. Overall, the results have been disappointing.

Currently, a second-generation anti-Aβ monoclonal antibody, solanezumab (LY2062430), which targets the central region of β-amyloid, is undergoing phase III clinical trials [41]. Further research into immunization against Aβ peptides is needed to refine our understanding of the complexities involved in γ-secretase-substrate recognition, γ-secretase-mediated βAPP cleavage, and the kinetics of Aβ peptide metabolism, translocation, proteolysis, clearance, and off-target effects.

Interestingly, antibodies targeting specific regions of Aβ peptides (e.g., residues 4–10) have shown promise in inhibiting cytotoxicity and fibrillogenesis. Recent advances include DNA immunization using HBsAg-based particles expressing a B cell epitope of Aβ, which has been shown to slow disease progression and extend survival in AD mouse models. Additionally, tau-targeted immunization has demonstrated the ability to reduce the progression of neurofibrillary tangles in aged P301L tau transgenic mice [8].

More recently, APP-C99 substrate-targeting monoclonal antibodies have emerged as potential novel immunotherapeutics and Notch-sparing agents, showing promise in lowering Aβ levels in mice. However, it remains uncertain whether these effects can be replicated in humans [34] (Table 13.1).

13.6 Future Prospects

Despite achieving significant reductions in Aβ levels with various anti-Aβ drugs during clinical trials, there has been limited progress in slowing cognitive decline. It is important to recognize that anti-amyloid therapy should not function as a standalone treatment; it must be combined with other agents that target different facets of the disease. Furthermore, extensive clinical trials focused on Aβ have uncovered several persistent challenges in developing effective disease-modifying treatments to slow Alzheimer's disease progression: (1) the absence of reliable early biomarkers, (2) difficulties in objectively and quantitatively measuring cognitive impairment, (3) the need to carefully consider risk factors linked to AD, such as APOE status and family history, (4) the timing of treatment initiation, and (5) the identification of suitable targets capable of effectively penetrating the brain.

A range of nanocarriers and nanoparticle prodrugs has been developed to enhance drug bioavailability and effectiveness. Techniques such as conventional adsorption and covalent conjugation are utilized to link anti-Aβ antibodies to these nanocarriers. Some methods can even inhibit the oligomerization and accumulation of Aβ, providing a significant benefit. Research indicates that clinically normal individuals with biomarker evidence of elevated brain amyloid (Aβ+) are prime candidates for early intervention, as they have an increased risk of progressing to mild cognitive impairment and dementia. Additionally, these high-risk individuals often carry the ApoE allele and tend to have a more pronounced family history of Alzheimer's disease.

Aβ deposition occurs well in advance of cognitive decline, indicating that initiating anti-Aβ treatment after the onset of symptoms is often too late. By that stage, significant neurodegeneration may have already become irreversible, and simply reducing amyloid levels may not restore neural networks. Thus, a paradigm shift from "treatment" to "prevention" in Alzheimer's disease management is crucial,

Table 13.1 Various amyloid β agents in clinical trials

Invested agent	Status of clinical trial	AD stage	Results	Clinical trial identifier (NCT number)
APP synthesis				
Phospiphen tartrate (Bunatanetap)	Phase I/II	Mild-moderate	Terminated (business decision)	04524351
β-secretase1 (BACE 1) inhibitor				
Verubecestat (MK-8931)	Phase II/III	Mild-moderate	No cognitive/functional benefits; more adverse effects	01739348
Umibecestat (CNP520)	Phase II/III	Mild or risk of AD with one allele of APOE4 and increased amyloid in the brain	Severe adverse effects; no cognitive effects	03131453
Atabecestat (JNJ-54861911)	Phase II	Mild-moderate	Terminated due to adverse effects; hepatotoxicity	02406027
Lanabecestat (LY3314814)	Phase II/II	Mild	Discontinued due to adverse effects; futility	02783573
Elenbecestat	Phase II	Early-mild	Discontinued due to unfavorable risk: Benefit outcomes	02322021
γ-secretase inhibitor				
Semagacestat	Phase III	Mild-moderate	Discontinued due to adverse cognitive effects; skin cancer	00762411
Avagacestat	Phase II	Prodromal	Discontinued due to adverse cognitive effects; skin cancer	00810147
γ-secretase modulator				
Tarenflurbil	Phase III	Mild	No functional benefits; poor brain penetration	00105547
Itanapraced (CHF 5074)	Phase II completed	Mild	Adverse effects but no reports after 2013	01602393
Pinitol	Phase II	Mild	Limitations not specified	00470418
α-secretase enhancer				
Etazolate	Phase II/III	Mild-moderate	No reports after 2009	00880412
Acitretin	Phase II	Mild-moderate	Second generation retinoid	01078168

(continued)

Table 13.1 (continued)

Invested agent	Status of clinical trial	AD stage	Results	Clinical trial identifier (NCT number)
Epigallocatechin-3-gallate		Apo-E carriers with subjective cognitive decline	No further reports since 2020	03978052
Bryostatin 1	Phase II	Mild	Natural molecule	02221947
APH-1105	Phase II	MCI due to AD		03806478
ID1201	Phase II	Mild	No result posted	03363269
Active immunization				
AN1792	Phase II	Mild-moderate	Cause meningoencephalitis	00021723
Affitope® AD02	Phase II	Mild-moderate	Safe and well tolerated; immunogenic; no clinical effect	00711321
CAD106	Phase II/III	Mild-moderate	Safe and well tolerated; immunogenic; no clinical effect	01097096
ACC-001	Phase II	Mild-moderate	Safe and well tolerated; immunogenic; no clinical effect	01284387
Lu AF20513	Phase I	Mild-moderate	Low titters observed	03668405
ACI-24	Phase II	Mild-moderate	Safe and well tolerated; poorly immunogenic	02388152
ABvac40	Phase II	Mild	Favorable safety profile	03113812
UB-311	Phase III	MCI/AD	Safe and well tolerated; immunogenic	00965588
Passive immunotherapy				
Bapineuzumab	Phase III	Mild-moderate	No cognitive effects; ARIA incidence is high	00574132
Solanezumab	Phase III	Mild-moderate	No cognitive/functional effects	00905373
Crenezumab	Phase III	Prodromal to mild	No efficacy	03491150
Gantenerumab	Phase III	Individuals at risk	No clinical benefits	03443973
Ponezumab	Phase II	Mild-moderate	Safe and well tolerated	01125631
Gammagard®	Phase III	Mild-moderate	Lack of efficacy	00818662
Octagam® 10%	Phase II	Mild-moderate	Lack of efficacy	00818662
SAR-228810	Phase I	Mild-moderate	Safe and well tolerated	01485302
MEDI-1814	Phase I	Mild-moderate	No further reports since 2019	02036645
Donanemab	Phase II	Mild	Less cognitive decline than placebo	01837641

(continued)

Table 13.1 (continued)

Invested agent	Status of clinical trial	AD stage	Results	Clinical trial identifier (NCT number)
Aducanumab (Aduhelm®)	Phase IV	Prodromal-mild	Accelerated approval; benefits outweigh risk	02782975
Receptor for advanced glycation end products (RAGE) antagonist				
Azeliragon		Mild AD with impaired glucose tolerance	Orphan drug designation	01766336
Inhibits Aβ aggregation				
Sycllo-inositol	Phase III	Mild-moderate	More adverse effects	01735630
GV-971	Phase III	Mild-moderate	Conditionally approved by China National Medical Product Administration	02986529
ALZT-OP1a	Phase III	Early	No results posted	02547818

emphasizing the need for therapeutic interventions during the asymptomatic preclinical phases. Efficient clearance of neuronal debris and Aβ peptides is vital for neuronal survival, as research has demonstrated that Aβ clearance is impaired in Alzheimer's and correlates with disease progression. Clinical trials are exploring drugs that modify Aβ misfolding, aggregation, metabolism, and clearance, representing promising avenues of investigation that could yield effective treatments for both preventing and addressing the disease.

References

1. Abramowski D, Wiederhold K-H, Furrer U, Jaton A-L, Neuenschwander A, Runser M-J, Danner S, Reichwald J, Ammaturo D, Staab D. Dynamics of Aβ turnover and deposition in different β-amyloid precursor protein transgenic mouse models following γ-secretase inhibition. J Pharmacol Exp Ther. 2008;327(2):411–24.
2. Amin F, Shamsi A, Asghar MN, Khaki PSS, Khan MS, Tabrez S, Zaidi SK, Khan W, Bano B. Alzheimer's: a progressive brain disease: causes, symptoms, and prevention. In: Biological, diagnostic and therapeutic advances in alzheimer's disease: non-pharmacological therapies for alzheimer's disease. Singapore: Springer; 2019. p. 31–51.
3. Atta AM, Elgawish MS, Moustafa SM. Recent insight into BACE1 as a potential target for treatment of Alzheimer's disease. Rec Pharm Biomedical Sci. 2021;5(Chemistry):100–11.
4. Bandyopadhyay S, Goldstein L, Lahiri D, Rogers J. Role of the APP non-amyloidogenic signaling pathway and targeting α-secretase as an alternative drug target for treatment of Alzheimer's disease. Curr Med Chem. 2007;14(27):2848–64.
5. Barage SH, Sonawane KD. Amyloid cascade hypothesis: pathogenesis and therapeutic strategies in Alzheimer's disease. Neuropeptides. 2015;52:1–18.
6. Berecek K, King S, Wu J. Angiotensin-converting enzyme and converting enzyme inhibitors. Cellular and molecular biology of the renin-angiotensin system. CRC Press; 2018. p. 183–220.

7. Bernstein H-G, Stricker R, Lendeckel U, Bertram I, Dobrowolny H, Steiner J, Bogerts B, Reiser G. Reduced neuronal co-localisation of nardilysin and the putative α-secretases ADAM10 and ADAM17 in Alzheimer's disease and down syndrome brains. Age. 2009;31:11–25.
8. Bi M, Ittner A, Ke YD, Götz J, Ittner LM. Tau-targeted immunization impedes progression of neurofibrillary histopathology in aged P301L tau transgenic mice. PLoS One. 2011;6(12):e26860.
9. Cai Y, Liu J, Wang B, Sun M, Yang H. Microglia in the neuroinflammatory pathogenesis of Alzheimer's disease and related therapeutic targets. Front Immunol. 2022;13:856376.
10. Cameron DJ, Galvin C, Alkam T, Sidhu H, Ellison J, Luna S, Ethell DW. Alzheimer's-related peptide amyloid-β plays a conserved role in angiogenesis. PLoS One. 2012;7(7):e39598.
11. Carson JA, Turner AJ. β-Amyloid catabolism: roles for neprilysin (NEP) and other metallopeptidases? J Neurochem. 2002;81(1):1–8.
12. Cavallucci V, D'Amelio M, Cecconi F. Aβ toxicity in Alzheimer's disease. Mol Neurobiol. 2012;45:366–78.
13. Chow VW, Mattson MP, Wong PC, Gleichmann M. An overview of APP processing enzymes and products. NeuroMolecular Med. 2010;12:1–12.
14. DaSilva KA, Shaw JE, McLaurin J. Amyloid-β fibrillogenesis: structural insight and therapeutic intervention. Exp Neurol. 2010;223(2):311–21.
15. De Strooper B, Karran E. New precision medicine avenues to the prevention of Alzheimer's disease from insights into the structure and function of γ-secretases. EMBO J. 2024;43(6):887–903.
16. De Strooper B, Iwatsubo T, Wolfe MS. Presenilins and γ-secretase: structure, function, and role in Alzheimer disease. Cold Spring Harb Perspect Med. 2012;2(1):a006304.
17. Dhanavade MJ, Sonawane KD. Amyloid beta peptide-degrading microbial enzymes and its implication in drug design. 3 Biotech. 2020;10(6):247.
18. Eckman EA, Reed DK, Eckman CB. Degradation of the Alzheimer's amyloid β peptide by endothelin-converting enzyme. J Biol Chem. 2001;276(27):24540–8.
19. Ghaffari D, Griffin JK, St George-Hyslop P. ABI3 deletion in TgCRND8 mice is associated with reduced amyloid plaque pathology and altered glial response. bioRxiv: 2023.2010. 2005.560956.
20. Ghosh AK, Brindisi M, Tang J. Developing β-secretase inhibitors for treatment of Alzheimer's disease. J Neurochem. 2012;120:71–83.
21. Goyal D, Shuaib S, Mann S, Goyal B. Rationally designed peptides and peptidomimetics as inhibitors of amyloid-β (Aβ) aggregation: potential therapeutics of Alzheimer's disease. ACS Comb Sci. 2017;19(2):55–80.
22. Green RC, Schneider LS, Amato DA, Beelen AP, Wilcock G, Swabb EA, Zavitz KH, T. P. S. Group. Effect of tarenflurbil on cognitive decline and activities of daily living in patients with mild Alzheimer disease: a randomized controlled trial. JAMA. 2009;302(23):2557–64.
23. Grüninger-Leitch F, Schlatter D, Kung E, Nelböck P, Döbeli H. Substrate and inhibitor profile of BACE (β-secretase) and comparison with other mammalian aspartic proteases. J Biol Chem. 2002;277(7):4687–93.
24. Gurses MS, Ural MN, Gulec MA, Akyol O, Akyol S. Pathophysiological function of ADAMTS enzymes on molecular mechanism of Alzheimer's disease. Aging Dis. 2016;7(4):479.
25. Hölscher C. Development of beta-amyloid-induced neurodegeneration in Alzheimer's disease and novel neuroprotective strategies. Rev Neurosci. 2005;16(3):181–212.
26. Hsu R-L, Lee K-T, Wang J-H, Lee LY-L, Chen RP-Y. Amyloid-degrading ability of nattokinase from Bacillus subtilis natto. J Agric Food Chem. 2009;57(2):503–8.
27. Hussain MS, Mishra AK. Advancements in understanding and addressing Alzheimer's disease: a comprehensive review. Int J Res Med Sci. 2024;12(5):1789.
28. Jalkute CB, Sonawane KD. Evaluation of a possible role of Stigmatella aurantiaca ACE in Aβ peptide degradation: a molecular modeling approach. J Mol Microbiol Biotechnol. 2015;25(1):26–36.
29. Kumar D, Sharma A, Sharma L. A comprehensive review of Alzheimer's association with related proteins: pathological role and therapeutic significance. Curr Neuropharmacol. 2020;18(8):674–95.

30. Leissring MA, Farris W, Chang AY, Walsh DM, Wu X, Sun X, Frosch MP, Selkoe DJ. Enhanced proteolysis of β-amyloid in APP transgenic mice prevents plaque formation, secondary pathology, and premature death. Neuron. 2003;40(6):1087–93.
31. Lichtenthaler SF. Alpha-secretase in Alzheimer's disease: molecular identity, regulation and therapeutic potential. J Neurochem. 2011;116(1):10–21.
32. Ling Y, Morgan K, Kalsheker N. Amyloid precursor protein (APP) and the biology of proteolytic processing: relevance to Alzheimer's disease. Int J Biochem Cell Biol. 2003;35(11):1505–35.
33. Lozupone M, Dibello V, Sardone R, Castellana F, Zupo R, Lampignano L, Bortone I, Stallone R, Altamura M, Bellomo A. Lessons learned from the failure of solanezumab as a prospective treatment strategy for Alzheimer's disease. Expert Opin Drug Discov. 2024;19(6):639–47.
34. Lukiw WJ. Amyloid beta (Aβ) peptide modulators and other current treatment strategies for Alzheimer's disease (AD). Expert Opin Emerg Drugs. 2012;17(1):43–60.
35. Menting KW, Claassen JA. Β-Secretase inhibitor; a promising novel therapeutic drug in Alzheimer's disease. Front Aging Neurosci. 2014;6:165.
36. Moraca F, Vespoli I, Mastroianni D, Piscopo V, Gaglione R, Arciello A, De Nisco M, Pacifico S, Catalanotti B, Pedatella S. Synthesis, biological evaluation and metadynamics simulations of novel N-methyl β-sheet breaker peptides as inhibitors of Alzheimer's β-amyloid fibrillogenesis. RSC Med Chem. 2024;15(7):2286–99.
37. Neaţu M, Covaliu A, Ioniţă I, Jugurt A, Davidescu EI, Popescu BO. Monoclonal antibody therapy in Alzheimer's disease. Pharmaceutics. 2023;16(1):60.
38. Nunan J, Small DH. Regulation of APP cleavage by α-, β-and γ-secretases. FEBS Lett. 2000;483(1):6–10.
39. Olagunju AS, Ahammad F, Alagbe AA, Otenaike TA, Teibo JO, Mohammad F, Alsaiari AA, Omotoso O, Talukder MEK. Mitochondrial dysfunction: a notable contributor to the progression of Alzheimer's and Parkinson's disease. Heliyon. 2023;9(3):e14387.
40. Oliver D, Reddy PH. Dynamics of dynamin-related protein 1 in Alzheimer's disease and other neurodegenerative diseases. Cells. 2019;8(9):961.
41. Plascencia-Villa G, Perry G. Lessons from antiamyloid-β immunotherapies in Alzheimer's disease. Handb Clin Neurol. 2023;193:267–91.
42. Postina R. Activation of α-secretase cleavage. J Neurochem. 2012;120:46–54.
43. Sciacca MF, La Rosa C, Milardi D. Amyloid-mediated mechanisms of membrane disruption. Biophysica. 2021;1(2):137–56.
44. Serneels L, Narlawar R, Perez-Benito L, Municoy M, Guallar V, T'Syen D, Dewilde M, Bischoff F, Fraiponts E, Tresadern G. Selective inhibitors of the PSEN1–gamma-secretase complex. J Biol Chem. 2023;299(6):104794.
45. Shen Y, Joachimiak A, Rich Rosner M, Tang W-J. Structures of human insulin-degrading enzyme reveal a new substrate recognition mechanism. Nature. 2006;443(7113):870–4.
46. Shih I-M, Wang T-L. Notch signaling, γ-secretase inhibitors, and cancer therapy. Cancer Res. 2007;67(5):1879–82.
47. Sun Y-X, Ji X, Mao X, Xie L, Jia J, Galvan V, Greenberg DA, Jin K. Differential activation of mTOR complex 1 signaling in human brain with mild to severe Alzheimer's disease. J Alzheimers Dis. 2014;38(2):437–44.
48. Takasugi N, Tomita T, Hayashi I, Tsuruoka M, Niimura M, Takahashi Y, Thinakaran G, Iwatsubo T. The role of presenilin cofactors in the γ-secretase complex. Nature. 2003;422(6930):438–41.
49. Taliyan R, Chandran SK, Kakoty V. Therapeutic approaches to Alzheimer's type of dementia: a focus on FGF21 mediated neuroprotection. Curr Pharm Des. 2019;25(23):2555–68.
50. Tanzi RE, Bertram L. Twenty years of the Alzheimer's disease amyloid hypothesis: a genetic perspective. Cell. 2005;120(4):545–55.
51. Vassar R. β-secretase (BACE) as a drug target for Alzheimer's disease. Adv Drug Deliv Rev. 2002;54(12):1589–602.
52. Vekrellis K, Ye Z, Qiu WQ, Walsh D, Hartley D, Chesneau V, Rosner MR, Selkoe DJ. Neurons regulate extracellular levels of amyloid β-protein via proteolysis by insulin-degrading enzyme. J Neurosci. 2000;20(5):1657–65.

53. Weggen S, Eriksen JL, Sagi SA, Pietrzik CU, Ozols V, Fauq A, Golde TE, Koo EH. Evidence that nonsteroidal anti-inflammatory drugs decrease amyloid β42 production by direct modulation of γ-secretase activity. J Biol Chem. 2003;278(34):31831–7.
54. Wolfe MS. Γ-Secretase inhibitors and modulators for Alzheimer's disease. J Neurochem. 2012;120:89–98.
55. Yoon S-S, Jo SA. Mechanisms of amyloid-β peptide clearance: potential therapeutic targets for Alzheimer's disease. Biomol Ther. 2012;20(3):245.
56. Yuan X-Z, Sun S, Tan C-C, Yu J-T, Tan L. The role of ADAM10 in Alzheimer's disease. J Alzheimers Dis. 2017;58(2):303–22.
57. Zou K, Yamaguchi H, Akatsu H, Sakamoto T, Ko M, Mizoguchi K, Gong J-S, Yu W, Yamamoto T, Kosaka K. Angiotensin-converting enzyme converts amyloid β-protein 1–42 (Aβ1–42) to Aβ1–40, and its inhibition enhances brain Aβ deposition. J Neurosci. 2007;27(32):8628–35.

A Comprehensive Review of the Ubiquitin-Proteasome System in Synapse Remodeling and Neurodegenerative Diseases

14

Meenakshi Kumari

14.1 Introduction

Alzheimer's disease (AD) disease is identified by memory loss and thinking disorders such as memory loss, not attention, reasoning, and problem-solving. This disorder affects a person's ability to think clearly, remember information, and perform daily activities. Cognitive disorders are variable in their causes, severity, and impact on daily functioning.

An excitatory neurotransmitter is a chemical messenger in the brain that increases the possibility that a neuron (nerve cell) will send a signal to other neurons. When an excitatory neurotransmitter binds to receptors on a neuron, it makes the neuron more likely to become active and transmit an electrical signal. Glutamate is an amino acid present in the neural center and works as a signal molecule. It helps neurons communicate and is essential for learning, memory, and normal brain function. It is primary stimulated chemical messenger in the central nervous system. But in excess, Nevertheless, excessive levels of glutamate can be detrimental. When there is an excess of glutamate or when it over-stimulates neurons, it can lead to a process called excitotoxicity, which can damage or kill neurons [50]. Glutamate is the predominant stimulating neurotransmitter in the brain, essential for functions such as memory, learning, and overall neural communication. It is too much excitatory activity and lead to excitotoxicity problems. Deu to this, damage or kill neurons and cause Alzheimer's disease (AD). This disease damage and death of brain cells, contributing to neurodegeneration. This disease occurs in 50–70% in old age and presenile period [40]. Parkinson's disease is a nerve degenerative condition affecting the central nervous system, marked by tremors, bradykinesia, stiffness, and postural instability, which usually progress gradually [46]. These non-motor symptoms are common in all neurodegeneration, contributing significantly to the

M. Kumari (✉)
Department of Botany, Career Point University, Kota, Rajasthan, India

N. Puranik (ed.), *Proteostasis*, https://doi.org/10.1007/978-981-96-6202-9_14

overall disease burden [46]. Huntington's disease (HD) is an inherited somatic disorder that affects nerve cells, causing both physical and emotional changes in individuals. This disease is also known as Huntington's chorea. It causes basal ganglia disease such as hyperkinetic movement disorder [19]. Amyotrophic Lateral Sclerosis (ALS), also called Lou Gehrig's disease, is a progressive neurodegenerative condition that impacts motor nerve cells in the brain and spinal cord. This results in muscle weakness, paralysis, and often death, usually occurring within 3–5 years of diagnosis [9]. ALS is classified into two forms: sporadic (90–95% of cases) and familial (5–10%) [97]. The characteristic features of ALS are the degeneration of cortical and spinal motor neurons. This neurodegeneration results in progressive muscle atrophy, spasticity, and the loss of voluntary motor control [92]. Frontotemporal Dementia (FTD) is an assembly of neurodegenerative disorders and identified by gradual deterioration of the frontal and temporal regions of the brain. This leads to symptoms like behavioral changes, language difficulties, and impaired executive functions. FTD contain several forms such as behavioral variant FTD (bvFTD) and primary progressive aphasia (PPA), which further highlights the diversity of this condition. Neuronal degeneration conditions are identified by the deterioration of neuronal function. They cause inflammation in the brain and eventually the death of neurons. Alzheimer's disease (AD), Parkinson's disease (PD), and amyotrophic lateral sclerosis (ALS) are neurodegenerative diseases that degeneration the structure and function of the central or peripheral nervous system [42]. Alzheimer's disease (AD), Parkinson's disease (PD), Huntington's disease (HD), amyotrophic lateral sclerosis (ALS), and frontotemporal dementia (FTD) are classified as cognitive decline disorders (Fig. 14.1).

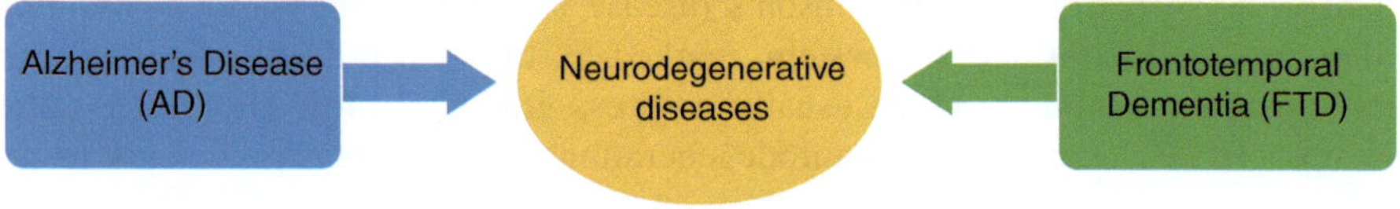

Fig. 14.1 Neurological diseases are a category of disorders marked by the gradual loss of structure or function of neurons, causing their eventual death. These conditions typically worsen over time and affect various functions of the brain and nervous system, contributing to cognitive decline, motor impairments, and other serious complications

14.2 Alzheimer Disease (AD)

Alzheimer's disease (AD) is identified by two hallmark features: Amyloid β plaques and Neurofibrillary aggregates. Amyloid β plaques are deposits of β-amyloid peptides that accumulate outside neurons, leading to disruption of normal brain function, inflammation, and oxidative stress, ultimately causing neuronal damage and death. Neurofibrillary aggregates, on the other hand, are warped strands of tau protein that aggregated inside neurons, dismantling normal cellular processes, causing neuronal damage and death, and contributing to cognitive decline. In addition to these hallmark features, Alzheimer's disease (AD) is also marked by other molecular alterations, including neuropil threading, where abnormal tau protein accumulates in the region between neurons, disrupting neural connections and impairing communication between neurons. Dystrophic neurites, abnormal, swollen, or fragmented nerve endings, indicate neuronal damage and degeneration, further disrupting neural connections. Neuroinflammation, the stimulating of antibody-producing cell and inflammation in neural center, also plays a significant role, contributing to neuronal damage and death, and disease progression. Cerebral amyloid angiopathy (CAA), the accumulation of beta-amyloid in blood vessel walls, leads to vascular damage and dysfunction, increasing the risk of microbleeds and hemorrhages. These molecular alterations collectively contribute to the cognitive decline, memory loss, and neuronal degeneration characteristic of Alzheimer's disease (AD) [86].

RNA metabolism plays an important role in the expansion of this disease. Garofalo et al. [28] investigated RNA sequencing from peripheral blood mononuclear cells (PBMCs) of patients with neurodegenerative diseases—specifically Amyotrophic Lateral Sclerosis (ALS), Alzheimer's Disease (AD), and Parkinson's Disease (PD)—along with healthy controls. The research revealed significant dysregulation of long non-coding RNAs (lncRNAs) and messenger RNAs (mRNAs). Amyotrophic Lateral Sclerosis (ALS) patients exhibited the most pronounced changes, with 293 differentially expressed (DE) lncRNAs and 87 mRNAs. In contrast, Alzheimer's Disease (AD) patients displayed moderate changes with 23 differentially expressed (DE) genes, comprising 19 protein-coding genes and 4 lncRNAs, while Parkinson's Disease (PD) patients showed minimal alterations with only 5 differentially expressed (DE) genes. The researchers employed the Kyoto Encyclopedia of Genes and Genomes (KEGG) and Gene Ontology (GO) analyses to investigate the affected pathways and biological processes. They identified common disruptions in immune response, inflammation, cell death and survival, and neuronal function in both Amyotrophic Lateral Sclerosis (ALS) and Alzheimer's Disease (AD). This groundbreaking study highlights the critical role of RNA metabolism in the pathogenesis of neurodegenerative diseases, opening avenues for the development of novel biomarkers and therapeutic targets, and offering hope for millions affected worldwide (Fig. 14.2).

In Alzheimer's Disease (AD), amyloid β plaques and tau tangles are hallmark protein aggregates that damage synapses and neurons. The UPS fails to degrade these pathological proteins efficiently, contributing to disease progression. Studies have shown that both ubiquitin-tagged proteins and proteasomal subunits

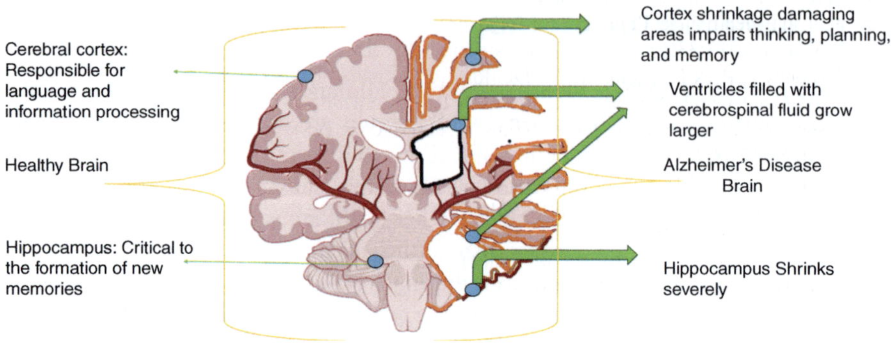

Fig. 14.2 Diagrammatic representation of a Healthy and Alzheimer Disease Brain. It is a permanent neurodegenerative condition that leads to a gradual deterioration in memory, language abilities, and impairments in judgment and comprehension

accumulate in AD brains, suggesting a compromised degradation system. Tau degradation: Hyperphosphorylated tau, which forms neurofibrillary tangles, is normally tagged for degradation by the UPS. Impaired UPS activity leads to tau aggregation and synaptic dysfunction, accelerating AD pathology [18, 68].

14.2.1 Ubiquitin–Proteasome System

The ubiquitin protein was discovered by Gideon Goldstein in 1975 and further characterized during the late 1970s and 1980s, revealing its essential roles in various cellular processes. It is a small regulatory biopolymer, with a molecular mass of 8.6 kDa, discovered in nearly all tissues of eukaryotic organisms, earning its name due to its ubiquitous presence. Ubiquitin B, Ubiquitin C, Ubiquitin A-52 residue, and Ribosomal Protein S27A are genes in the human genome that code for ubiquitin. They play a vital role in maintaining cellular homeostasis, protein degradation, recycling, cell cycle regulation, DNA repair, signal transduction, and apoptosis. The dysregulation of ubiquitin-related processes has been associated with cancer and neurodegenerative diseases, and metabolic disorders, highlighting the importance of ubiquitin in maintaining cytological health. Through its widespread presence and diverse functions, ubiquitin has become a fundamental focus of research in understanding cellular biology and disease mechanisms. The ubiquitin-proteasome system (UPS) is a key cellular network responsible for protein degradation and homeostasis. Ubiquitin E3 ligases, recognize target proteins that need to be degraded, while the 26S proteasome catalyzes the degradation itself. The UPS not only removes damaged or misfolded proteins but also actively participates in synaptic remodeling by regulating the levels of synaptic proteins like receptors, scaffolding proteins, and other signaling molecules that are crucial for synaptic function [25, 37] (Fig. 14.3).

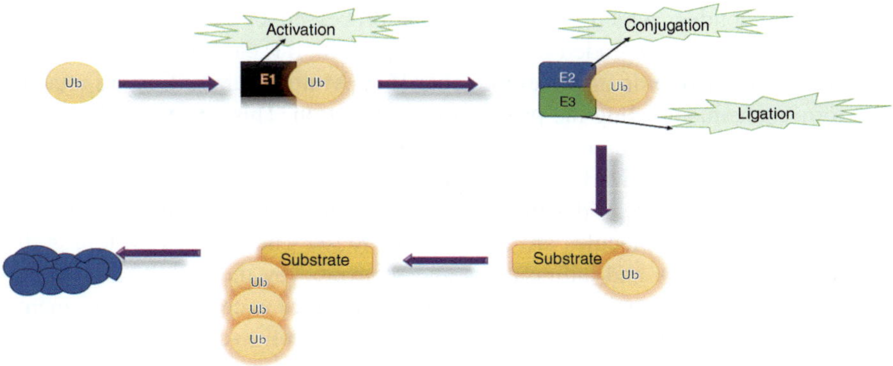

Fig. 14.3 The image shows the process of ubiquitination, a key cellular mechanism for targeting proteins for degradation via the proteasome. E1 protein is attached with ubiquitin and activates ubiquitin. It is an alarm clock. Conjugation is the intermediate step. Conjugating enzymes grab the ubiquitin molecule and create E2-Ubiquitin conjugate intermediate. E3 ubiquitin enzyme is the final step. This is the anchor of the UPS, handing off and finally linking the ubiquitin to the protein

Synaptic remodeling is the process by which synapses strengthen, weaken, or are eliminated in response to activity, environmental changes, or developmental cues. The E1 enzyme (Ubiquitin-activating enzyme) activated and binds to ubiquitin in an ATP-driven process. The binding forms a dynamic thioester bond interaction of E1 enzyme with ubiquitin, stimulating ubiquitin for the upcoming step. The turning on ubiquitin is subsequently moved from the Ubiquitin-conjugating enzyme E1–E2. E2 carries ubiquitin and, in collaboration with another enzyme (E3), support the conjugation of ubiquitin to the target protein. The E3 enzyme (Ubiquitin ligase) recognizes and binds the target protein (often referred to as a "substrate"). The E3 enzyme aids in the transfer of ubiquitin from the E2 enzyme to the target protein, forming a covalent bond between ubiquitin and the target protein. E3 ubiquitin ligase enzyme identifies the specific protein substrate to be degraded and catalyzes the conveyance of ubiquitin from E2 to the substrate. E3 enzymes are crucial for substrate focus and ensure that the correct proteins are tagged for degradation. Ubiquitin molecules are bound to the lysine residues of the target protein, frequently creating a chain of ubiquitin molecules (polyubiquitination). This chain serves as a signal that the protein is destined for degradation. E3 enzymes provide substrate specificity, meaning they help identify which proteins need to be degraded [37].

14.2.2 Polyubiquitination

This process often repeats, adding multiple ubiquitin molecules in a chain (polyubiquitination) to the substrate protein. Polyubiquitinated proteins are marked for degradation by the protein degrading complex. Previously, protein is polyubiquitin-tagged, it is recognized and targeted to the proteasome and plays a crucial role in degrading damaged proteins. The proteasome breaks down the ubiquitinated protein into small peptides, which are then further processed or recycled by the cell. Ubiquitin itself is not degraded with the target protein. It is cleaved off and recycled for use in further rounds of protein tagging [64]. Multiple ubiquitins are attached and form a chain, signaling the protein for degradation. This protein is directed to the proteasome for degradation, while ubiquitin is recycled. Once a protein is polyubiquitinated, it is recognized and directed to the proteasome, a large protease complex. The proteasome degrades the tagged protein into smaller peptide fragments, while the ubiquitin molecules are recycled for reuse. The reduced protein is dismantled into peptides and amino acids and recovered by the cell. The ubiquitin-proteasome system is vital for sustaining protein balance in cells. It is instrumental in regulating numerous cellular functions, including cell cycle regulation, signal transduction, and immune responses.

p53 is a tumor suppressor protein. Under normal conditions, p53 is kept at reduced intensity in the cell by an E3 enzyme called Mouse Double Minute 2, which tags p53 with ubiquitin, marking it for proteasomal degradation. However, in response to DNA damage, p53 is stabilized (i.e., not degraded), allowing it to accumulate and activate the expression of genes related to cell cycle arrest, DNA repair, and programmed cell death. UPS plays a critical role in cellular regulation, as misregulation of p53 can lead to cancer development. Alzheimer's and Parkinson's, have been linked to dysfunction in the UPS, as the improper degradation of misfolded proteins can result in toxic aggregates [31]. This process is essential for regulating protein levels and quality control within the cell, playing a vital role in many cellular mechanisms.

14.2.3 Ubiquitin-Proteasome System in Synapse Remodeling in Alzheimer's Disease (AD)

The UPS is crucial in maintaining this synaptic plasticity by modulating the availability of key proteins at the synapse. The absorption and breakdown α-amino-3-hydroxy-5-methyl-4-isoxazolepropionic acid (AMPA) receptors are vital for excitatory synaptic communication. Postsynaptic density protein 95 and other postsynaptic scaffolding proteins are also regulated by the UPS. By degrading these proteins, the UPS influences the stability and plasticity of synapses. Synaptic pruning is done by forming an excess of synaptic connections during development. The UPS help in elimination of redundant synapses, it is a process crucial for proper neural circuit formation. Disruption in this pruning process is associated with autism spectrum disorders (ASDs) and schizophrenia [16, 36]. Proteolysis process is done by ubiquitin-proteasome pathway. It represents a novel molecular mechanism that governs numerous essential functions for the nervous system, including synaptic connectivity and synaptic plasticity. This process sculpts the nervous system and controls the axonal growth and dendritic pruning and also regulates the synaptic size and number [24, 47].

Glial cells are non-neuronal cells in the brain that provide support, protection, and nourishment to neurons. These cells do not participate directly in synaptic transmission, but they play critical roles in preserving homeostasis, producing myelin, and offering support and insulation for neurons, and contributing to immune responses within the brain. When neurodegenerative diseases occur, glial cells can become "reactive," a process known as reactive gliosis. This means they take on a more active role in the inflammatory response, which can affect how they handle protein degradation. Interestingly, while neurons often show a buildup of these damaged proteins in neurodegenerative diseases, glial cells don't seem to accumulate them as much. This suggests that glial cells might be better at maintaining "protein homeostasis," which means they can manage and clear out proteins more efficiently. One reason might be that glial cells produce a special type of proteasome called the immunoproteasome during inflammation, which might help break down disease-related proteins more effectively [39]. Alzheimer's disease accumulates in the brain. Ubiquitin is a small protein that marks them for degradation by the proteasome, a structure that breaks them down. The failure of this process (the ubiquitin-proteasome system, or UPS) is thought to contribute to this disease because the buildup of damaged proteins can harm cells. In Alzheimer's disease, there is proteasomal dysfunction, leading to the accumulation of abnormal proteins, specifically amyloid-beta (Aβ) and tau proteins. These proteins are toxic to neurons and contribute to the neuronal cell death seen in Alzheimer's disease. Proteasomal dysfunction results in the failure to degrade misfolded or abnormal proteins. This dysfunction leads to the accumulation of proteins like Aβ and tau, which is associated with Alzheimer's. Ultimately, the failure to remove these proteins triggers neuronal cell death (Fig. 14.4).

In Alzheimer's disease (AD) the UPS malfunctions, leading to the accumulation of misfolded amyloid-β (Aβ) and tau proteins. These proteins aggregate into plaques and tangles, causing neurotoxicity and eventually leading to neuronal cell death

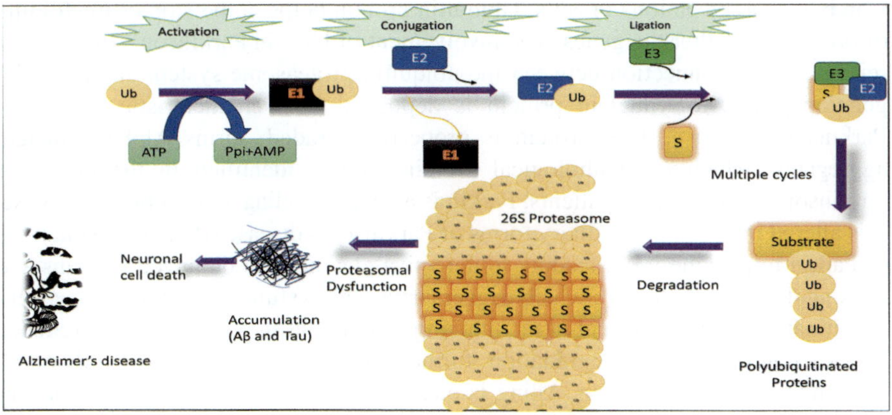

Fig. 14.4 The image appears to illustrate the polyubiquitinated proteins and its relation to Alzheimer's disease. The diagram visually connects the UPS system to how it malfunctions in Alzheimer's disease, showing the breakdown of protein regulation and leading to toxic accumulations that cause neurodegeneration

[29]. The proteasome's dysfunction prevents the clearance of these pathological proteins, exacerbating disease progression. Ubiquitinated proteins are found in both Aβ plaques and tau tangles, and a mutation in ubiquitin-related genes contributes to Alzheimer's disease (AD) [84, 98].

14.3 Parkinson's Disease (PD)

The substantia nigra or midbrain region is an area of the brain and is vital for controlling movement. It is a component of the basal ganglia and is split into two primary sections: the pars compacta and the pars reticulata. The substantia nigra pars compacta are particularly important for generating dopamine, a neurotransmitter that modulates motor function. The degeneration of dopamine-producing neurons in this region is a key feature of Parkinson's disease, resulting in motor symptoms such as tremors, stiffness, and slowed movement (bradykinesia) linked to the condition. Parkinson's disease (PD) is marked by the loss of dopaminergic neurons in the substantia nigra and the buildup of alpha-synuclein aggregates (Lewy bodies). Mutations in genes encoding UPS elements, such as parkin (an E3 ligase), lead to hereditary forms disease. Loss of parkin activity leads to the buildup of aberrant protein mitochondrial dysfunction, contributing to neuronal death. The 26S proteasome's function is reduced in Parkinson's disease (PD), leading to a buildup of toxic alpha-synuclein aggregates that disrupt normal neuronal function [32].

14.3.1 Ubiquitin-Proteasome System in Synapse Remodeling in Parkinson's Disease (PD)

It is a cellular and molecular disorder. Proteins are dysfunction and make the inclusion bodies aggregations in cells. Ubiquitin protein helps in the stop of misfolding of proteins and alter the genes. The involvement of UPS in Parkinson's disease has reinforced the connection between the ubiquitin-proteasome system and neurodegeneration [6]. α-Synuclein protein is deposited in a characteristic feature of Parkinson's disease. This protein is properly degraded, forms Alpha-synuclein aggregates, which are pathological protein clumps identified in the brains of Parkinson's disease (PD) patients. Parkin is a gene encoding an E3 ubiquitin ligase. It is altered in specific hereditary forms of Parkinson's disease (PD) [17]. Mutations in Parkin impair the UPS, leading to defective clearance of damaged proteins and causing their accumulation, which triggers neuronal dysfunction. UCH-L1 is a deubiquitinating enzyme critical in maintaining free ubiquitin pools. It is also reduced the efficiency of protein degradation, further contributing to proteasome dysfunction and protein aggregation. Aging is one of the most significant risk factors for Parkinson's disease (PD), as it leads to a natural decline in the efficiency of proteostasis systems, including the UPS [12]. Oxidative stress arises from a disparity between the generation of free radicals and the cell's capacity to neutralize these reactive species (Fig. 14.5).

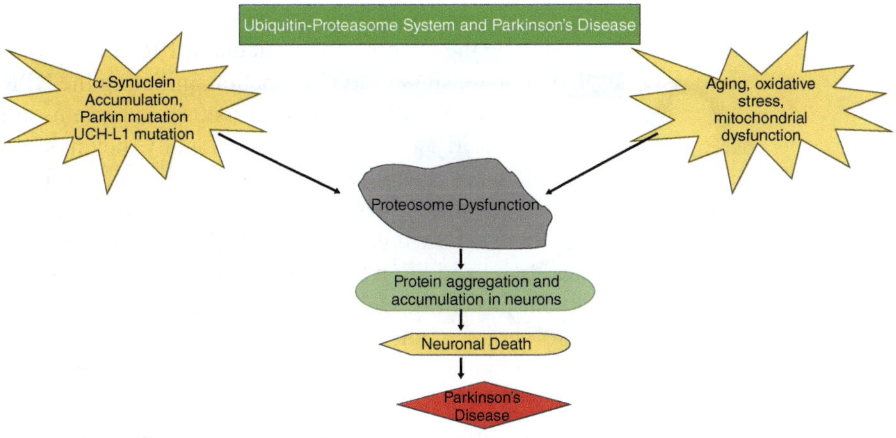

Fig. 14.5 This diagram shows that multiple genetic, environmental, and cellular factors converge to cause proteasome dysfunction in Parkinson's disease, resulting in the toxic aggregate of proteins like α-synuclein, the infected of neurons, and the eventual development of Parkinson's disease (PD) symptoms. The dysfunction of the UPS is a central feature of PD pathology, making it a target for therapeutic intervention

In Parkinson's disorder, oxidative damage affects proteins and contributes to mitochondrial dysfunction, exacerbating UPS impairment [23, 55]. Mitochondrial dysfunction further accelerates proteasome inefficiency by reducing cellular energy production, causing the aggregation of aberrant proteins and increasing neuronal vulnerability. Proteasome Dysfunction is the central component of the diagram that shows the breakdown of the UPS [81]. When the UPS is unable to degrade misfolded or damaged proteins, it results in their accumulation within neurons. Misfolded proteins, such as α-synuclein, begin to aggregate in neurons. These aggregates are toxic to neurons and can disrupt intracellular processes. The accumulation of protein aggregates induces neuronal stress, causing the stimulate the program cell death. The culmination of these processes—proteasome dysfunction, protein aggregation, and neuronal death—leads to the medical manifestation of Parkinson's disease [56]. The progressive degeneration of neurons in critical brain regions results in the motor and cognitive symptoms seen in Parkinson's disease (PD) patients [58]. Ubiquitination is a protein modification that involves attaching ubiquitin (a small regulatory protein) to substrate proteins, either marking them for degradation or regulating their cellular function. This process requires three enzymes with activation conjugation and ligation [37]. Lysine, represented by "K" in biochemical shorthand, contains an amine group that forms bonds with other molecules, making it a key site for post-translational modifications like ubiquitination. Depending on the lysine used for attachment, the function of the polyubiquitin chain differs. K48-attached sequence signals protein degradation, while K63-attached sequence is typically engaged in signaling pathways or other non-degradative roles. K48-linked polyubiquitin chains tag proteins for break down by the 26S proteasome, a process crucial for maintaining protein homeostasis by

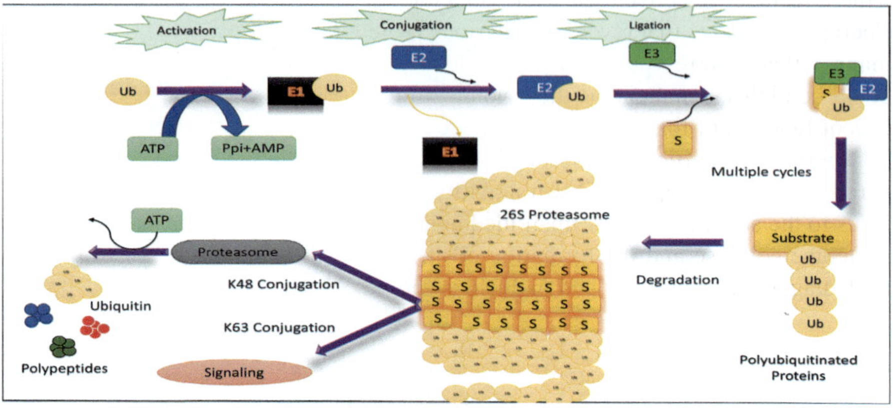

Fig. 14.6 The image depicts the ubiquitination pathway, a process by which proteins are tagged for degradation or signaling using ubiquitin molecules

degrading misfolded unneeded proteins. Once ubiquitinated with a K48 chain, the protein is recognized by the proteasome, which unfolds and degrades it into smaller peptides. This process regulates key cellular functions, incorporating cell cycle oversight, programmed apoptosis, and response to redox imbalance [10]. On the other hand, K63-linked polyubiquitin chains are involved in non-proteolytic functions, such as protein signaling, DNA repair, and endocytosis. K63 ubiquitin chains act as scaffolds to recruit proteins to specific locations, such as DNA damage sites or membrane-bound receptors during signal transduction. This type of ubiquitination is critical for processes like activation of the NF-κB signaling pathway, essential for immune responses [21] (Fig. 14.6).

14.4 Huntington's Disease (HD)

The huntingtin protein is large, located on chromosome 4. It plays a vital role in several cellular processes, including vesicle trafficking, transcriptional regulation, and synaptic function. Genetic alteration in the HTT gene causes an abnormal expansion of CAG trinucleotide sequence, leading to an elongated polyglutamine (polyQ) sequence in the huntingtin protein. Normally, huntingtin contains fewer than 36 glutamine repeats, but in individuals with Huntington's Disease (HD), the number of repeats exceeds this threshold, leading to the formation of a mutated huntingtin protein with toxic properties [99]. It is inherited in an autosomal dominant fashion, meaning a single copy of the mutated gene is sufficient to trigger the disorder [82]. It is demonstrated with a combination of neuromuscular symptoms, cognitive deficits, and emotional symptoms. Motor symptoms include chorea (involuntary, jerky movements) and dystonia, whereas decline typically affects managerial functions [78]. Psychiatric disturbances include depression, irritability, and, in some cases, severe psychiatric disorders like psychosis [82]. It is identified by neurodegeneration, especially in the striatal region and cortex, due to the toxic effects of the mutated huntingtin protein [4]. This neurodegeneration underlies movement disorders, cognitive impairments, and emotional disturbances. The diagnosis of this disease is confirmed via genetic testing, which detects the Cytosine-Adenine-Guanine repeat in the huntingtin gene. Neuroimaging, such as MRI, often reveals brain atrophy, especially in the striatum, as the condition advances [99]. Current clinically trial is focus on managing symptoms. Tetrabenazine is commonly used to treat chorea, while antipsychotic medications can help with psychiatric symptoms [82]. Physical, occupational, and speech therapy also contribute to maintaining patients' quality of life. Recent research efforts focus on gene-silencing therapies and CRISPR-Cas9 gene-editing technologies. These approaches aim to either suppress or repair the mutant HTT gene, offering hope for future treatments [88].

Huntington's Disease (HD) is a fatal genetic disorder caused by a mutated huntingtin protein, leading to protein misfolding and aggregation, proteostasis network collapse, and organelle dysfunction. The disease is characterized by a polyglutamine expansion exceeding 40 repeats, resulting in proteolytic cleavage and accumulation of mutant huntingtin (mHtt). Although mHtt is expressed from early life, symptoms typically appear in mid-adulthood due to aging-related factors. The pathogenesis of HD involves the disruption of cellular proteostasis, where chronic production of misfolded mHtt overwhelms the chaperone machinery, diverting other misfolded proteins to the proteasome and autophagy pathways. Mitochondrial dysfunction plays a crucial role, as sustained mitochondrial function is necessary for ATP-dependent chaperones and proteolytic machinery. Recent studies highlight the importance of mitochondria-endoplasmic reticulum interactions and dedicated proteostasis machinery in mitochondria. Potential treatments aim to enhance cytosolic proteostasis pathways and modulate proteostasis network components. Research has shown that pharmacological and genetic modulation of these components can impact disease phenotypes in cellular and in vivo HD models. Understanding the complex interplay between mHtt, proteostasis, and organelle dysfunction is crucial for developing effective therapeutic strategies to combat HD.

14.4.1 Ubiquitin-Proteasome System in Synapse Remodeling in Huntington's Disease (HD)

Ubiquitin (Ub) is a small regulatory protein that needs to be activated before being attached to a substrate protein [41]. The ubiquitination process commences with the formation of a thioester bond between the ubiquitin-activating enzyme, E1, and ubiquitin, fueled by the energy released from ATP hydrolysis to ADP and a phosphate group and make the activated ubiquitin-E1 complex. The activated ubiquitin is transferred to the ubiquitin-conjugating enzyme, E2, via another thioester bond, and contains four classes—Classes I, II, III, and IV—based on subcellular localization, interaction with E1, and their ability to modulate E3 enzyme activity factors.

This intricate process sets the stage for protein degradation or modification, highlighting the crucial roles of E1, E2, and ubiquitin in regulating protein fate. The transfer involves the formation of an E2 ~ Ub complex, where ubiquitin is bound to the E2 enzyme. E3 ubiquitin ligase catalyzes the transfer of ubiquitin from E2 to the target protein. Directly transfers ubiquitin to the target protein. RBR (RING-between-RING) transfers ubiquitin through an intermediate. U-box and RING ligases coordinate the transfer by binding both E2 and the substrate. The Cys residue on the target protein interacts with ubiquitin, allowing the transfer of ubiquitin to bind covalently between ubiquitin and the ligand protein. Ubiquitin can be attached to proteins at specific lysine (K) residues, such as K6, K9, and K15 [87]. This process can result in either monoubiquitination (Attachment of a single ubiquitin molecule) or polyubiquitination (Formation of a ubiquitin chain) [101]. Additional ubiquitins are attached to lysine residues on the already attached ubiquitin. Ubiquitination can target a protein for degradation, alter its location, affect its activity, or promote/inhibit protein interactions [37]. After ubiquitination, deubiquitinating enzymes (DUBs) reverse the process by removing ubiquitin molecules through various cleavage mechanisms. This crucial step regulates protein fate and recycles ubiquitin molecules [73] (Figs. 14.7 and 14.8).

Deubiquitination, the removal of ubiquitin chains from substrate proteins, shows different cleavage types involved in this process. It is mediated by deubiquitinating enzymes (DUBs), which regulate ubiquitin-dependent processes by trimming or removing ubiquitin chains.

I. Exo cleavage is removing a single ubiquitin molecule from the end of a ubiquitin chain.
II. Endo cleavage involves breaking the ubiquitin chain at internal points, leaving smaller chains intact.

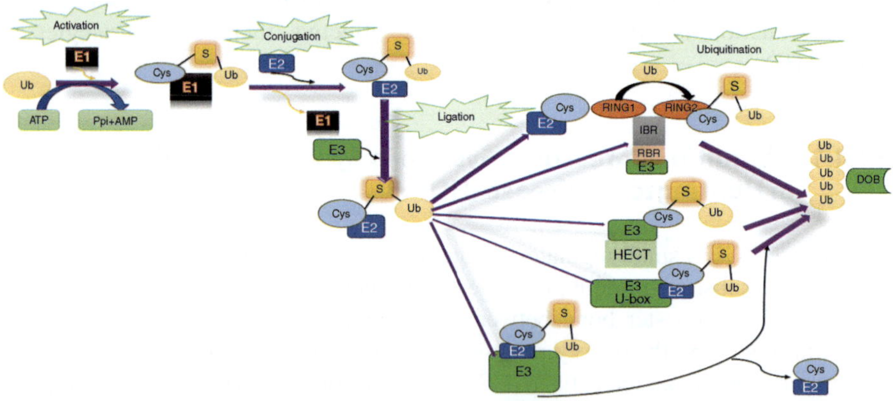

Fig. 14.7 This image illustrates the process of protein ubiquitination, which is a series of biochemical reactions. Ubiquitin molecules tag target proteins for degradation or functional modification, signaling the proteasome to break them down or altering their cellular behavior

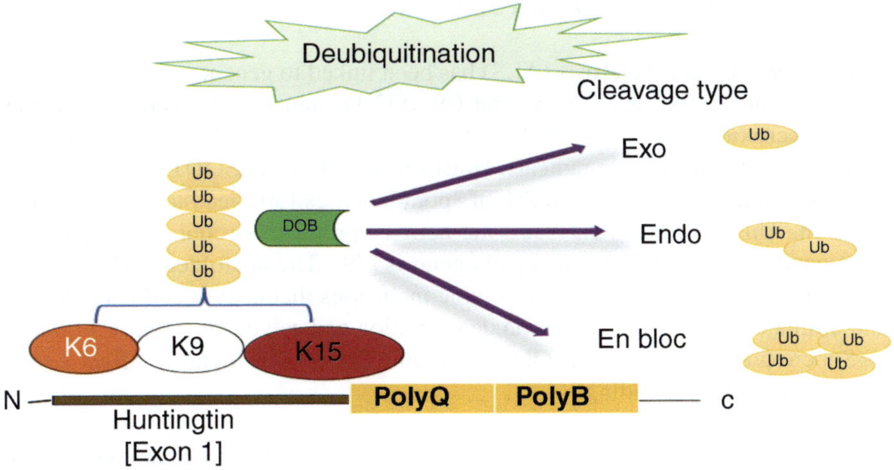

Fig. 14.8 The potential removal of ubiquitin through deubiquitination. After ubiquitination, proteins can be deubiquitinated by DUB (Deubiquitinating enzymes). The target protein may be sent for degradation or have its function modulated depending on the specific ubiquitination pattern

III. En bloc cleavage is a bulk removal of ubiquitin chains is shown, where a whole polyubiquitin chain is detached from the substrate. These cleavages regulate different cellular processes, including protein degradation, signal transduction, and cellular stress responses.
IV. The bottom part of the image refers to Huntingtin (Exon-1), a specific protein with polyQ (polyglutamine) and polyB (polyubiquitin) domains, suggesting a connection to diseases like Huntington's, where the misregulation of ubiquitination and deubiquitination plays a role in protein aggregation and neuronal dysfunction [15].

Deubiquitination, a vital cellular process, involves the enzymatic removal of ubiquitin chains from proteins, primarily facilitated by deubiquitinating enzymes (DUBs). These enzymes ensure that ubiquitin-dependent signaling pathways are reversible, thereby maintaining cellular homeostasis. As depicted in the image, there are three main types of deubiquitination: exo-cleavage, where a single ubiquitin molecule is removed from the end of a chain; endo-cleavage, which splits the chain at internal positions; and en bloc cleavage, which removes an entire ubiquitin chain in one step. These different modes of deubiquitination regulate various functions like protein degradation, endocytosis, and cell signaling [87]. Huntingtin (Exon-1), a protein linked to Huntington's disease. In this context, polyglutamine (PolyQ) expansions in Huntingtin are associated with neurodegenerative disorders, where misfolded proteins accumulate due to disruptions in ubiquitination and deubiquitination. Dysregulation in these processes can lead to protein accumulation, as seen in Huntington's and other neurodegenerative diseases, affecting cellular function [51].

14.5 Amyotrophic Lateral Sclerosis (ALS)

Amyotrophic Lateral Sclerosis (ALS) has been linked to genetic variations, including mutations in the SOD1 gene and C9orf72 hexanucleotide repeat expansions [76]. Superoxide dismutase 1 (SOD1) typically protects cells from oxidative stress by neutralizing superoxide radicals. However, mutations in SOD1, responsible for 20% of familial ALS cases and 2% of sporadic cases [30], lead to a toxic gain of function. This aberrant SOD1 enzyme activity promotes protein misfolding and aggregation, contributing to ALS pathogenesis [79]. The misfolded SOD1 proteins accumulate in neurons, forming insoluble inclusions that disrupt cellular functions and ultimately induce motor neuron degeneration and cell death [100].

Superoxide Dismutase 1 (SOD1) mutations have been found to affect several cellular pathways, including:

I. SOD1 is typically an antioxidant enzyme, its mutated forms contribute to increased oxidative damage by either reducing its protective capacity or by directly generating reactive oxygen species (ROSs) [48].
II. Misfolded SOD1 proteins exhibit a propensity for aggregation, resulting in the creation of harmful protein inclusions that disrupt cellular transport and impair neuronal function [75].
III. The mutant SOD1 protein localizes to mitochondria, causing structural damage and disrupting energy production, thereby contributing to mitochondrial dysfunction, a hallmark of ALS pathology [43].
IV. Aggregates of mutant SOD1 can interfere with axonal transport, disrupting the communication between neurons and muscles, which accelerates motor neuron degeneration [63].
V. Dysfunction of the Ubiquitin-Proteasome System (UPS) occurs due to mutant SOD1, hindering the degradation of misfolded proteins and leading to the accumulation of toxic protein aggregates within motor neurons [52].
VI. SOD1 Mouse Models Mouse models expressing mutant SOD1 have been widely used to study ALS. These models replicate many of the pathological features seen in human ALS, including protein aggregation, motor neuron death, and muscle atrophy. Studies using SOD1 mouse models have greatly advanced our understanding of ALS pathogenesis and have been instrumental in developing potential therapeutic strategies [94].
VII. Therapeutic Approaches Targeting SOD1: Efforts to develop therapies for SOD1-related ALS have focused on reducing the levels of mutant SOD1 or correcting its toxic gain of function. Some approaches include:
VIII. Gene Silencing: Antisense oligonucleotides (ASOs) are being investigated to specifically target and reduce the expression of mutant SOD1, offering a promising strategy for patients with SOD1 mutations [80]. Small Molecule Inhibitors: Compounds that prevent SOD1 misfolding and aggregation are being explored to inhibit the toxic effects of mutant SOD1 [70].

IX. Immunotherapy: Targeted immunotherapy aimed at clearing misfolded SOD1 proteins is another avenue of research, as reducing SOD1 aggregates could slow disease progression [27].
X. Clinical Presentation: Patients with ALS often present with asymmetrical limb weakness, progressing to difficulties in swallowing, speaking, and breathing. Bulbar onset ALS, which affects speech and swallowing first, has a worse prognosis than limb onset [11].
XI. Current Treatments: There is no cure for ALS, but treatments aim to slow progression and manage symptoms. Riluzole, a glutamate inhibitor, is the only FDA-approved drug proven to extend survival, albeit modestly [60]. Newer treatments, such as edaravone, an antioxidant, have shown promise in reducing oxidative stress in ALS patients [1]. Despite substantial progress in elucidating the genetic and molecular mechanisms underlying ALS, a cure remains elusive. Ongoing research into neuroprotective agents and gene therapies holds promise for potential breakthroughs in future treatments.

14.5.1 Ubiquitin-Proteasome System in Synapse Remodeling in Amyotrophic Lateral Sclerosis (ALS)

The ubiquitin–proteasome system (UPS) is essential controller of protein turnover and cellular balance. In ALS, dysfunction of the UPS is implicated in the accumulation of misfolded proteins, contributing to synaptic damage and neurodegeneration [52]. The UPS's role in synapse remodeling is especially crucial because synaptic changes are early markers of ALS progression [2]. Synaptic remodeling is essential for maintaining neural plasticity and proper motor function. The UPS regulates synaptic proteins, including those involved in neurotransmitter release, synaptic growth, and maintenance [35]. Through targeted degradation of these proteins, the UPS modulates synaptic strength and structure, processes that are disrupted in ALS [83]. Mutations in genes like SOD1, TDP-43, and C9orf72, which are associated with ALS, can interfere with the ubiquitination and subsequent degradation of synaptic proteins [85]. Accumulation of harmful protein clusters disrupts synaptic function and triggers motor neuron degeneration [89]. Additionally, proteasomal impairment has been linked to increased oxidative stress and mitochondrial dysfunction in ALS, further exacerbating synaptic damage [14]. In ALS, early synaptic dysfunction precedes motor neuron loss, suggesting that targeting synaptic changes might offer therapeutic potential [61]. The UPS is central to this remodeling process, controlling the turnover of key synaptic proteins. Dysfunctional UPS activity in ALS may lead to the inability to clear damaged proteins at the synapse, accelerating disease progression [93]. Given the role of the UPS in ALS, therapeutic strategies that enhance proteasomal activity or correct UPS dysfunction could slow synaptic degeneration and improve motor function. Recent studies have explored proteasome activators and autophagy-inducing drugs as potential treatments to restore protein homeostasis and protect synapses in ALS [52]. The UPS plays a vital role in synapse remodeling, and its dysfunction is essential for ALS pathogenesis. Further

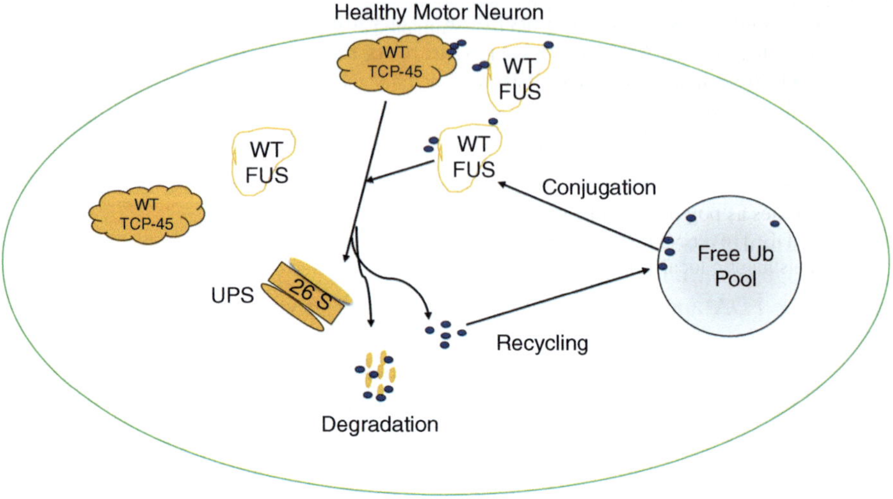

Fig. 14.9 This image compares the function of motor neurons in healthy individuals versus those affected by Amyotrophic Lateral Sclerosis (ALS), with a focus on the protein proteasomal degradation pathway (UPS)

understanding of the relationship between the UPS and synaptic changes could lead to new therapeutic interventions aimed at preserving synaptic integrity and slowing ALS progression (Figs. 14.9 and 14.10).

I. Normal neuronal function relies on efficient UPS-mediated regulation of proteins, including WT TDP-43 and FUS, where misfolded or unnecessary proteins are ubiquitinated and degraded.
II. The ubiquitination process directs damaged proteins to the proteasome, ensuring timely degradation and preventing protein buildup that could disrupt neuronal function.
III. Free Ubiquitin Pool: After degradation, the ubiquitin molecules are recycled back to a free pool, maintaining a balance in the cellular environment.
IV. In ALS Motor Neuron mutant TDP-43 and FUS: In ALS-affected neurons, mutant forms of TDP-43 and FUS disrupt the UPS.
V. Protein Aggregation: Instead of being degraded, these mutant proteins aggregate and sequester ubiquitin, forming toxic clumps.
VI. UPS Dysfunction: The dysfunction in the UPS prevents proper protein degradation. As a result, the free ubiquitin pool is reduced, leading to further protein mismanagement.
VII. Accumulation of toxic protein clusters contributes to the progressive decline and eventual loss of motor neurons, typifying ALS pathology.

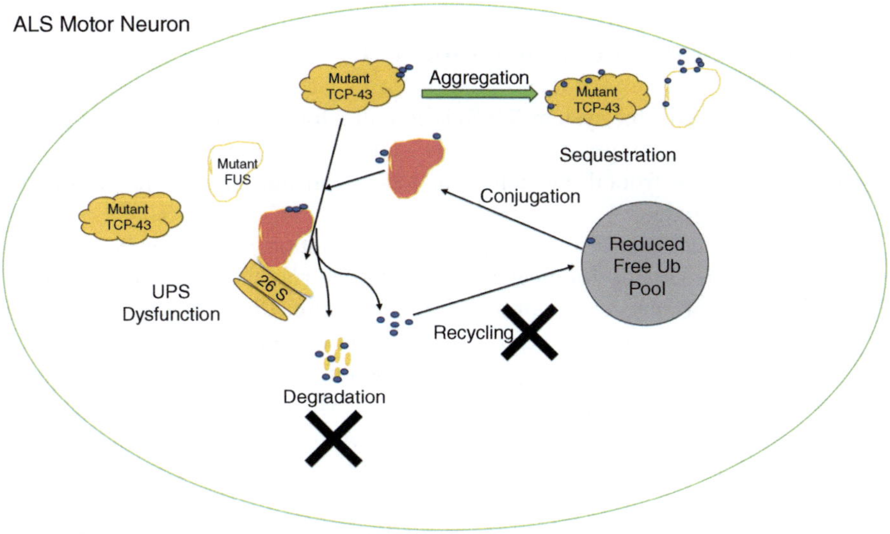

Fig. 14.10 This highlights the critical role of protein mismanagement in ALS pathogenesis, where UPS dysfunction contributes to neurodegenerative processes

14.6 Frontotemporal Dementia (FTD)

Frontotemporal dementia (FTD) is characterized by the pathological accumulation of ubiquitinated proteins in neurons, indicating impaired ubiquitin-proteasome system (UPS) function, which normally clears misfolded or damaged proteins. In the context of synapse remodeling, the UPS is essential for regulating the turnover of synaptic proteins, maintaining synaptic plasticity, and modulating neurotransmission. In FTD, dysfunction in the UPS may lead to impaired synaptic function, contributing to the cognitive and behavioral symptoms observed in these patients. Synapse remodeling, the process by which synaptic connections are altered, pruned, or strengthened, is critical for learning, memory, and adaptation to new environments. The UPS helps control this by regulating the degradation of synaptic proteins, such as synaptophysin and postsynaptic density proteins [77]. In FTD, mutations in genes like C9orf72, TARDBP, and GRN have been linked to defects in the UPS, causing the accumulation of ubiquitinated peptides and the dysfunction of normal neural transmission. The interaction between the UPS and synapse remodeling in FTD is an emerging area of research. Defective protein clearance mechanisms likely contribute to synaptic dysfunction, which in turn underlies the cognitive and behavioral symptoms of the disease. Understanding this process better could open up new therapeutic avenues targeting protein degradation pathways in FTD.

14.6.1 Ubiquitin-Proteasome System in Synapse Remodeling in Frontotemporal Dementia (FTD)

The ubiquitin-proteasome system (UPS) is essential for synaptic remodeling, which is disrupted in Frontotemporal Dementia (FTD), a group of neurodegenerative disorders affecting the frontal and temporal lobes, causing changes in personality, behavior, and language.

The pathological features of FTD are the disruption of protein homeostasis, where the UPS is heavily implicated. The UPS controls the degradation of damaged, misfolded, or unnecessary proteins at the synapse. Ubiquitin tags proteins for degradation, which are then broken down by the proteasome. This process is essential for maintaining synaptic integrity and function. In FTD, dysfunction in this system leads to the accumulation of toxic proteins, which can interfere with synapse structure and function. Synapse remodeling is an activity-dependent process, where proteins are constantly synthesized and degraded to regulate synaptic strength. The UPS ensures proper turnover of synaptic proteins involved in neurotransmitter release, receptor trafficking, and cytoskeletal dynamics. In FTD, altered UPS function can impair synaptic plasticity, affecting cognitive and behavioral processes. In many cases of FTD, mutations in genes related to the UPS, such as TARDBP (encoding TDP-43) and GRN (encoding progranulin), lead to misregulation of ubiquitin-tagging and protein clearance. TDP-43 inclusions, a hallmark of FTD, are often ubiquitinated but not efficiently degraded, disrupting neuronal function and synaptic connections. During development and neuroplasticity, the UPS is involved in synaptic pruning, where unnecessary or weak synapses are removed. Defects in this process can contribute to synaptic loss, one of the early signs of neurodegeneration in FTD. Impaired synaptic pruning due to UPS dysfunction can lead to abnormal network connectivity in the brain, exacerbating the symptoms of FTD. UPS maintaining synaptic health, targeting components of this system may offer therapeutic strategies for FTD. Enhancing UPS function or reducing the accumulation of toxic proteins could help preserve synaptic function and slow disease progression (Fig. 14.11).

Polyphenols influence the UPS by regulating protein degradation, enhancing UPS enzyme activity, promoting autophagy and mitophagy, reducing oxidative stress, and modulating inflammation. Specifically, polyphenols have been shown to reduce protein aggregation, a hallmark of neurodegenerative diseases. For instance, in AD, polyphenols reduce Aβ aggregation and enhance UPS function, mitigating cognitive decline. In PD, polyphenols protect against α-synuclein aggregation and promote UPS-mediated clearance, alleviating motor symptoms. The ubiquitin-proteasome system (UPS) plays a vital role in maintaining protein homeostasis and preventing neurodegenerative diseases. These are as followes:

I. The UPS involves a complex process of tagging and degrading damaged proteins, facilitated by three key enzymes: E1, E2, and E3. Impairments in the UPS have been linked to various neurodegenerative diseases, including Alzheimer's, Parkinson's, Huntington's, ALS, and Frontotemporal Dementia.

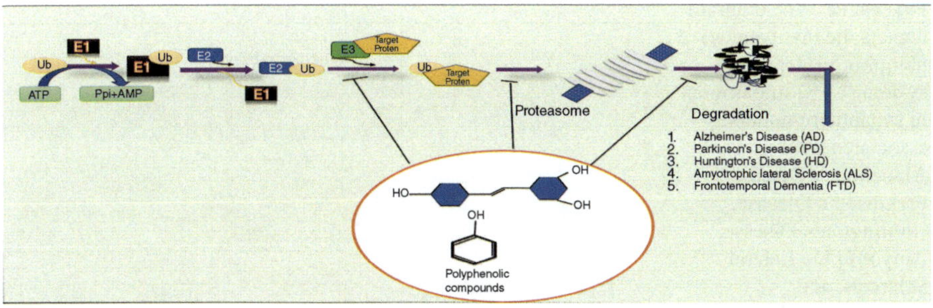

Fig. 14.11 This figure emphasizes the relationship between the UPS dysfunction and neurodegenerative diseases and highlights the therapeutic potential of polyphenolic compounds in modulating the system. Proper UPS function is essential for protein quality control. Dysregulation results in toxic protein aggregation, a hallmark of neurodegenerative diseases. Polyphenols' interaction with the UPS suggests potential for alleviating these conditions

II. Polyphenolic compounds, found in plant-based foods and natural products, have shown potential neuroprotective effects by enhancing UPS function and inhibiting protein aggregation. Further research is necessary to elucidate the mechanisms underlying polyphenol-mediated neuroprotection and to develop innovative therapies for neurodegenerative diseases.
III. Ubiquitinated proteins are selectively degraded by the proteasome, a multisubunit complex, into peptides that can be either reused for protein synthesis or further degraded.
IV. Impairments in the ubiquitin-proteasome pathway have been linked to various neurodegenerative disorders, including Alzheimer's Disease (AD), Parkinson's Disease (PD), Huntington's Disease (HD), Amyotrophic Lateral Sclerosis (ALS), and Frontotemporal Dementia (FTD).
V. In these diseases, the proteasome may fail to clear misfolded or aggregated proteins, contributing to neurodegeneration. Polyphenolic compounds (depicted in the central circle of the diagram) have potential protective roles in neurodegenerative diseases. These compounds, typically found in various plants, may help inhibit the formation of protein aggregates or enhance the function of the proteasome, thereby promoting the clearance of harmful proteins in these diseases. Two chemical structures in the circle likely represent specific types of polyphenols that could have neuroprotective effects.

Frontotemporal Dementia (FTD) is the root condition, and it forms the core of the graphic. It involves progressive damage to the frontal and temporal lobes, causing significant declines in behavioral regulation, language processing, and motor coordination [95] (Fig. 14.12).

Behavioral Variant Frontotemporal Dementia (bvFTD), the most prevalent form of FTD, is marked by significant changes in personality, behavior, and social interactions, often manifesting as apathy, impulsivity, and diminished empathy [57]. Primary Progressive Aphasia (PPA) is a type of FTD that affects language skills. It

Fig. 14.12 The diagram depicts the involvement of ubiquitin-proteasome system (UPS) impairment in synaptic remodeling associated with Alzheimer's Disease, Parkinson's Disease, Huntington's Disease, Amyotrophic Lateral Sclerosis, and Frontotemporal Dementia

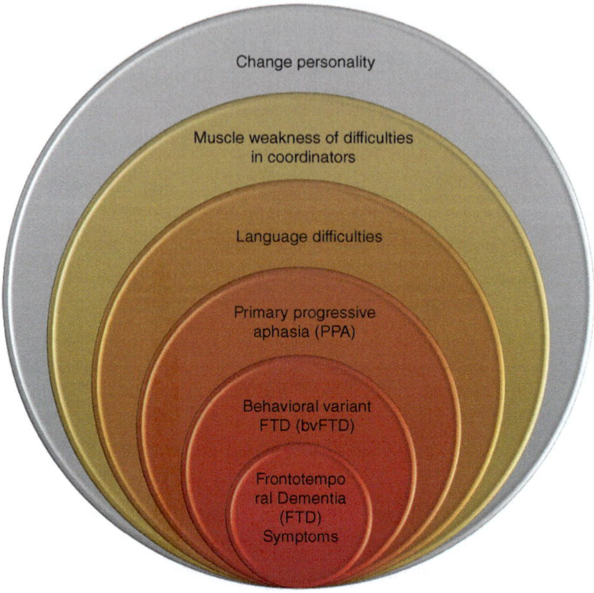

leads to difficulties in speaking, understanding, reading, and writing, while other cognitive functions remain relatively intact, especially in the early stages [38]. Language Difficulties is the layer emphasizes that as FTD progresses, language issues become more pronounced. People may struggle with forming coherent sentences, understanding speech, and word finding. As FTD progresses further, individuals may develop motor symptoms, such as weakness or difficulty coordinating muscle movements. This can resemble other neurodegenerative conditions like ALS (Amyotrophic Lateral Sclerosis), which sometimes overlaps with FTD [8]. The outermost layer represents one of the hallmark symptoms of FTD. The disease often starts with personality changes, where individuals may become more irritable, impulsive, or lose social inhibitions. These personality changes typically occur before motor and language issues [20].

14.7 Comparative Analysis: Ubiquitin-Proteasome System (UPS) in Synapse Remodeling Across Neurodegenerative Diseases

Neurodegenerative diseases, including Alzheimer's Disease (AD), Parkinson's Disease (PD), Huntington's Disease (HD), Amyotrophic Lateral Sclerosis (ALS), and Frontotemporal Dementia (FTD), share a common pathology: accumulation of misfolded proteins overwhelming the Ubiquitin-Proteasome System (UPS). This study examines the role of UPS in synapse remodeling and its implications in these diseases (Table 14.1):

Table 14.1 UPS dysfunction in neurodegenerative diseases: a comparative overview

Aspect	Synapse remodeling	Alzheimer's disease (AD)	Parkinson's disease (PD)	Huntington's disease (HD)	Amyotrophic lateral sclerosis (ALS)	Frontotemporal dementia (FTD)	References
Key proteins	1. AMPA/NMDA receptors 2. Scaffolding proteins (PSD-95).	1. Ubiquitination of Tau and Amyloid-beta (Aβ) 2. Failure to degrade these proteins leads to their accumulation and aggregation (neurofibrillary tangles and plaques)	1. α-Synuclein is a key protein involved in PD 2. Its accumulation forms Lewy bodies 3. Leading to neuronal damage and synaptic dysfunction.	1. Mutant Huntingtin (mHTT) protein is ubiquitinated but not efficiently degraded 2. Leading to its aggregation and synaptic dysfunction.	Ubiquitinated inclusions are characteristic of ALS pathology, often involving proteins like TDP-43, SOD1, and FUS.	Key proteins involved in FTD include TDP-43, FUS, and progranulin, all of which are found ubiquitinated in neuronal inclusions.	1. Tai and Human [89]. 2. Lim and Tan [54]. 3. Li and Li [53]. 4. Peters et al. [71]. 5. Neumann et al. [65, 66].

(continued)

Table 14.1 (continued)

Aspect	Synapse remodeling	Alzheimer's disease (AD)	Parkinson's disease (PD)	Huntington's disease (HD)	Amyotrophic lateral sclerosis (ALS)	Frontotemporal dementia (FTD)	References
Role of UPS	1. Synaptic plasticity 2. Protein turnover 3. Synapse formation/ elimination by degrading misfolded or damaged proteins.	1. Accumulation of Aβ and Tau 2. Disrupting synapse function and contributing to neuronal death.	1. Impaired UPS leads to α-synuclein accumulation and the formation of toxic aggregates 2. Disrupting dopamine signaling and synaptic function.	1. Accumulation of mHTT aggregates disrupts synaptic function 2. Contributing to neuronal toxicity.	Impaired UPS in ALS leads to the accumulation of ubiquitinated aggregates, disrupting motor neurons and synaptic function.	Impaired UPS in FTD leads to the accumulation of ubiquitinated protein inclusions, which disrupt synaptic and neuronal function.	1. Dib-Hajj et al. [24], Kauffman et al. [47]. 2. McNaught et al. [59]. 3. Zheng and Wang [102]. 4. Tashiro et al. [91]. 5. Urushitani et al. [96].
Proteasome activity	1. Proteasomes degrade polyubiquitinated proteins 2. Maintaining protein homeostasis in synapses.	1. Reduced proteasomal activity leads to the accumulation of neurotoxic proteins (Aβ, Tau) 2. Promoting neuroinflammation and synaptic dysfunction.	1. Reduced proteasomal activity 2. Allowing toxic α-synuclein aggregates to accumulate.	1. Mutant Huntingtin inhibits proteasomal activity 2. Reducing the degradation of misfolded proteins and causing toxic aggregate buildup.	Decreased proteasome activity in ALS contributes to the accumulation of misfolded proteins like SOD1, leading to motor neuron death.	Reduced proteasome activity in FTD, especially in cases with TDP-43 pathology, leads to an accumulation of abnormal synaptic proteins.	1. Bedford et al. [5]. 2. Betarbet et al. [7]. 3. Jana et al. [44]. 4. Kabashi et al. [45]. 5. DeJesus-Hernandez et al. [20].

Synaptic dysfunction	Synaptic proteins are ubiquitinated for turnover, contributing to synaptic plasticity and remodeling.	1. Aβ plaques and Tau tangles interfere with synaptic plasticity 2. Leading to synapse loss 3. Impaired cognition, and neuronal degeneration.	1. α-Synuclein aggregation impairs synaptic vesicle recycling and disrupts synaptic transmission 2. Contributing to motor deficits.	1. mHTT disrupts synaptic transmission by altering glutamate receptors 2. Reducing synaptic plasticity, and impairing cognitive function.	Misfolded TDP-43 and SOD1 impair synaptic transmission in ALS, leading to motor neuron degeneration and muscle atrophy.	TDP-43 or FUS aggregates disrupt synaptic function, leading to the loss of synaptic plasticity and cognitive decline in FTD patients.	1. Gong et al. [33]. 2. Lim and Tan [54]. 3. Li and Li [53]. 4. Philips and Rothstein [72]. 5. Neumann et al. [67].
UPS components implicated	1. E3 ligases (e.g., Parkin, Mdm2) 2. Regulate synaptic protein ubiquitination 3. Determining degradation or recycling.	1. E3 ligase UCHL1 and other UPS components are impaired in AD 2. Affecting the clearance of toxic proteins, leading to protein aggregation.	1. Parkin, an E3 ubiquitin ligase, is mutated in familial PD 2. Leading to impaired ubiquitination and protein degradation.	1. Mutant Huntingtin interferes with the function of E3 ligases 2. Leading to reduced ubiquitination and degradation of synaptic proteins.	ALS mutations in genes like UBQLN2 and VCP interfere with the ubiquitination process, causing defective protein clearance.	Mutations in genes like GRN (progranulin) and UBQLN2 impair protein ubiquitination and degradation, leading to FTD pathology.	1. Keller et al. [49]. 2. Moore et al. [62]. 3. Glickman and Ciechanover [31]. 4. Deng et al. [22]. 5. Baker et al. [3].

(continued)

Table 14.1 (continued)

Aspect	Synapse remodeling	Alzheimer's disease (AD)	Parkinson's disease (PD)	Huntington's disease (HD)	Amyotrophic lateral sclerosis (ALS)	Frontotemporal dementia (FTD)	References
Disease mechanism	Proteostasis in synapses is crucial for plasticity and synaptic health.	1. Accumulation of misfolded proteins (Aβ, Tau) 2. Triggering synaptic dysfunction 3. Neuronal death, and memory loss.	1. α-Synuclein accumulation 2. Leading to mitochondrial dysfunction, oxidative stress, and synaptic degeneration.	1. mHTT aggregates block proteasome function, preventing the degradation of damaged proteins, resulting in synaptic loss and neurodegeneration.	UPS dysfunction allows the accumulation of ubiquitinated protein inclusions, leading to synaptic loss and progressive neurodegeneration.	Dysfunctional Dysfunctional UPS leads to the accumulation of toxic protein clusters like TDP-43 and ubiquitinated inclusions, causing neuronal death.	1. Tan et al. [90]. 2. Chung et al. [13]. 3. Ortega et al. [69]. 4. Blokhuis et al. [8]. 5. Rabinovich-Toidman et al. [74].
Therapeutic Potential	Modulation of UPS activity can help maintain synaptic plasticity and prevent neurodegeneration.	1. Proteasome activation or 2. Inhibitors targeting deubiquitinases (e.g., UCHL1) are explored as potential therapies for AD to prevent toxic protein accumulation.	1. Targeting Parkin and proteasome activity with small molecules may restore proper protein degradation and synaptic function in PD.	1. Targeting UPS components, such as enhancing proteasome activity or promoting clearance of mHTT aggregates, is explored as a therapy.	Enhancing UPS function, reducing misfolded protein aggregates, or modulating proteasome activity are potential therapeutic approaches in ALS.	Enhancing proteasome function or reducing protein aggregates (e.g., TDP-43) is being explored as potential therapies in FTD.	1. Grune et al. [34]. 2. McNaught and Jenner [57]. 3. Forman et al. [26].

Proper synaptic function relies on the ubiquitin-proteasome system (UPS), which eliminates misfolded proteins and governs protein degradation. In neurodegenerative disorders, including AD, PD, HD, ALS, and FTD, UPS dysfunction triggers protein accumulation, synaptic impairment, and neuronal loss. In ALS, ubiquitinated inclusions, typically involving proteins like TDP-43, SOD1, and FUS, accumulate in motor neurons. These aggregates are not efficiently cleared due to impaired UPS function, which results in progressive synaptic loss and motor neuron degeneration. Decreased proteasome activity in ALS further exacerbates this issue, preventing the degradation of misfolded proteins, contributing to neuronal toxicity, muscle atrophy, and motor deficits. Mutations in genes such as UBQLN2 and VCP disrupt the ubiquitination process, leading to defective protein clearance and the formation of toxic aggregates. Across neurodegenerative diseases, similar mechanisms involving key proteins and UPS components contribute to disease pathology. In AD, the accumulation of Tau and Amyloid-beta (Aβ), in PD the aggregation of α-synuclein, and in HD the build-up of mutant Huntingtin (mHTT) similarly disrupt proteostasis and synaptic function. Targeting UPS components and enhancing proteasome activity hold therapeutic potential across these disorders, with efforts underway to modulate the clearance of toxic aggregates. In ALS and FTD, therapies aimed at boosting proteasome function or promoting the degradation of misfolded proteins may help mitigate disease progression by preserving synaptic health and preventing neurodegeneration.

14.8 Conclusion

Neurodegenerative diseases, including Alzheimer's, Parkinson's, Huntington's, Amyotrophic Lateral Sclerosis, and Frontotemporal Dementia, share a common underlying mechanism: abnormal protein accumulation and processing leading to neuronal dysfunction and cell death. This protein misfolding disorder results from disrupted protein homeostasis, highlighting the need for novel therapeutic strategies. Recent research focuses on neuronal translation and the ubiquitin-proteasome system (UPS), critical for regulating synaptic plasticity, neuronal function, and protein degradation. However, in neurodegenerative diseases, the UPS becomes overwhelmed, leading to protein accumulation and cellular dysfunction. To address this, therapeutic strategies aim to regulate disrupted key factors, alleviate translational repression, and reduce protein overload effects. Modulating UPS activity offers potential for restoring synaptic function and slowing disease progression. Continued research seeks to identify effective treatments by understanding the shared mechanisms underlying neurodegenerative diseases. By targeting these commonalities, researchers can develop innovative treatments to combat these devastating disorders. The ubiquitin-proteasome system's role in protein degradation and synaptic function makes it a promising therapeutic target, holding hope for improving the lives of those affected by neurodegenerative diseases.

References

1. Abe K, Ito H, Sobue G. Edaravone: a new drug approved for ALS. Lancet Neurol. 2017;16(1):15–6. https://doi.org/10.1016/S1474-4422(16)30341-3.
2. Bae JR, Kim SH. Synapses in neurodegenerative diseases. Banchōng Neurosci. 2018;122(5):1519–31. https://doi.org/10.1016/j.neuro.2018.01.010.
3. Baker M, Mackenzie I, Pickering-Brown S, et al. Mutations in progranulin cause tau-negative frontotemporal dementia linked to chromosome 17. Nature. 2006;442:916–919. https://doi.org/10.1038/nature05016.
4. Bates GP. Hunting aggregation and toxicity in Huntington's disease. Lancet. 2003;361(9369):1642–4. https://doi.org/10.1016/S0140-6736(03)13306-0.
5. Bedford L, Paine S, Sheppard PW, Mayer RJ, Roelofs J. Assembly, structure, and function of the 26S proteasome. Trends Cell Biol. 2020;30(12):991–1003. https://doi.org/10.1016/j.tcb.2020.09.001.
6. Behi T, Kumar S, Althafar ZM, Sehgal A, Singh S, Sharma N, Bhatia S. Exploring the role of ubiquitin–proteasome system in Parkinson's disease. Mol Neurobiol. 2022;59(6):4257–73. https://doi.org/10.1007/s12035-022-02851-1.
7. Betarbet R, Sherer TB, MacKenzie G, Garcia-Osuna M, Panov AV, Greenamyre JT. Chronic systemic pesticide exposure reproduces features of Parkinson's disease. Nat Neurosci. 2005;3(12):1301–6. https://doi.org/10.1038/nn1103.
8. Blokhuis AM, Groen EJ, Koppers M, van den Berg LH, Pasterkamp RJ. Protein aggregation in amyotrophic lateral sclerosis. Acta Neuropathol. 2013;125(6):777–94. https://doi.org/10.1007/s00401-013-1125-6.
9. Brown RH, Al-Chalabi A. Amyotrophic lateral sclerosis. N Engl J Med. 2017;377(2):162–72. https://doi.org/10.1056/NEJMra1603471.
10. Chau V, Tobias JW, Bachmair A, Marriott D, Ecker DJ, Gonda DK, Varshavsky A. A multiubiquitin chain is confined to specific lysine in a targeted short-lived protein. Science. 1989;243(4898):1576–83. https://doi.org/10.1126/science.2538923.
11. Chiò A, Logroscino G, Traynor BJ, Collins J, Simeone JC, Goldstein LA, White LA. Global epidemiology of amyotrophic lateral sclerosis: a systematic review of the published literature. Neuroepidemiology. 2011;36(2):123–39. https://doi.org/10.1159/000323927.
12. Chung KKK, Dawson VL. The role of the ubiquitin–proteasomal pathway in Parkinson's disease and other neurodegenerative disorders. Trends Neurosci. 2004;27(11):597–602. https://doi.org/10.1016/j.tins.2004.07.008.
13. Chung KKK, Dawson VL, Dawson TM. The role of the ubiquitin-proteasomal pathway in Parkinson's disease and other neurodegenerative disorders. Trends Neurosci. 2001;24(11):S7–S14. https://doi.org/10.1016/s0166-2236(00)01998-6.
14. Ciechanover A, Kwon YT. Proteolysis: ubiquitin-proteasome system and lysosome. Cold Spring Harb Perspect Biol. 2015;7(11):a016337. https://doi.org/10.1101/cshperspect.a016337.
15. Clague MJ, Urbé S, Komander D. Breaking the chains: deubiquitylating enzyme specificity begets function. Nat Rev Mol Cell Biol. 2019;20:338–52. https://doi.org/10.1038/s41580-019-0099-.
16. Colledge M, Snyder EM, Crozier RA, Soderling JA, Jin Y, Langeberg LK, Lu H, Bear MF, Scott JD. Ubiquitin-mediated proteasomal degradation of AMPA receptor subunits. Cell. 2003;114(4):443–55. https://doi.org/10.1016/S0092-8674(03)00697-9.
17. Cookson MR. The role of leucine-rich repeat kinase 2 (LRRK2) in Parkinson's disease. Nat Rev Neurosci. 2010;11(12):791–7. https://doi.org/10.1038/nrn2935.
18. Dantuma NP, Bott LC. The ubiquitin-proteasome system in neurodegenerative diseases: precipitating factor, yet part of the solution. Front Mol Neurosci. 2014;7:70. https://doi.org/10.3389/fnmol.2014.00070.
19. Dayalu P, Albin RL. Huntington disease: pathogenesis and treatment. Neurol Clin. 2015;33(1):101–14. https://doi.org/10.1016/j.ncl.2014.09.003.

20. DeJesus-Hernandez M, Mackenzie IRA, Boeve BF, Boxer AL, Baker M, Rutherford NJ, Nicholson AM, Finch NA, Flynn H, Adamson J, Kouri N, Wojtas A, Sengdy P, Hsiung GYR, Karydas A, Seeley WW, Josephs KA, Coppola G, Geschwind DH, Rademakers R. Expanded GGGGCC hexanucleotide repeat in the noncoding region of C9ORF72 causes chromosome 9p-linked FTD and ALS. Neuron. 2011;72(2):245–56. https://doi.org/10.1016/j.neuron.2011.09.011.
21. Deng L, Wang C, Spencer E, Yang L, Braun A, You J, Slaughter C, Pickart C, Chen ZJ. Activation of the IκB kinase complex by TRAF6 requires a dimeric ubiquitin-conjugating enzyme complex and a unique polyubiquitin chain. Cell. 2000;103(2):351–61. https://doi.org/10.1016/S0092-8674(00)00126-4.
22. Deng HX, Chen W, Hong ST, et al. Mutations in UBQLN2 cause dominant X-linked juvenile and adult-onset ALS and ALS/dementia. Nature. 2011;477:211–5. https://doi.org/10.1038/nature10353.
23. Dias V, Junn E, Mouradian MM. The role of oxidative stress in Parkinson's disease. J Parkinsons Dis. 2013;3(4):461–91. https://doi.org/10.3233/JPD-130230.
24. Dib-Hajj SD, Cummins TR, Black JA, Waxman SG. From genes to pain: Nav1.7 and human pain disorders. Trends Neurosci. 2007;30(11):555–63. https://doi.org/10.1016/j.tins.2007.08.004.
25. Ding Q, Shen HM. The role of ubiquitin-proteasome pathway in synaptic plasticity and memory. Neuroscience. 2007;148(3):515–29. https://doi.org/10.1016/j.neuroscience.2007.06.023.
26. Forman M, Trojanowski J, Lee VY. Neurodegenerative diseases: a decade of discoveries paves the way for therapeutic breakthroughs. Nat Med. 2004;10:1055–63. https://doi.org/10.1038/nm1113.
27. Forsberg K, Andersen PM, Marklund SL, Brännström T. Glial nuclear aggregates of superoxide dismutase-1 are regularly present in patients with amyotrophic lateral sclerosis. Acta Neuropathol. 2010;121(5):623–34. https://doi.org/10.1007/s00401-010-0780-5.
28. Garofalo M, Pandini C, Bordoni M, Pansarasa O, Rey F, Costa A, Minafra B, Diamanti L, Zucca S, Carelli S, Cereda C, Gagliardi S. Alzheimer's, Parkinson's disease and amyotrophic lateral sclerosis gene expression patterns divergence reveals different grades of RNA metabolism involvement. Int J Mol Sci. 2020;21(24):9500. https://doi.org/10.3390/ijms21249500.
29. Gentier RJ, van Leeuwen FW. Misframed ubiquitin and impaired protein quality control: an early event in Alzheimer's disease. Front Mol Neurosci. 2015;8:47. https://doi.org/10.3389/fnmol.2015.00047.
30. Ghasemi M, Brown RH. Genetics of amyotrophic lateral sclerosis. Cold Spring Harb Perspect Med. 2018;8(5):a024125. https://doi.org/10.1101/cshperspect.a024125.
31. Glickman MH, Ciechanover A. The ubiquitin-proteasome proteolytic pathway: destruction for the sake of construction. Physiol Rev. 2002;82(2):373–428. https://doi.org/10.1152/physrev.00027.2001.
32. Goldberg MS, Fleming SM, Palacino JJ, Cepeda C, Lam HA, Bhatnagar A, Meloni EG, Wu N, Ackerson LC, Klapstein GJ, Gajendiran M, Roth BL, Chesselet MF, Maidment NT, Shen J. The role of parkin in ubiquitin-proteasomal degradation of damaged mitochondria. J Biol Chem. 2003;278(33):31177–87. https://doi.org/10.1074/jbc.M300451200.
33. Gong C, Liu F, Grundke-Iqbal I, Iqbal K. Post-translational modifications of tau protein in Alzheimer's disease. J Neural Transm. 2004;112(6):813–38. https://doi.org/10.1007/s00702-004-0221-0.
34. Grune T, Jung T, Merker K, Davies KJA. Decreased proteolysis caused by protein aggregates, inclusion bodies, plaques, lipofuscin, ceroid, and "aggresomes" during oxidative stress, aging, and disease. Int J Biochem Cell Biol. 2004;36(12):2519–30. https://doi.org/10.1016/j.biocel.2004.04.020.
35. Hegde AN. The ubiquitin-proteasome pathway and synaptic plasticity. Learn Mem. 2010;17(7):314–27. https://doi.org/10.1101/lm.1812910.
36. Hegde AN, DiAntonio A. Ubiquitin and the synapse. Nat Rev Neurosci. 2002;3(11):854–61. https://doi.org/10.1038/nrn959.

37. Hershko A, Ciechanover A. The ubiquitin system. Annu Rev Biochem. 1998;67:425–79. https://doi.org/10.1146/annurev.biochem.67.1.425.
38. Hipp MS, Kasturi P, Hartl FU. The proteostasis network and its decline in ageing. Nat Rev Mol Cell Biol. 2012;13(9):668–79. https://doi.org/10.1038/nrm3454.
39. Hoflinger P, Rodríguez J. The ubiquitin proteasome system in glia and its role in neurodegenerative diseases. Front Mol Neurosci. 2014;7:Article 73. https://doi.org/10.3389/fnmol.2014.00073.
40. Huang N, Zhang L, Li X, Wu H, Zhang X, Li L, Zhang L. Tanshinone IIA-incubated mesenchymal stem cells inhibit lipopolysaccharide-induced inflammation of N9 cells through TREM2 signaling pathway. Stem Cells Int. 2022;2022:Article 9977610. https://doi.org/10.1155/2022/9977610.
41. Husnjak K, Dikic I. Ubiquitin-binding proteins: decoders of ubiquitin-mediated cellular functions. Annu Rev Biochem. 2012;81:291–322. https://doi.org/10.1146/annurev-biochem-051810-094654.
42. Hussain R, Zubair H, Pursell S, Shahab M. Neurodegenerative diseases: regenerative mechanisms and novel therapeutic approaches. Brain Sci. 2018;8(9):177. https://doi.org/10.3390/brainsci8090177.
43. Israelson A, Yamanaka K, Cleveland DW. Mutant superoxide dismutase 1 and mitochondria in ALS pathogenesis: a destructive alliance. Free Radic Biol Med. 2010;48(6):707–14. https://doi.org/10.1016/j.freeradbiomed.2009.12.030.
44. Jana NR, Zemskov EA, Wang G, Nukina N. Altered proteasomal function due to the expression of polyglutamine-expanded truncated N-terminal huntingtin induces apoptosis by caspase activation through mitochondrial cytochrome c release. Hum Mol Genet. 2001;10(10):1049–59. https://doi.org/10.1093/hmg/10.10.1049.
45. Kabashi E, Valdmanis PN, Dion P, Spiegelman D, McConkey BJ, Velde CV, Rouleau GA. TARDBP mutations in individuals with sporadic and familial amyotrophic lateral sclerosis. Nat Genet. 2011;40(5):572–4. https://doi.org/10.1038/ng.132.
46. Kalia LV, Lang AE. Parkinson's disease. Lancet. 2015;386(9996):896–912. https://doi.org/10.1016/S0140-6736(14)61393-3.
47. Kauffman AS, Clifton DK, Steiner RA. Emerging ideas about kisspeptin– GPR54 signaling in the neuroendocrine regulation of reproduction. Trends Neurosci. 2007;30(10):504–11. https://doi.org/10.1016/j.tins.2007.08.001.
48. Kaur SJ, McKeown SR, Rashid S. Mutant SOD1 mediated pathogenesis of amyotrophic lateral sclerosis. Gene Ther. 2021;28(1–2):1–21. https://doi.org/10.1038/s41434-020-0142-z.
49. Keller JN, Hanni KB, Markesbery WR. Impaired proteasome function in Alzheimer's disease. J Neurochem. 2000;75(1):436–9. https://doi.org/10.1046/j.1471-4159.2000.0750436.x.
50. Kocahan S, Akillioglu K. Effects of NMDA receptor blockade during the early development period on the retest performance of adult Wistar rats in the elevated plus maze. Neurochem Res. 2013;38(8):1496–500. https://doi.org/10.1007/s11064-013-1070-7.
51. Komander Renton D, Rape M. The ubiquitin code. Annu Rev Biochem. 2012;81:203–29. https://doi.org/10.1146/annurev-biochem-060310-170328.
52. Leigh PN, Anderton BH, Shaw CE. Ubiquitin-proteasome dysfunction in ALS and frontotemporal lobar degeneration: a new therapeutic target. Trends Neurosci. 2020;43(9):647–59. https://doi.org/10.1016/j.tins.2020.06.005.
53. Li S-H, Li XJ. Huntingtin-protein interactions and the pathogenesis of Huntington's disease. Trends Genet. 2004;20(3):146–54. https://doi.org/10.1016/j.tig.2004.01.008.
54. Lim K-L, Tan JMM. Role of the ubiquitin-proteasome system in Parkinson's disease. BMC Biochem. 2007;8:S13. https://doi.org/10.1186/1471-2091-8-S1-S13.
55. Lim KL, Zhang CW. Molecular events underlying Parkinson's disease – an interwoven tapestry. Front Neurol. 2013;4:33. https://doi.org/10.3389/fneur.2013.00033.
56. Longobardi A, Catania M, Geviti A, Salvi E, Vecchi ER, Bellini S, Saraceno C, Nicsanu R, Squitti R, Binetti G. Autophagy markers are altered in Alzheimer's disease, dementia

with Lewy bodies and frontotemporal dementia. Int J Mol Sci. 2024;25(2):1125. https://doi.org/10.3390/ijms25021125.
57. McNaught KS, Jenner P. Proteasomal function is impaired in substantia nigra in Parkinson's disease. Neurosci Lett. 2001;297(3):191–4. https://doi.org/10.1016/S0304-3940(00)01732-9.
58. McNaught KS, Olanow CW. Protein aggregation in the pathogenesis of familial and sporadic Parkinson's disease. Neurobiol Aging. 2006;27(4):530–45. https://doi.org/10.1016/j.neurobiolaging.2005.03.016.
59. McNaught KSP, Belizaire R, Isacson O, Jenner P, Olanow CW. Altered proteasomal function in sporadic Parkinson's disease. Exp Neurol. 2001;179(1):38–46. https://doi.org/10.1006/exnr.2001.7864.
60. Miller RG, Mitchell JD, Moore DH. Riluzole for amyotrophic lateral sclerosis (ALS)/motor neuron disease (MND). Cochrane Database Syst Rev. 2012;3(3):CD001447. https://doi.org/10.1002/14651858.CD001447.pub3.
61. Moloney EB, de Winter F, Verhaagen J. ALS as a distal axonopathy: molecular mechanisms affecting neuromuscular junction stability in the presymptomatic stages of the disease. Front Neurosci. 2014;8(252):1–12. https://doi.org/10.3389/fnins.2014.00252.
62. Moore DJ, Dawson VL, Dawson TM. Role for the Ubiquitin-Proteasome system in Parkinson's disease and other neurodegenerative brain amyloidoses. Neuromolecular Med. 2003;4(1–2):95–108. https://doi.org/10.1385/nmm:4:1-2:95.
63. Münch C, O'Brien J, Bertolotti A. Aberrant disulfide bonding in proteins and neurodegeneration: the example of ALS. Cell Death Differ. 2011;18(9):1391–400. https://doi.org/10.1038/cdd.2011.43.
64. Nalepa G, Rolfe M, Harper JW. Drug discovery in the ubiquitin-proteasome system. Nat Rev Drug Discov. 2006;5(7):596–613. https://doi.org/10.1038/nrd2056.
65. Neumann M, Mackenzie IRA, Cairns NJ, Boyer PJ, Markesbery WR, Smith CD, Trojanowski JQ. TDP-43 in the ubiquitin pathology of frontotemporal dementia with ubiquitin-positive inclusions and amyotrophic lateral sclerosis. J Neuropathol Exp Neurol. 2006;65(7):635–44. https://doi.org/10.1097/01.jnen.0000228130.51254.7f.
66. Neumann M, Sampathu DM, Kwong LK, Truax AC, Micsenyi MC, Chou TT, Bruce J, Schuck T, Grossman M, Clark CM, McCluskey LF, Miller BL, Masliah E, Mackenzie IRA, Feldman H, Feiden W, Kretzschmar HA, Trojanowski JQ, Lee VMY. Ubiquitinated TDP-43 in frontotemporal lobar degeneration and amyotrophic lateral sclerosis. Science. 2006;314(5796):130–3. https://doi.org/10.1126/science.1134108.
67. Neumann M, Rademakers R, Roeber S, Baker M, Kretzschmar HA, Mackenzie IRA. A new subtype of frontotemporal lobar degeneration with FUS pathology. Brain. 2009;132(11):2922–31. https://doi.org/10.1093/brain/awp214.
68. Oddo S, Caccamo A, Shepherd JD, Murphy MP, Golde TE, Kayed R, Metherate R, Mattson MP, Akbari Y, LaFerla FM. Amyloid deposition precedes tangle formation in a triple-transgenic model of Alzheimer's disease. Neurobiol Aging. 2006;27(9):1199–208. https://doi.org/10.1016/j.neurobiolaging.2005.05.022.
69. Ortega Z, Díaz-Hernández M, Lucas JJ. Is the ubiquitin-proteasome system impaired in Huntington's disease? Cell. Mol. Life Sci. 2007;64:2245–57. https://doi.org/10.1007/s00018-007-7222-8.
70. Pehar M, Beeson G, Beeson CC, Johnson JA, Vargas MR. Mitochondria-targeted catalase reverts the neurotoxicity of hSOD1G93A astrocytes without extending the survival of ALS-linked mutant hSOD1 mice. PLoS One. 2014;9(6):e100783. https://doi.org/10.1371/journal.pone.0100783.
71. Peters OM, Ghasemi M, Brown RH. Emerging mechanisms of molecular pathology in ALS. J Clin Invest. 2015;125(5):1767–79. https://doi.org/10.1172/JCI71601.
72. Philips T, Rothstein JD. Glial cells in amyotrophic lateral sclerosis. Exp Neurol. 2014;262:111–20. https://doi.org/10.1016/j.expneurol.2014.01.011.
73. Pickart CM. Mechanisms underlying ubiquitination. Annu Rev Biochem. 2001;70:503–33. https://doi.org/10.1146/annurev.biochem.70.1.503.

74. Rabinovich-Toidman P, Rabinovich-Nikitin I, Solomon B, Priel E. Ubiquitin-proteasome system regulation in amyotrophic lateral sclerosis: implications for therapeutic intervention. Neurobiol Dis. 2015;77:135–44. https://doi.org/10.1016/j.nbd.2015.02.015.
75. Redler RL, Dokholyan NV. The complex molecular biology of amyotrophic lateral sclerosis (ALS). Prog Mol Biol Transl Sci. 2012;107(1):215–62. https://doi.org/10.1016/B978-0-12-385883-2.00002-1.
76. Renton AE, Chio A, Traynor BJ. State of play in amyotrophic lateral sclerosis genetics. Nat Neurosci. 2014;17(1):17–23. https://doi.org/10.1038/nn.3584.
77. Richard D, Arjmand P, Berksu Ozkara B, Radmard M, Gad M, Sheikhy A, Wintermark M, Yedavalli V, Sair HI, Luna LP. Diagnostic performance of ASL-MRI and FDG-PET in frontotemporal dementia: a systematic review and meta-analysis. Am J Neuroradiol. 2024;46(2):341–8. https://doi.org/10.3174/ajnr.A8440.
78. Roos RA. Huntington's disease: a clinical review. Orphanet J Rare Dis. 2010;5(1):40. https://doi.org/10.1186/1750-1172-5-40.
79. Rosen DR, Siddique T, Patterson D, Figlewicz DA, Sapp P, Hentati A, Brown RH. Mutations in Cu/Zn superoxide dismutase gene are associated with familial amyotrophic lateral sclerosis. Nature. 1993;362(6415):59–62. https://doi.org/10.1038/362059a0.
80. Rosenblum LT, Kosciuk MC, Lewis JS, Pytel P. Antisense oligonucleotides: therapy for SOD1-linked amyotrophic lateral sclerosis. Annu Rev Med. 2017;68(1):217–29. https://doi.org/10.1146/annurev-med-050214-013459.
81. Ross CA, Poirier MA. Protein aggregation and neurodegenerative disease. Nat Med. 2004;10(7):S10–7. https://doi.org/10.1038/nm1066.
82. Ross CA, Tabrizi SJ. Huntington's disease: from molecular pathogenesis to clinical treatment. Lancet Neurol. 2011;10(1):83–98. https://doi.org/10.1016/S1474-4422(10)70245-3.
83. Schoen M. Synaptic dysfunction in ALS: unraveling the connection to the UPS. Neurodegener Dis. 2020;19(1–2):55–64. https://doi.org/10.1159/000507000.
84. Schwartz AL, Ciechanover A. Targeting proteins for destruction by the ubiquitin system: implications for human pathobiology. Annu Rev Pharmacol Toxicol. 2009;49(1):73–96. https://doi.org/10.1146/annurev.pharmtox.051208.165340.
85. Scotter EL, Chen HJ, Shaw CE. TDP-43 proteinopathy and ALS: insights into disease mechanisms and therapeutic targets. Neurotherapeutics. 2015;12(2):352–63. https://doi.org/10.1007/s13311-014-0310-1.
86. Serrano-Pozo A, Frosch MP, Masliah E, Hyman BT. Neuropathological alterations in Alzheimer disease. Cold Spring Harb Perspect Med. 2011;1:a006189. https://doi.org/10.1101/cshperspect.a006189.
87. Swatek KN, Komander D. Ubiquitin modifications. Cell Res. 2016;26(4):399–422. https://doi.org/10.1038/cr.2016.39.
88. Tabrizi SJ, Ghosh R, Leavitt BR. Huntingtin lowering strategies for disease modification in Huntington's disease. Neuron. 2019;101(5):801–19. https://doi.org/10.1016/j.neuron.2019.02.033.
89. Tai HC, Schuman EM. Ubiquitin, the proteasome and synaptic plasticity. Nat Rev Neurosci. 2008;9(11):826–38. https://doi.org/10.1038/nrn2499.
90. Tan JM, Wong ES, Kirkpatrick DS, Pletnikova O, Ko HS, Tay S, Ho MW, Troncoso J, Gygi SP, Lee MK, Dawson VL, Dawson TM, Lim K. Lysine 63-linked ubiquitination promotes the formation and autophagic clearance of protein inclusions associated with neurodegenerative diseases. Hum Mol Genet. 2007;17(3):431–9. https://doi.org/10.1093/hmg/ddm320.
91. Tashiro E, Maruki H, Minato Y, Imoto M. Ubiquitin-proteasome system as a therapeutic target for diseases related to cell cycle and apoptosis. Curr Med Chem. 2012;19(3):402–11. https://doi.org/10.2174/092986712803414188.
92. Taylor JP, Brown RH, Cleveland DW. Decoding ALS: from genes to mechanism. Nature. 2016;539(7628):197–206. https://doi.org/10.1038/nature20413.
93. Todi SV, Paulson HL. Balancing act: deubiquitinating enzymes in the nervous system. Trends Neurosci. 2011;34(8):370–82. https://doi.org/10.1016/j.tins.2011.04.001.

94. Turner MR, Talbot K. SOD1 gene therapy for ALS: super, or only so-so? Brain. 2008;131(6):1535–7. https://doi.org/10.1093/brain/awn091.
95. Upadhya SC, Hegde AN. Role of the ubiquitin-proteasome system in Alzheimer's disease. BMC Biochem. 2007;8(1):1–12. https://doi.org/10.1186/1471-2091-8-S1-S12.
96. Urushitani M, Takahashi R. ALS pathogenesis: protein quality control, stress granules, and beyond. Trends Neurosci. 2010;34(1):24–33. https://doi.org/10.1016/j.tins.2010.09.003.
97. Van Es MA, Hardiman O, Chio A, Al-Chalabi A, Pasterkamp RJ, Veldink JH, van den Berg LH. Amyotrophic lateral sclerosis. Lancet. 2017;390(10107):2084–98. https://doi.org/10.1016/S0140-6736(17)31287-4.
98. Van Tijn P, Kamphuis W, van de Berg WD, Hol EM, Van Leeuwen FW. Ubiquitin B and its mutant UBB+1 in Alzheimer's and other neurodegenerative diseases. Front Neurosci. 2011;5:20. https://doi.org/10.3389/fnins.2011.00020.
99. Walker FO. Huntington's disease. Lancet. 2007;369(9557):218–28. https://doi.org/10.1016/S0140-6736(07)60111-1.
100. Wei Y, Zhong S, Yang H, Wang X, Lv B, Bian Y, Pei Y, Xu C, Zhao Q, Wu Y, Luo D, Wang F, Sun H, Chen Y. Current therapy in amyotrophic lateral sclerosis (ALS): a review on past and future therapeutic strategies. Eur J Med Chem. 2024;272:116496., ISSN 0223-5234. https://doi.org/10.1016/j.ejmech.2024.116496.
101. Wilkinson KD. Ubiquitination and deubiquitination: targeting of proteins for degradation by the proteasome. Semin Cell Dev Biol. 2000;11(3):141–8. https://doi.org/10.1006/scdb.2000.0167.
102. Zheng Z, Wang H. The ubiquitin-proteasome system in viral infection: novel targets for drug development. Drug Resist Updat. 2003;6(1):1–5. https://doi.org/10.1016/S1368-7646(03)00002-5.

Endoplasmic Reticulum Stress and Protein Misfolding in Neurodegenerative Diseases

15

Suman Samaddar, Moqbel Ali Moqbel Redhwan, and Mohan Muttanahally Eraiah

15.1 Introduction

Degenerative neurologic diseases, including Parkinson's disease (PD), Alzheimer's disease (AD), amyotrophic lateral sclerosis (ALS), Huntington's disease (HD), and prion diseases, are characterized by a shared pathological hallmark: the buildup and aggregation of misfolded proteins [1–3]. These misfolded proteins initiate a range of cellular responses, notably the endoplasmic reticulum (ER) stress response, which protects the cell from the toxic accumulation of these proteins [4–7]. However, when misfolded protein accumulation surpasses the capacity of cellular quality control systems, these protective mechanisms fail, resulting in organelle and cell dysfunction [8–10].

The ubiquitin-proteasome system (UPS) is essential for the degradation of cellular proteins and is intimately linked with ER-associated degradation (ERAD) that targets misfolded proteins for dislocation from ER to the cytosol for proteasomal degradation [3, 9, 10]. Disruptions in ER protein retro translocation or proteasome activity can result in a buildup of misfolded proteins in the ER, triggering an ER stress response. Misfolded proteins activate this response regardless of location within the ER, cytosol, or nucleus, and propagate harmful signals across different cellular compartments.

S. Samaddar (✉)
Research Institute, BGS Global Institute of Medical Sciences,
BGS Health & Education City, Bengaluru, Karnataka, India
e-mail: sumansamaddar@bgsgims.edu.in

M. A. M. Redhwan
Department of Pharmacology, KLE College of Pharmacy, Bengaluru, Karnataka, India

KLE Academy of Higher Education and Research, Belagavi, Karnataka, India

M. M. Eraiah
Department of Medicine, BGS Global Institute of Medical Sciences,
Bengaluru, Karnataka, India
e-mail: principal@bgsgims.edu

N. Puranik (ed.), *Proteostasis*, https://doi.org/10.1007/978-981-96-6202-9_15

Misfolded proteins can form plaques inside cells or in extracellular, where they interact with various cellular components, leading to toxicity [1]. This toxicity can arise through several mechanisms: inhibition of synaptic function, loss of synapses disrupting neuronal communication, isolation of essential cellular chaperones and transcription factors, disruption of signal transduction pathways and calcium homeostasis, generation of oxidative stress through free radicals, impairment of the UPS, and activation of cell-death pathways, ultimately resulting in apoptosis.

This chapter will explore recent advancements in understanding ER stress and unfolded protein response (UPR) and their implications for neuronal diseases and apoptosis associated with protein misfolding. By elucidating these complex regulatory mechanisms, we aim to provide insights into potential therapeutic strategies for these debilitating disorders.

15.2 ER Function

ER is a critical organelle in eukaryotic cells, known for its extensive roles in the synthesis of steroids and lipids, calcium homeostasis, metabolism of carbohydrates, and protein synthesis. Its unique structure includes the nuclear envelope (NE), peripheral ER cisternae, and an intricate network of tubules, facilitating its multifunctional nature. ER forms multiple contact points with the plasma membrane (PM), Golgi apparatus, mitochondria, peroxisomes, and endosomes. These connections are crucial for cellular homeostasis and signaling [11, 12].

15.3 ER and Plasma Membrane

The ER is the primary calcium reservoir in cells, establishing an intricate network of contacts with the PM to regulate calcium exchange. These contact sites are pivotal for store-operated calcium entry (SOCE). When calcium levels within the ER fall, stromal-interacting molecule (STIM) proteins undergo a structural transformation, repositioning the ER's tubular structures to the ER–PM junctions. This shift enables the activation of Orai, the protein that forms the pore of the calcium release-activated calcium (CRAC) channel, allowing for calcium influx into the cell [13].

Moreover, ER–PM contact sites play a vital role in the metabolism of phosphatidylinositol. The oxysterol-binding homology (Osh) protein family, along with vesicle-associated membrane protein-associated protein (VAP), modulates the lipid signaling molecule phosphatidylinositol 4-phosphate (PI4P) at these junctions. In yeast, the absence of Osh proteins leads to a marked increase in PI4P levels, while the introduction of recombinant Osh3 diminishes the levels of proteins that activate Sac1 phosphatase, highlighting the regulatory function of Osh proteins in controlling PI4P levels at ER–PM contact sites [14].

15.4 ER and Mitochondria

The interaction between the ER and mitochondria, often referred to as mitochondria-associated ER membranes (MAMs), is crucial for the proper functioning and survival of cells. These specialized contact sites enable numerous essential processes such as lipid exchange, autophagosome formation, mitochondrial division, calcium homeostasis, and programmed cell death. Calcium influx into mitochondria occurs at these contact sites via ER channels that interact with mitochondrial membranes, requiring IP3 receptor (IP3R) and voltage-dependent anion channel 1 (VDAC1) interactions. These calcium dynamics influence apoptosis, mitochondrial division, motility, and the activity of mitochondrial calcium-binding proteins [15].

MAMs are also essential for lipid biosynthesis. Phosphatidylserine (PS) synthesis occurs on the ER membrane, followed by its transformation into phosphatidylethanolamine (PE) on mitochondrial proteins and then to phosphatidylcholine (PC) using ER enzymes. This lipid movement between ER and mitochondrial membranes is necessary for each conversion step. Additionally, ER tubules have been found to play a role in positioning the mitochondrial fission machinery, aiding in mitochondrial biogenesis, particularly in yeast and mammalian cells. Molecular bridges, such as the ER-mitochondria encounter structure (ERMES) in yeast and mitofusin 2 (Mfn2) in mammals, mediate these interactions and likely adjust based on the functional demands of ER-mitochondria communications [16, 17].

Disruptions in ER-mitochondria signaling can lead to mitochondrial damage, calcium dyshomeostasis, ER stress, lipid metabolism defects, autophagy issues, decreased respiratory chain efficiency, and oxidative phosphorylation impairments. Such disruptions are often observed in neurodegenerative diseases, highlighting the critical role of ER–mitochondria interactions in maintaining cellular and neurological health [17].

15.5 ER Stress and UPR

The ER is essential for protein biosynthesis, post-translational modifications, and proper folding of newly synthesized proteins. However, the accumulation of misfolded or unfolded proteins can compromise its structure and function, leading to ER stress. This stress triggers a protective mechanism known as UPR, which aims to restore the balance within the ER [18].

The UPR is orchestrated by three main sensor proteins: PKR-like ER kinase (PERK), inositol-requiring enzyme 1a (IRE1α), and activating transcription factor 6 (ATF6). In normal conditions, these proteins are bound to the molecular chaperone BiP/GRP78, keeping them in an inactive state. During ER stress, BiP detaches from the sensors, leading to their activation. PERK activation causes phosphorylation of the eukaryotic translation initiation factor 2α (eIF2α), reducing general protein synthesis but promoting the translation of ATF4, a transcription factor that enhances the expression of genes linked to apoptosis, such as C/EBP homologous protein (CHOP).

IRE1α, possessing kinase and endoribonuclease activities, oligomerizes and autophosphorylates upon ER stress, splicing XBP1 mRNA to produce the active transcription factor XBP1s. XBP1s enhances the expression of genes related to UPR- and ER-associated degradation (ERAD). Prolonged IRE1α activation can lead to apoptosis by activating ASK1 and downstream kinases JNK and p38 MAPK. ATF6, a type II transmembrane protein, moves to the Golgi under ER stress. There, it is cleaved by proteases S1P and S2P, releasing a cytosolic fragment that enters the nucleus to induce UPR target gene expression, including those encoding BiP, CHOP, and XBP1 [19–21].

15.6 Interaction between UPR, Protein Aggregation, and Neurodegeneration

Protein misfolding can lead to the loss of normal protein function, the acquisition of neurotoxic properties, and chronic neuroinflammation [22]. Neurons are susceptible to the harmful effects of protein misfolding and aggregation, which can disrupt synaptic functions, induce apoptosis, and cause selective neuronal death [23].

ER stress has been linked to abnormal neuronal death in psychiatric disorders such as schizophrenia, depression, and PTSD [24]. Studies using transgenic animal models, along with genetic and neuropathological research, suggest that neurodegenerative diseases are often caused by the misfolding, aggregation, and buildup of certain proteins in the brain.

There is growing evidence that mediators of the UPR play a critical role in the pathophysiology of protein misfolding disorders (PMDs) [25]. ER stress markers have been identified in various neurodegenerative diseases (Fig. 15.1), and interestingly, these markers often co-localize with protein aggregates in the brains of

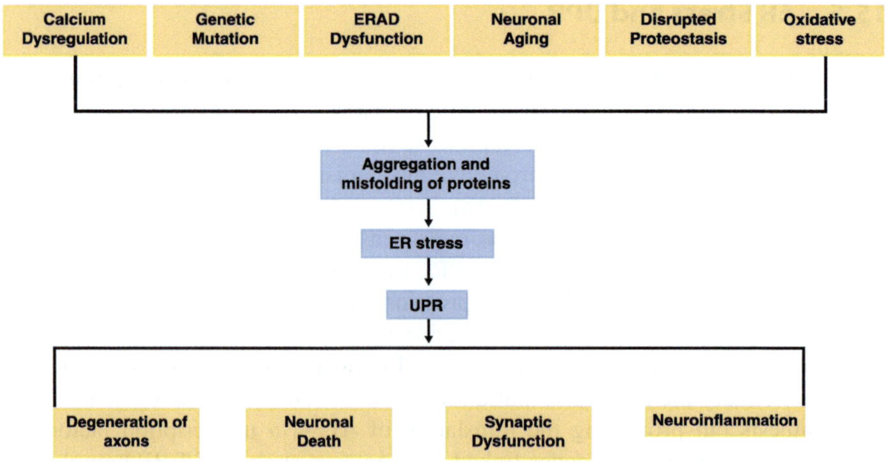

Fig. 15.1 ER stress, UPR, and neurodegeneration

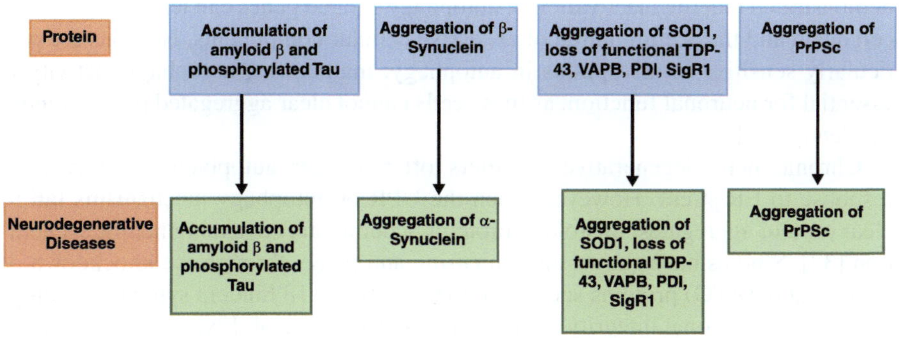

Fig. 15.2 Protein misfolding and aggregation in the pathogenesis of neurodegenerative diseases

affected individuals [26, 27]. UPR markers such as p-IRE1, active ATF6α, p-PERK, p-eIF2α, BiP, and CHOP have been detected in patients with AD [28], PD [29], ALS [30], and prion diseases [31]. Postmortem brain tissues of patients with neurodegenerative diseases have shown the presence of p-PERK and p-eIF2α, which are associated with the accumulation of misfolded and aggregated proteins. In particular, phosphorylated forms of PERK, IRE1α, and eIF2α have been found in neurons affected by Alzheimer's disease and in the substantia nigra of Parkinson's disease patients [32]. Furthermore, UPR markers are overexpressed in the spinal cords of ALS patients and the striatum, parietal cortex, and caudate putamen of prion disease patients.

Protein aggregates in neurons lose their normal function, contributing to the development of degenerative disorders. The accumulation of these aggregates triggers ER stress, which in turn activates the UPR. Initially, the UPR aims to alleviate ER protein overload and maintain neuronal viability. However, if ER stress persists, it can lead to ER stress-mediated cell death and neuronal loss. Notably, each type of protein aggregate operates through distinct mechanisms, leading to various disruptions in ER proteostasis and consequently to neurodegeneration (Fig. 15.2) [24].

Despite the involvement of misfolded proteins in neurodegenerative diseases, the precise relationship between these proteins and disease pathogenesis remains unclear. Additionally, the role of ER stress in neurodegenerative diseases is still debated, with some researchers suggesting that ER stress may play only a minor role in neurodegeneration mechanisms [33, 34].

15.7 Interaction Between ER Stress, Autophagy, and Neurodegeneration

Autophagy is a crucial cellular process that targets damaged organelles and misfolded proteins for degradation within the lysosome, facilitating the recycling of macromolecules [35, 36]. This mechanism is vital for tissue remodeling, cell survival, and regeneration [36]. Research across various model organisms has demonstrated that the inactivation of autophagy-related genes can lead to significant cell death and the accumulating of excessive cellular components. Neurons are particularly sensitive to disruptions in autophagy; maintaining autophagic activity is essential for neuronal function, as these cells cannot clear aggregated proteins independently [35].

Chronic neurodegenerative disorders often trigger autophagic activity as a response to ER stress. However, when the UPR or autophagy mechanisms fail to clear protein aggregates, misfolded proteins accumulate, advancing neurodegeneration [37]. Studies have shown that impairing autophagy through the knockdown of AuTophaGy (ATG) proteins such as ATG1, -2, -6, or -18 hinders synaptic development, disrupts axonal integrity by affecting microtubule stability, and contributes to the onset and progression of various neurodegenerative diseases [38, 39]. Autophagy-deficient mice exhibit neuronal defects, and deficiencies in autophagy are linked to neurodegenerative diseases like ALS, Parkinson's disease, Alzheimer's disease, and prion diseases [37].

ER stress can modulate autophagy through the PERK/eIF2α/ATF4 signaling pathway [40]. ATG5 and ATG7 have been shown to link autophagy with ER stress via PERK signaling [41]. Under ER stress conditions, ATF4 promotes the transcription of autophagy-related genes, including Beclin-1 (BECN1), microtubule-associated protein 1 light chain 3 beta (MAP 1LC3B), ATG5, ATG7, and ATG12 [42]. The eIF2α-kinases, GCN2 and PERK, along with ATF4 and CHOP, enhance the transcription of genes essential for autophagosome formation and function [42]. Inhibition of eIF2α dephosphorylation has been found to accelerate cell death in an ER stress and autophagy-dependent manner, highlighting the critical role of the PERK pathway in autophagy regulation [43].

The IRE1a pathway has also been explored in the context of autophagy and neurodegenerative diseases. The IRE1/XBP1 pathway may impact neurodegenerative disease pathogenesis, with XBP1s acting as a transcription factor that activates Beclin-1, a key gene in autophagy [44]. The interaction between XBP1 and FoxO1 (Forkhead box protein O1) regulates ER stress-induced autophagy [45]. Blocking IRE1 or ATG7 expression is associated with dopaminergic neuronal loss, locomotor impairment, reduced lifespan, and the progression of PD, indicating that the IRE1 pathway links ER stress to autophagy-dependent neuronal death [46]. Additionally, the IRE1/XBP1s pathway and the IRE1/TRAF2/ASK1/JNK (c-Jun N-terminal kinase) pathway have been found to interact with autophagy. Recent studies have shown that in Parkinson's-like disorders, the IRE1 signaling pathway activates JNK signaling via the ASK1-TRAF2 complex, initiating autophagy [47].

The ATF6 pathway is involved in interferon-gamma (IFN-ϒ)-induced activation of death-associated kinase 1 (DAPK1), a regulator of cell death and autophagy. ATF6-mediated upregulation of CHOP, XBP1, and GRP78 further contributes to autophagy induction. However, the role of the ATF6 pathway in autophagy has received less attention and warrants further investigation [48].

Emerging evidence suggests that communication between the ER and other organelles at membrane contact sites is closely linked to neurodegenerative disorders such as PD, AD, and ALS (Table 15.1) [49].

Table 15.1 Key pathways and molecular interactions between ER stress, autophagy, and neurodegeneration

Pathway	Key molecules	Role in autophagy and neurodegeneration
PERK/eIF2α/ATF4	PERK, eIF2α, ATF4, ATG5, ATG7, Beclin-1, MAP1LC3B	Modulates autophagy under ER stress. ATF4 promotes the transcription of autophagy-related genes, and PERK signaling links autophagy with ER stress. Inhibition of eIF2α dephosphorylation leads to cell death.
IRE1/XBP1	IRE1, XBP1s, FoxO1, Beclin-1, ATG7	IRE1/XBP1 pathway impacts neurodegeneration by activating autophagy-related genes. XBP1 regulates autophagy, while IRE1 activation links ER stress to autophagy-dependent neuronal death.
IRE1/TRAF2/ASK1/JNK	IRE1, TRAF2, ASK1, JNK	IRE1 activation triggers JNK signaling through the ASK1-TRAF2 complex, initiating autophagy in Parkinson's-like disorders.
ATF6	ATF6, CHOP, XBP1, GRP78, DAPK1	ATF6 activates autophagy-related genes under IFN-? Signaling, with CHOP and XBP1 contributing to autophagy induction. DAPK1 regulates cell death and autophagy.
eIF2α-Kinases (GCN2 & PERK)	GCN2, PERK, ATF4, CHOP	These kinases enhance the transcription of autophagosome-related genes, crucial for autophagy.
Communication with organelles	ER, mitochondria, lysosomes	Membrane contact sites between ER and organelles are linked to the progression of neurodegenerative diseases like PD, AD, and ALS.

15.8 ER Stress in Neurodegenerative Diseases

15.8.1 ER Stress in AD

AD is a serious neurodegenerative disorder that affects over 25 million people worldwide, leading to a progressive decline in cognitive functions. At the cellular level, AD is associated with several harmful processes, including excitotoxicity, oxidative stress within neurons, disrupted intracellular signaling pathways, synaptic dysfunction, and widespread neurodegeneration in key brain areas such as the frontal cortex, hippocampus, and basal forebrain [50]. AD development involves altered synaptic communication, dysregulated calcium homeostasis, increased ER stress, and persistent neuroinflammation. The hallmark neuropathological characteristics of AD include the accumulation of hyperphosphorylated tau proteins and the formation of extracellular amyloid-β (Aβ) plaques (Fig. 15.3) [51].

Tau protein, which normally stabilizes neuronal microtubules, becomes pathogenic when hyperphosphorylated, forming neurofibrillary tangles. Amyloid β-precursor protein (β-APP) is processed by β-secretase (BACE1) and γ-secretase complexes to produce Aβ peptides [52]. These peptides can accumulate into insoluble plaques and soluble oligomers, both of which are toxic to neurons and contribute to neurodegeneration and apoptosis [53, 54]. The γ-secretase complex is a

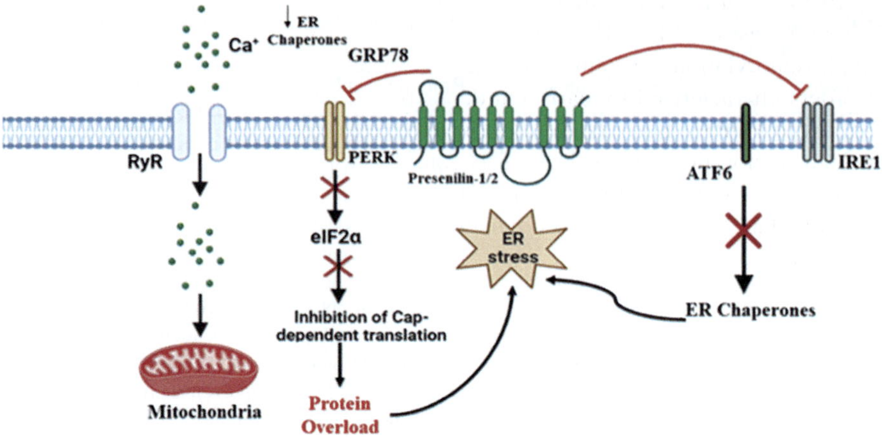

Fig. 15.3 Schematic diagrams showing links between ER stress, the UPR, and AD

membrane-bound assembly that includes presenilin-1/2 (PS1/2), nicastrin, Pen2, and Aph1 with the catalytic activity residing in PS1/2 [55]. Mutations in APP and presenilin genes increase Aβ peptide aggregation in the brain, leading to AD.

Continuous buildup of Aβ and p-tau disrupts calcium balance in the ER, encourages improper protein folding, and triggers ER stress. Specifically, tau protein can inhibit the ER-associated degradation (ERAD) pathway, resulting in the accumulation of misfolded proteins within the ER lumen [56]. Aβ oligomers interact with neuronal NMDA receptors, disrupting cytosolic calcium balance, causing ER stress-dependent cell death, impaired synaptic function, and losing dendritic spines. The neurotoxicity of Aβ peptides is linked to ER stress-mediated apoptosis, involving the activation of ASK1 and JNK pathways [57].

Studies have indicated abnormal ER stress levels in AD brains. Under mild ER stress, the UPR serves a protective role, but prolonged ER stress activates pro-apoptotic UPR pathways, contributing to neurodegeneration. Chaperones such as BiP, Hsp72, Hsp73, Grp94, PDI, and calreticulin are upregulated in AD cerebrospinal fluid and brain tissue [58]. Phosphorylated forms of PERK and eIF2α are significantly elevated in AD brains, particularly in the hippocampus and frontal cortex, often activated by tau aggregation [59]. PERK activation increases the expression of transcription factors ATF4 and BACE1, enhancing Aβ peptide production [60]. Inhibition of PERK reduces BACE1 expression and Aβ levels in AD mouse models [61]. Salubrinal treatment, which inhibits eIF2α dephosphorylation, increases BACE1 and Aβ production in neurons [62]. Aβ oligomers in the hippocampus are associated with elevated ATF4 expression, crucial for neuronal plasticity and long-term memory [63].

IRE1 activation correlates with AD progression. Genetic deletion of IRE1's RNAse domain in the nervous system reduces amyloid deposition, Aβ oligomers, and neuronal loss, improving cognitive functions in AD models. A polymorphism in the XBP1 promoter is a risk factor for AD. XBP1 influences ADAM10 transcription, reducing Aβ plaques and enhancing ERAD machinery components [64].

Recent findings suggest that ATF6 plays a role in AD. Reduced ATF6 expression in AD models correlates with lower APP and BACE1 levels, suppressing Aβ production and preserving spatial memory. However, prolonged ER stress elevates pro-apoptotic UPR components in AD brains, likely driving neurodegeneration [65]. CHOP, a key ER stress-related apoptotic factor, is upregulated in AD, along with caspase-12 and GADD34. CHOP activation generates reactive oxygen species (ROS), exacerbating oxidative stress, inflammation, and neuronal apoptosis [66]. Despite evidence linking UPR to AD, its role remains controversial. For instance, in 5XFAD transgenic mice, UPR markers such as BiP, p-IRE1, p-eIF2α, ATF4, and CHOP are not significantly elevated, suggesting ER stress might not be a universal feature of AD [33]. Further studies are necessary to clarify the involvement of UPR in AD pathogenesis.

15.8.2 ER Stress, Neuroinflammation, and AD

A bidirectional relationship exists between ER stress and inflammation, significantly impacting neurodegeneration and AD [66]. The UPR encompasses three pathways that facilitate the production and secretion of pro-inflammatory factors, including the transcription factor NF-κB. Activation of NF-κB leads to the expression of pro-inflammatory cytokines such as TNFα, IL-1, IL-8, and IL-6. Additionally, NF-κB regulates genes like the nuclear proto-oncogene SET, which is found at elevated levels and mislocalized within the neuronal cytoplasm in AD brains, thereby contributing directly to the disease's progression [67]. In this process, phosphorylated IRE1 (p-IRE1) interacts with TRAF-2, recruiting the IκB kinase (IKK) complex. This interaction results in the phosphorylation and degradation of IkB, subsequently activating NF-κB [68]. Additionally, the TRAF2-IRE1 complex can bind ASK-1, activating the JNK pathway and promoting inflammatory signaling through the transcription factor AP-1. The JNK-AP1 pathway is implicated in Aβ-induced neuroinflammation in AD-affected brains [69]. Furthermore, phosphorylated PERK (p-PERK), through phosphorylated eIF2α (p-eIF2α), inhibits IκB expression, further enhancing NF-κB activation [66]. In mouse models, ER stress-induced activation of the JAK1/STAT3 pathway via p-PERK also leads to neuroinflammation (Table 15.2) [70].

PERK can directly phosphorylate and activate glycogen synthase kinase 3β (GSK-3β), resulting in the hyperphosphorylation of tau, increased Aβ production, and consequent deficits in learning and memory, coupled with neurodegeneration [71, 72]. The role of ATF6 in pro-inflammatory responses within brain pathology is less documented. However, ATF6 siRNA has been shown to mitigate ER stress-mediated pro-inflammatory effects via the Akt/NF-κB pathway and the CREBH (cyclic adenosine monophosphate (cAMP)-responsive element-binding protein H) transcription factor [57, 66].

Table 15.2 Molecular pathways linking ER stress, neuroinflammation, and AD

Pathway/mechanism	Key players	Outcome/effect
NF-κB activation via UPR	p-IRE1, TRAF-2, IKK complex, NF-κB, IκB	Activation of NF-κB, leading to increased pro-inflammatory cytokines (TNFα, IL-1, IL-8, IL-6) and mislocalized SET in AD brains
TRAF2-IRE1 complex and JNK pathway	TRAF-2, IRE1, ASK-1, JNK, AP-1	Promotes Aβ-induced neuroinflammation via AP-1 transcription factor
PERK-eIF2α pathway	p-PERK, p-eIF2α, NF-κB	Inhibits IκB, enhancing NF-κB activation, contributing to neuroinflammation
PERK-JAK1/STAT3 pathway	p-PERK, JAK1, STAT3	Activation of neuroinflammation in mouse models
PERK-GSK-3β interaction	PERK, GSK-3β, tau, Aβ	Hyperphosphorylation of tau, increased Aβ production, memory deficits, neurodegeneration
ATF6 and pro-inflammatory response	ATF6, Akt, NF-κB, CREBH	Less documented, but ATF6 siRNA reduces pro-inflammatory effects via the Akt/NF-κB pathway and CREBH transcription factor

15.8.3 ER Stress in PD

PD is a progressive neurodegenerative condition that is often identified in its later stages. It ranks as the second most prevalent neurodegenerative disorder after AD [73], predominantly affecting people over the age of 40. Globally, the prevalence is around 2% in men and 1.3% in women [74–76]. Key symptoms of PD include tremors, impaired balance, posture, and movement, which worsen as the disease progresses. While most PD cases are sporadic, 10–15% are linked to mutations in various genes, such as SNCA/PARK1 (encoding α-synuclein (α-SYN)), Parkin (PRKN), PARK9/ATP13A2, PARK7/DJ1, LRRK2, and PINK1 [77].

PD is characterized by two primary pathological features: the selective degeneration of dopaminergic neurons in the substantia nigra pars compacta (SNpc), which regulates movement and muscle control, and the accumulation of misfolded α-synuclein (α-SYN) fibrils. These fibrils accumulate in neuronal cell bodies (Lewy bodies) or within axons and dendrites (Lewy neurites) [78]. α-SYN is predominantly present in presynaptic terminals of neurons in the central nervous system but is also detected in peripheral tissues and blood [79, 80]. Mutations in the SNCA gene, along with gene duplications and triplications, are hereditary causes of PD and function as dominant variants that trigger the disease [81]. α-SYN monomers aggregate into β-sheet structures, forming insoluble protofibrils and fibrils, spreading through interconnected neural circuits, causing neuronal damage, and being implicated in various neurodegenerative diseases known as α-synucleinopathies [82].

ER stress plays a critical role in the toxicity of α-SYN and the degeneration of dopaminergic neurons. Research indicates that α-SYN aggregates bind to the ER chaperone BiP, which triggers the UPR pathway [83, 84]. Evidence of ER stress, including increased levels of phosphorylated PERK and eIF2α, has been observed in the dopaminergic neurons of PD patients, particularly in the substantia nigra, but not in healthy control tissues. Additionally, α-SYN is found in higher

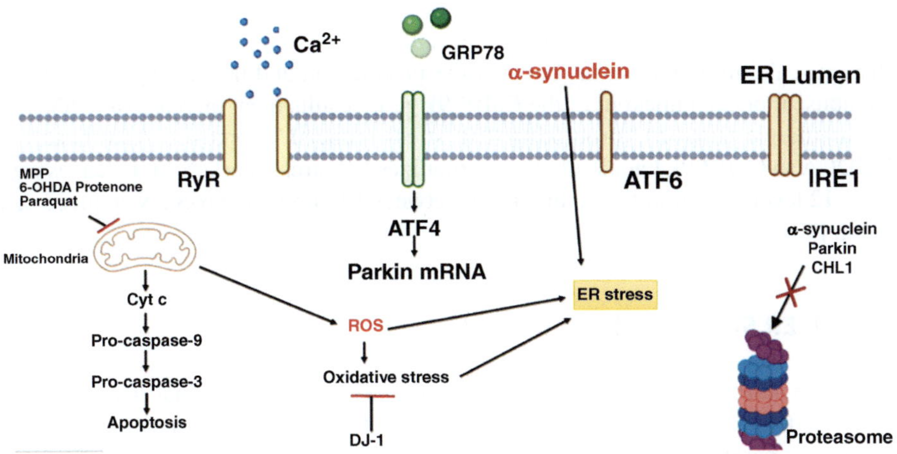

Fig. 15.4 Schematic diagrams showing links between ER stress, the UPR, and PD

concentrations in the ER/microsomal fractions of PD brains compared to controls [85]. In A53T transgenic mice, α-synucleinopathy coincides with increased ER chaperones and abnormal UPR in affected neurons, with further support from the accumulation of polyubiquitin chains and caspase-12 activation [86]. ER stress inhibitors like salubrinal significantly ameliorate α-synucleinopathy onset and dopaminergic neurodegeneration in A53T models, protecting against cell death and reducing A53T cytotoxicity (Fig. 15.4) [87, 88].

In differentiated rat sympathetic-like neuron cells (PC12), the overexpression of the A53T α-SYN leads to increased ROS activity, decreased proteasome activity, and activation of ER stress as a final defense mechanism before triggering caspase-dependent apoptosis. The onset of ER stress is marked by increased phosphorylation of eIF2 and the upregulation of ER stress-associated genes, such as GRP17 and GADD153. In dopaminergic SH-SY5Y cells, glucose deprivation leads to α-SYN/ER stress-mediated cell death, where α-SYN functions as a cellular stress sensor. Furthermore, elevated expression of ATF4/CREB-2 has been observed in the substantia nigra of SYN120 transgenic mice, connecting ER stress with the pathogenesis of PD [84].

The involvement of the IRE1–XBP1 pathway in PD remains debated. In animal models treated with 6-hydroxydopamine, gene therapy that activates XBP1 has demonstrated neuroprotective effects [89]. However, in Drosophila melanogaster, IRE1 activation causes cell loss in photoreceptor neurons independently of XBP1, preventable by autophagy inhibition [46].

Research suggests that the accumulation of α-SYN leads to ER stress by disrupting ER-Golgi trafficking, potentially through interactions with RAB1 GTPase or ATF6 [90, 91]. Co-expression of RAB1 with α-SYN can prevent the loss of dopaminergic neurons in animal models. α-SYN inhibits ATF6 activation via COPII-mediated ER-Golgi transit, reducing its cytoprotective effect and promoting apoptosis [91].

The aggregation of α-SYN may lead to ER stress and disturb calcium balance within the ER. This aggregation stimulates the ER calcium pump, SERCA, which disrupts calcium homeostasis, boosts ROS production, and triggers apoptosis [92]. In mouse models, knockout of the CaBP-9k gene results in an increase of α-SYN in dopaminergic neurons, promoting ER stress-induced apoptosis. Administering TUDCA to CaBP-9k knockout mice normalizes ER stress markers and cleaves caspase-12 levels, highlighting a direct link between ER stress and α-SYN-induced cell death [93].

15.8.4 ER Stress in ALS

ALS is a serious neurodegenerative disease marked by the gradual loss of motor neurons in the brain, spinal cord, and brainstem. This degeneration results in muscle weakness, paralysis, atrophy, and spasticity [94]. Most ALS cases are sporadic, with familial ALS (fALS) constituting only 5–10% of cases.

Several genes have been implicated in ALS, including protein disulfide isomerase (PDI), superoxide dismutase 1 (SOD1), TAR DNA-binding protein (TARDBP or TDP-43), fused in sarcoma (FUS), chromosome 9 open reading frame 72 (C9ORF72), vesicle-associated membrane protein-associated protein B (VAPB), and Sigma-1 receptor (SIGR1) [26, 94]. Mutations in these genes disrupt mRNA metabolism, protein homeostasis, and protein aggregation, leading to neurotoxicity and neuroinflammation, contributing to ALS pathogenesis. ER stress has been widely reported in ALS, with upregulated UPR markers detected in both patients and animal models [95–97].

Approximately 20% of familial ALS cases and 1–2% of all ALS cases are linked to mutations in the SOD1 gene [98]. Interestingly, female ALS patients with SOD1 mutations have better survival rates than males [99]. Transcriptional analysis of motor neurons derived from induced pluripotent stem cells (iPSCs) from patients with SOD1 mutations has shown high ER stress and upregulated UPR markers [100]. Examination of lumbar spinal cord sections in G93A SOD1 mice revealed upregulation of PERK, IRE1α, and ATF6 [100]. In post-mortem tissues, mutant SOD1 co-localizes with UPR markers like BiP, calnexin, and PDI, potentially sequestering these chaperones and disrupting ER function [24, 101]. Structural alterations in the ER, such as rough ER fragmentation and ribosome detachment, have been observed in ALS patients and SOD1 mouse models [102]. Mutant SOD1 interacts with Derlin-1, an ER-associated degradation (ERAD) pathway component, triggering ER stress and motor neuron death [103].

Initial studies indicated that PERK haploinsufficiency speeds up the progression of ALS in SOD1 transgenic mice by reducing the ability to manage misfolded SOD1, leading to early UPR overload [104]. Genetic deletion of GADD34 and the inhibition of eIF2α phosphatase have been found to delay the onset of disease and extend motor neuron survival [105, 106]. However, more recent research suggests that neither PERK haploinsufficiency nor genetic UPR enhancement through GADD34 deletion confers benefits in motor neuron disease caused by mutant SOD1

[107]. The deletion of ATF4 in SOD1 G85R mutant mice protected ALS, possibly due to reduced levels of apoptotic components like CHOP [108].

The involvement of the IRE1 pathway in ALS has been explored by examining the role of homeodomain-interacting protein kinase 2 (HIPK2), a crucial part of the IRE1-ASK1 apoptotic pathway that activates JNK under ER stress conditions [109]. In SOD1 G93A mice, the loss of HIPK2 has been associated with delayed disease onset, decreased spinal motor neuron cell death, and improved overall survival [109]. XBP1 ablation reduced motor neuron death and mutant SOD1 aggregation, likely due to a balance between UPR and autophagy pathways [96]. Although the role of ATF6 in ALS is less studied, elevated ATF6 levels in SOD1 G93A mouse models suggest its involvement in ALS progression [110].

ER chaperones like PDI play a broad protective role in ALS. Increased PDI levels have been detected in the cerebrospinal fluid and blood of ALS patients, indicating that PDI dysfunction contributes to ALS pathology and can aid in diagnosis [111]. Multiple PDIA1 and PDIA3 missense variants have been identified in ALS patients, with single nucleotide polymorphisms (SNPs) in these genes enriched in ALS cases [112]. In motoneuron cell culture models, PDI variants induce motor defects and disrupt motor neuron connectivity and dendritic outgrowth [113]. Additionally, overexpression of mutant TDP-43 induces ER stress pathways in neuroblastoma cells, with interactions between PDI and TDP-43 observed in transfected cell lysates and spinal cords of mutant TDP-43 transgenic mice [114]. Abnormal eIF2α phosphorylation by TDP-43 aggregates has been noted in Drosophila models, with therapeutic modulation attenuating TDP-43 toxicity in ALS [115].

In ALS models, FUS, normally located in the nucleus, translocates to the cytoplasm and forms inclusions linked to ER stress. Mutant FUS colocalizes with PDI in human ALS lumbar spinal cords and familial ALS tissues [116]. The common familial ALS mutation in C9ORF72 is associated with UPR activation and vulnerability to Ca^{2+}-permeable AMPA receptor-mediated excitotoxicity [117, 118]. Mutations in ER-related genes, such as E102Q in the SIGR1 gene, lead to ER stress-mediated protein homeostasis defects and dysregulated RNA-binding proteins [119]. Mutant VAPB P56S affects intracellular Ca^{2+} storage and signaling, inhibiting ATF6 and XBP1, leading to chronic ER stress, synaptic loss, and cell death in ALS [120].

Further research is needed to fully understand these mechanisms and their implications for ALS treatment strategies.

15.9 ER Stress in Prion Diseases

Prion diseases, also known as transmissible spongiform encephalopathies (TSEs), are a group of rare, rapidly progressing neurodegenerative disorders that impact the nervous systems of both humans and animals. In humans, these diseases include Creutzfeldt–Jakob disease (CJD), Gerstmann–Sträussler–Scheinker syndrome, and fatal familial insomnia. In animals, they manifest as scrapie in sheep and goats, bovine spongiform encephalopathy (BSE, commonly known as "mad cow disease") in cattle, and chronic wasting disease (CWD) in deer and elk. Prion diseases can occur through infectious transmission, arise sporadically without apparent cause, or be genetically inherited. Despite their different modes of occurrence, all prion diseases share a common feature: the abnormal folding of the prion protein (PrP). This misfolded protein becomes resistant to protease enzymes, which normally break down proteins, leading to its accumulation in brain tissue. The buildup of these abnormal proteins causes brain damage, resulting in the characteristic sponge-like appearance of brain tissue seen in affected individuals [121].

In prion diseases, the fundamental molecular event involves the transformation of the normal prion protein (PrP^{C}), which primarily has an α-helical structure, into its misfolded counterpart, PrP^{Sc}, which is rich in β-sheets. This misfolded PrP^{Sc} accumulates in the brain, forming aggregates that are linked to the disease's pathology [122]. While the presence of PrP^{Sc} serves as an indicator of prion infection, its accumulation alone does not directly cause toxicity [123]. Instead, the harmful effects are due to PrP^{Sc} inducing the misfolding of additional PrP^{C} molecules, which leads to the activation of various toxic cellular pathways. These pathways include caspase activation, ER stress, autophagy dysregulation, and disturbances in calcium homeostasis [124, 125].

ER stress plays a crucial role in the development of prion diseases, contributing to the buildup of PrP^{Sc} and increasing the likelihood of PrP^{C} misfolding [126, 127]. This connection is supported by the observed upregulation of ER chaperones such as Grp58/ERp57 in mice infected with scrapie and Grp78/BiP, Grp94, PDIA1, and Grp58/ERp57 in the cerebral cortex of patients with both variant and CJD [128, 129]. Moreover, heat shock protein 70 (HSP70) levels are elevated in CJD patients and rodents infected with scrapie. The absence of HSP70 has been shown to hasten the progression of prion diseases in mice, suggesting that HSP70 may provide a protective mechanism against these conditions [130].

The UPR pathways involving PERK, IRE1α, and ATF6 are implicated in prion diseases. Modulating these pathways has shown potential in preventing neuronal loss and extending survival in prion-infected mice [131]. Specifically, research has demonstrated that prion-infected astrocytes exhibit elevated levels of p-PERK and associated neuronal degeneration. In the hippocampus of prion-infected mice and mice overexpressing PrP^{C}, activation of the PERK pathway results in increased p-eIF2α, leading to a decrease in the expression of synaptic proteins, disruption of synaptic function, and subsequent neuronal loss [132]. Inhibiting the p-PERK signaling pathway selectively in astrocytes has been shown to prevent neuronal degeneration and improve survival outcomes in prion disease models, suggesting a neuroprotective effect of UPR over-activation.

Therapeutic interventions targeting the UPR pathways have shown promise. Gene therapy to deliver eIF2α phosphatase, oral treatment with a PERK inhibitor [133], and treatment with ISRIB (a compound that blocks p-eIF2α consequences) have all provided neuroprotection and slowed disease progression in prion-infected animals [132, 134]. Additionally, prion infection induces XBP1 mRNA splicing and activates stress kinases JNK and ERK via IRE1α. Overexpression of XBP1s in cellular models of prion disease prevented PrP misfolding, while dominant-negative forms of XBP1 and IRE1a increased PrP aggregation [135, 136]. Despite this, mice lacking XBP1 in the brain showed normal prion replication and neuropathology, indicating compensatory mechanisms from other UPR pathways [135].

ATF4 and ATF6 also play roles in prion disease. Overexpression of ATF4 or an active mutant form of ATF6 in cellular models prevented PrP aggregation [137]. These findings highlight the critical involvement of ER stress and UPR pathways in prion diseases, particularly the PERK pathway. Further research is needed to fully understand the roles of IRE1 and ATF6 in prion disease pathogenesis and to explore potential therapeutic targets.

15.10 Conclusion

ER stress plays a significant role in the development of numerous neurodegenerative diseases, including AD, PD, ALS, and prion diseases. These conditions are often marked by the accumulation of misfolded proteins and aggregates. Despite extensive research, the exact involvement of the UPR and the specific mechanisms through which ER stress contributes to disease progression remains unclear, as they may produce varying or even contradictory effects. This complexity arises from the intricate interactions between ER stress, UPR pathways, energy regulation, proteostasis, and neuroinflammation. As highlighted throughout this work, both in vivo and in vitro models of neurodegenerative diseases demonstrate that the accumulation of disease-specific misfolded proteins leads to neuronal and synaptic dysfunction. Targeting the UPR offers a promising strategy for treating a variety of these disorders. However, fully elucidating the signaling pathways and biological roles of UPR-associated molecules is essential for the development of novel therapeutic interventions. Identifying new compounds that can modulate UPR signaling and target genes linked to ER stress in animal and cellular models is crucial, as it will provide valuable insights into the role of the UPR in the progression of protein misfolding disorders.

References

1. Selkoe DJ. Folding proteins in fatal ways. Nature. 2003;426:900–4. https://doi.org/10.1038/nature02264.
2. Taylor JP, Hardy J, Fischbeck KH. Toxic proteins in neurodegenerative disease. Science (1979). 2002;296:1991–5. https://doi.org/10.1126/science.1067122.
3. Kopito RR, Ron D. Conformational disease. Nat Cell Biol. 2000;2:E207–9. https://doi.org/10.1038/35041139.
4. Dobson CM. Protein folding and misfolding. Nature. 2003;426:884–90. https://doi.org/10.1038/nature02261.
5. Kopito RR. Aggresomes, inclusion bodies and protein aggregation. Trends Cell Biol. 2000;10:524–30. https://doi.org/10.1016/S0962-8924(00)01852-3.

6. Kaufman RJ. Stress signaling from the lumen of the endoplasmic reticulum: coordination of gene transcriptional and translational controls. Genes Dev. 1999;13:1211–33. https://doi.org/10.1101/gad.13.10.1211.
7. Dobson CM, Ellis RJ. Protein folding and misfolding inside and outside the cell. EMBO J. 1998;17:5251–4. https://doi.org/10.1093/emboj/17.18.5251.
8. Sitia R, Braakman I. Quality control in the endoplasmic reticulum protein factory. Nature. 2003;426:891–4. https://doi.org/10.1038/nature02262.
9. Goldberg AL. Protein degradation and protection against misfolded or damaged proteins. Nature. 2003;426:895–9. https://doi.org/10.1038/nature02263.
10. Sherman MY, Goldberg AL. Cellular defenses against unfolded proteins. Neuron. 2001;29:15–32. https://doi.org/10.1016/S0896-6273(01)00177-5.
11. Yang J, Griffin A, Qiang Z, Ren J. Organelle-targeted therapies: a comprehensive review on system design for enabling precision oncology. Signal Transduct Target Ther. 2022;7:379. https://doi.org/10.1038/s41392-022-01243-0.
12. Schwarz DS, Blower MD. The endoplasmic reticulum: structure, function and response to cellular signaling. Cell Mol Life Sci. 2016;73:79–94. https://doi.org/10.1007/s00018-015-2052-6.
13. Burgoyne T, Patel S, Eden ER. Calcium signaling at ER membrane contact sites. Biochim Biophys Acta. 2015;1853:2012–7. https://doi.org/10.1016/j.bbamcr.2015.01.022.
14. Stefan CJ, Manford AG, Baird D, Yamada-Hanff J, Mao Y, Emr SD. Osh proteins regulate phosphoinositide metabolism at ER-plasma membrane contact sites. Cell. 2011;144:389–401. https://doi.org/10.1016/j.cell.2010.12.034.
15. Area-Gomez E, de Groof A, Bonilla E, Montesinos J, Tanji K, Boldogh I, Pon L, Schon EA. A key role for MAM in mediating mitochondrial dysfunction in Alzheimer disease. Cell Death Dis. 2018;9:335. https://doi.org/10.1038/s41419-017-0215-0.
16. Gómez-Suaga P, Bravo-San Pedro JM, González-Polo RA, Fuentes JM, Niso-Santano M. ER–mitochondria signaling in Parkinson's disease. Cell Death Dis. 2018;9:337. https://doi.org/10.1038/s41419-017-0079-3.
17. Friedman JR, Lackner LL, West M, DiBenedetto JR, Nunnari J, Voeltz GK. ER tubules mark sites of mitochondrial division. Science (1979). 2011;334:358–62. https://doi.org/10.1126/science.1207385.
18. Uddin MS, Tewari D, Sharma G, Kabir MT, Barreto GE, Bin-Jumah MN, Perveen A, Abdel-Daim MM, Ashraf GM. Molecular mechanisms of ER stress and UPR in the pathogenesis of Alzheimer's disease. Mol Neurobiol. 2020;57:2902–19. https://doi.org/10.1007/s12035-020-01929-y.
19. Carrara M, Prischi F, Nowak PR, Kopp MC, Ali MM. Noncanonical binding of BiP ATPase domain to Ire1 and Perk is dissociated by unfolded protein CH1 to initiate ER stress signaling. elife. 2015;4:e03522. https://doi.org/10.7554/eLife.03522.
20. Wu J, Rutkowski DT, Dubois M, Swathirajan J, Saunders T, Wang J, Song B, Yau GD-Y, Kaufman RJ. ATF6α optimizes long-term endoplasmic reticulum function to protect cells from chronic stress. Dev Cell. 2007;13:351–64. https://doi.org/10.1016/j.devcel.2007.07.005.
21. Hillary RF, FitzGerald U. A lifetime of stress: ATF6 in development and homeostasis. J Biomed Sci. 2018;25:48. https://doi.org/10.1186/s12929-018-0453-1.
22. Soto C. Unfolding the role of protein misfolding in neurodegenerative diseases. Nat Rev Neurosci. 2003;4:49–60. https://doi.org/10.1038/nrn1007.
23. Ashraf G, Greig N, Khan T, Hassan I, Tabrez S, Shakil S, Sheikh I, Zaidi S, Akram M, Jabir N, Firoz C, Naeem A, Alhazza I, Damanhouri G, Kamal M. Protein misfolding and aggregation in Alzheimer's disease and type 2 diabetes mellitus. CNS Neurol Disord Drug Targets. 2014;13:1280–93. https://doi.org/10.2174/1871527313666140917095514.
24. Hetz C, Saxena S. ER stress and the unfolded protein response in neurodegeneration. Nat Rev Neurol. 2017;13:477–91. https://doi.org/10.1038/nrneurol.2017.99.
25. Sweeney P, Park H, Baumann M, Dunlop J, Frydman J, Kopito R, McCampbell A, Leblanc G, Venkateswaran A, Nurmi A, Hodgson R. Protein misfolding in neurodegenerative dis-

eases: implications and strategies. Transl Neurodegener. 2017;6:6. https://doi.org/10.1186/s40035-017-0077-5.
26. Hughes D, Mallucci GR. The unfolded protein response in neurodegenerative disorders – therapeutic modulation of the PERK pathway. FEBS J. 2019;286:342–55. https://doi.org/10.1111/febs.14422.
27. Ghemrawi R, Khair M. Endoplasmic reticulum stress and unfolded protein response in neurodegenerative diseases. Int J Mol Sci. 2020;21:6127. https://doi.org/10.3390/ijms21176127.
28. Gerakis Y, Hetz C. Emerging roles of ER stress in the etiology and pathogenesis of Alzheimer's disease. FEBS J. 2018;285:995–1011. https://doi.org/10.1111/febs.14332.
29. Colla E. Linking the endoplasmic reticulum to Parkinson's disease and alpha-synucleinopathy. Front Neurosci. 2019;13:560. https://doi.org/10.3389/fnins.2019.00560.
30. Prell T, Stubendorff B, Le TT, Gaur N, Tadić V, Rödiger A, Witte OW, Grosskreutz J. Reaction to endoplasmic reticulum stress via ATF6 in amyotrophic lateral sclerosis deteriorates with aging. Front Aging Neurosci. 2019;11:5. https://doi.org/10.3389/fnagi.2019.00005.
31. Lindholm D, Korhonen L, Eriksson O, Kõks S. Recent insights into the role of unfolded protein response in ER stress in health and disease. Front Cell Dev Biol. 2017;5:48. https://doi.org/10.3389/fcell.2017.00048.
32. Stutzbach LD, Xie SX, Naj AC, Albin R, Gilman S, Lee VMY, Trojanowski JQ, Devlin B, Schellenberg GD. The unfolded protein response is activated in disease-affected brain regions in progressive supranuclear palsy and Alzheimer's disease. Acta Neuropathol Commun. 2013;1:31. https://doi.org/10.1186/2051-5960-1-31.
33. Sadleir KR, Popovic J, Vassar R. ER stress is not elevated in the 5XFAD mouse model of Alzheimer's disease. J Biol Chem. 2018;293:18434–43. https://doi.org/10.1074/jbc.RA118.005769.
34. Zhao L, Ackerman SL. Endoplasmic reticulum stress in health and disease. Curr Opin Cell Biol. 2006;18:444–52. https://doi.org/10.1016/j.ceb.2006.06.005.
35. Son JH, Shim JH, Kim K-H, Ha J-Y, Han JY. Neuronal autophagy and neurodegenerative diseases. Exp Mol Med. 2012;44:89. https://doi.org/10.3858/emm.2012.44.2.031.
36. Tettamanti G, Carata E, Montali A, Dini L, Fimia GM. Autophagy in development and regeneration: role in tissue remodelling and cell survival. Eur Zool J. 2019;86:113–31. https://doi.org/10.1080/24750263.2019.1601271.
37. Cai Y, Arikkath J, Yang L, Guo M-L, Periyasamy P, Buch S. Interplay of endoplasmic reticulum stress and autophagy in neurodegenerative disorders. Autophagy. 2016;12:225–44. https://doi.org/10.1080/15548627.2015.1121360.
38. Birdsall V, Waites CL. Autophagy at the synapse. Neurosci Lett. 2019;697:24–8. https://doi.org/10.1016/j.neulet.2018.05.033.
39. Negrete-Hurtado A, Overhoff M, Bera S, De Bruyckere E, Schätzmüller K, Kye MJ, Qin C, Lammers M, Kondylis V, Neundorf I, Kononenko NL. Autophagy lipidation machinery regulates axonal microtubule dynamics but is dispensable for survival of mammalian neurons. Nat Commun. 2020;11:1535. https://doi.org/10.1038/s41467-020-15287-9.
40. Liu C, Yan D, Wang C, Ma Z, Deng Y, Liu W, Xu B. Manganese activates autophagy to alleviate endoplasmic reticulum stress–induced apoptosis via PERK pathway. J Cell Mol Med. 2020;24:328–41. https://doi.org/10.1111/jcmm.14732.
41. Zheng W, Xie W, Yin D, Luo R, Liu M, Guo F. ATG5 and ATG7 induced autophagy interplays with UPR via PERK signaling. Cell Commun Signal. 2019;17:42. https://doi.org/10.1186/s12964-019-0353-3.
42. B'chir W, Maurin A-C, Carraro V, Averous J, Jousse C, Muranishi Y, Parry L, Stepien G, Fafournoux P, Bruhat A. The eIF2α/ATF4 pathway is essential for stress-induced autophagy gene expression. Nucleic Acids Res. 2013;41:7683–99. https://doi.org/10.1093/nar/gkt563.
43. Yu C-L, Yang S-F, Hung T-W, Lin C-L, Hsieh Y-H, Chiou H-L. Inhibition of eIF2α dephosphorylation accelerates pterostilbene-induced cell death in human hepatocellular carcinoma cells in an ER stress and autophagy-dependent manner. Cell Death Dis. 2019;10:418. https://doi.org/10.1038/s41419-019-1639-5.

44. Hosoi T, Nomura J, Tanaka K, Ozawa K, Nishi A, Nomura Y. Link between endoplasmic reticulum stress and autophagy in neurodegenerative diseases. Endoplasm Reticul Stress Dis. 2017;4:37–45. https://doi.org/10.1515/ersc-2017-0004.
45. Kishino A, Hayashi K, Hidai C, Masuda T, Nomura Y, Oshima T. XBP1-FoxO1 interaction regulates ER stress-induced autophagy in auditory cells. Sci Rep. 2017;7:4442. https://doi.org/10.1038/s41598-017-02960-1.
46. Yan C, Liu J, Gao J, Sun Y, Zhang L, Song H, Xue L, Zhan L, Gao G, Ke Z, Liu Y, Liu J. IRE1 promotes neurodegeneration through autophagy-dependent neuron death in the Drosophila model of Parkinson's disease. Cell Death Dis. 2019;10:800. https://doi.org/10.1038/s41419-019-2039-6.
47. Liu C, Yan D-Y, Wang C, Ma Z, Deng Y, Liu W, Xu B. IRE1 signaling pathway mediates protective autophagic response against manganese-induced neuronal apoptosis in vivo and in vitro. Sci Total Environ. 2020;712:136480. https://doi.org/10.1016/j.scitotenv.2019.136480.
48. Mei Y, Thompson MD, Cohen RA, Tong X. Endoplasmic reticulum stress and related pathological processes. J Pharmacol Biomed Anal. 2013;1:1000107.
49. Wang P, Kou D, Le W. Roles of VMP1 in autophagy and ER–membrane contact: potential implications in neurodegenerative disorders. Front Mol Neurosci. 2020;13:42. https://doi.org/10.3389/fnmol.2020.00042.
50. Lindholm D, Wootz H, Korhonen L. ER stress and neurodegenerative diseases. Cell Death Differ. 2006;13:385–92. https://doi.org/10.1038/sj.cdd.4401778.
51. Mattsson N, Zetterberg H, Janelidze S, Insel PS, Andreasson U, Stomrud E, Palmqvist S, Baker D, Tan Hehir CA, Jeromin A, Hanlon D, Song L, Shaw LM, Trojanowski JQ, Weiner MW, Hansson O, Blennow K. Plasma tau in Alzheimer disease. Neurology. 2016;87:1827–35. https://doi.org/10.1212/WNL.0000000000003246.
52. Walsh DM, Selkoe DJ. Aβ oligomers – a decade of discovery. J Neurochem. 2007;101:1172–84. https://doi.org/10.1111/j.1471-4159.2006.04426.x.
53. Zhang H, Ma Q, Zhang Y, Xu H. Proteolytic processing of Alzheimer's β-amyloid precursor protein. J Neurochem. 2012;120:9–21. https://doi.org/10.1111/j.1471-4159.2011.07519.x.
54. Gendron TF, Petrucelli L. The role of tau in neurodegeneration. Mol Neurodegener. 2009;4:13. https://doi.org/10.1186/1750-1326-4-13.
55. Haass C. Take five—BACE and the γ-secretase quartet conduct Alzheimer's amyloid β-peptide generation. EMBO J. 2004;23:483–8. https://doi.org/10.1038/sj.emboj.7600061.
56. Abisambra JF, Jinwal UK, Blair LJ, O'Leary JC, Li Q, Brady S, Wang L, Guidi CE, Zhang B, Nordhues BA, Cockman M, Suntharalingham A, Li P, Jin Y, Atkins CA, Dickey CA. Tau accumulation activates the unfolded protein response by impairing endoplasmic reticulum-associated degradation. J Neurosci. 2013;33:9498–507. https://doi.org/10.1523/JNEUROSCI.5397-12.2013.
57. Song J, Park K, Lee W, Lee J. Apoptosis signal regulating kinase 1 (ASK1): potential as a therapeutic target for Alzheimer's disease. Int J Mol Sci. 2014;15:2119–29. https://doi.org/10.3390/ijms15022119.
58. Lin Q, Cao Y, Gao J. Serum calreticulin is a negative biomarker in patients with Alzheimer's disease. Int J Mol Sci. 2014;15:21740–53. https://doi.org/10.3390/ijms151221740.
59. Sharma V, Ounallah-Saad H, Chakraborty D, Hleihil M, Sood R, Barrera I, Edry E, Kolatt Chandran S, Ben Tabou de Leon S, Kaphzan H, Rosenblum K. Local inhibition of PERK enhances memory and reverses age-related deterioration of cognitive and neuronal properties. J Neurosci. 2018;38:648–58. https://doi.org/10.1523/JNEUROSCI.0628-17.2017.
60. Hugon J, Mouton-Liger F, Dumurgier J, Paquet C. PKR involvement in Alzheimer's disease. Alzheimers Res Ther. 2017;9:83. https://doi.org/10.1186/s13195-017-0308-0.
61. Devi L, Ohno M. PERK mediates eIF2α phosphorylation responsible for BACE1 elevation, CREB dysfunction and neurodegeneration in a mouse model of Alzheimer's disease. Neurobiol Aging. 2014;35:2272–81. https://doi.org/10.1016/j.neurobiolaging.2014.04.031.
62. O'Connor T, Sadleir KR, Maus E, Velliquette RA, Zhao J, Cole SL, Eimer WA, Hitt B, Bembinster LA, Lammich S, Lichtenthaler SF, Hébert SS, De Strooper B, Haass C, Bennett DA, Vassar R. Phosphorylation of the translation initiation factor eIF2α increases BACE1

levels and promotes amyloidogenesis. Neuron. 2008;60:988–1009. https://doi.org/10.1016/j.neuron.2008.10.047.
63. Pasini S, Corona C, Liu J, Greene LA, Shelanski ML. Specific downregulation of hippocampal ATF4 reveals a necessary role in synaptic plasticity and memory. Cell Rep. 2015;11:183–91. https://doi.org/10.1016/j.celrep.2015.03.025.
64. Duran-Aniotz C, Cornejo VH, Espinoza S, Ardiles ÁO, Medinas DB, Salazar C, Foley A, Gajardo I, Thielen P, Iwawaki T, Scheper W, Soto C, Palacios AG, Hoozemans JJM, Hetz C. IRE1 signaling exacerbates Alzheimer's disease pathogenesis. Acta Neuropathol. 2017;134:489–506. https://doi.org/10.1007/s00401-017-1694-x.
65. Du Y, Liu X, Zhu X, Liu Y, Wang X, Wu X. Activating transcription factor 6 reduces Aβ1–42 and restores memory in Alzheimer's disease model mice. Int J Neurosci. 2020;130:1015–23. https://doi.org/10.1080/00207454.2020.1715977.
66. Santos LE, Ferreira ST. Crosstalk between endoplasmic reticulum stress and brain inflammation in Alzheimer's disease. Neuropharmacology. 2018;136:350–60. https://doi.org/10.1016/j.neuropharm.2017.11.016.
67. Feng Y, Li X, Zhou W, Lou D, Huang D, Li Y, Kang Y, Xiang Y, Li T, Zhou W, Song W. Regulation of SET gene expression by NFkB. Mol Neurobiol. 2017;54:4477–85. https://doi.org/10.1007/s12035-016-9967-2.
68. Mohammed-Ali Z, Cruz GL, Dickhout JG. Crosstalk between the unfolded protein response and NF-κB-mediated inflammation in the progression of chronic kidney disease. J Immunol Res. 2015;2015:1–10. https://doi.org/10.1155/2015/428508.
69. Vukic V, Callaghan D, Walker D, Lue L-F, Liu QY, Couraud P-O, Romero IA, Weksler B, Stanimirovic DB, Zhang W. Expression of inflammatory genes induced by beta-amyloid peptides in human brain endothelial cells and in Alzheimer's brain is mediated by the JNK-AP1 signaling pathway. Neurobiol Dis. 2009;34:95–106. https://doi.org/10.1016/j.nbd.2008.12.007.
70. Meares GP, Liu Y, Rajbhandari R, Qin H, Nozell SE, Mobley JA, Corbett JA, Benveniste EN. PERK-dependent activation of JAK1 and STAT3 contributes to endoplasmic reticulum stress-induced inflammation. Mol Cell Biol. 2014;34:3911–25. https://doi.org/10.1128/MCB.00980-14.
71. Hooper C, Killick R, Lovestone S. The GSK3 hypothesis of Alzheimer's disease. J Neurochem. 2008;104:1433–9. https://doi.org/10.1111/j.1471-4159.2007.05194.x.
72. Llorens-Marítin M, Jurado J, Hernández F, Ávila J. GSK-3β, a pivotal kinase in Alzheimer disease. Front Mol Neurosci. 2014;7:46. https://doi.org/10.3389/fnmol.2014.00046.
73. Hirtz D, Thurman DJ, Gwinn-Hardy K, Mohamed M, Chaudhuri AR, Zalutsky R. How common are the "common" neurologic disorders? Neurology. 2007;68:326–37. https://doi.org/10.1212/01.wnl.0000252807.38124.a3.
74. Elbaz A, Carcaillon L, Kab S, Moisan F. Epidemiology of Parkinson's disease. Rev Neurol (Paris). 2016;172:14–26. https://doi.org/10.1016/j.neurol.2015.09.012.
75. Tan LCS, Venketasubramanian N, Jamora RDG, Heng D. Incidence of Parkinson's disease in Singapore. Parkinsonism Relat Disord. 2007;13:40–3. https://doi.org/10.1016/j.parkreldis.2006.07.003.
76. Zhang Z-X, Roman GC, Hong Z, Wu C-B, Qu Q-M, Huang J-B, Zhou B, Geng Z-P, Wu J-X, Wen H-B, Zhao H, Zahner GE. Parkinson's disease in China: prevalence in Beijing, Xian, and Shanghai. Lancet. 2005;365:595–7. https://doi.org/10.1016/S0140-6736(05)17909-4.
77. Corti O, Lesage S, Brice A. What genetics tells us about the causes and mechanisms of Parkinson's disease. Physiol Rev. 2011;91:1161–218. https://doi.org/10.1152/physrev.00022.2010.
78. Goedert M, Spillantini MG, Del Tredici K, Braak H. 100 years of Lewy pathology. Nat Rev Neurol. 2013;9:13–24. https://doi.org/10.1038/nrneurol.2012.242.
79. Malek N, Swallow D, Grosset KA, Anichtchik O, Spillantini M, Grosset DG. Alpha-synuclein in peripheral tissues and body fluids as a biomarker for Parkinson's disease – a systematic review. Acta Neurol Scand. 2014;130:59–72. https://doi.org/10.1111/ane.12247.

80. Witt SN. Molecular chaperones, alpha-synuclein, and neurodegeneration. Mol Neurobiol. 2013;47:552–60. https://doi.org/10.1007/s12035-012-8325-2.
81. Krüger R, Kuhn W, Müller T, Woitalla D, Graeber M, Kösel S, Przuntek H, Epplen JT, Schols L, Riess O. AlaSOPro mutation in the gene encoding α-synuclein in Parkinson's disease. Nat Genet. 1998;18:106–8. https://doi.org/10.1038/ng0298-106.
82. Soto C. Transmissible proteins: expanding the prion heresy. Cell. 2012;149:968–77. https://doi.org/10.1016/j.cell.2012.05.007.
83. Bellani S, Mescola A, Ronzitti G, Tsushima H, Tilve S, Canale C, Valtorta F, Chieregatti E. GRP78 clustering at the cell surface of neurons transduces the action of exogenous alpha-synuclein. Cell Death Differ. 2014;21:1971–83. https://doi.org/10.1038/cdd.2014.111.
84. Bellucci A, Navarria L, Zaltieri M, Falarti E, Bodei S, Sigala S, Battistin L, Spillantini M, Missale C, Spano P. Induction of the unfolded protein response by α-synuclein in experimental models of Parkinson's disease. J Neurochem. 2011;116:588–605. https://doi.org/10.1111/j.1471-4159.2010.07143.x.
85. Hoozemans JJM, van Haastert ES, Eikelenboom P, de Vos RAI, Rozemuller JM, Scheper W. Activation of the unfolded protein response in Parkinson's disease. Biochem Biophys Res Commun. 2007;354:707–11. https://doi.org/10.1016/j.bbrc.2007.01.043.
86. Colla E, Coune P, Liu Y, Pletnikova O, Troncoso JC, Iwatsubo T, Schneider BL, Lee MK. Endoplasmic reticulum stress is important for the manifestations of α-synucleinopathy *in vivo*. J Neurosci. 2012;32:3306–20. https://doi.org/10.1523/JNEUROSCI.5367-11.2012.
87. Boyce M, Bryant KF, Jousse C, Long K, Harding HP, Scheuner D, Kaufman RJ, Ma D, Coen DM, Ron D, Yuan J. A selective inhibitor of eIF2α dephosphorylation protects cells from ER stress. Science (1979). 2005;307:935–9. https://doi.org/10.1126/science.1101902.
88. Smith WW, Jiang H, Pei Z, Tanaka Y, Morita H, Sawa A, Dawson VL, Dawson TM, Ross CA. Endoplasmic reticulum stress and mitochondrial cell death pathways mediate A53T mutant alpha-synuclein-induced toxicity. Hum Mol Genet. 2005;14:3801–11. https://doi.org/10.1093/hmg/ddi396.
89. Valdés P, Mercado G, Vidal RL, Molina C, Parsons G, Court FA, Martinez A, Galleguillos D, Armentano D, Schneider BL, Hetz C. Control of dopaminergic neuron survival by the unfolded protein response transcription factor XBP1. Proc Natl Acad Sci. 2014;111:6804–9. https://doi.org/10.1073/pnas.1321845111.
90. Gitler AD, Bevis BJ, Shorter J, Strathearn KE, Hamamichi S, Su LJ, Caldwell KA, Caldwell GA, Rochet J-C, McCaffery JM, Barlowe C, Lindquist S. The Parkinson's disease protein α-synuclein disrupts cellular Rab homeostasis. Proc Natl Acad Sci. 2008;105:145–50. https://doi.org/10.1073/pnas.0710685105.
91. Cooper AA, Gitler AD, Cashikar A, Haynes CM, Hill KJ, Bhullar B, Liu K, Xu K, Strathearn KE, Liu F, Cao S, Caldwell KA, Caldwell GA, Marsischky G, Kolodner RD, LaBaer J, Rochet J-C, Bonini NM, Lindquist S. α-Synuclein blocks ER-Golgi traffic and Rab1 rescues neuron loss in Parkinson's models. Science (1979). 2006;313:324–8. https://doi.org/10.1126/science.1129462.
92. Betzer C, Lassen LB, Olsen A, Kofoed RH, Reimer L, Gregersen E, Zheng J, Calì T, Gai W, Chen T, Moeller A, Brini M, Fu Y, Halliday G, Brudek T, Aznar S, Pakkenberg B, Andersen JP, Jensen PH. Alpha-synuclein aggregates activate calcium pump SERCA leading to calcium dysregulation. EMBO Rep. 2018;19:e44617. https://doi.org/10.15252/embr.201744617.
93. Jung E-M, Yoo Y-M, Park SY, et al. Calbindin-D9k is a novel risk gene for neurodegenerative disease. Cell Physiol Biochem. 2020;54:438–56. https://doi.org/10.33594/000000229.
94. Matus S, Valenzuela V, Medinas DB, Hetz C. ER dysfunction and protein folding stress in ALS. Int J Cell Biol. 2013;2013:1–12. https://doi.org/10.1155/2013/674751.
95. Dion PA, Daoud H, Rouleau GA. Genetics of motor neuron disorders: new insights into pathogenic mechanisms. Nat Rev Genet. 2009;10:769–82. https://doi.org/10.1038/nrg2680.
96. Ito Y, Yamada M, Tanaka H, Aida K, Tsuruma K, Shimazawa M, Hozumi I, Inuzuka T, Takahashi H, Hara H. Involvement of CHOP, an ER-stress apoptotic mediator, in both human sporadic ALS and ALS model mice. Neurobiol Dis. 2009;36:470–6. https://doi.org/10.1016/j.nbd.2009.08.013.

97. Hetz C, Thielen P, Matus S, Nassif M, Court F, Kiffin R, Martinez G, Cuervo AM, Brown RH, Glimcher LH. XBP-1 deficiency in the nervous system protects against amyotrophic lateral sclerosis by increasing autophagy. Genes Dev. 2009;23:2294–306. https://doi.org/10.1101/gad.1830709.
98. Chen D, Wang Y, Chin ER. Activation of the endoplasmic reticulum stress response in skeletal muscle of G93A*SOD1 amyotrophic lateral sclerosis mice. Front Cell Neurosci. 2015;9:170. https://doi.org/10.3389/fncel.2015.00170.
99. Tang L, Ma Y, Liu X-L, Chen L, Fan D-S. Better survival in female SOD1-mutant patients with ALS: a study of SOD1-related natural history. Transl Neurodegener. 2019;8:2. https://doi.org/10.1186/s40035-018-0142-8.
100. Kikuchi H, Almer G, Yamashita S, Guégan C, Nagai M, Xu Z, Sosunov AA, McKhann GM, Przedborski S. Spinal cord endoplasmic reticulum stress associated with a microsomal accumulation of mutant superoxide dismutase-1 in an ALS model. Proc Natl Acad Sci. 2006;103:6025–30. https://doi.org/10.1073/pnas.0509227103.
101. Lautenschlaeger J, Prell T, Grosskreutz J. Endoplasmic reticulum stress and the ER mitochondria calcium cycle in amyotrophic lateral sclerosis. Amyotroph Lateral Scler. 2012;13:166–77. https://doi.org/10.3109/17482968.2011.641569.
102. Nishitoh H, Kadowaki H, Nagai A, Maruyama T, Yokota T, Fukutomi H, Noguchi T, Matsuzawa A, Takeda K, Ichijo H. ALS-linked mutant SOD1 induces ER stress- and ASK1-dependent motor neuron death by targeting Derlin-1. Genes Dev. 2008;22:1451–64. https://doi.org/10.1101/gad.1640108.
103. Wang L, Popko B, Roos RP. The unfolded protein response in familial amyotrophic lateral sclerosis. Hum Mol Genet. 2011;20:1008–15. https://doi.org/10.1093/hmg/ddq546.
104. Wang L, Popko B, Roos RP. An enhanced integrated stress response ameliorates mutant SOD1-induced ALS. Hum Mol Genet. 2014;23:2629–38. https://doi.org/10.1093/hmg/ddt658.
105. Jiang H-Q, Ren M, Jiang H-Z, Wang J, Zhang J, Yin X, Wang S-Y, Qi Y, Wang X-D, Feng H-L. Guanabenz delays the onset of disease symptoms, extends lifespan, improves motor performance and attenuates motor neuron loss in the SOD1 G93A mouse model of amyotrophic lateral sclerosis. Neuroscience. 2014;277:132–8. https://doi.org/10.1016/j.neuroscience.2014.03.047.
106. Dzhashiashvili Y, Monckton CP, Shah HS, Kunjamma RB, Popko B. The UPR-PERK pathway is not a promising therapeutic target for mutant SOD1-induced ALS. Neurobiol Dis. 2019;127:527–44. https://doi.org/10.1016/j.nbd.2019.03.024.
107. Matus S, Lopez E, Valenzuela V, Nassif M, Hetz C. Functional contribution of the transcription factor ATF4 to the pathogenesis of amyotrophic lateral sclerosis. PLoS One. 2013;8:e66672. https://doi.org/10.1371/journal.pone.0066672.
108. Lee S, Shang Y, Redmond SA, Urisman A, Tang AA, Li KH, Burlingame AL, Pak RA, Jovičić A, Gitler AD, Wang J, Gray NS, Seeley WW, Siddique T, Bigio EH, Lee VM-Y, Trojanowski JQ, Chan JR, Huang EJ. Activation of HIPK2 promotes ER stress-mediated neurodegeneration in amyotrophic lateral sclerosis. Neuron. 2016;91:41–55. https://doi.org/10.1016/j.neuron.2016.05.021.
109. Clark EM, Nonarath HJT, Bostrom JR, Link BA. Establishment and validation of an endoplasmic reticulum stress reporter to monitor zebrafish ATF6 activity in development and disease. Dis Model Mech. 2020;13:dmm041426. https://doi.org/10.1242/dmm.041426.
110. Walker AK, Atkin JD. Mechanisms of neuroprotection by protein disulphide isomerase in amyotrophic lateral sclerosis. Neurol Res Int. 2011;2011:1–7. https://doi.org/10.1155/2011/317340.
111. Woehlbier U, Colombo A, Saaranen MJ, Pérez V, Ojeda J, Bustos FJ, Andreu CI, Torres M, Valenzuela V, Medinas DB, Rozas P, Vidal RL, Lopez-Gonzalez R, Salameh J, Fernandez-Collemann S, Muñoz N, Matus S, Armisen R, Sagredo A, Palma K, Irrazabal T, Almeida S, Gonzalez-Perez P, Campero M, Gao F, Henny P, van Zundert B, Ruddock LW, Concha ML, Henriquez JP, Brown RH, Hetz C. ALS-linked protein disulfide isomerase variants cause motor dysfunction. EMBO J. 2016;35:845–65. https://doi.org/10.15252/embj.201592224.

112. Ogura J, Ruddock LW, Mano N. Cysteine 343 in the substrate binding domain is the primary S-nitrosylated site in protein disulfide isomerase. Free Radic Biol Med. 2020;160:103–10. https://doi.org/10.1016/j.freeradbiomed.2020.07.029.
113. Walker AK, Soo KY, Sundaramoorthy V, Parakh S, Ma Y, Farg MA, Wallace RH, Crouch PJ, Turner BJ, Horne MK, Atkin JD. ALS-associated TDP-43 induces endoplasmic reticulum stress, which drives cytoplasmic TDP-43 accumulation and stress granule formation. PLoS One. 2013;8:e81170. https://doi.org/10.1371/journal.pone.0081170.
114. Kim H-J, Raphael AR, LaDow ES, McGurk L, Weber RA, Trojanowski JQ, Lee VM-Y, Finkbeiner S, Gitler AD, Bonini NM. Therapeutic modulation of eIF2α phosphorylation rescues TDP-43 toxicity in amyotrophic lateral sclerosis disease models. Nat Genet. 2014;46:152–60. https://doi.org/10.1038/ng.2853.
115. Farg MA, Soo KY, Walker AK, Pham H, Orian J, Horne MK, Warraich ST, Williams KL, Blair IP, Atkin JD. Mutant FUS induces endoplasmic reticulum stress in amyotrophic lateral sclerosis and interacts with protein disulfide-isomerase. Neurobiol Aging. 2012;33:2855–68. https://doi.org/10.1016/j.neurobiolaging.2012.02.009.
116. Selvaraj BT, Livesey MR, Zhao C, Gregory JM, James OT, Cleary EM, Chouhan AK, Gane AB, Perkins EM, Dando O, Lillico SG, Lee Y-B, Nishimura AL, Poreci U, Thankamony S, Pray M, Vasistha NA, Magnani D, Borooah S, Burr K, Story D, McCampbell A, Shaw CE, Kind PC, Aitman TJ, Whitelaw CBA, Wilmut I, Smith C, Miles GB, Hardingham GE, Wyllie DJA, Chandran S. C9ORF72 repeat expansion causes vulnerability of motor neurons to Ca2+-permeable AMPA receptor-mediated excitotoxicity. Nat Commun. 2018;9:347. https://doi.org/10.1038/s41467-017-02729-0.
117. Dreser A, Vollrath JT, Sechi A, Johann S, Roos A, Yamoah A, Katona I, Bohlega S, Wiemuth D, Tian Y, Schmidt A, Vervoorts J, Dohmen M, Beyer C, Anink J, Aronica E, Troost D, Weis J, Goswami A. The ALS-linked E102Q mutation in Sigma receptor-1 leads to ER stress-mediated defects in protein homeostasis and dysregulation of RNA-binding proteins. Cell Death Differ. 2017;24:1655–71. https://doi.org/10.1038/cdd.2017.88.
118. Gregory JM, Livesey MR, McDade K, Selvaraj BT, Barton SK, Chandran S, Smith C. Dysregulation of AMPA receptor subunit expression in sporadic ALS post-mortem brain. J Pathol. 2020;250:67–78. https://doi.org/10.1002/path.5351.
119. Gkogkas C, Middleton S, Kremer AM, Wardrope C, Hannah M, Gillingwater TH, Skehel P. VAPB interacts with and modulates the activity of ATF6. Hum Mol Genet. 2008;17:1517–26. https://doi.org/10.1093/hmg/ddn040.
120. Hall CE, Yao Z, Choi M, Tyzack GE, Serio A, Luisier R, Harley J, Preza E, Arber C, Crisp SJ, Watson PMD, Kullmann DM, Abramov AY, Wray S, Burley R, Loh SHY, Martins LM, Stevens MM, Luscombe NM, Sibley CR, Lakatos A, Ule J, Gandhi S, Patani R. Progressive motor neuron pathology and the role of astrocytes in a human stem cell model of VCP-related ALS. Cell Rep. 2017;19:1739–49. https://doi.org/10.1016/j.celrep.2017.05.024.
121. Mays CE, Soto C. The stress of prion disease. Brain Res. 2016;1648:553–60. https://doi.org/10.1016/j.brainres.2016.04.009.
122. Hughes D, Halliday M. What is our current understanding of PrPSc-associated neurotoxicity and its molecular underpinnings? Pathogens. 2017;6:63. https://doi.org/10.3390/pathogens6040063.
123. Chakrabarti O, Ashok A, Hegde RS. Prion protein biosynthesis and its emerging role in neurodegeneration. Trends Biochem Sci. 2009;34:287–95. https://doi.org/10.1016/j.tibs.2009.03.001.
124. Thapa S, Abdelaziz DH, Abdulrahman BA, Schatzl HM. Sephin1 reduces prion infection in prion-infected cells and animal model. Mol Neurobiol. 2020;57:2206–19. https://doi.org/10.1007/s12035-020-01880-y.
125. Orsi A, Fioriti L, Chiesa R, Sitia R. Conditions of endoplasmic reticulum stress favor the accumulation of cytosolic prion protein. J Biol Chem. 2006;281:30431–8. https://doi.org/10.1074/jbc.M605320200.
126. Nunziante M, Ackermann K, Dietrich K, Wolf H, Gädtke L, Gilch S, Vorberg I, Groschup M, Schätzl HM. Proteasomal dysfunction and endoplasmic reticulum stress enhance traf-

ficking of prion protein aggregates through the secretory pathway and increase accumulation of pathologic prion protein. J Biol Chem. 2011;286:33942–53. https://doi.org/10.1074/jbc.M111.272617.

127. Rane NS, Kang S-W, Chakrabarti O, Feigenbaum L, Hegde RS. Reduced translocation of nascent prion protein during ER stress contributes to neurodegeneration. Dev Cell. 2008;15:359–70. https://doi.org/10.1016/j.devcel.2008.06.015.
128. Hetz C, Russelakis-Carneiro M, Maundrell K, Castilla J, Soto C. Caspase-12 and endoplasmic reticulum stress mediate neurotoxicity of pathological prion protein. EMBO J. 2003;22:5435–45. https://doi.org/10.1093/emboj/cdg537.
129. Mays CE, Armijo E, Morales R, Kramm C, Flores A, Tiwari A, Bian J, Telling GC, Pandita TK, Hunt CR, Soto C. Prion disease is accelerated in mice lacking stress-induced heat shock protein 70 (HSP70). J Biol Chem. 2019;294:13619–28. https://doi.org/10.1074/jbc.RA118.006186.
130. Park K-W, Eun Kim G, Morales R, Moda F, Moreno-Gonzalez I, Concha-Marambio L, Lee AS, Hetz C, Soto C. The endoplasmic reticulum chaperone GRP78/BiP modulates prion propagation in vitro and in vivo. Sci Rep. 2017;7:44723. https://doi.org/10.1038/srep44723.
131. Smith HL, Freeman OJ, Butcher AJ, Holmqvist S, Humoud I, Schätzl T, Hughes DT, Verity NC, Swinden DP, Hayes J, de Weerd L, Rowitch DH, Franklin RJM, Mallucci GR. Astrocyte unfolded protein response induces a specific reactivity state that causes non-cell-autonomous neuronal degeneration. Neuron. 2020;105:855–866.e5. https://doi.org/10.1016/j.neuron.2019.12.014.
132. Moreno JA, Halliday M, Molloy C, Radford H, Verity N, Axten JM, Ortori CA, Willis AE, Fischer PM, Barrett DA, Mallucci GR. Oral treatment targeting the unfolded protein response prevents neurodegeneration and clinical disease in prion-infected mice. Sci Transl Med. 2013;5:206ra138. https://doi.org/10.1126/scitranslmed.3006767.
133. Halliday M, Radford H, Sekine Y, Moreno J, Verity N, le Quesne J, Ortori CA, Barrett DA, Fromont C, Fischer PM, Harding HP, Ron D, Mallucci GR. Partial restoration of protein synthesis rates by the small molecule ISRIB prevents neurodegeneration without pancreatic toxicity. Cell Death Dis. 2015;6:e1672. https://doi.org/10.1038/cddis.2015.49.
134. Urano F, Wang X, Bertolotti A, Zhang Y, Chung P, Harding HP, Ron D. Coupling of stress in the ER to activation of JNK protein kinases by transmembrane protein kinase IRE1. Science. 2000;287:664–6. https://doi.org/10.1126/science.287.5453.664.
135. Casas-Tinto S, Zhang Y, Sanchez-Garcia J, Gomez-Velazquez M, Rincon-Limas DE, Fernandez-Funez P. The ER stress factor XBP1s prevents amyloid-beta neurotoxicity. Hum Mol Genet. 2011;20:2144–60. https://doi.org/10.1093/hmg/ddr100.
136. Hetz C, Lee A-H, Gonzalez-Romero D, Thielen P, Castilla J, Soto C, Glimcher LH. Unfolded protein response transcription factor XBP-1 does not influence prion replication or pathogenesis. Proc Natl Acad Sci USA. 2008;105:757–62. https://doi.org/10.1073/pnas.0711094105.
137. Sasaki S. Endoplasmic reticulum stress in motor neurons of the spinal cord in sporadic amyotrophic lateral sclerosis. J Neuropathol Exp Neurol. 2010;69:346–55. https://doi.org/10.1097/NEN.0b013e3181d44992.

16 Ameliorative Effects of Heat Shock Proteins in Therapeutic Implications of Neurodegenerative Diseases

Divya Bisht and Deena Prakash

Abbreviations

AAV	Adeno-associated viral
AIF	Apoptosis-inducing factor
Aβ	Amyloid-beta
BBB	Blood–brain barrier
CMA	Chaperone-mediated autophagy
ERAD	Endoplasmic reticulum-associated degradation
GGA	Geranylgeranylacetone
HSF1	Heat shock factor 1
HSPs	Heat Shock Proteins
LTP	Long-term potentiation
NDs	Neurodegenerative diseases
PQC	Protein quality control
ROS	Reactive oxygen species
sHSPs	Small Heat Shock Proteins
UPR	Unfolded protein response
UPS	Ubiquitin–proteasome system

D. Bisht (✉) · D. Prakash
Biochemistry Discipline, School of Sciences, Indira Gandhi National Open University, New Delhi, India
e-mail: divyabisht489@gmail.com

N. Puranik (ed.), *Proteostasis*, https://doi.org/10.1007/978-981-96-6202-9_16

16.1 Introduction

Neurodegenerative diseases (NDs) are a group of disorders characterized by progressive dysfunction and death of neurons, resulting in cognitive decline, motor dysfunction, and other neurological impairments causing difficulty in the normal functioning of the patient. Major NDs include Alzheimer's disease, Parkinson's disease, Huntington's disease, and Amyotrophic Lateral Sclerosis, each with unique signs and symptoms but sharing several common pathological mechanisms [31].

Alzheimer's disease is the most prevalent ND characterized by progressive memory loss and cognitive decline due to amyloid-beta (Aβ) plaques and tau protein tangles, which disrupt neuron communication and lead to extensive neuronal death, particularly in regions critical for memory, such as the hippocampus and cortex. Parkinson's disease is the second most common ND, primarily affecting body movement characterized by tremors, rigidity, and bradykinesia. Its hallmark pathology includes the accumulation of Lewy bodies (aggregates of misfolded alpha-synuclein protein) in dopaminergic neurons of the substantia nigra, a brain region responsible for movement coordination. Huntington's disease is an autosomal dominant genetic disorder, results in involuntary movements, cognitive decline, and psychiatric symptoms due to the accumulation of mutated huntingtin protein, primarily affecting the basal ganglia. Amyotrophic Lateral Sclerosis, also known as Lou Gehrig's disease, leads to rapid muscle weakness, paralysis, and eventual respiratory failure, caused by the loss of motor neurons in the brain and spinal cord. All these NDs have own characteristic impacts on the body; however, each of these NDs involves protein misfolding and aggregation disrupting cellular function, contributing to oxidative stress and mitochondrial dysfunction and ultimately resulting in cell death. Neuroinflammation, driven by activated microglia and astrocytes, also plays a key role in disease progression. Moreover, impairment in the autophagy–lysosome and ubiquitin–proteasome pathways, responsible for clearing damaged proteins, is a common mechanism across these diseases [13, 27, 31].

Neurodegenerative diseases represent a prominent public health challenge due to their progressive nature and lack of effective treatments. While each disease affects different regions of the brain and leads to distinct clinical symptoms, they share common pathological features, including protein misfolding, oxidative stress, and neuroinflammation [27].

Heat Shock Proteins (HSPs) are a family of proteins with essential functioning of cellular homeostasis and protection against various stress conditions. Firstly named after their response to heat, HSPs are also known to react to various other stressors, including oxidative stress, inflammation, ischemia, and toxin exposure. These proteins primarily work as molecular chaperones managing correct protein folding, preventing the accumulation of misfolded proteins, and facilitating the repair or degradation of damaged proteins. By modulating various cellular pathways, HSPs not only prevent cell damage but also promote the recovery and survival of cells in stressful conditions. HSPs also play a role in immune regulation acting as danger signals when released extracellularly during cell stress or damage, stimulating both innate and adaptive immune responses. Overall, HSPs serve as protectors of cellular integrity under both normal and stress conditions. By ensuring correct protein folding, preventing toxic aggregation, inhibiting apoptosis, and regulating immune responses, HSPs are fundamental for maintaining cellular homeostasis and soldiers in the face of environmental and physiological challenges. The function of HSPs extends beyond simple stress response in the body, as they contribute to broader cellular functions that make them essential for the survival and function of cells across diverse biological contexts [32].

HSPs are also essential in addressing the molecular pathology of NDs due to their main pathological features, such as protein misfolding, oxidative stress, mitochondrial dysfunction, and chronic neuroinflammation, which contribute to neuronal loss. As molecular chaperones, HSPs, particularly HSP70 and HSP90, check and prevent the toxic accumulation of misfolded proteins and promote correct folding and direct damaged proteins to degradation pathways to reduce harmful aggregates. HSPs also inhibit and reduce oxidative stress by stabilizing antioxidant enzymes and repairing mitochondrial damage, thus helping in preservation of neuronal function. Additionally, HSPs modulate inflammatory responses by acting as signaling molecules to manage immune activity, reducing the damaging effects of chronic neuroinflammation [17].

This chapter delves into the fundamental characteristics of HSPs, covering their various types, molecular mechanisms, and critical roles in cellular function. It further explores how these proteins act as protective agents, with a focus on their potential to counteract the cellular and molecular dysfunctions that drive neurodegenerative diseases. Through examining their role in protein folding, oxidative stress response, inflammation modulation, and apoptosis inhibition, this chapter highlights the promising therapeutic potential of Heat Shock Proteins in mitigating the progression and impact of conditions such as Alzheimer's, Parkinson's, Huntington's, and Amyotrophic Lateral Sclerosis.

16.2 Heat Shock Proteins: Structure and Function

Heat Shock Proteins (HSPs) are a highly conserved family of proteins with primary role in maintaining cellular homeostasis, especially under stress conditions as shown in Fig. 16.1. The structure of HSPs is characterized by the presence of

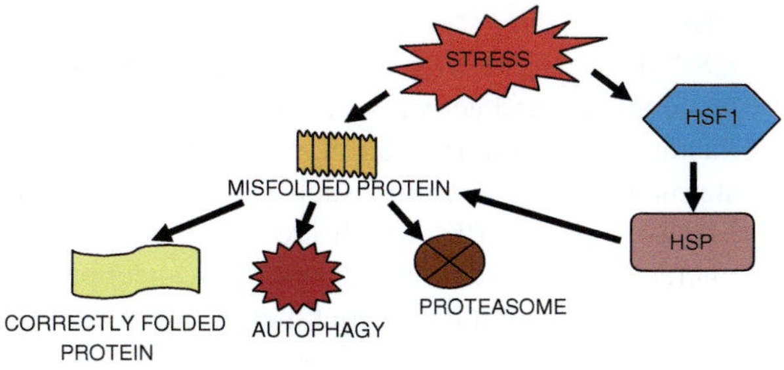

Fig. 16.1 HSPs and stress response

specific domains to bind with and stabilize other proteins, assisting efficient protein folding, assembly, and function. This structural organization varies across different HSP families, categorized by molecular weight into various classes. Each class is distinguished by unique structural features and chaperone functions tailored to specific cellular needs [15, 32].

16.2.1 Types of Heat Shock Proteins (HSPs)

- HSP27 (Small Heat Shock Proteins, sHSPs)
  HSP27 is a small heat shock protein with a molecular weight of ~27 kDa. It protects cells under stress by preventing misfolded protein aggregation, stabilizing the cytoskeleton, and inhibiting apoptosis through caspase-3 suppression. It also regulates oxidative stress and inflammation, making it vital for cellular resilience.
- HSP40 (DNAJ Proteins)
  HSP40 are also known as DNAJ proteins with the molecular weight ranging from 40 to 50 kDa. These primarily function as co-chaperone to HSP70 by recognizing misfolded proteins and delivering them to HSP70 for refolding. They also regulate HSP70's ATPase activity, enhancing its refolding capacity. Together, HSP40 and HSP70 prevent protein aggregation and maintain cellular proteostasis.
- HSP60 (Chaperonins)
  HSP60 is a chaperonin with a molecular weight of approximately 60 kDa. It is located mainly in mitochondria, facilitating the correct folding of mitochondrial proteins crucial for energy production and cellular respiration. HSP60 prevents dysfunction and aids in repairing damaged proteins in response to stress ensuring proper functioning of mitochondria and preventing cellular apoptosis.
- HSP70
  HSP70 is among the most extensively studied member of the HSP family with a molecular weight of approximately 70 kDa. It plays a central role in protein folding, assembly, translocation, and degradation by binding to nascent polypeptides, preventing premature folding and aggregation. Under stress, HSP70 assists in refolding denatured proteins and promotes their degradation when repair is not possible. Additionally, it has antiapoptotic functions by inhibiting key components of the cell death machinery, such as apoptosis-inducing factor (AIF) and caspases.
- HSP90
  HSP90 has a molecular weight of approximately 90 kDa and plays a crucial role in stabilizing and activating a variety of client proteins involved in signal transduction, cell cycle regulation, and apoptosis. It is crucial for the maturation of steroid hormone receptors, such as estrogen and androgen receptors, and kinases like Akt and MAPK, which are vital for cellular growth and survival. It also interacts with proteins associated with cancer progression and neurodegenerative diseases, making it a potential therapeutic target. HSP90 also plays a key role in

protein triage decisions, determining whether damaged proteins should be refolded or degraded. Structure of one such HSP90 protein is shown in Fig. 16.2.

- HSP100
  HSP100 (approximately 100 kDa) specializes in preventing protein aggregation and assisting in refolding them into their functional conformations. These work in conjunction with HSP70 and other chaperones to dissolve large protein aggregates, thereby restoring protein homeostasis under stress conditions. HSP100 is crucial in cells exposed to extreme stress, such as heat shock, where the likelihood of protein aggregation increases significantly.
- Ubiquitin
  Ubiquitin has a molecular weight of approximately 8.5 kDa and plays a vital role in the proteostasis network by tagging damaged or misfolded proteins for degradation via the proteasome. Being smaller than most of the HSPs, ubiquitin collaborates closely with them to ensure that irreparably damaged proteins are eliminated maintaining cellular health by clearing potentially toxic proteins that could accumulate and disrupt normal cellular functions.

As discussed, each HSP type has specialized functions, yet together they form a network that protects cells from stress-induced damage. By facilitating protein folding, preventing aggregation, aiding in protein degradation, and protecting against apoptosis, HSPs are essential for cellular homeostasis.

16.2.2 Mechanism of Protein Quality Control and Prevention of Protein Misfolding

Protein quality control (PQC) is essential for ensuring proper folding of proteins, maintaining their functional conformations, degrading damaged or misfolded proteins, and eliminating them from the cell. This system is crucial for cellular homeostasis, as misfolded proteins lead to toxic aggregates associated with neurodegenerative diseases like Alzheimer's, Parkinson's, and Huntington's. PQC

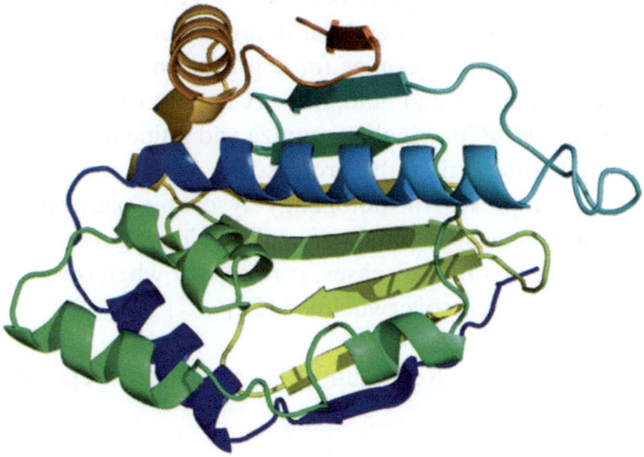

Fig. 16.2 Structure of the HSP90AA1 protein

upholds proteostasis through various mechanisms involving molecular chaperones, degradation pathways, and cellular stress responses [19, 24].

- Molecular Chaperones

Molecular chaperones, including Heat Shock Proteins (HSPs), are essential for preventing all sorts of protein misfolding by crucially facilitating efficient protein folding, refolding, and preventing aggregation. HSPs bind to nascent polypeptide chains during translation ensuring that newly synthesized proteins achieve their correct three-dimensional structure. HSP70 and HSP90 are key chaperone systems where HSP70 binds to unfolded or partially folded proteins to prevent aggregation and promotes correct folding through cycles of binding and release powered by ATP hydrolysis. On the other hand, HSP90 stabilizes and matures the proteins involved in cell signaling. Together, HSP70 AND HSP90 form a critical defense against protein misfolding and aid in refolding misfolded proteins resulting from stress conditions such as heat shock and oxidative stress [7].

- Proteasomal Degradation (Ubiquitin–Proteasome System)

In extreme cases where it is difficult to refold the misfolded proteins, these faulty proteins are usually targeted for degradation to prevent their accumulation in the cell. The ubiquitin–proteasome system (UPS) serves as the primary mechanism for this degradation. This process involves small ubiquitin molecules which get attached to irreparably misfolded proteins. Such ubiquitinated proteins are directed to the proteasome, a large proteolytic complex that breaks them down into smaller peptides. This system ensures that improperly folded or damaged proteins are eliminated, thereby maintaining the health and functionality of the cellular proteome [29].

- Autophagy–Lysosome Pathway

When misfolded proteins accumulate into larger aggregates that the proteasome cannot efficiently degrade, the autophagy–lysosome pathway is activated for their clearance. Autophagy is a cellular process that targets these larger aggregates, damaged organelles, and cellular debris by enclosing them in autophagosomes, which then fuse with lysosomes for degradation. A specific type of autophagy, known as chaperone-mediated autophagy (CMA), involves heat shock proteins (HSPs), particularly HSP70, which recognize and guide misfolded proteins to the lysosome for degradation. Unlike the proteasome, which mainly handles individual proteins or small complexes, the autophagy–lysosome pathway effectively degrades large protein aggregates, making it crucial for clearing the protein aggregates associated with neurodegenerative diseases, particularly when other protein quality control mechanisms have been overwhelmed [22].

- Endoplasmic Reticulum (ER)-Associated Degradation (ERAD)

The endoplasmic reticulum (ER) is critical cellular organelle in protein folding process, particularly for secreted and membrane-bound proteins. When proteins misfold in the ER, the ER-associated degradation (ERAD) pathway is activated which retrotranslocates misfolded proteins back into the cytosol, where they are ubiquitinated and degraded by the proteasome. ERAD maintain proteostasis within the ER and preventing ER stress, which can activate the unfolded protein response (UPR). The UPR is a cellular stress response that increases the production of chaperones and other PQC components to manage the accumulation of misfolded proteins [24].

- Stress Response Pathways

Cells have evolved multiple stress response pathways, including HSR and UPR, to enhance protein quality control under conditions like heat shock and oxidative stress which can trigger protein misfolding. In HSR, heat shock factor 1 (HSF1) is activated during stress, translocating to the nucleus to induce the expression of HSPs and other chaperones to improve protein refolding or degradation of misfolded proteins.

Meanwhile, the UPR is triggered by the accumulation of misfolded proteins in ER, enhancing the protein folding capacity of the endoplasmic reticulum by increasing chaperone production and reducing the influx of new proteins, while also boosting ER-associated degradation (ERAD) activity and potentially inducing autophagy if protein aggregation persists, ensuring overall cellular health [15].

- Prevention of Protein Aggregation

PQC mechanisms not only assist in protein folding and degradation but also actively prevent the formation of harmful protein aggregates. Chaperones such as HSP70, HSP90, and sHSPs bind to misfolded proteins, while HSP100 proteins specialize in separating these large aggregates to restore protein functionality or facilitate degradation [5].

Therefore, a protein quality control system is vital for preventing misfolding and maintaining cellular proteostasis through chaperones, degradation pathways, and stress responses. Dysfunction in these systems can result in the accumulation of misfolded proteins, contributing to various diseases, especially neurodegenerative disorders.

16.2.3 Heat Shock Factor (HSF) Pathway and Cellular Stress Response

The Heat Shock Factor (HSF) pathway is an important cellular mechanism for the regulation of proteotoxic stress response as well as the promotion of the heat shock response (HSR) to protect cells from environmental and physiological challenges. HSFs, mainly HSF1, are transcription factors that activate heat shock proteins

(HSPs) during the accumulation of misfolded proteins under any stress condition. Therefore, HSF1 remains dormant in normal conditions binding with HSP70 and HSP90. However during stress, misfolded proteins displace HSF1 from these chaperones, promoting its trimerization and nuclear translocation to bind heat shock elements (HSEs) in gene promoters. This binding initiates the HSPs transcription (HSP70, HSP90, small HSPs (e.g., HSP27), HSP60, and HSP100), facilitating protein folding, inhibit aggregation, and manage the degradation of irreparable proteins. On alleviation of stress, the HSF1 pathway is downregulated through dephosphorylation and feedback mechanisms involving increased HSP levels that return HSF1 to its inactive form, ensuring a transient response tailored to cellular protection [25, 30].

- Role of HSF1 in Diseases and Aging

HSF1 and the heat shock response are vital for both acute stress management and the long-term health of cellular system, playing a key role in preventing various diseases. Dysregulation of the HSF1 pathway is linked to several pathological conditions, including neurodegenerative diseases, cancer, and aging. In neurodegenerative diseases like Alzheimer's, Parkinson's, and Huntington's, impaired HSF1 activation and reduced HSP expression lead to the accumulation of misfolded proteins, making it a potential target for therapeutic enhancement. In the case of cancer, HSF1 is often hyperactivated, promoting cancer cell survival by increasing HSP expression and stabilizing oncogenic proteins, which is why inhibitors of HSF1 and HSP90 are being investigated as cancer treatments. In addition to its role in neurodegenerative diseases and cancer, aging is associated with a decline in the heat shock response, resulting in decreased chaperone function and an increased risk of protein misfolding, contributing to age-related diseases [12].

The Heat Shock Factor (HSF) pathway is crucial for cellular stress response, enabling cells to recover from protein misfolding and damage by activating heat shock proteins. Proper regulation of HSF1 is vital for maintaining proteostasis and cellular health, with its dysregulation linked to diseases like neurodegeneration, cancer, and aging. Targeting the HSF1 pathway presents promising therapeutic opportunities for stress-related conditions.

16.3 Molecular Pathogenesis of Neurodegenerative Diseases

Neurodegenerative diseases (NDs) encompass a diverse range of disorders marked by the progressive degeneration of the nervous system, with notable examples including Alzheimer's, Parkinson's, Huntington's, and amyotrophic lateral sclerosis. Their molecular pathogenesis is intricate and multifactorial, involving genetic, environmental, and biochemical factors. Key mechanisms contributing to NDs include protein misfolding and aggregation, oxidative stress, mitochondrial dysfunction, neuroinflammation, and impaired protein quality control [4, 17].

- Protein Misfolding and Aggregation

Neurodegenerative diseases are characterized by the accumulation of misfolded proteins that form toxic aggregates, adversely affecting neuronal function. In Alzheimer's disease, amyloid-beta (Aβ) peptides aggregate into plaques, while hyperphosphorylated tau proteins form neurofibrillary tangles, both contributing to neuronal death. Parkinson's disease is marked by the misfolding of alpha-synuclein into Lewy bodies, which disrupts synaptic function. In Huntington's disease, an expanded CAG repeat in the HTT gene results in a polyglutamine (polyQ) tract in the huntingtin protein, leading to misfolding and aggregation that impairs cellular processes. The aggregation mechanisms are influenced not only by the proteins' intrinsic properties but also by the cellular environment, including chaperone availability and cellular stressors [35].

- Oxidative Stress

Oxidative stress occurs when there is an imbalance between the production of reactive oxygen species (ROS) and the body's ability to detoxify them or repair the resulting damage. In neurodegenerative diseases, elevated ROS levels cause the oxidation of proteins, lipids, and nucleic acids, leading to cellular injury and death, with neurons being particularly vulnerable due to their high metabolic rate and low antioxidant capacity. In Alzheimer's disease, oxidative stress is linked to amyloid plaque accumulation, resulting in neuronal damage and impaired protein function, which can further promote misfolding and aggregation. Similarly, in Parkinson's disease, the degeneration of dopaminergic neurons is associated with oxidative stress stemming from mitochondrial dysfunction and increased ROS production. Consequently, therapeutic strategies aimed at enhancing antioxidant defenses are being investigated to reduce oxidative damage in these diseases [18].

- Mitochondrial Dysfunction

Mitochondria are essential for energy production and are pivotal in cellular metabolism, apoptosis, and calcium homeostasis. Mitochondrial dysfunction is a hallmark of various neurodegenerative diseases, contributing to neuronal cell death through decreased ATP production, increased ROS generation, and the release of proapoptotic factors, ultimately leading to neuronal apoptosis. In Alzheimer's disease, mitochondrial dysfunction is linked to reduced electron transport chain activity and changes in mitochondrial dynamics. In Parkinson's disease, mutations in genes such as PINK1 and Parkin, which are crucial for mitochondrial quality control, disrupt mitophagy—the process that selectively degrades damaged mitochondria—resulting in the accumulation of dysfunctional mitochondria and worsening neurodegeneration [9, 17].

- Neuroinflammation

Neuroinflammation is a crucial aspect of the molecular pathogenesis of neurodegenerative diseases, characterized by the activation of glial cells, such as microglia and astrocytes, in response to neuronal injury. While neuroinflammation can act as a protective response, chronic inflammation often becomes harmful, leading to neuronal dysfunction and death. In Alzheimer's disease, activated microglia cluster around amyloid plaques, releasing proinflammatory cytokines that exacerbate neuronal injury and create a chronic inflammatory state, which recruits additional immune cells and perpetuates neuroinflammation and neurodegeneration. Similarly, in Parkinson's disease, inflammation is believed to contribute to the loss of dopaminergic neurons, with activated microglia producing neurotoxic substances that further damage surrounding neurons [10].

- Impaired Protein Quality Control

Maintaining proteostasis is vital for neuronal health, as impaired protein quality control mechanisms, such as those involving molecular chaperones and proteasomal degradation pathways, can result in the accumulation of misfolded proteins and contribute to neurodegeneration. Many neurodegenerative diseases exhibit decreased expression and function of chaperones, hindering the refolding and removal of misfolded proteins from cells. The ubiquitin–proteasome system (UPS) serves as the primary mechanism for degrading damaged or misfolded proteins; dysregulation of this system is linked to various neurodegenerative diseases, leading to the accumulation of toxic proteins. For instance, in Huntington's disease, impaired proteasomal function results in the buildup of polyglutamine-expanded huntingtin, exacerbating neuronal toxicity.

- Genetic and Environmental Factors

Genetic predispositions significantly influence the pathogenesis of neurodegenerative diseases, with mutations in specific genes such as APP, PSEN1, and PSEN2 linked to Alzheimer's disease, and SNCA associated with Parkinson's disease identified as key risk factors. Additionally, polygenic risk factors contribute to overall disease susceptibility. Environmental factors, including exposure to toxins, oxidative stress, and lifestyle choices, further interact with genetic susceptibility to affect the onset and progression of these diseases. For example, exposure to pesticides and heavy metals has been correlated with an increased risk of developing Parkinson's disease [34].

The molecular pathogenesis of neurodegenerative diseases involves a complex interplay of mechanisms, including protein misfolding, oxidative stress, mitochondrial dysfunction, neuroinflammation, and impaired protein quality control. Understanding these pathways is vital for developing targeted therapies to prevent or slow disease progression. Advances in molecular biology and genetics hold promise for identifying new therapeutic targets and improving disease management. Ongoing research into the mechanisms of neurodegeneration may lead to innovative interventions that enhance neuronal survival and function [4, 18].

16.4 Mechanism of Action: HSPs in Neuroprotection

Heat Shock Proteins (HSPs) are highly conserved molecular chaperones that play a vital role in cellular protection, especially in neurodegenerative diseases. Induced by stressors like heat, oxidative stress, and inflammation, HSPs assist in proper protein folding, prevent aggregation, and promote the degradation of misfolded proteins. Their neuroprotective mechanisms are multifaceted, involving key pathways such as protein quality control, antiapoptotic effects, modulation of inflammation, and enhancement of cellular resilience [21].

- Protein Quality Control

Heat Shock Proteins (HSPs) play a crucial role in maintaining proteostasis, particularly in neurons that are sensitive to misfolded proteins and aggregates. They assist in proper protein folding by binding to nascent polypeptides and partially folded proteins, with HSP70 and HSP90 preventing aggregation through ATP-dependent cycles of binding and release. Additionally, HSPs like HSP100 can disaggregate protein aggregates, restoring protein function and mitigating the toxicity associated with aggregated proteins, such as Aβ plaques in Alzheimer's disease and Lewy bodies in Parkinson's disease. Furthermore, HSPs facilitate the degradation of misfolded proteins by directing them to the ubiquitin–proteasome pathway, helping to prevent cellular toxicity [7].

- Antiapoptotic Effects

Heat Shock Proteins (HSPs) exhibit significant antiapoptotic properties, protecting neurons from programmed cell death in response to stress. They inhibit the activation of caspases, proteases that are crucial in the apoptotic process, thereby promoting neuronal survival under stress conditions. HSPs also interact with members of the Bcl-2 family of proteins, which regulate apoptosis; for instance, HSP70 can bind to and inhibit proapoptotic proteins like Bad, enhancing cell survival. Additionally, HSPs help maintain mitochondrial function and integrity, which are vital for neuronal health. By preventing mitochondrial permeability transition and supporting the proper functioning of the electron transport chain, HSPs enable neurons to withstand metabolic stress [29].

- Modulation of Inflammation

Chronic neuroinflammation significantly contributes to neurodegenerative diseases, and Heat Shock Proteins (HSPs) play a role in modulating inflammatory responses. HSPs, such as HSP70, can influence the activation state of microglia, the resident immune cells in the central nervous system, promoting an anti-inflammatory phenotype that reduces neuroinflammation and its harmful effects on neurons. Additionally, HSPs regulate the production of proinflammatory cytokines and chemokines, helping to modulate the inflammatory environment in the brain. By limiting excessive inflammation, HSPs help protect neurons from inflammatory damage [10].

- Enhancement of Cellular Resilience

Heat Shock Proteins (HSPs) enhance the resilience of neurons against stressors, promoting cell survival and function. Their upregulation during stress conditions increases the cellular pool of chaperones, enabling neurons to better manage protein misfolding and aggregation, thereby maintaining functional integrity. HSPs also participate in various signaling pathways that promote cellular survival; for instance, HSP70 interacts with kinases involved in survival signaling to support neuroprotective pathways. Additionally, HSPs play a role in synaptic plasticity and memory formation, with HSP70 being essential for long-term potentiation (LTP), a critical process for learning and memory. This involvement in synaptic function highlights the broader significance of HSPs in maintaining neuronal health [3].

- Clinical Implications and Therapeutic Potential

Given their critical roles in neuroprotection, Heat Shock Proteins (HSPs) have emerged as potential therapeutic targets for treating neurodegenerative diseases. Researchers are exploring compounds that induce HSP expression, such as geranylgeranylacetone (GGA) and other small molecules, for their neuroprotective effects, which can enhance HSP levels and improve cellular resilience in various neurodegeneration models. Gene therapy approaches aim to boost the expression of specific HSPs in neurons, potentially providing enhanced neuroprotection in conditions like Alzheimer's and Parkinson's diseases. Additionally, developing therapies that directly target specific HSPs or their pathways could open new avenues for intervention in these neurodegenerative disorders [6, 8, 21].

Heat Shock Proteins (HSPs) are crucial for neuroprotection, maintaining protein quality control, inhibiting apoptosis, modulating inflammation, and enhancing cellular resilience. As research uncovers their complex roles in the nervous system, the therapeutic potential of HSPs in neurodegenerative diseases becomes clear. Harnessing their protective functions may lead to innovative strategies to mitigate neurodegeneration and promote neuronal health, offering hope for those affected by these conditions.

16.5 Therapeutic Implications of HSPs in Neurodegenerative Diseases

Heat Shock Proteins (HSPs) play a crucial role in maintaining proteostasis and neuroprotection in neurodegenerative diseases (NDs) such as Alzheimer's, Parkinson's, and Huntington's disease, where misfolded proteins lead to neuronal degeneration. Their upregulation helps manage the accumulation of toxic proteins like amyloid-beta and alpha-synuclein, making HSPs promising therapeutic targets. Strategies to enhance HSP expression include pharmacological induction with compounds like geranylgeranylacetone (GGA) and quercetin, natural methods such as exercise and heat stress, and gene therapy for direct HSP gene delivery. HSP-based therapeutic agents, including small molecule chaperones and peptide therapeutics, are being developed to aid protein folding and reduce aggregation. Clinical applications are expanding, with HSP levels serving as potential biomarkers for disease progression, the possibility of combination therapies, and personalized medicine approaches based on individual HSP expression profiles. However, challenges such as the specificity of HSP activation, efficient delivery to the central nervous system, and long-term safety remain to be addressed. Overall, advancements in HSP modulation present significant potential for improving treatment outcomes in neurodegenerative diseases.

Extracellular heat shock proteins (HSPs), particularly Hsp70, exhibit protective effects in neurodegenerative conditions such as amyotrophic lateral sclerosis (ALS) and Alzheimer's disease (AD) by interacting with synapses, masking toxic extracellular proteins, and maintaining trophic support to prevent synaptic dysfunction. Hsp70's affinity for phosphatidylserine and complement proteins may inhibit microglial pruning, and it might modulate immune responses to reduce glial activation or promote adaptive immunity. Although research on the therapeutic use of exogenous Hsp70 is in the early stages, challenges like differences between HSP isoforms and delivery methods remain. For example, while Hsp70 demonstrated efficacy in SOD1G93A mice, Hsc70 did not, emphasizing the importance of selecting specific HSP isoforms for treatment. Successful central nervous system penetration of Hsp70 through intranasal administration contrasts with less effective intraperitoneal injections, and targeted delivery methods could enhance outcomes for peripheral neuron diseases. Additionally, extracellular HSPs interact with Toll-like receptors, potentially triggering immune responses; thus, ensuring endotoxin-free preparations and optimized delivery could unlock HSPs' neuroprotective potential. HSP90, which stabilizes proteins associated with neurodegeneration, is another therapeutic target, as its inhibitors can upregulate Hsp70 and improve neurological outcomes. HSPs also play critical roles in stabilizing tau and amyloid-beta proteins in AD and in maintaining protein homeostasis in Parkinson's disease by refolding misfolded proteins and aiding autophagy. The pharmacological induction of Hsp70 shows promise for neuroprotection, with agents like geranylgeranylacetone (GGA) demonstrating neuroprotective effects. Overall, the multifaceted roles of Hsp70 and HSP90 in modulating neurodegeneration highlight their potential as therapeutic targets in neurodegenerative diseases, warranting further research to address current challenges and optimize therapeutic strategies [2, 10, 11, 14, 21, 23].

16.6 HSP Modulation: Potential Therapeutic Strategies

Heat Shock Proteins (HSPs) are essential molecular chaperones that play a critical role in cellular protection and homeostasis, particularly under stress conditions, making them vital in neurodegenerative diseases (NDs). Their ability to assist in protein folding, prevent aggregation, and promote the degradation of misfolded proteins has led to the exploration of HSP modulation as a therapeutic strategy. Various approaches are being investigated, including pharmacological agents that induce HSP expression, such as geranylgeranylacetone (GGA) and quercetin, which activate Heat Shock Factor 1 (HSF1) to enhance resilience against cellular stress. Gene therapy using adeno-associated viral (AAV) vectors allows for the direct delivery of HSP genes to neurons, promoting sustained production and countering neurodegenerative processes. Lifestyle interventions, like regular exercise and controlled heat exposure, can also elevate HSP levels and bolster cellular resilience. Additionally, the development of HSP-based drugs, including small molecule chaperones and peptide therapeutics, aims to enhance HSP functions, aiding protein folding and preventing aggregation. Combination therapies, integrating HSP modulation with conventional treatments, may provide synergistic neuroprotective effects. However, challenges such as ensuring specificity in targeting beneficial HSPs, long-term safety, and effective delivery mechanisms must be addressed. Overall, the modulation of HSPs presents a promising avenue for improving cellular resilience, preventing protein misfolding, and providing neuroprotection in neurodegenerative diseases, with ongoing research crucial for translating these strategies into clinical practice [5, 16, 20, 26, 33].

16.7 Challenges and Future Directions

Heat Shock Proteins (HSPs) have emerged as promising therapeutic targets for neurodegenerative diseases (NDs) due to their crucial roles in maintaining cellular proteostasis and protecting neurons from stress. However, several challenges must be addressed before HSP modulation can be effectively translated into clinical practice. The complexity of HSP functions complicates the identification of specific HSP targets for various NDs, and the risks associated with indiscriminate activation may disrupt normal cellular processes. Additionally, delivering HSP-modulating agents across the blood–brain barrier (BBB) and ensuring long-term safety and efficacy in diverse patient populations remain significant hurdles.

Future research should focus on elucidating the specific roles of individual HSPs, developing targeted therapies, and exploring innovative drug delivery systems to enhance efficacy while minimizing side effects. Investigating combination therapies and lifestyle interventions that naturally influence HSP expression may also provide synergistic benefits. Ultimately, advancing our understanding of HSP biology and integrating personalized medicine approaches will be crucial for optimizing therapeutic strategies and improving outcomes for individuals affected by neurodegenerative diseases [1, 28, 30].

16.8 Conclusion

The exploration of Heat Shock Proteins (HSPs) as therapeutic targets in neurodegenerative diseases (NDs) has opened new avenues for interventions aimed at mitigating the complex pathophysiological processes underlying these conditions. HSPs are essential for protein homeostasis, cellular protection, and stress response, enhancing neuronal resilience against misfolded proteins, oxidative stress, and neuroinflammation characteristic of Alzheimer's, Parkinson's, and Huntington's diseases. Researchers are pursuing various strategies, including pharmacological agents, gene therapy, lifestyle interventions, and HSP-based drugs, to harness these protective effects, aiming to enhance HSP expression while ensuring targeted delivery to affected neurons. However, challenges remain, such as the complexity of HSP functions, risks of indiscriminate activation, and effective delivery to the central nervous system. Future research should focus on understanding the specific roles of different HSPs, advancing drug development, and integrating lifestyle factors that influence HSP expression. As our understanding of HSP biology deepens, the potential for translating HSP modulation into effective therapeutic strategies becomes increasingly promising, offering hope for improved outcomes and quality of life for affected individuals. Ultimately, the ongoing exploration of HSP modulation represents a crucial step toward developing innovative therapies that could alter the course of neurodegenerative diseases and provide a beacon of hope for patients and their families.

References

1. Akhtar A, Andleeb A, Waris TS, Bazzar M, Moradi AR, Awan NR, Yar M. Neurodegenerative diseases and effective drug delivery: a review of challenges and novel therapeutics. J Control Release. 2021;330:1152–67.
2. Alberti G, Paladino L, Vitale AM, Caruso Bavisotto C, Conway de Macario E, Campanella C, Macario AJ, Marino Gammazza A. Functions and therapeutic potential of extracellular hsp60, hsp70, and hsp90 in neuroinflammatory disorders. Appl Sci. 2021;11(2):736.
3. Almalki AFY, Arabdin M, Khan A. The role of heat shock proteins in cellular homeostasis and cell survival. Cureus. 2021;13(9):e18316.
4. Armstrong R. What causes neurodegenerative disease? Folia Neuropathol. 2020;58(2):93–112.
5. Bose S, Cho J. Targeting chaperones, heat shock factor-1, and unfolded protein response: promising therapeutic approaches for neurodegenerative disorders. Ageing Res Rev. 2017;35:155–75.
6. Campanella C, Pace A, Caruso Bavisotto C, Marzullo P, Marino Gammazza A, Buscemi S, Palumbo Piccionello A. Heat shock proteins in Alzheimer's disease: role and targeting. Int J Mol Sci. 2018;19(9):2603.
7. Dubrez L, Causse S, Borges Bonan N, Dumétier B, Garrido C. Heat-shock proteins: chaperoning DNA repair. Oncogene. 2020;39(3):516–29.
8. Dutta N, Pal K, Pal M. Heat shock factor 1 and its small molecule modulators with therapeutic potential. In: Heat shock proteins in inflammatory diseases. Cham: Springer International Publishing; 2020. p. 69–88.
9. Eldeeb MA, Thomas RA, Ragheb MA, Fallahi A, Fon EA. Mitochondrial quality control in health and in Parkinson's disease. Physiol Rev. 2022;102(4):1721–55.

10. Gao C, Jiang J, Tan Y, Chen S. Microglia in neurodegenerative diseases: mechanism and potential therapeutic targets. Signal Transduct Target Ther. 2023;8(1):359.
11. Ghit A. Protective role of heat shock proteins during neurodegeneration in Parkinson's disease. In: Heat shock proteins in inflammatory diseases. Cham: Springer; 2020. p. 33–51.
12. Gonzales MM, Garbarino VR, Pollet E, Palavicini JP, Kellogg DL, Kraig E, Orr ME. Biological aging processes underlying cognitive decline and neurodegenerative disease. J Clin Invest. 2022;132(10):e158453.
13. Guo T, Zhang D, Zeng Y, Huang TY, Xu H, Zhao Y. Molecular and cellular mechanisms underlying the pathogenesis of Alzheimer's disease. Mol Neurodegener. 2020;15:1–37.
14. Gupta A, Bansal A, Hashimoto-Torii K. HSP70 and HSP90 in neurodegenerative diseases. Neurosci Lett. 2020;716:134678.
15. Hu C, Yang J, Qi Z, Wu H, Wang B, Zou F, Mei H, Liu J, Wang W, Liu Q. Heat shock proteins: biological functions, pathological roles, and therapeutic opportunities. MedComm. 2022;3(3):e161.
16. Hussain R, Zubair H, Pursell S, Shahab M. Neurodegenerative diseases: regenerative mechanisms and novel therapeutic approaches. Brain Sci. 2018;8(9):177.
17. Johnson J, Mercado-Ayon E, Mercado-Ayon Y, Dong YN, Halawani S, Ngaba L, Lynch DR. Mitochondrial dysfunction in the development and progression of neurodegenerative diseases. Arch Biochem Biophys. 2021;702:108698.
18. Jurcau A. Insights into the pathogenesis of neurodegenerative diseases: focus on mitochondrial dysfunction and oxidative stress. Int J Mol Sci. 2021;22(21):11847.
19. Luh LM, Bertolotti A. Potential benefit of manipulating protein quality control systems in neurodegenerative diseases. Curr Opin Neurobiol. 2020;61:125–32.
20. Lyon MS, Milligan C. Extracellular heat shock proteins in neurodegenerative diseases: new perspectives. Neurosci Lett. 2019;711:134462.
21. Mishra PS, Singh A. Heat shock proteins in neural signaling: implications in health and disease. In: Heat shock proteins in signaling pathways. Cham: Springer; 2019. p. 459–78.
22. Park H, Kang JH, Lee S. Autophagy in neurodegenerative diseases: a hunter for aggregates. Int J Mol Sci. 2020;21(9):3369.
23. Patel F, Mandal P. Neurodegenerative diseases and their therapeutic approaches. In: Neurons: dendrites and axons. IntechOpen; 2019.
24. Phillips BP, Gomez-Navarro N, Miller EA. Protein quality control in the endoplasmic reticulum. Curr Opin Cell Biol. 2020;65:96–102.
25. Pincus D. Regulation of Hsf1 and the heat shock response. In: HSF1 and molecular chaperones in biology and cancer. Cham: Springer; 2020. p. 41–50.
26. Pratt WB, Gestwicki JE, Osawa Y, Lieberman AP. Targeting Hsp90/Hsp70-based protein quality control for treatment of adult onset neurodegenerative diseases. Annu Rev Pharmacol Toxicol. 2015;55(1):353–71.
27. Rauf A, Badoni H, Abu-Izneid T, Olatunde A, Rahman MM, Painuli S, Semwal P, Wilairatana P, Mubarak MS. Neuroinflammatory markers: key indicators in the pathology of neurodegenerative diseases. Molecules. 2022;27(10):3194.
28. Santha M, Durham HD, Vigh L, Prodromou C. The role of heat shock proteins in neuroprotection. Front Pharmacol. 2020;11:1227.
29. Shan Q, Ma F, Wei J, Li H, Ma H, Sun P. Physiological functions of heat shock proteins. Curr Protein Pept Sci. 2020;21(8):751–60.
30. Singh MK, Shin Y, Ju S, Han S, Choe W, Yoon KS, Kim SS, Kang I. Heat shock response and heat shock proteins: current understanding and future opportunities in human diseases. Int J Mol Sci. 2024;25(8):4209.
31. Teleanu DM, Niculescu AG, Lungu II, Radu CI, Vladâcenco O, Roza E, Costăchescu B, Grumezescu AM, Teleanu RI. An overview of oxidative stress, neuroinflammation, and neurodegenerative diseases. Int J Mol Sci. 2022;23(11):5938.
32. Tutar L, Tutar Y. Heat shock proteins; an overview. Curr Pharm Biotechnol. 2010;11(2):216–22.
33. Vendredy L, Adriaenssens E, Timmerman V. Small heat shock proteins in neurodegenerative diseases. Cell Stress Chaperones. 2020;25(4):679–99.

34. Wang H, Yang F, Zhang S, Xin R, Sun Y. Genetic and environmental factors in Alzheimer's and Parkinson's diseases and promising therapeutic intervention via fecal microbiota transplantation. NPJ Parkinsons Dis. 2021;7(1):70.
35. Wilson DM, Cookson MR, Van Den Bosch L, Zetterberg H, Holtzman DM, Dewachter I. Hallmarks of neurodegenerative diseases. Cell. 2023;186(4):693–714.

Degradation of Misfolded Proteins in Neurodegenerative Diseases: Therapeutic Targets and Strategies

17

Aradhana Rajak

17.1 Introduction

Proteostasis is the mechanism that maintains a functional proteome by precisely coordinating the molecular machinery that supports a protein from its synthesis to its demise [1]. The quality control of cellular protein promises accurate conformational folding of nascent proteins, maintaining unfolding, refolding, and/or degradation of misfolded proteins. As we age, exposure to both internal and external stresses hinders the proteostasis network and causes aberrant proteins to build up or the components of proteostasis to decrease [2]. An overview of current developments in our knowledge of the regulation of proteostasis and its function in aging and various diseases, such as infectious and monogenetic disorders. We mention a few of the open-ended and developing issues surrounding proteostasis to understand diseases linked to damaged proteins and to encourage aging in a healthful manner [3]. Proteostasis is a vital process in nonmitotic cells such as neurons, and their protein homeostasis systems are diminished with aging, resulting in the accumulation of defective organelles and misfolded protein debris [4]. The two central cellular degradation systems are the ubiquitin–proteasomal system (UPS) and autophagy–lysosomal system, targeting degradation of organelles, single functional and nonfunctional proteins, large molecules, and aggregates of proteins, respectively [5].

A. Rajak (✉)
Bharathiar University, Coimbatore, Tamil Nadu, India
e-mail: aradhana.rajak@gmail.com

N. Puranik (ed.), *Proteostasis*, https://doi.org/10.1007/978-981-96-6202-9_17

17.2 Conformational Folding, Refolding, and Misfolding of Proteins

The efficient folding and escape from aggregation and misfolding of the nascent polypeptide chains is ensured with the expression of classes of molecular chaperones proteins [6]. Chaperones guide the nascent polypeptide along the productive folding pathway, by either reverting misfolding or avoiding aggregation [7]. In a nascent but misfolded protein to retain its biological properties, first chaperones try to repair the unfolded protein; if this doesn't work, they can trigger a variety of cellular programs, such as the heat shock response (HSR), ubiquitin–proteasome system (UPS), unfolded protein response (UPR), and endoplasmic reticulum-associated degradation (ERAD), which can take more extreme actions to either repair the unfolded protein or eliminate it completely [8]. Protein misfolding is inherent due to processes known to occur when residues are searching for stable native-like interactions. Even though tiny proteins can fold quickly and completely in vitro [9], for larger proteins, complete yield with regular folding is frequently inefficient due to off-pathway aggregation. Therefore, a significant biological and medical issue arises when proteins fail to folding or stay folded physiologically [6]. Aggregations, mostly caused by hydrophobic interactions that are nonspecific and exposed to prevent undesirable interactions with charged or polar moieties and water. Chaperones selectively bind to nonpolar hydrophobic regions within nascent unfolded chains to shield nonspecific aggregation [10]. These molecular machinery controls the quality of protein folding [11]. Chaperones have been demonstrated to perform novel roles such as breakdown of preexisting aggregates and cover their hydrophobic regions or transformation into huge masses leading suppression of their toxicity [12]. Unfolded proteins accumulate in the endoplasmic reticulum (ER) after pathogenic insults that cause the chaperone machinery to malfunction, initiating the UPR pathway. Two mechanisms can lead to misfolded proteins. First, some proteins are translated correctly from their amino acid sequences, but occasionally they misfold because they discover a different, more stable conformation [13]. Alternatively, genomic mutations can lead to protein fold errors and dysfunctions a single faulty amino acid can lead to the misfolding of a whole protein, which can cause cellular disaster and protein aggregation [14]. In addition to the potential for a genetic mutation in a disease-specified protein, protein misfolding disorders are also related to changes in the cellular protein folding machinery [15]. While protein mutations are associated with degradation occasionally allowing survival of misfolded proteins and eventually aggregation to form fibrils. Mutational chaperones promote proteins proper translation and folding into various conformations easily [4]. Moreover, protein misfolding is linked to physiological disturbances such as inflammation, calcium dysregulation, and mitochondrial dysfunction. More than 40 human diseases have been linked directly to the deposition of such aggregates in tissue, most of which are age-dependent and appear to be caused by a decline in the proteostasis network's capacity that happens with aging [16]. This includes neurodegenerative diseases. A crucial protein is generated and misfolded frequently in each of these diseases, eluding cellular degradation mechanisms and protein folding machinery

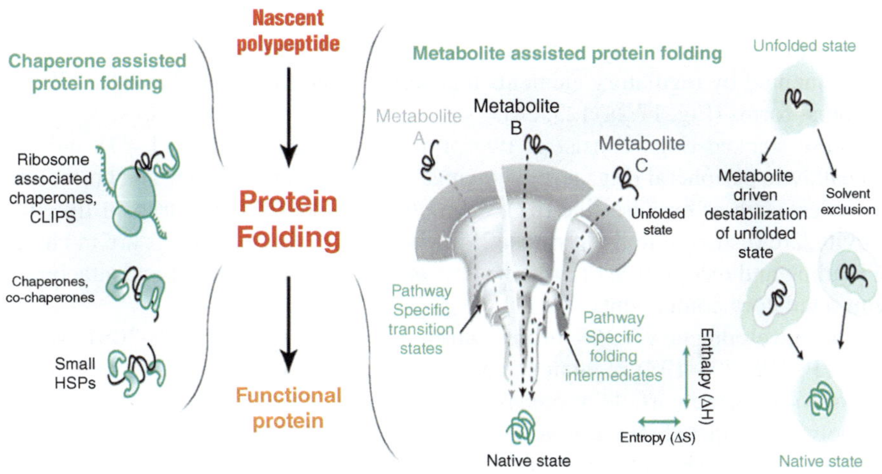

Fig. 17.1 A diagram illustrating protein folding. It shows the transition from the nascent polypeptide to the functional protein, highlighting chaperone and metabolite-assisted protein folding. Relevant molecules and pathways are labeled. On the right, the three metabolic-assisted folding components of protein folding, i.e., A, B, and C. The extreme right panel shows the molecular mechanism of metabolite-based protein folding. The grey and green represent solvent and metabolite, respectively. CLIPS, chaperones linked to protein synthesis; HSP, heat shock protein. (Adapted from Verma et al. [54])

alike. This misfolded protein accumulates and eventually nucleates as a massive fibrillar aggregates [17] (Fig. 17.1).

17.3 Ubiquitin–Proteasome System (UPS)

A key proteolytic mechanism that targets a significant number of intracellular proteins and is necessary for regular cell function is the ubiquitin–proteasome system (UPS). It is responsible for the proteolytic breakdown of 80–90% of all cellular proteins, including many regulated, short-lived, misfolded/damaged, or otherwise outdated proteins [21]. It is composed of around 850 distinct components [18–20]. The core particle of the 20S proteasome, where the proteolysis occurs, is made up of four stacked rings that combine to form a barrel-shaped structure with a central cavity. With the addition of regulatory elements affecting the 20S proteasome's selectivity and activity, this structure serves as the functional center of all proteasomal forms [18]. These stacked rings consist of two catalytic inner rings called β-rings and two noncatalytic outer rings termed α-rings (with subunits α1–α7) [19]. The bulk of intracellular proteins are targeted by the central proteolytic mechanism known as the ubiquitin–proteasome system (UPS), which is necessary for regular cell function. Comprising around 850 distinct elements [20], it is accountable for the proteolytic breakdown of 80–90% of all cellular proteins, encompassing numerous regulated, transient, misfolded/damaged proteins [21]. The 20S proteosome is barrel shaped molecule, that comprises the central core, where the proteolysis occurs [22]. The selectivity and activity of molecules have a central cavity and are supplemented by regulatory elements that serve as the functional basis for all proteasomal forms (Fig. 17.2a) [23, 24].

These stacked rings consist of two proteolytic inner β-rings (β1–β7) and two noncatalytic peripheral rings termed α-rings (with subunits α1–α7). The proteolytic activities, such as caspase-like (β1), chymotrypsin-like (β5), and trypsin-like (β2) activity, are restricted to the β-rings [25], whereas the interior α-rings are in charge of binding and recognition of protein substrate to the inner chamber and attachment with regulatory components [26].

The development of 26S proteasomes with one (RP1CP; 19S-20S) or two (RP2CP; 19S-20S-19S) regulatory complex particles. A 19S particle binds to the α-rings of the core particle, it consists of a lid made up of eight subunits to detect the polyubiquitinated target protein and a base of six ATPases that exerts a chaperone-like activity. The entry gate of 20S proteasome is opened when 19S proteasome binds and unfolds polyubiquitinated proteins [27]. The protein ubiquitin, covalently labels and signals the target proteins for breakdown [28]. The enzyme E1, which activates ubiquitin, breaks down ATP to create a thioester-linked conjugate with ubiquitin called ubiquitin activator enzyme (E1). The enzyme E2 called ubiquitin conjugating enzyme, conjugates ubiquitin by receiving ubiquitin from E1 and creates a thioester intermediate with it. Finally, the enzyme E3, also known as ubiquitin ligase, binds both E2 and the substrate, transferring the ubiquitin to the substrate [29]. Occasionally, a polyubiquitin chain needs to be extended using a combination of three ubiquitination enzymes and a fourth enzyme called E4 (ubiquitin chain elongation factor) [30, 31]. Polyubiquitinated substrate proteins are subsequently

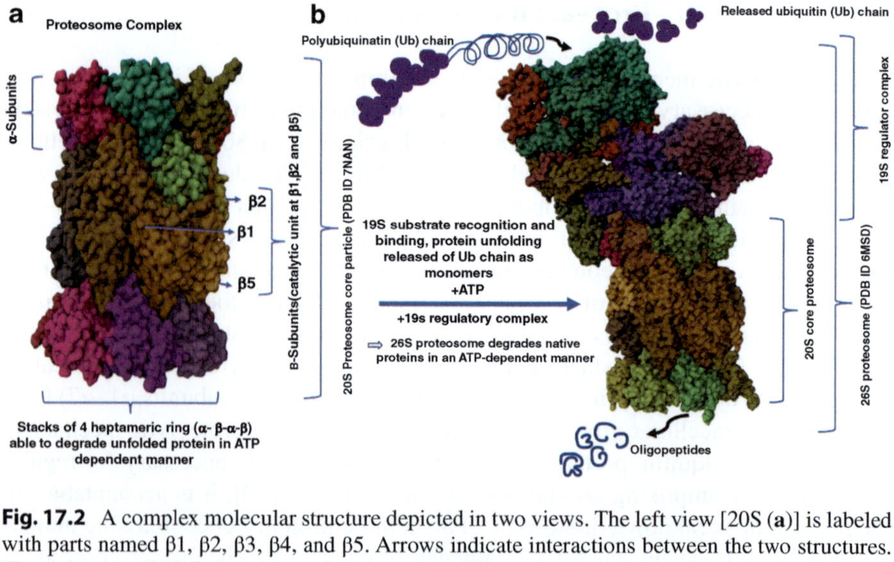

Fig. 17.2 A complex molecular structure depicted in two views. The left view [20S (**a**)] is labeled with parts named β1, β2, β3, β4, and β5. Arrows indicate interactions between the two structures. The right view [26S (**b**)] Various colors represent different components of the molecules

transported to the 26S proteasome for oligopeptide degradation. Additionally, ubiquitin recycling enzymes liberate ubiquitin molecules, which can then be used in subsequent polyubiquitination activities [32] (Fig. 17.2b). Although the 26S proteasome has a high selectivity for polyubiquitinated proteins, it is not very effective at breaking down oxidized proteins [33–35], which are not ubiquitinated primarily [36–38]. Selective breakdown of oxidatively damaged proteins does not require an intact ubiquitination system [39], since unfolded proteins can enter the inner chambers of proteasome on their own lacking the aid of ubiquitination or regulator particles [40].

17.4 Immunoproteasome

Immunoproteosome called the most activated form of proteosome molecule [41]. Immunoproteosome can be assembled by stimulants like lipopolysaccharide, tumor necrosis factor-α, and interferons and oxidative stress [42]. This is achieved by substituting the immunoproteosomal subunits for the constitutively expressed catalytic subunits: multi-catalytic endopeptidase complex-like 1 (MECL1 or β2i), low molecular weight protein 2 (LMP2 or β1i), and low molecular weight protein 7 (LMP7 or β5i) [43]. Furthermore, the Proteosome activator 28 kDa, regulator binds with the 20S core unit. Immunoproteosomes are primarily investigated concerning the processing of oligopeptides for the presentation of antigens to the adaptive immune response. The immunoproteasome is crucial for the removal of oxidized proteins in addition to its function in the immune system, the 26S proteasome is reported for poor efficiency in degrading oxidized protein [44]. The result of the dissembling of 26S proteosome under oxidative stress, there are more 20S proteasomes available for use, boosting the cellular ability to digest oxidized protein [45]. The 11S regulators indicate improved binding to 20S proteasomes after oxidative stress, which increases the capacity to preferentially degrade proteins that have been oxidatively damaged [43, 46]. Since the 11S regulator binds to immunoproteasomes [43, 44], the selectivity for oxidized protein has been improved, indicating that immunoproteasomes play a significant biological role in addition to immunological responses [47, 48] enhancing the selectivity for oxidized protein as 11S regulator bind to immunoproteasomes [43, 44]. The upregulation of 20S, immunoproteosome, and 11S regulator levels is demonstrated by low level of H_2O_2, and it does not always involve protein damage [49]. This proves that the mild stimuli activate the signal-transduction pathways to prime the cell to withstand oxidative assaults [50]. Diseases like cancer, inflammatory conditions, and neurological illnesses have also been connected to immunoproteasome expression and activity [51]. It has been found that amyloid plaques have increased immunoproteasome expression [52]. Furthermore, it was discovered that astrocytes and microglia from ALS mice models have elevated immunoproteasomes [53] and that suppressing immunoproteasome induction lowers survival and higher expression, limits the severity of disease [54]. All things considered, a working UPS is crucial, and any changes made to its parts could have harmful effects [55] (Fig. 17.3).

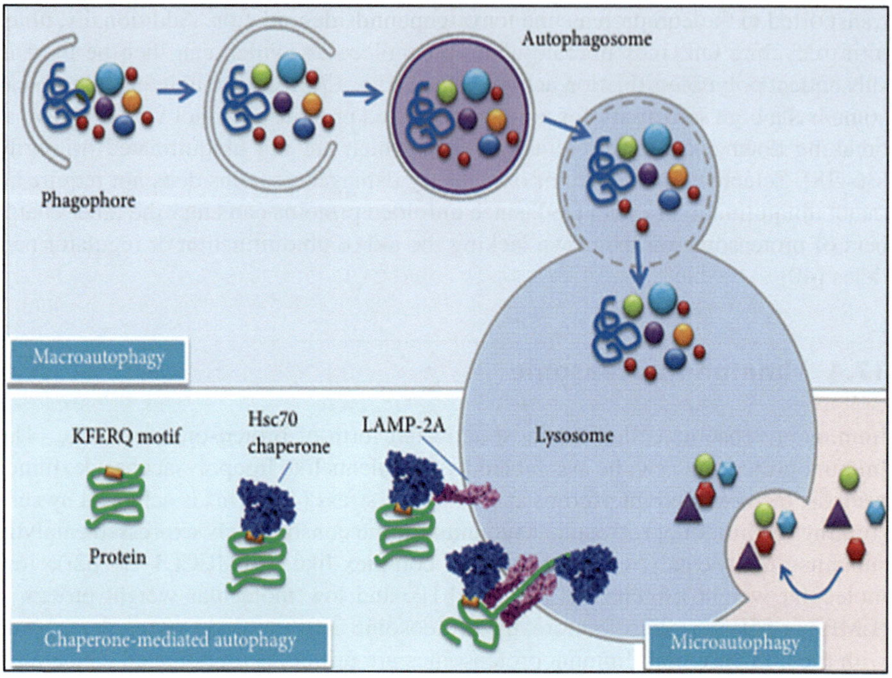

Fig. 17.3 Diagram illustrating vesicle formation in mammalian cells, showing exocytosis and endocytosis processes with labeled proteins and cellular structures. Includes phases of vesicle development, including phagosome and autophagosome formation. (Adapted from Höhn et al. [131])

17.5 Endoplasmic Reticulum-Associated Protein Degradation (ERAD)

The nascent proteins were kept at the endoplasmic reticulum (ER) after synthesis till the maturation of proteins. Homeostasis depends on proper protein conformation folding, yet despite all the effort put into the process, a sizable fraction of nascent polypeptides ready to enter at ER are unable to secure a native conformation [56]. The ER protein quality control systems monitor the protein folding accuracy. The process known as ER-associated protein degradation (ERAD), which is in charge of removing potentially hazardous proteins from the ER, is applied to proteins that are unable to fold or assemble correctly [57]. Retrotranslocation, ubiquitination, proteasomal degradation, and detection and targeting of substrates are the first steps in the intricate, multistep process known as ERAD. Owing to its intricacy, this quality-control procedure is only skimmed over here, with references to studies that exclusively concentrate on ERAD and provide detailed descriptions of the stages [58]. Typically, maturation tags resulting from posttranslational modifications allow ERAD substrates to be identified [59]. Additionally, misfolded proteins are drawn to the ERAD complex by chaperones and other recognition factors such as EDEM1 (ER degradation enhancing α-mannosidase like protein 1) BiP/GRP78, and the lectins OS9 (osteosarcoma amplified 9) and XTP3B [60–62]. These factors can attach to misfolded proteins and transfer them to the Sel1L-Hrd1 ERAD complex so they can be transported to the cytosol. Furthermore, misfolded glycoproteins must be de-mannosylated for degradation [63]. After the termination of the mannose, α1,6-bonded mannose residue is exposed, serving as a signal for recognition of OS9 and XTP3-B. Misfolded luminal proteins require full retrotraslocation [64], and membrane proteins only require the transport of specific domains. The substrate becomes accessible to the cytoplasm during the late retrotranslocation phase, which results in luminal substrate ubiquitination [57]. Cytosolic proteins acting as ubiquitin receptors must recognize ubiquitinated substrates in order to couple retrotranslocation with degradation [65]. Proteins with ubiquitin-binding domains can transfer ubiquitinated proteins from the ER membrane's retro translocation complex to the proteasome [66]. ERAD activates at higher levels of proteotoxic stress and at low levels under basal conditions. Brain samples from AD patients show greater ER stress; these samples also show increased BiP/GRP78 as well as activation of UPR at activation at hippocampus and temporal cortex, which are regularly correlated with neurofibrillary tangles NFTs and the accumulation of Aβ plaque [67]. ER stress is caused by accumulation of misfolded/unfolded protein in the ER lumen, lowering the level of free chaperone [68]. Which causes reduced protein synthesis and activated UPR therefore reducing the rate at which developing proteins enter the ER [69]. By upregulating ERAD genes, UPR also increases ERAD capability, ensuring that faulty proteins are degraded in the event that folding attempts are unsuccessful [69]. Last, UPR also triggers cell death processes like autophagy or apoptosis in the presence of steadily rising cellular stress [67].

17.6 Autophagy–Lysosome System

The lysosomal digestive system, which has an almost infinite capacity for degradation, is a significant cellular "degradation mechanism" aside from the proteasomal pathway. Enclosed within membranes, lysosomes are multienzyme-containing organelles that contain proteases, lipases, hydrolases, glycosidases, nucleases, peptidases, sulfatases, and phosphatases. Extracellular substances may enter the body through receptor, phago, or pinocytosis-mediated endocytosis. When intracellular organelles and proteins were observed to be ensnared by double-membrane vesicles in the 1960s, the phrase autophagy, or "self-eating," was first used [70]. Through a process known as autophagy the soluble, big protein aggregates, prolonged proteins, extracellular substances even organelles are directed toward the lysosome [18, 20]. The autophagic system targets membrane-associated proteins or large aggregates and is made up of about 500 distinct components [71, 72]. It is essential for maintaining the metabolic balance of cells and ensuring cell viability [73]. In order to preserve cellular nutrition levels and support the regulation of intracellular organelle homeostasis, autophagy functions as a starvation response [74, 75]. The change is linked to a number of human illnesses, including cancer, lysosomal storage disorders, and neurodegenerative diseases. It is essential for the elimination of aggregated/toxic proteins and compromised cellular organelles that might harm cells during cellular stress condition [76, 77]. In addition to inducing type II apoptosis, disruption of autophagy is crucial for the etiology of a number of illnesses. There are three main categories of autophagy that are commonly recognized: chaperone-mediated autophagy (CMA), microautophagy, and macroautophagy (see Fig. 17.3).

Although their contents are delivered for lysosomal breakdown via all three channels, their mechanisms differ [78]. The most well-studied process that is essential to preserving cellular and organismal homeostasis is macroautophagy, sometimes known as autophagy [79]. Numerous autophagy-related (Atg) proteins are involved in this evolutionarily conserved system, which coordinates vesicle creation through three distinct phases (initiation, elongation, and maturation). The vesicles are then fused with lysosomes to produce autolysosomes [73]. A double-membrane vesicle called an autophagosome is used during macroautophagy to trap pathogenic organisms, damaged mitochondria, misfolded protein aggregates, and excess organelles [79]. The production of phagophore is the initiation process, phagophore is a major cytoplasmic double membrane compartment.

By the combination of the autophagosome's outer membrane with the lysosomal membrane, the phagophore expands to form a circular autophagosome. Fusion of autophagosome with lysosome produces an autolysosome. The autophagosome cargo interacts with lysosomal hydrolases in this last stage, facilitating its breakdown. When the autophagosome inner membrane is revealed to the resident hydrolases and inner lysosomal acidic lumen degrading autophagic cargo. The constituent elements are then recycled back into the cytosol to power additional cellular processes [80]. Two similar mechanisms to ubiquitination are needed for the elongating step: LC3 modification and Atg12-Atg5-conjugation [81].

The Atg12 (ubiquitin-like protein) is triggered by ATG7 and transported to ATG10 in the first conjugation system (Atg12-Atg5). The Atg12, Atg5, Atg16, and Atg31 can form a noncovalent protein complex outside the developing autophagosome [82]. This process also encourages the formation of LC3I (microtubule-associated-protein-light-chain-3)-PE (phosphatidylethanolamine), which is similar to the action of E3 enzymes of the ubiquitin conjugation complex. The Atg4 removes the C-terminal tail of LC3 in the second conjugation mechanism, revealing glycine residue (G116) which is activated by tg7. Membrane elongation is made possible by the LC3-PE complex known as LC3II, which is located at the autophagosomal membrane. Furthermore, LC3II can attach to the adaptor protein sequestosome 1 (SQSTM1/p62), which assists in the direction and identification of ubiquitinated protein aggregates toward selective autophagic destruction. According to newly available data, ubiquitin controls how the UPS and autophagy system break down cellular targets. The seven lysine residues that construct ubiquitin—K6, K11, K27, K29, K33, K48, and K6—can participate in further ubiquitination. The autophagosomal pathway preferentially degrades K63-linked chains of monoubiquitinated substrates, whereas the proteasomal system recognizes K48-linked polyubiquitin chains [83]. Ubiquitin is used in autophagosomal and proteasomal pathways because

SQSTM1/p62 has incomplete specificity, even though p62 has a higher affinity for K63-linked polyubiquitin chains [84]. K48-linked polyubiquitin chains are recognized by the proteasomal system, while the K63-linked chains of monoubiquitinated substrates are preferentially degraded by autophagosomal pathway [83].

17.7 Degradation of Misfolded Protein in Neurodegenerative Disease: Therapeutic Targets and Strategies

The misfolded proteins are found as unfolded, intermediately folded, and correctly folded species in cells [1]. Healthy active cells utilize chaperone proteins in misfolded proteins for trafficking, intermediate stabilization, and degradation/refolding [1]. It is confident now that not all but many proteins can form amyloid fibrils under appropriate biochemical conditions [2, 3]. However, many amyloidogenic proteins associated with the disorder have extensive disorder in their soluble forms due to the internal amino acid sequences for aggregation [3–5]. The cytotoxic amyloid fibrils are the protein fragments that are liberated from nondiseased proteins [2, 5]. The Proteasomal system degrades fully or partially unfolded proteins only, which are monomeric [6]. The native proteins can be converted into amyloid proteins (to seed into prion-like propagation) as the amyloidal forms are thermodynamically more stable and possess extensive contacts between polymeric peptides [7].

The mutations, cellular aging, and proteotoxic stress can lead the proteins to escape the cells' quality control system and start aggregating to form nonconventional structural motifs. These structures can range from oligomeric to amorphous clusters to highly organized plaques and amyloid fibrils (Table 17.1: enlisting disease associated with protein misfolding and aggregation). The misfolded proteins are continuously originated due to physiological stressors, cellular mutations, and biogenesis. The refolding and destruction of misfolded proteins are handled by particular intracellular organelles, such as aggresomes or various kinds of inclusion bodies.

Table 17.1 The misfolded proteins associated with neurodegenerative disorder

Proteinopathy	Aggregating polypeptide(s)
Alzheimer's disease	Amyloid beta (Ab) peptide; Tau
Parkinson's disease	α-synuclein
Multiple tauopathies	Tau protein (microtubule-associated)
Huntington's disease	Huntingtin with tandem glutamine repeats
Amyotrophic lateral sclerosis	Superoxide dismutase 1
Spongiform encephalopathies	Prion proteins
Familial amyloidotic polyneuropathy	Transthyretin (mutant forms)

Chaperone proteins bind with the naïve polypeptides and assist in proper folding. Alternatively targets the degradation of misfolded proteins via ubiquitin–proteasome system or autophagy pathway [8, 9]. Proteotoxic stressors like heat shock and unfolded protein responses (UPR) involve rearrangement via proteostasis regulation. This is regulated by both chaperones (e.g., Hsp70, Hsp90) and non-chaperone proteins (including transcription factors, cell cycle regulators, signaling proteins, and receptors) [17]. Due to aging or diseased misfolded proteins, cells may undergo proteostasis system collapse. This system collapse is related to the buildup of ubiquitinated inclusion bodies (IBs) [9]. These ubiquitinated aggregates lead to inhibition or clogging of the proteasomal system [19, 20].

17.8 Misfolded Protein Toxicity

Neuronal apoptosis is the outcome of all neurodegenerative diseases. The misfolded proteins induce synaptic dysfunction and loss of function in neurons. The specific upstream process of various misfolded disease proteins that result in neurotoxicity is unknown and depends on the species of involved proteins. These disease-inducing misfolded proteins primarily acted by dominant-negative effects or toxic gain-of-function. A few loss-of-function effects are also reported. When neurons are dealt with purified oligomers and transfected with expression vectors, it has been noted that defective proteins have a detrimental effect on neuronal function. For example, β-amyloid, α-synuclein, and tau all of them disrupt synaptic communication [30–32]. Neuronal transport pathways and microtubule function are disrupted by a tau protein mutation [32, 33]. The import of mitochondrial proteins is distorted by an alpha-synuclein [32, 34]. Furthermore, by attaching to and ensnaring other cytosolic proteins, large misfolded protein aggregates use harmful consequences.

The result of proteomic studies confirms that when artificial proteins intended to form amyloid-like fibrils. Causes the toxicity, which results in aggregate formation and engagement into aberrant interactions thereby distorting cytosolic stress responses [35]. Misfolded proteins generally have a distorted ability for proteostasis. The aggregates of multiple different proteins at cytosol that include artificial β-sheets, TAR DNA binding protein-43 (TDP-43), and fragmented mutant huntingtin are reported to distort nucleocytoplasmic transport of RNA and proteins [36]. Along with synaptic dysfunction, numerous cellular changes are associated with neurodegenerative diseases especially neuroinflammation, oxidative stress, calcium signaling abnormalities, and mitochondrial dysfunction. These cellular distress signals appear in the early process of disease and are both the expected cause and consequence of neurodegeneration. The relationship between misfolded protein accumulation and cellular distress signs is bidirectional and mutually aggravated. The misfolded proteins (amyloid-β, a-synuclein, and mHtt) all result in severe oxidative stress in astrocytes and impair the anti-oxidant responses of the astroglial [37–40]. Conversely, oxidative stress leads to aging and pathogenic-related proteostatic breakdown, as well as increasing the aggregation of disease proteins [41, 42]. The relationship between folding errors in proteins and neuroinflammation, which

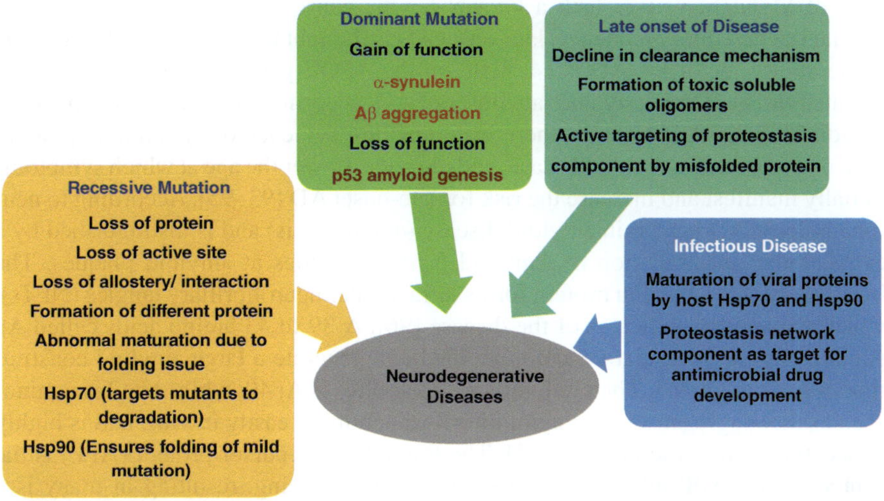

Fig. 17.4 Flowchart illustrating factors contributing to neurodegenerative diseases. Shows three main causes: Dominant mutation, late-onset disease, and recessive mutation, each with subpoints, leading to Neurodegenerative Diseases at the center

is being most thoroughly investigated in AD, also seems to follow an ascending spiraling cycle. It has been demonstrated that insoluble Aβ aggregates and soluble Aβ oligomers bind to and activate astrocytes and microglia, causing a persistently low level of neuroinflammation [43]. Though potentially beneficial in the near term, a number of lines of evidence indicate that Aβ's pro-inflammatory actions eventually compromise microglial and astrocyte function, including their capacity to eliminate Aβ and other misfolded proteins [38, 43–45]. Neuroinflammation has damaging effects that are due to misfolded proteins and leads to an ongoing age-related immune system senescence [46, 47] (Fig. 17.4).

17.9 Alzheimer Disease

Alzheimer's disease (AD) is a neurodegenerative, unalterable brain ailment that gradually deteriorates cognitive function, leading psychosis, amnesia, language difficulties, and ultimately the inability to perform even the most basic tasks [85]. The number of people with AD along with additiona l dementias rose by twofold globally between 1990 and 2016, mostly due to population aging and growth [86]. AD has two forms an inherent early onset and periodic late onset. Genetics primarily determines familial AD. Thus far, causative mutations have been found in three genes: t PSEN2 gene (chromosome 1) [88], PSEN1 gene (chromosome 14) [87], and APP gene (chromosome 21) [87–89]. These mutations cause early-onset AD, which is primarily inherited autosomally dominantly. These mutations only cause 5–10% of cases of early onset. The majority of sporadic instances and more than 50% of Mendelian cases lack a known genetic cause [90–92]. Despite having a substantial hereditary tendency, late-onset AD is thought to be complex. The genetic vulnerability factor for AD is ε4 allele on the APOE gene, which is linked to an elevated threat in the early and late onset of AD, according to a genome-wide association analysis. Additionally, there is compelling evidence of systemic inflammation, obesity, cardiovascular disease, and diabetes. Lower the age at which symptoms initially manifest and increase the risk for late-onset AD [93–95]. According to neuropathology, AD is a brain amyloid disease (amyloidosis) and is distinguished by a extracellular accumulation of Amyloid β (Aβ) peptides at amyloid plaques. The hyperphosphorylated Tau protein forms intracellular neurofibrillary tangles (NFTs), which are another indicator of the disease [96]. A 39 to 43 amino acids called Aβ peptide turn self-associative and misfolded and generate a large amyloid constitution in patients of AD. The predominant Aβ species is Aβ40, while Aβ42 is a minor species that aggregates into amyloid fibrils much more easily in vitro and is highly concentrated in senile plaques [97]. The β-amyloid precursor protein (APP) is the source of Aβ. APP mRNA undergoes alternative splicing, resulting in many isoforms of varying lengths. The human APP gene, resent at chromosome 21 has three major isoforms due to alternate splicing [98], which are APP695, APP770 and APP751 (where the 695, 770 and 751is amino acids number). After being sequentially cleaved by β- and γ-secretase, these isoforms can all produce Aβ [99, 100]. The cleavage of APP at the Aβ peptide location thus α-Secretase stops the production of Aβ. The Golgi/trans-Golgi network (TGN) apparatus carries long APP peptide, which has generated at ER to undergo post-translational changes throughout maturation [101]. In TGN-derived secretory vesicles, a complete APP, the α-, β-, and γ-secretase are delivered at plasma membrane. α-secretase primarily cleaves APP at the plasma membrane to create non-amyloidogenic sAPPα [102]. The intact APP and secretases at the cell surface undergo disintegration by endosome and recycled back, when delivered to the lysosomal system. Because endosomes and lysosomes provide the ideal acidic condition for the activity of β-secretase and increased collection of APP in these structures encourages Aβ formation and amyloidogenic processing [103]. Aβ can either be eliminated by proteolysis or transferred from the brain into the peripheral blood circulations. Aβ has been shown to be degraded by the insulin-degrading enzyme (IDE), neprilysin, angiotensin-converting enzyme (ACE), and endothelin-converting enzyme (ECE) [104, 105]. Furthermore, the UPS is crucial in controlling the deposition of Aβ at neuronal cells by either reducing Aβ synthesis or encouraging proteolytic breakdown [106]. Thus, any malfunction inside UPS functioning might potentially contribute to Aβ buildup in the neuronal cytoplasm, favoring the buildup of Aβ plaque [107]. Additionally, dysregulation of the ER's proteostatic systems, including UPR and ERAD, is frequently observed in neurodegenerative illnesses. The ERAD is significant for the pathophysiology of AD. Over the past 20 years, the amyloid cascade hypothesis has gained more traction [108]. Specifically, the focus now lies on smaller, soluble oligomers called protofibrils rather than amyloid fibrils, which are still the most common harmful form of Aβ [109, 110]. This reevaluation has been somewhat spurred by the discovery that there is no correlation between the quantity of aged plaques and the degree of memory loss, and smaller solublized oligomers are relatively hazardous in contrast to fibrillar Aβ. Tau is a structural protein linked to microtubules that are involved in axonal transport and is the second protein linked to AD [111, 112]. It has been demonstrated that the primary constituent of NFTs in AD is filamentous tau that is hyperphosphorylated and insoluble [113]. Tau inclusions are observed in brain regions with impaired function, contributing to the pathogenesis of AD. Furthermore, the development of NFTs is associated with the severity of AD and its pathological progression [114]. In vitro studies has shown that Tau protein (unstructured polypeptide) can be degraded by 20S proteosome, without ubiquitination [113]. Tau protein is expressed in a lesser degree at the kidney, lung, testis, and central and peripheral neurological systems [115]. It is most commonly detected in neuronal somatodendritic compartments, oligodendrocytes, and neuronal axons [116–118]. Studies on tau turnover suggest that autophagy and UPS, with selectivity for different pools, may contribute to its destruction [119, 120]. For instance, UPS preferentially degrades hypoacetylated tau [121]. Another characteristic of AD that has been found is the buildup of advanced glycation end products (AGEs) [122]. Increased oxidative stress and inflammation are linked to insoluble protein aggregates formed by AGEs [123]. Research has shown that AGEs raise the generation of ROS, which activates downstream pathways involved in the processing of APP and the synthesis of Aβ [124]. Furthermore, the immunoproteasome has been found to express itself more when it is present in amyloid plaques [54], which are the main site of AGEs. It has previously been demonstrated that the build-up of AGEs activates the AGE-receptor (RAGE), which in turn causes phosphorylation and a significant increase in activity and expression of Jak1/2 of STAT1 in immunoproteasome [29].

17.10 Parkinson's Disease

The degeneration of dopaminergic neurons in the substantia nigra and other brain regions with aging is known as Parkinson's disease (PD). The cytoplasmic deposition of amyloid-like aggregates known as Lewy bodies or fibrous inclusions primarily made of the protein α-synuclein (αS) is the clinical hallmark of Parkinson's disease (PD) [125, 126]. Similar to other neurodegenerative illnesses, the majority of experimental data indicate that protein accumulation due to gain-of-toxic-function fundamental in pathophysiology [15, 127], although toxic species are still not defined yet. Although some investigations have revealed that αS fibrils can also be very hazardous. The reports recommend that a specific oligomeric class of αS is its extremely cytostatic form and crucial in disease pathophysiology [128–132]. However, the causative factor of Parkinson's disease (PD) is still mostly unclear, and 90% of instances are thought to be random, with 10–15% of individuals associated with a positive ancestry of the disease [133]. Aging is a significant challenging factor for Parkinson's disease (PD), as the disease frequency rises rapidly with age and peaks at 80 years of age [134]. Nonetheless, it has been evident over the past 20 years that exposure to various environmental factors, either alone or in conjunction with genetic factors, may play a significant role in late-onset sporadic Parkinson's disease cases. Mutations in the Parkin, PINK1, and DJ-1 genes have been linked to recessive autosomal Parkinson's disease type [135]. Monogenic types of recessive Parkinson's are a significant source for the early onset of disease (before the age of 40 years), whereas the age of onset for dominant autosomal Parkinsonism is close to that of sporadic PD [136]. Parkin protein participates in the UPS-mediated degradation, as it is an E3 ubiquitin ligase [137]. Therefore, mutations associated with the Parkin-encoding gene are linked to dysfunctional ubiquitin ligase activity, which in turn leads to the aggregation of neurotoxic proteins, especially in the substantia nigra [138]. Numerous investigations validated this theory, exhibiting that the proteasome degrades αS via ubiquitin-dependent [139–141], as well as ubiquitin-independent [142, 143] pathways. Lysosomes are the main degradation agents of αS. It was discovered that PINK1 encodes the serine/threonine protein kinase known as PTEN-induced putative kinase 1, which interacts with PARKIN to promote the selective autophagy of injured mitochondria [144]. The breakdown of both cytosolic and vesicular αS is linked to the autophagic pathway. Misfold αS aggregates, in particular, exhibit strong colocalization with both early and late endosomal/lysosomal markers [145–147]. This suggests that αS is subjected to lysosomal degradation, despite some contradictory reports regarding the kinetics of this process [116], possibly due to variations in the experimental methods used to prepare the aggregates and deliver them to cells. Pathogenic αS can be released from the neuron by emptying the targeted vesicular bodies for exocytosis to release the lysosomal degradation pathways. In vitro, the role of autophagy in the clearance of αS has also been shown using 3-methyladenine (3-MA), an autophagy inhibitor that significantly increased the amounts of soluble and aggregated αS in nonneuronal cells [148]. Moreover, neuroblastoma cells were shown to accumulate A53T oligomers when the macroautophagic process was inhibited both pharmacologically and molecularly [149]. A different lysosomal degradation process most likely plays a role in the clearance of αS, as there is contradicting research showing that neither animal models nor neural cultured cells buildup of any insoluble αS as a result of 3-MA inhibition [150]. By blocking Rab1a protein, a crucial early mediator of the process, αS overexpression drastically reduced macroautophagy in both transgenic mice models and mammalian nonneuronal cells. This resulted in the mislocalization of the autophagy protein Atg9 and a decrease in autophagy some maturation [151]. It has been suggested that fibrillar assemblies can be broken down by macroautophagy, while oligomeric species are more likely to be broken down by CMA. In fact, it was demonstrated that CMA was essential for controlling intracellular αS levels [152]. Numerous investigations demonstrated the significance of this method for the in vivo breakdown of distinct lysosomes, mono and dimeric αS, and the potent inhibition of CMA by A30P and A53T αS mutations [153]. Moreover, a notable rise in aggregated αS was seen upon downregulating LAMP-2A, the CMA pathway's rate-limiting step, indicating its pivotal function in αS breakdown [148]. A tiny quantity of misfolded αS can elude the breakdown process by lysosomal rupture, so liberating itself in the cytosol [154]. Furthermore, the lysosomal inhibitor chloroquine causes a significant rise in cytosolic misfolded αS [155]. Crucially, age is found to significantly affect lysosomal functionality, which may be one of the most pertinent ways in which aging influences αS disease [124, 154]. Hence, as demonstrated by studies conducted on cultured cells and animal models, deficiencies in the elimination of αS lead to the build-up of harmful aggregates [156, 157]. After entering the cells, αS forms perinuclear thread-like inclusions that undergo various modifications, including phosphorylation and ubiquitination. The accumulation of pathogenic αS can persist for months, according to studies done on the brains of mice [158], but the cells may also release small oligomeric clusters that come from these substantial inclusions. Ultimately, other proteases, including calpain, neurosin (kallikrein-6), and matrix metalloproteinases [159–161], can also be activated in vitro and in vivo to cause αS breakdown. Such cleavage results in shortened protein forms that can cause the full-length endogenous protein to congregate. These forms, along with aggregated αS, have been identified as parts of neurites and Lewy bodies [162].

17.11 Amyotrophic Lateral Sclerosis

A devastating neurodegenerative disease amyotrophic lateral sclerosis (ALS) is characterized by a progressive loss of motor neurons in the central nervous system (CNS), which results in stiffness, wasting, and paralysis of the muscles. The primary nuclear TAR DNA-binding protein 43 (TDP-43) was identified in 2006 as a significant component of the insoluble and ubiquitinated inclusions observed in the brains of patients with frontotemporal lobar degeneration (FTLD or FTLD–TDP) and amyotrophic lateral sclerosis (ALS) [163, 164]. Gain-of-function (GOF) and loss-of-function (LOF) pathways combine to cause neurodegeneration when nuclear native TDP-43 is depleted, which happens concurrently with the cytoplasmic accumulation of TDP-43 [165–167]. It is well recognized that cytosolic aggregates are inherently hazardous [165, 168] and that they can attract nuclear TDP-43, which amplifies their negative consequences by raising the nuclear LOF [169–171]. Proteostasis systems and other physiological activities seem to be compromised in ALS [172], indicating that neurons are essential for maintaining TDP-43 protein homeostasis through protein degradation pathways [173–175]. It has been noted that one of the main factors leading to the start of ALS in the elderly is a steady decline in the effectiveness of cellular proteostasis [172]. Genetic data supports this concept as well. For example, several of the mutations linked to ALS affect genes like VCP, CHMP2B, progranulin (PGRN), OPTN, TMEM106B, SQSTM1, TBK1, UBQLN2, ALS2, and C9ORF72 that are involved in UPS- or autophagy-mediated degradation [79, 106, 176–184]. Even though ALP contributes to the breakdown of TDP-43 stubborn aggregates, recent studies have demonstrated that UPS is primarily responsible for the cytosolic accumulation of TDP-43 [174, 185]. Consequently, cytosolic diffuse TDP-43 was found to assemble in a variety of misfolded species, that vary from soluble monomers to nondegradable macroaggregates [185], even though a collection of TDP-43 inclusions is unable to separate [185, 186]. Specifically, the report suggests that of the different TDP-43 aggregates that can be distinguished based on how susceptible they are to different protein degradation systems, the nondisposable fraction is approximately 49%, while the aggregates that can be broken down by UPS and ALP are approximately 24% and 26%, respectively [186]. In fact, under conditions where both routes are operating, the inhibition of one system increased the pace of degradation even though the quantity of TDP 43 that has been degraded is smaller. These results support the idea that there is cross-talk between the two clearance systems and have validated earlier findings that blocking one clearance pathway increases the effectiveness of the other [131, 187]. One of the main characteristics of pathogenic TDP-43 that has been discovered in the brain and spinal cord of ALS patients is ubiquitination of TDP-43 [133]. As a result, the mechanism of disease of ubiquitinated TDP-43 in ALS, notably its origin and allocation, has been extensively studied during the last 10 years.

It has been demonstrated that the ubiquitin lysines K-48 and K-63 allow the E3 ubiquitin ligase (Parkin) to ubiquitinate TDP-43. This promotes the cytoplasmic accumulation of TDP-43 into inclusions without leaving any traces of its protein breakdown visible [188, 189]. Further studies have shown that K-48 and K-63-linked polyubiquitin chains label TDP-43 particles from complete proteins and that these particles are destined for different outcomes: UPS degradation of TDP-43 for the K-48-linked polyubiquitin chains and autophagic removal of the TDP-43 containing K-63-linked polyubiquitin chains [174].

TDP-43's insoluble high molecular weight aggregates were shown to be more readily ubiquitinated and accumulated when it interacted with the ubiquitin-conjugating enzyme UBE2E3 and the ubiquitin-isopeptidase Y (UBPY) [191]. Additionally, it was discovered that mutations at the ubiquitination sites in TDP-43 reduced the protein's accumulation, indicating the ubiquitination in adjusting the aggregation of TDP-43 [190]. Specifically, a TDP-43 linked to FTLD and carrying the K263E mutation was shown to be highly ubiquitinated, potentially as a result of its folding incorrectly because the lysine residue that is positively charged and negatively charged the RRM2 domain's aspartate residue [191]. Furthermore, mutations in the vicinity of the TDP-43's ubiquitination sites It was also discovered that the RRM1 domain reduced the TDP-43's buildup, indicating that ubiquitination is involved in adjusting the aggregation of TDP-43 [190]. Autophagy's function in preserving TDP-43-associated Conflict suggests that the process of toxicity may be complicated. Data indicating that autophagy can proceed at a faster or slower rate down illness [190].

The involvement of autophagy in reversing TDP-43-associated toxicity may be a complicated process, as seen by conflicting studies that suggest autophagy may either accelerate or halt the progression of sickness [173]. It has been discovered that the endo-lysosomal pathways and the vacuolar fusion machinery are essential for both preserving cell survival and TDP-43 clearance. The toxicity of TDP-43 in ALS may be due to defective endocytosis. It has been discovered that aberrant TDP-43 levels, in cellular and yeast models as well as in the frontal cortex tissue of an ALS patient, inhibit endocytosis by colocalizing with the endocytosis-related proteins [191]. TDP-43 aggregation is associated with impaired endocytosis, even if TDP-43 toxicity and motor neuron dysfunction can be reversed by increasing endocytosis [191].

17.12 Mutant Huntingtin Disease (HD)

Huntington's disease (HD) is a dominant autosomal neurodegenerative disorder [192]. The disease patients experience growing motor and cognitive functional decrease with loss of awareness, dementia, depression, and anxiety [193]. The illness is brought on by the accumulation of mutant huntingtin (mHTT) proteins. In a wild-type huntingtin protein (HTT), a polyQ tract, and a stretch of glutamine residue are found, this is encrypted by a repetitive CAG codon found within exon 1 of the HTT gene [194]. The length of the CAG repeats ranging from 16 to 20 and varies among individuals. The patients usually have more than 35 CAG repeats [195]. Thereby, producing an elongated polyQ tract of mHTT proteins, which is aggregation-prone and neurologically toxic [196]. The PolyQ stretch is highly abundant in amyloid proteins, enriched with β-sheets, and low solubility with detergent. The PolyQ inclusions in mHTT are cytotoxic and neurodegenerative [197].

The mHTT is crucial in the pathological process of HD. However, the mechanism of cytotoxic mHTT removal has not been discovered so far. A defective substrate for every known proteolytic pathway is the mHTT. and functions as an inhibitor for proteolytic machinery, i.e., UPS, macroautophagy, and CMA [198]. For instance, the inclusion of mHTT in the brains of HD mice and HD patients is augmented with UPS components such as Ub and ubiquitinated HTT. The mHTT can be labeled with Ub but acts as faulty substrate for proteasomal degradation [199]. The build-up of mHTT is not caused by direct obstruction of proteasomal machinery but by a failed protein quality control system along with sequestration of molecular chaperones [200].

The wild-type HTT are targeted for degraded through CMA, the Hsc70 molecules recognizes the two types of KFERQ-like motifs at KDRVN [99–103] and NEIKV [245–248] residue, respectively. Just as wild HTT, mHTT can be targeted for CMA degradation by Hsc70 molecules [201]. The mHTT shows a high affinity for Hsc70 and LAMP-2A therefore, mHTT delivery is delayed for the lysosomal membrane [202]. This delay in delivery of primarily targeted mHTT to the lysosomal membrane can prevent CMA-dependent autophagic breakdown, which could cause a little disruption in proteostasis [203]. The accumulation of nondegraded mHTT evolves into deposition of intranuclear inclusions and perinuclear cytoplasmic aggregates at neurons of HD patients [204].

Major elements of macroautophagy, including LC3, are often increased in neuronal and nonneuronal cells in HD patients as well as in a variety of HD mouse models [205]. The overproduction of cargo-free vacuoles in autophagic cells is linked to the apparent elevation of macroautophagy, presumably as a result of compromised cargo delivery to autophagic vacuoles [206]. It was discovered that elements of macroautophagy, including p62, LC3-II, mTOR, and Beclin-1, are deposited in the striatum of HD transgenic mice when the autophagic flux is decreased [207]. Increased synthesis of autophagic core components is facilitated by the sequestration of autophagic regulators in mHTT inclusions, including mTOR [208]. Therefore, decreased macroautophagy activities linked to HTT suppression of macroautophagy worsen the course of HD disease in an age-dependent way [209].

17.13 Therapeutic Targets

17.13.1 Targeting Production, Misfolding, and Aggregation

The drugs targeted against misfolded proteins or aggregated proteins are challenging. The main challenge is determining whether the type or forms of a particular disease protein are mainly accountable for illness. The β-amyloid (Aβ) was studied originally as fibrilar plaques and were the disease-causing species in Alzheimer's disease. However present reports suggest that intermediate aggregates (oligomers and soluble protofibrils) are primarily accountable for AD. The varying species of α-synuclein in Parkinson's disease have also supported a similar discovery [210]. The current reports suggest that accumulated fibrils of α-synuclein, Aβ, and tau is found in various conformational "strains" or variants, with dissimilar propagational properties and different extents of neurotoxicity [211, 212]. Detection and characterization of toxic misfolded protein species by the conventional biochemical analytical method to characterize intermediate species. The denaturing SDS–PAGE technique changes the oligomeric state of Aβ42 oligomers [213]. Currently, nuclear magnetic resonance (NMR) spectroscopy and ion mobility spectrometry-mass spectrometry (IMS–MS) had been employed to analyze the aggregation and conformation of amyloid proteins and their inhibitors of the processes in solution form [214, 215]. Stabilizing adult fibrils to stop prion-like proliferation is another strategy now being investigated instead of halting the early misfolding and aggregation of amyloids [216], the protein production pathway is intervened therapeutically as far upstream is practical and would be one strategy, i.e., at the stage of protein translation, cleavage, or post-translational modification, in situations where the relative toxicity of different misfolded protein species is unknown [217]. The theory, focusing on the initial stages of the processing pathway, would yield the maximum level of therapeutic specificity, eradicate harmful gains, or function losses brought on by misfolded or aggregated protein forms, and stop aberrant folding and aggregation from spreading [218, 219]. Targeting the protein cleavage and post-translational modification tau and Aβ are being actively explored as inhibitors of tau phosphorylation (such as glycogen synthase 3β inhibitors) and acetylation, as well as Beta Amyloid Cleaving Enzyme (BACE) inhibitors for Aβ, a few of that are presently undergoing clinical trials [220].

17.13.2 Focusing Chaperones

Another potential target class for treatment in diseases involving protein misfolding is chaperones. Chaperones offer a feasible therapeutic entryway signal for each phase of a pathogenic protein's processing because they are engaged in each component of proteostasis [221]. The mammalian brain produces more than 200 chaperone proteins, and several cells express distinct protein combinations of chaperones and co-chaperones exclusively [222]. Additionally In addition to offering prospects for the development of medications that target the chaperones particular to neurons or glial cells, cell type-specific expression of chaperone subgroups may help to explain why some misfolded proteins are harmful in one cell type but not in others [223]. However, we still have little insight into which particular chaperones regulate specific disease proteins and how the enormous quantity of chaperones and the variety of their activity methods pose a hurdle to developing various treatments [224]. Links between inherited variants of some neurodegenerative illnesses and mutations in particular chaperones have provided some indications. For instance, Parkinson's disease has been associated with mutations in Hsp70 and Hsp40, while ALS has been associated with mutations in the co-chaperone valosin-containing protein (VCP) [225]. Another potential target class for treatment in illness situations involving protein misfolding is chaperones. Chaperones provide putative therapeutic input points to every stage of a pathogenic protein's synthesis because they are engaged in every facet of proteostasis; diverse cell types express diverse sets of chaperones and co-chaperones [227], and the mammalian brain expresses more than 200 distinct chaperone proteins [223]. In addition to offering prospects for the development of medications that target the chaperones specific to neurons or glial cells, cell type-specific expression of chaperone groups might also assist in clarifying why some misfolded proteins are harmful in one cell type but not in others [228]. However, the development of treatments is further hampered by the immense number of chaperones and the variety of their modes of action. Our understanding of how one chaperones interact with which one disease proteins and how is currently quite restricted. Links between inherited variants of some neurodegenerative illnesses and mutations in particular chaperones have provided some hints [222]. For instance, Parkinson's disease has been associated with mutations in Hsp70 and Hsp40, whereas ALS has been connected to mutations in the co-chaperone valosin-containing protein (VCP) [221].

Furthermore, in cell or animal models of diverse neurodegenerative disorders, overexpression of more than 20 distinct chaperones has been identified to give neuroprotection; in several instances, particular chaperones appear to provide protection against multiple disease proteins [228]. Creating small chemical inhibitors or activators of particular chaperones has been one method of therapeutically addressing the chaperone system. Hsp70 and Hsp90 have been the most extensively researched chaperones associated with neurodegeneration [229]. Hsp70 and Hsp90 have different impacts on the stability of client proteins. While Hsp90 fixes client proteins and prevents their ubiquitination, Hsp70 encourages their destruction through the UPS system.

Furthermore, in cell or animal models of diverse neurodegenerative disorders, overexpression of more than 20 distinct chaperones has been identified to give neuroprotection; in many instances, individual chaperones present and save multiple pathogenic proteins [228]. Creating small chemical inhibitors or activators of particular chaperones has been one method of therapeutically addressing the chaperone system. The two chaperones that have been investigated the most among those related to neurodegeneration are Hsp70 and Hsp90. Hsp70 and Hsp90 have different impacts on the stability of client proteins. While Hsp90 stabilizes client proteins and prevents their ubiquitination, Hsp70 encourages their destruction through the UPS system. HSF1 provides a close connection between Hsp90 and Hsp70 activity. HSF1 is usually activated by Hsp90 inhibitors, and this causes Hsp70 to be induced [229]. In cellular and animal models of AD, PD, and HD, a number of small molecule medications that either activate Hsp70, inhibit Hsp90, or the both, been demonstrated to improve neurological phenotypes, decrease cellular toxicity, and mitigate the buildup of pathogenic protein aggregates [226]. Although there is ongoing work in this field, because of issues with peripheral cytotoxicity and/or low brain penetration, neither of these drugs have likely to move into clinical trials for treatment in neurodegenerative diseases [230]. Protein engineering has been used as an additional approach to produce chaperone-based treatments. For instance, it was found that the yeast breakdown Hsp104 can break down in vitro fibrils made of tau, polyglutamine, Aβ42, α-synuclein, and prion protein, among other proteins seen in neurological diseases [231]. To dissolve these proteins, however, comparatively high Hsp104 concentrations are required. Proteins with significantly increased breakdown activity and lower toxicity were produced by making minor modifications to the sequence of Hsp104, including variations that lessened neurodegeneration in a C. elegans model of Parkinson's disease [232]. This strategy may also be used for other chaperones, such as metazoan chaperones with disaggregase activity (e.g., Hsp110, Hsp70, and Hsp40) and the yeast chaperonin TRiC, which has the unique capacity to transfer through cell membranes and demonstrated to survive against HTT toxicity [233, 234].

17.13.3 Autophagy Targeted

Neurodegenerative disorders frequently exhibit defects in the UPS and autophagy processes of protein breakdown [235, 236]. For instance, PINK-1, Parkin (a ubiquitin ligase), UCH-L1 (Ub carboxy terminal hydrolase L1), DJ-1 (PARK7), and LRRK2/PRAK8 are among the proteins encoded by several gene mutations that result in familial Parkinson's disease (PD) and are involved in the UPS and/or autophagy [237, 238]. Similar to the chaperone system, the number of proteins involved (about 500–1000 is linked alone to the UPS system) makes it difficult to select viable therapeutic targets from the UPS or autophagy pathways. Furthermore, the form or forms of the illness protein that are largely responsible for cellular toxicity are unknown for most diseases. Targeting protein breakdown is crucial since the process by which a particular protein is broken down (such by UPS versus autophagy) might differ based on the protein's soluble or fibrillar state and the particular post-translational changes it carries [239].

17.13.4 Targeting Degradation

Neurodegenerative disorders frequently exhibit errors in the UPS and autophagy processes of protein breakdown [237, 238]. For instance, PINK-1, Parkin (a ubiquitin ligase), UCH-L1 (Ub carboxy terminal hydrolase L1), DJ-1 (PARK7), and LRRK2/PRAK8 are among the proteins encoded by several gene mutations that result in familial Parkinson's disease (PD) and are involved in the UPS and/or autophagy [237, 239]. Similar to the chaperone system, the amount of proteins involved (about 500–1000 are linked alone to the UPS system) makes it difficult to select viable therapeutic targets from the UPS or autophagy pathways. Furthermore, the form or forms of the illness protein that are primarily accountable for cellular toxicity are unknown for the majority of diseases. This problem is crucial when focusing on protein degradation since the process by which a particular protein is broken down (such as UPS versus autophagy) might change based on the protein's soluble or fibrillar state, and the particular post-translational modifications it carries [238]. (Table 17.2 enlisting potential drug target associated with protein misfolding and neurodisorder).

Table 17.2 Reported drug-specific interactions for neurological diseases associated with misfolded proteins

Neurodisorder	Drug class	Drug	Regulatory mechanism
Alzheimer disease	Amyloid-Directed	Aducanumab (Aduhelm) IgG1 monoclonal antibody	Acts by targeting and removing amyloid-beta plaques
	Cholinesterase Inhibitors	Donepezil, rivastigmine, galantamine	Prevent the knockdown of acetylcholine
	Glutamate regulators	Memantine	Antagonize N-methyl-D-aspartate (NMDA) receptor to improve signal-to-noise ratio of glutamatergic transmission
Parkinson disease	Dopamine supplements	Levodopa	Replenish the decreased dopamine levels
	Decarboxylase inhibitors	Carbidopa	Prevent peripheral breakdown of levodopa
	Dopamine agonist	Apomorphine hydrochloride, ropinirole hydrochloride, rotigotine, pergolide, pramipexole dihydrochloride	Produces dopamine-like effects
Amyotrophic Lateral Sclerosis	Glutamate-receptor antagonist	Riluzole	Inhibits glutamate receptors
	Free-radical scavenger	Edaravone	Scavenges free radicals

17.14 Preclinical Challenges to Clinical Trials

Preclinical research has discovered promising possibilities for Huntington's disease, Parkinson's disease, ALS, and Alzheimer's disease; however, only a small number of them have demonstrated appreciable advantages in clinical trials [240]. The development of treatments for neurological disorders in general has been hampered by this "failure to translate"; the probable causes of this have been covered in length elsewhere [241–243]. The absence of strong targets whose modification yields a therapeutic benefit is one of the main causes [244]. Preclinical research in animal models lacks predictive validity due to the uncertainty surrounding what molecule or pathway should be targeted and the consequent inability to show target engagement. Furthermore, finding translatable biomarkers in animal models for clinical research is crucial [245]. Pharmacokinetics and drug safety provide major obstacles to effective medication development for misfolded protein disease [246]. However, recent structural investigations of the interactions between antibodies and particular amyloid-β molecule epitopes, as well as the correlation between these contacts and clinical results, have made significant progress in the field of Alzheimer's disease [245, 247] (Table 17.3 enlists the drug compound and their mode of action).

Table 17.3 Enlisting the drug compound and their mode of action

Compound/company name	Identified disease	Mechanism of action	Status
TRx0237/TauRx therapeutics	Alzheimer Disease	Tau aggregation inhibitor	Phase II clinical trial complicated
AADvac1/Axon nueroscience SE	Alzheimer Disease	Active Tau-based immunotherapy	Phase I clinical trial complicated
ACI-35/AC immune AG	Alzheimer Disease	Phaspho-tau vaccine	Phase I clinical trial active
Armicomol / Orphazyme ApS	Amyotrophic lateral sclerosis	HSP activation	Phase I/II active
Nuedexta/Avanir pharamaceuticals	Amyotrophic lateral sclerosis- PBA Symptom treatment	Unknown for PBA, NNDA receptor antagonist	FDA approved
Deferiprone/Generic	Parkinson's Disease	Iron chelator	Phase II recruiting
Istradefylline/Kyowa Hakko Kirin	Parkinson's Disease	Adnosine A2A receptor antagonist	Approved in Japan, not FDA approval
APH-1105 (an α-secretase modulator)	Mild-to-moderate Alzheimer's Disease, dementia	Lipid nanoparticle	Phase II
CNM-Au8 (Nanocrystalline gold)	Amyotrophic lateral sclerosis	Gold nanocrystals	Phase II
CNM-Au8 (Nanocrystalline gold)	Parkinson's Disease	Gold nanocrystals	Phase II

17.15 Conclusion

The inability of protein quality control to break down or eliminate misfolded proteins in elderly people's brains is a common molecular process behind the majority of neurodegenerative diseases. Throughout aging, post-translational modifications (like endoproteolytic cleaves and phosphorylation) of native proteins (like tau and amyloid β in AD) or genetic mutations of otherwise non-pathogenic proteins (like HTT in HD, α-synuclein in PD, PrPC in prion disease, and SOD1 and TDP-43 in ALS) produce the disease-causing misfolded proteins. The clinical characteristics of many neurodegenerative illnesses are these pathogenic agents' propensity to aggregate into oligomers with enhanced β-sheet composition, which can then develop into fibrillar inclusion bodies or extracellular plaques. Through their interactions with different cellular molecules, including essential elements of proteolytic pathways, β-Sheet-enriched aggregates can either directly or indirectly affect the UPS, CMA, and macroautophagy. As a result, protein quality control is less effective, which speeds up the formation of harmful aggregates. Because aging inherently reduces the ability of these postmitotic cells to handle such challenges, the cycle of misfolded proteins and protein quality control is especially harmful to aged neurons. Global failures of all proteolytic pathways are commonly linked to neurological illnesses as a result of these regrettable occurrences.

A sizable fraction of cellular proteins misfold either spontaneously or in response to physiological stressors, either during translation/folding or when functioning as folded proteins. The UPS, CMA, and macroautophagy can effectively eliminate the majority of aberrantly folded cellular proteins from the human proteome. On the other hand, the pathogenic proteins listed above are frequently resistant to those proteolytic pathways, possibly due to the fact that molecular chaperones find it challenging to release their β-sheet-enriched folds. When these substrates are not fully unfolded, they may become trapped in the small cylinder of the proteasome, be unable to be fed into the proteasomal cylinder, or be difficult to separate from the CMA components (like LAMP-2) during delivery over the lysosomal membrane. As evidenced by a study where the overexpression of the molecular chaperone Hsp70 expedited the proteasomal degradation of tau, increasing the activity or levels of molecular chaperones involved in the UPS is one way to improve the clearance of pathogenic proteins. Since several studies have demonstrated that increasing CMA improved the elimination of harmful misfolded proteins, another tactic is to activate the molecular chaperones (like Hsc70), carriers (like histone deacetylase 6), and adaptors (like LAMP-2). The requirement that substrates be at least partially or fully unfolded into nascent polypeptides before to being fed into the proteasome or lysosome is a common restriction of the UPS and CMA.

Since there is no ATP-dependent unfolding step involved in the degradation by macroautophagy, lysosomal proteolysis offers a perfect quality control method for misfolded proteins that are prone to aggregation. Furthermore, numerous studies have demonstrated that altering autophagic regulators like mTOR completely restored the autophagic flux, indicating that while autophagic flux is frequently decreased in affected neurons in the majority of neurodegenerative diseases, the functions of core autophagic machinery appear to remain largely intact. In order to promote macroautophagy and improve the removal of harmful protein aggregates, numerous small-molecule drugs were created.

In the brains of several transgenic mouse models, known mTOR-dependent autophagic inducers have been effectively used to improve behavior and cognition, reduce neurodegeneration, and increase the clearance of different pathogenic protein aggregates. This is because the mTOR pathway is becoming a more promising therapeutic target.

An mTOR-independent autophagic inducer and rapamycin have synergistic effects, and other autophagy regulators, including the ULK1 kinase AMPK, are also being intensively pursued as possible therapeutic targets. It should be mentioned that the majority of these research employ transgenic mice that overexpress pathogenic proteins that have already developed large amounts of insoluble inclusions, despite the fact that it is becoming more and more obvious that autophagy inducers have therapeutic potential to eliminate protein aggregates. More research should be done in more physiologically realistic settings to examine the effects of these chemicals on a variety of biological processes, including off-target effects.

References

1. Sontag EM, Samant RS, Frydman J. Mechanisms and functions of spatial protein quality control. Annu Rev Biochem. 2017;86(1):97–122.
2. Mymrikov EV, Daake M, Richter B, Haslbeck M, Buchner J. The chaperone activity and substrate spectrum of human small heat shock proteins. J Biol Chem. 2017;292(2):672–84.
3. Balch WE, Morimoto RI, Dillin A, Kelly JW. Adapting proteostasis for disease intervention. Science. 2008;319(5865):916–9.
4. Labbadia J, Morimoto RI. The biology of proteostasis in aging and disease. Annu Rev Biochem. 2015;84(1):435–64.
5. Wolynes PG. Recent successes of the energy landscape theory of protein folding and function. Q Rev Biophys. 2005;38(4):405–10.
6. Hipp MS, Park SH, Hartl FU. Proteostasis impairment in protein-misfolding and-aggregation diseases. Trends Cell Biol. 2014;24(9):506–14.
7. Brockwell DJ. Probing the mechanical stability of proteins using the atomic force microscope. Biochem Soc Trans. 2007;35(6):1564–8.
8. Nillegoda NB, Bukau B. Metazoan Hsp70-based protein disaggregases: emergence and mechanisms. Front Mol Biosci. 2015;2:57.
9. Beeg M, Stravalaci M, Romeo M, Carrá AD, Cagnotto A, Rossi A, Diomede L, Salmona M, Gobbi M. Clusterin binds to Aβ1–42 oligomers with high affinity and interferes with peptide aggregation by inhibiting primary and secondary nucleation. J Biol Chem. 2016;291(13):6958–66.
10. Fernandez-Funez P, Sanchez-Garcia J, De Mena L, Zhang Y, Levites Y, Khare S, Golde TE, Rincon-Limas DE. Holdase activity of secreted Hsp70 masks amyloid-β42 neurotoxicity in Drosophila. Proc Natl Acad Sci. 2016;113(35):E5212–21.
11. Ojha J, Karmegam RV, Masilamoni JG, Terry AV Jr, Cashikar AG. Behavioral defects in chaperone-deficient Alzheimer's disease model mice. PLoS One. 2011;6(2):e16550.
12. Mannini B, Cascella R, Zampagni M, van Waarde-Verhagen M, Meehan S, Roodveldt C, Campioni S, Boninsegna M, Penco A, Relini A, Kampinga HH. Molecular mechanisms

used by chaperones to reduce the toxicity of aberrant protein oligomers. Proc Natl Acad Sci. 2012;109(31):12479–84.
13. Chiti F, Dobson CM. Protein misfolding, amyloid formation, and human disease: a summary of progress over the last decade. Annu Rev Biochem. 2017;86(1):27–68.
14. Douglas PM, Dillin A. Protein homeostasis and aging in neurodegeneration. J Cell Biol. 2010;190(5):719–29.
15. Rochet JC, Lansbury PT Jr. Amyloid fibrillogenesis: themes and variations. Curr Opin Struct Biol. 2000;10(1):60–8.
16. Nijman SM, Luna-Vargas MP, Velds A, Brummelkamp TR, Dirac AM, Sixma TK, Bernards R. A genomic and functional inventory of deubiquitinating enzymes. Cell. 2005;123(5):773–86.
17. Varshavsky A. The ubiquitin system, an immense realm. Annu Rev Biochem. 2012;81(1):167–76.
18. Sowa ME, Bennett EJ, Gygi SP, Harper JW. Defining the human deubiquitinating enzyme interaction landscape. Cell. 2009;138(2):389–403.
19. Rock KL, Gramm C, Rothstein L, Clark K, Stein R, Dick L, Hwang D, Goldberg AL. Inhibitors of the proteasome block the degradation of most cell proteins and the generation of peptides presented on MHC class I molecules. Cell. 1994;78(5):761–71.
20. Grune T. Oxidative stress, aging and the proteasomal system. Biogerontology. 2000;1:31–40.
21. Davies KJ. Degradation of oxidized proteins by the 20S proteasome. Biochimie. 2001;83(3–4):301–10.
22. Jung T, Grune T. Structure of the proteasome. Prog Mol Biol Transl Sci. 2012;109:1–39.
23. Groll M, Huber R. Substrate access and processing by the 20S proteasome core particle. Int J Biochem Cell Biol. 2003;35(5):606–16.
24. Köhler A, Bajorek M, Groll M, Moroder L, Rubin DM, Huber R, Glickman MH, Finley D. The substrate translocation channel of the proteasome. Biochimie. 2001;83(3–4):325–32.
25. George DE, Tepe JJ. Advances in proteasome enhancement by small molecules. Biomol Ther. 2021;11(12):1789.
26. Lim KL, Tan JM. Role of the ubiquitin proteasome system in Parkinson's disease. BMC Biochem. 2007;8:S13.
27. Grimm S, Höhn A, Grune T. Oxidative protein damage and the proteasome. Amino Acids. 2012;42:23–38.
28. Koegl M, Hoppe T, Schlenker S, Ulrich HD, Mayer TU, Jentsch S. A novel ubiquitination factor, E4, is involved in multiubiquitin chain assembly. Cell. 1999;96(5):635–44.
29. Layfield R, Tooth D, Landon M, Dawson S, Mayer J, Alban A. Purification of poly-ubiquitinated proteins by S5a-affinity chromatography. Proteomics: Int Ed. 2001;1(6):773–7.
30. Kandel R, Jung J, Neal S. Proteotoxic stress and the ubiquitin proteasome system. Semin Cell Dev Biol. 2024;156:107–20. Academic Press.
31. Tzounakas VL, Dzieciatkowska M, Anastasiadi AT, Karadimas DG, Vergaki A, Siourounis P, Stamoulis K, Papassideri IS, Kriebardis AG, D'Alessandro A, Antonelou MH. Red cell proteasome modulation by storage, redox metabolism and transfusion. Blood Transfus. 2022;20(1):27.
32. Reeg S, Jung T, Castro JP, Davies KJ, Henze A, Grune T. The molecular chaperone Hsp70 promotes the proteolytic removal of oxidatively damaged proteins by the proteasome. Free Radic Biol Med. 2016;99:153–66.
33. Raynes R, Pomatto LC, Davies KJ. Degradation of oxidized proteins by the proteasome: distinguishing between the 20S, 26S, and immunoproteasome proteolytic pathways. Mol Asp Med. 2016;50:41–55.
34. Zhong Q, Xiao X, Qiu Y, Xu Z, Chen C, Chong B, Zhao X, Hai S, Li S, An Z, Dai L. Protein posttranslational modifications in health and diseases: functions, regulatory mechanisms, and therapeutic implications. MedComm. 2023;4(3):e261.
35. Türker F, Cook EK, Margolis SS. The proteasome and its role in the nervous system. Cell Chem Biol. 2021;28(7):903–17.

36. Cetin G, Klafack S, Studencka-Turski M, Kruger E, Ebstein F. The ubiquitin-proteasome system in immune cells. Biomolecules. 2021;11(1):60.
37. Shringarpure R, Grune T, Mehlhase J, Davies KJ. Ubiquitin conjugation is not required for the degradation of oxidized proteins by proteasome. J Biol Chem. 2003;278(1):311–8.
38. Qiu M, Chen J, Li X, Zhuang J. Intersection of the ubiquitin–proteasome system with oxidative stress in cardiovascular disease. Int J Mol Sci. 2022;23(20):12197.
39. Höhn A, Jung T, Grimm S, Catalgol B, Weber D, Grune T. Lipofuscin inhibits the proteasome by binding to surface motifs. Free Radic Biol Med. 2011;50(5):585–91.
40. Kloetzel PM. Antigen processing by the proteasome. Nat Rev Mol Cell Biol. 2001;2(3):179–88.
41. Shachar I, Karin N. The dual roles of inflammatory cytokines and chemokines in the regulation of autoimmune diseases and their clinical implications. J Leukoc Biol. 2013;93(1):51–61.
42. Pickering AM, Davies KJ. Differential roles of proteasome and immunoproteasome regulators Pa28αβ, Pa28γ and Pa200 in the degradation of oxidized proteins. Arch Biochem Biophys. 2012;523(2):181–90.
43. Pickering AM, Koop AL, Teoh CY, Ermak G, Grune T, Davies KJ. The immunoproteasome, the 20S proteasome and the PA28αβ proteasome regulator are oxidative-stress-adaptive proteolytic complexes. Biochem J. 2010;432(3):585–95.
44. Grune T, Catalgol B, Licht A, Ermak G, Pickering AM, Ngo JK, Davies KJ. HSP70 mediates dissociation and reassociation of the 26S proteasome during adaptation to oxidative stress. Free Radic Biol Med. 2011;51(7):1355–64.
45. Aiken CT, Kaake RM, Wang X, Huang L. Oxidative stress-mediated regulation of proteasome complexes. Mol Cell Proteomics. 2011;10(5):R110.006924.
46. Hussong SA, Kapphahn RJ, Phillips SL, Maldonado M, Ferrington DA. Immunoproteasome deficiency alters retinal proteasome's response to stress. J Neurochem. 2010;113(6):1481–90.
47. Seifert U, Bialy LP, Ebstein F, Bech-Otschir D, Voigt A, Schröter F, Prozorovski T, Lange N, Steffen J, Rieger M, Kuckelkorn U. Immunoproteasomes preserve protein homeostasis upon interferon-induced oxidative stress. Cell. 2010;142(4):613–24.
48. Pickering AM, Staab TA, Tower J, Sieburth D, Davies KJ. A conserved role for the 20S proteasome and Nrf2 transcription factor in oxidative stress adaptation in mammals, Caenorhabditis elegans and Drosophila melanogaster. J Exp Biol. 2013;216(4):543–53.
49. Johnston-Carey HK, Pomatto LC, Davies KJ. The Immunoproteasome in oxidative stress, aging, and disease. Crit Rev Biochem Mol Biol. 2016;51(4):268–81.
50. Mishto M, Bellavista E, Santoro A, Stolzing A, Ligorio C, Nacmias B, Spazzafumo L, Chiappelli M, Licastro F, Sorbi S, Pession A. Immunoproteasome and LMP2 polymorphism in aged and Alzheimer's disease brains. Neurobiol Aging. 2006;27(1):54–66.
51. Puttaparthi K, Elliott JL. Non-neuronal induction of immunoproteasome subunits in an ALS model: possible mediation by cytokines. Exp Neurol. 2005;196(2):441–51.
52. Ahtoniemi T, Goldsteins G, Keksa-Goldsteine V, Malm T, Kanninen K, Salminen A, Koistinaho J. Pyrrolidine dithiocarbamate inhibits induction of immunoproteasome and decreases survival in a rat model of amyotrophic lateral sclerosis. Mol Pharmacol. 2007;71(1):30–7.
53. Holmberg CI, Staniszewski KE, Mensah KN, Matouschek A, Morimoto RI. Inefficient degradation of truncated polyglutamine proteins by the proteasome. EMBO J. 2004;23(21):4307–18.
54. Verma K, Verma M, Chaphalkar A, Chakraborty K. Recent advances in understanding the role of proteostasis. Fac Rev. 2021;10:72.
55. Meusser B, Hirsch C, Jarosch E, Sommer T. ERAD: the long road to destruction. Nat Cell Biol. 2005;7(8):766–72.
56. Christianson JC, Shaler TA, Tyler RE, Kopito RR. OS-9 and GRP94 deliver mutant α1-antitrypsin to the Hrd1–SEL1L ubiquitin ligase complex for ERAD. Nat Cell Biol. 2008;10(3):272–82.
57. Bhamidipati A, Denic V, Quan EM, Weissman JS. Exploration of the topological requirements of ERAD identifies Yos9p as a lectin sensor of misfolded glycoproteins in the ER lumen. Mol Cell. 2005;19(6):741–51.

58. Molinari M, Calanca V, Galli C, Lucca P, Paganetti P. Role of EDEM in the release of misfolded glycoproteins from the calnexin cycle. Science. 2003;299(5611):1397–400.
59. Plemper RK, Böhmler S, Bordallo J, Sommer T, Wolf DH. Mutant analysis links the translocon and BiP to retrograde protein transport for ER degradation. Nature. 1997;388(6645):891–5.
60. Groisman B, Shenkman M, Ron E, Lederkremer GZ. Mannose trimming is required for delivery of a glycoprotein from EDEM1 to XTP3-B and to late endoplasmic reticulum-associated degradation steps. J Biol Chem. 2011;286(2):1292–300.
61. Burr ML, van den Boomen DJ, Bye H, Antrobus R, Wiertz EJ, Lehner PJ. MHC class I molecules are preferentially ubiquitinated on endoplasmic reticulum luminal residues during HRD1 ubiquitin E3 ligase-mediated dislocation. Proc Natl Acad Sci. 2013;110(35):14290–5.
62. Lee JG, Ye Y. Bag6/Bat3/Scythe: a novel chaperone activity with diverse regulatory functions in protein biogene sis and degradation. BioEssays. 2013;35(4):377–85.
63. Li G, Zhou X, Zhao G, Schindelin H, Lennarz WJ. Multiple modes of interaction of the deglycosylation enzyme, mouse peptide N-glycanase, with the proteasome. Proc Natl Acad Sci. 2005;102(44):15809–14.
64. Scheper W, Hoozemans JJ. The unfolded protein response in neurodegenerative diseases: a neuropathological perspective. Acta Neuropathol. 2015;130:315–31.
65. Boelens J, Lust S, Offner F, Bracke ME, Vanhoecke BW. The endoplasmic reticulum: a target for new anticancer drugs. In Vivo. 2007;21(2):215–26.
66. Kim I, Xu W, Reed JC. Cell death and endoplasmic reticulum stress: disease relevance and therapeutic opportunities. Nat Rev Drug Discov. 2008;7(12):1013–30.
67. De Duve C, Wattiaux R. Functions of lysosomes. Annu Rev Physiol. 1966;28(1):435–92.
68. Ciechanover A, Kwon YT. Protein quality control by molecular chaperones in neurodegeneration. Front Neurosci. 2017;11:185.
69. Menzies FM, Fleming A, Rubinsztein DC. Compromised autophagy and neurodegenerative diseases. Nat Rev Neurosci. 2015;16(6):345–57.
70. Ghavami S, Shojaei S, Yeganeh B, Ande SR, Jangamreddy JR, Mehrpour M, Christoffersson J, Chaabane W, Moghadam AR, Kashani HH, Hashemi M. Autophagy and apoptosis dysfunction in neurodegenerative disorders. Prog Neurobiol. 2014;112:24–49.
71. García-Prat L, Martínez-Vicente M, Perdiguero E, Ortet L, Rodríguez-Ubreva J, Rebollo E, Ruiz-Bonilla V, Gutarra S, Ballestar E, Serrano AL, Sandri M. Autophagy maintains stemness by preventing senescence. Nature. 2016;529(7584):37–42.
72. Varshavsky A. The ubiquitin system, autophagy, and regulated protein degradation. Annu Rev Biochem. 2017;86(1):123–8.
73. Mizushima N, Levine B, Cuervo AM, Klionsky DJ. Autophagy fights disease through cellular self-digestion. Nature. 2008;451(7182):1069–75.
74. Dasuri K, Zhang L, Keller JN. Oxidative stress, neurodegeneration, and the balance of protein degradation and protein synthesis. Free Radic Biol Med. 2013;62:170–85.
75. Yang Z, Klionsky DJ. Mammalian autophagy: core molecular machinery and signaling regulation. Curr Opin Cell Biol. 2010;22(2):124–31.
76. Chow CY, Landers JE, Bergren SK, Sapp PC, Grant AE, Jones JM, Everett L, Lenk GM, McKenna-Yasek DM, Weisman LS, Figlewicz D. Deleterious variants of FIG 4, a phosphoinositide phosphatase, in patients with ALS. Am J Hum Genet. 2009;84(1):85–8.
77. Yorimitsu T, Klionsky DJ. Autophagy: molecular machinery for self-eating. Cell Death Differ. 2005;12(2):1542–52.
78. Tanida I, Ueno T, Kominami E. LC3 conjugation system in mammalian autophagy. Int J Biochem Cell Biol. 2004;36(12):2503–18.
79. Nakatogawa H. Two ubiquitin-like conjugation systems that mediate membrane formation during autophagy. Essays Biochem. 2013;55:39–50.
80. Lilienbaum A. Relationship between the proteasomal system and autophagy. Int J Biochem Mol Biol. 2013;4(1):1.
81. Erpapazoglou Z, Walker O, Haguenauer-Tsapis R. Versatile roles of k63-linked ubiquitin chains in trafficking. Cells. 2014;3(4):1027–88.

82. Ihara Y, Morishima-Kawashima M, Nixon R. The ubiquitin–proteasome system and the autophagic–lysosomal system in Alzheimer disease. Cold Spring Harb Perspect Med. 2012;2(8):a006361.
83. Nichols E, Szoeke CE, Vollset SE, Abbasi N, Abd-Allah F, Abdela J, Aichour MT, Akinyemi RO, Alahdab F, Asgedom SW, Awasthi A. Global, regional, and national burden of Alzheimer's disease and other dementias, 1990–2016: a systematic analysis for the Global Burden of Disease Study 2016. Lancet Neurol. 2019;18(1):88–106.
84. Kang J, Lemaire HG, Unterbeck A, Salbaum JM, Masters CL, Grzeschik KH, Multhaup G, Beyreuther K, Müller-Hill B. The precursor of Alzheimer's disease amyloid A4 protein resembles a cell-surface receptor. Nature. 1987;325(6106):733–6.
85. Sherrington R, Rogaev EI, Liang YA, Rogaeva EA, Levesque G, Ikeda M, Chi H, Lin C, Li G, Holman K, Tsuda T. Cloning of a gene bearing missense mutations in early-onset familial Alzheimer's disease. Nature. 1995;375(6534):754–60.
86. Levy-Lahad E, Wasco W, Poorkaj P, Romano DM, Oshima J, Pettingell WH, Yu CE, Jondro PD, Schmidt SD, Wang K, Crowley AC. Candidate gene for the chromosome 1 familial Alzheimer's disease locus. Science. 1995;269(5226):973–7.
87. Wingo TS, Lah JJ, Levey AI, Cutler DJ. Autosomal recessive causes likely in early-onset Alzheimer disease. Arch Neurol. 2012;69(1):59–64.
88. Janssen JC, Beck JA, Campbell TA, Dickinson A, Fox NC, Harvey RJ, Houlden H, Rossor MN, Collinge J. Early onset familial Alzheimer's disease: mutation frequency in 31 families. Neurology. 2003;60(2):235–9.
89. Brouwers N, Sleegers K, Van Broeckhoven C. Molecular genetics of Alzheimer's disease: an update. Ann Med. 2008;40(8):562–83.
90. Hildreth KL, Van Pelt RE, Schwartz RS. Obesity, insulin resistance, and Alzheimer's disease. Obesity (Silver Spring, Md). 2012;20(8):1549.
91. Stewart R. Cardiovascular factors in Alzheimer's disease. J Neurol Neurosurg Psychiatry. 1998;65(2):143–7.
92. de Toledo Ferraz T, Alves TC, Ferreira LK, Wajngarten M, Busatto GF. Cardiac disorders as risk factors for Alzheimer's disease. J Alzheimers Dis. 2010;20(3):749–63.
93. Lamptey RN, Chaulagain B, Trivedi R, Gothwal A, Layek B, Singh J. A review of the common neurodegenerative disorders: current therapeutic approaches and the potential role of nanotherapeutics. Int J Mol Sci. 2022;23(3):1851.
94. Goate A, Chartier-Harlin MC, Mullan M, Brown J, Crawford F, Fidani L, Giuffra L, Haynes A, Irving N, James L, Mant R. Segregation of a missense mutation in the amyloid precursor protein gene with familial Alzheimer's disease. Nature. 1991;349(6311):704–6.
95. Sisodia SS, Koo EH, Hoffman PN, Perry G, Price DL. Identification and transport of full-length amyloid precursor proteins in rat peripheral nervous system. J Neurosci. 1993;13(7):3136–42.
96. Neve RL, Finch EA, Dawes LR. Expression of the Alzheimer amyloid precursor gene transcripts in the human brain. Neuron. 1988;1(8):669–77.
97. Xu H, Sweeney D, Wang R, Thinakaran G, Lo AC, Sisodia SS, Greengard P, Gandy S. Generation of Alzheimer β-amyloid protein in the trans-Golgi network in the apparent absence of vesicle formation. Proc Natl Acad Sci. 1997;94(8):3748–52.
98. Sisodia SS. Beta-amyloid precursor protein cleavage by a membrane-bound protease. Proc Natl Acad Sci. 1992;89(13):6075–9.
99. Jiang S, Li Y, Zhang X, Bu G, Xu H, Zhang YW. Trafficking regulation of proteins in Alzheimer's disease. Mol Neurodegener. 2014;9:1–3.
100. Pivovarova O, Höhn A, Grune T, Pfeiffer AF, Rudovich N. Insulin-degrading enzyme: new therapeutic target for diabetes and Alzheimer's disease? Ann Med. 2016;48(8):614–24.
101. Eckman EA, Reed DK, Eckman CB. Degradation of the Alzheimer's amyloid β peptide by endothelin-converting enzyme. J Biol Chem. 2001;276(27):24540–8.
102. Cirulli ET, Lasseigne BN, Petrovski S, Sapp PC, Dion PA, Leblond CS, Couthouis J, Lu YF, Wang Q, Krueger BJ, Ren Z. Exome sequencing in amyotrophic lateral sclerosis identifies risk genes and pathways. Science. 2015;347(6229):1436–41.

103. Tramutola A, Di Domenico F, Barone E, Perluigi M, Butterfield DA. It is all about ubiquitin: role of altered ubiquitin-proteasome system and UCHL1 in Alzheimer's disease. Oxidative Med Cell Longev. 2016;2016(1):2756068.
104. Hardy JA, Higgins GA. Alzheimer's disease: the amyloid cascade hypothesis. Science. 1992;256(5054):184–5.
105. Mroczko B, Groblewska M, Litman-Zawadzka A, Kornhuber J, Lewczuk P. Amyloid β oligomers (AβOs) in Alzheimer's disease. J Neural Transm. 2018;125:177–91.
106. Hayden EY, Teplow DB. Amyloid β-protein oligomers and Alzheimer's disease. Alzheimers Res Ther. 2013;5:1–1.
107. Takei Y, Teng J, Harada A, Hirokawa N. Defects in axonal elongation and neuronal migration in mice with disrupted tau and map1b genes. J Cell Biol. 2000;150(5):989–1000.
108. Iqbal K, Liu F, Gong CX, Grundke-Iqbal I. Tau in Alzheimer disease and related tauopathies. Curr Alzheimer Res. 2010;7(8):656–64.
109. Kondo J, Honda T, Mori H, Hamada Y, Miura R, Ogawara M, Ihara Y. The carboxyl third of tau is tightly bound to paired helical filaments. Neuron. 1988;1(9):827–34.
110. Giannakopoulos P, Herrmann FR, Bussière T, Bouras C, Kovari E, Perl DP, Morrison JH, Gold G, Hof P. Tangle and neuron numbers, but not amyloid load, predict cognitive status in Alzheimer's disease. Neurology. 2003;60(9):1495–500.
111. Gu Y, Oyama F, Ihara Y. τ is widely expressed in rat tissues. J Neurochem. 1996;67(3):1235–44.
112. Luk KC, Song C, O'Brien P, Stieber A, Branch JR, Brunden KR, Trojanowski JQ, Lee VM. Exogenous α-synuclein fibrils seed the formation of Lewy body-like intracellular inclusions in cultured cells. Proc Natl Acad Sci. 2009;106(47):20051–6.
113. Klein C, Krämer EM, Cardine AM, Schraven B, Brandt R, Trotter J. Process outgrowth of oligodendrocytes is promoted by interaction of fyn kinase with the cytoskeletal protein tau. J Neurosci. 2002;22(3):698–707.
114. Tashiro K, Hasegawa M, Ihara Y, Iwatsubo T. Somatodendritic localization of phosphorylated tau in neonatal and adult rat cerebral cortex. Neuroreport. 1997;8(12):2797–801.
115. Schmidt M, Finley D. Regulation of proteasome activity in health and disease. Biochim Biophys Acta-Mol Cell Res. 2014;1843(1):13–25.
116. Lee MJ, Lee JH, Rubinsztein DC. Tau degradation: the ubiquitin–proteasome system versus the autophagy-lysosome system. Prog Neurobiol. 2013;105:49–59.
117. Crippa V, Sau D, Rusmini P, Boncoraglio A, Onesto E, Bolzoni E, Galbiati M, Fontana E, Marino M, Carra S, Bendotti C. The small heat shock protein B8 (HspB8) promotes autophagic removal of misfolded proteins involved in amyotrophic lateral sclerosis (ALS). Hum Mol Genet. 2010;19(17):3440–56.
118. Vitek MP, Bhattacharya K, Glendening JM, Stopa E, Vlassara H, Bucala R, Manogue K, Cerami A. Advanced glycation end products contribute to amyloidosis in Alzheimer disease. Proc Natl Acad Sci. 1994;91(11):4766–70.
119. Bierhaus A, Humpert PM, Morcos M, Wendt T, Chavakis T, Arnold B, Stern DM, Nawroth PP. Understanding RAGE, the receptor for advanced glycation end products. J Mol Med. 2005;83:876–86.
120. Flavin WP, Bousset L, Green ZC, Chu Y, Skarpathiotis S, Chaney MJ, Kordower JH, Melki R, Campbell EM. Endocytic vesicle rupture is a conserved mechanism of cellular invasion by amyloid proteins. Acta Neuropathol. 2017;134:629–53.
121. Goedert M. Alpha-synuclein and neurodegenerative diseases. Nat Rev Neurosci. 2001;2(7):492–501.
122. Mori F, Piao YS, Hayashi S, Fujiwara H, Hasegawa M, Yoshimoto M, Iwatsubo T, Takahashi H, Wakabayashi K. α-Synuclein accumulates in Purkinje cells in Lewy body disease but not in multiple system atrophy. J Neuropathol Exp Neurol. 2003;62(8):812–9.
123. Roberts RF, Wade-Martins R, Alegre-Abarrategui J. Direct visualization of alpha-synuclein oligomers reveals previously undetected pathology in Parkinson's disease brain. Brain. 2015;138(6):1642–57.

124. Mor DE, Tsika E, Mazzulli JR, Gould NS, Kim H, Daniels MJ, Doshi S, Gupta P, Grossman JL, Tan VX, Kalb RG. Dopamine induces soluble α-synuclein oligomers and nigrostriatal degeneration. Nat Neurosci. 2017;20(11):1560–8.
125. Ibanez P, Bonnet AM, Debarges B, Lohmann E, Tison F, Agid Y, Dürr A, Brice A, Pollak P. Causal relation between α-synuclein locus duplication as a cause of familial Parkinson's disease. Lancet. 2004;364(9440):1169–71.
126. Tokuda T, Qureshi MM, Ardah MT, Varghese S, Shehab SA, Kasai T, Ishigami N, Tamaoka A, Nakagawa M, El-Agnaf OM. Detection of elevated levels of α-synuclein oligomers in CSF from patients with Parkinson disease. Neurology. 2010;75(20):1766–70.
127. Fusco G, Chen SW, Williamson PT, Cascella R, Perni M, Jarvis JA, Cecchi C, Vendruscolo M, Chiti F, Cremades N, Ying L. Structural basis of membrane disruption and cellular toxicity by α-synuclein oligomers. Science. 2017;358(6369):1440–3.
128. Cascella R, Perni M, Chen SW, Fusco G, Cecchi C, Vendruscolo M, Chiti F, Dobson CM, De Simone A. Probing the origin of the toxicity of oligomeric aggregates of α-synuclein with antibodies. ACS Chem Biol. 2019;14(6):1352–62.
129. Papapetropoulos S, Adi N, Ellul J, Argyriou AA, Chroni E. A prospective study of familial versus sporadic Parkinson's disease. Neurodegener Dis. 2007;4(6):424–7.
130. Driver JA, Logroscino G, Gaziano JM, Kurth T. Incidence and remaining lifetime risk of Parkinson disease in advanced age. Neurology. 2009;72(5):432–8.
131. Höhn A, Tramutola A, Cascella R. Proteostasis failure in neurodegenerative diseases: focus on oxidative stress. Oxidative Med Cell Longev. 2020;2020(1):5497046.
132. Schrag A. Quality of life and depression in Parkinson's disease. J Neurol Sci. 2006;248(1–2):151–7.
133. Webb JL, Ravikumar B, Atkins J, Skepper JN, Rubinsztein DC. α-Synuclein is degraded by both autophagy and the proteasome. J Biol Chem. 2003;278(27):25009–13.
134. Kalinderi K, Bostantjopoulou S, Fidani L. The genetic background of Parkinson's disease: current progress and future prospects. Acta Neurol Scand. 2016;134(5):314–26.
135. Imai Y, Soda M, Takahashi R. Parkin suppresses unfolded protein stress-induced cell death through its E3 ubiquitin-protein ligase activity. J Biol Chem. 2000;275(46):35661–4.
136. Bennett JP. Free radicals, oxidative stress and the origin of Parkinson's disease. J Neurol Sci. 1999;170(2):75–6.
137. Tofaris GK, Layfield R, Spillantini MG. α-Synuclein metabolism and aggregation is linked to ubiquitin-independent degradation by the proteasome. FEBS Lett. 2001;509(1):22–6.
138. Nakajima T, Takauchi S, Ohara K, Kokai M, Nishii R, Maeda S, Takanaga A, Tanaka T, Takeda M, Seki M, Morita Y. α-Synuclein-positive structures induced in leupeptin-infused rats. Brain Res. 2005;1040(1–2):73–80.
139. Seirafi M, Kozlov G, Gehring K. Parkin structure and function. FEBS J. 2015;282(11):2076–88.
140. Lee HJ, Suk JE, Bae EJ, Lee JH, Paik SR, Lee SJ. Assembly-dependent endocytosis and clearance of extracellular a-synuclein. Int J Biochem Cell Biol. 2008;40(9):1835–49.
141. Desplats P, Lee HJ, Bae EJ, Patrick C, Rockenstein E, Crews L, Spencer B, Masliah E, Lee SJ. Inclusion formation and neuronal cell death through neuron-to-neuron transmission of α-synuclein. Proc Natl Acad Sci. 2009;106(31):13010–5.
142. Brahic M, Bousset L, Bieri G, Melki R, Gitler AD. Axonal transport and secretion of fibrillar forms of α-synuclein, Aβ42 peptide and HTTExon 1. Acta Neuropathol. 2016;131:539–48.
143. Vogiatzi T, Xilouri M, Vekrellis K, Stefanis L. Wild type α-synuclein is degraded by chaperone-mediated autophagy and macroautophagy in neuronal cells. J Biol Chem. 2008;283(35):23542–56.
144. Yu J, Lyubchenko YL. Early stages for Parkinson's development: α-synuclein misfolding and aggregation. J Neuroimmune Pharmacol. 2009;4:10–6.
145. Klucken J, Poehler AM, Ebrahimi-Fakhari D, Schneider J, Nuber S, Rockenstein E, Schlötzer-Schrehardt U, Hyman BT, McLean PJ, Masliah E, Winkler J. Alpha-synuclein aggregation involves a bafilomycin A1-sensitive autophagy pathway. Autophagy. 2012;8(5):754–66.
146. Winslow AR, Chen CW, Corrochano S, Acevedo-Arozena A, Gordon DE, Peden AA, Lichtenberg M, Menzies FM, Ravikumar B, Imarisio S, Brown S. α-Synuclein impairs macroautophagy: implications for Parkinson's disease. J Cell Biol. 2010;190(6):1023–37.

147. Cuervo AM, Stefanis L, Fredenburg R, Lansbury PT, Sulzer D. Impaired degradation of mutant α-synuclein by chaperone-mediated autophagy. Science. 2004;305(5688):1292–5. Mak SK, McCormack AL, Manning-Boğ AB, Cuervo AM, Di Monte DA. Lysosomal degradation of α-synuclein in vivo. J Biol Chem. 2010;285(18):13621–9.
148. Jiang P, Gan M, Yen SH, McLean PJ, Dickson DW. Impaired endo-lysosomal membrane integrity accelerates the seeding progression of α-synuclein aggregates. Sci Rep. 2017;7(1):1–3.
149. Karpowicz RJ, Haney CM, Mihaila TS, Sandler RM, Petersson EJ, Lee VM. Selective imaging of internalized proteopathic α-synuclein seeds in primary neurons reveals mechanistic insight into transmission of synucleinopathies. J Biol Chem. 2017;292(32):13482–97.
150. Lee HJ, Khoshaghideh F, Patel S, Lee SJ. Clearance of α-synuclein oligomeric intermediates via the lysosomal degradation pathway. J Neurosci. 2004;24(8):1888–96.
151. Ravikumar B, Sarkar S, Rubinsztein DC. Clearance of mutant aggregate-prone proteins by autophagy. In: Autophagosome and phagosome. Totowa: Humana Press; 2008. p. 195–211.
152. Osterberg VR, Spinelli KJ, Weston LJ, Luk KC, Woltjer RL, Unni VK. Progressive aggregation of alpha-synuclein and selective degeneration of lewy inclusion-bearing neurons in a mouse model of Parkinsonism. Cell Rep. 2015;10(8):1252–60.
153. Dufty BM, Warner LR, Hou ST, Jiang SX, Gomez-Isla T, Leenhouts KM, Oxford JT, Feany MB, Masliah E, Rohn TT. Calpain-cleavage of α-synuclein: connecting proteolytic processing to disease-linked aggregation. Am J Pathol. 2007;170(5):1725–38.
154. Iwata A, Maruyama M, Akagi T, Hashikawa T, Kanazawa I, Tsuji S, Nukina N. Alpha-synuclein degradation by serine protease neurosin: implication for pathogenesis of synucleinopathies. Hum Mol Genet. 2003;12(20):2625–35.
155. Choi DH, Kim YJ, Kim YG, Joh TH, Beal MF, Kim YS. Role of matrix metalloproteinase 3-mediated α-synuclein cleavage in dopaminergic cell death. J Biol Chem. 2011;286(16):14168–77.
156. Liu CW, Giasson BI, Lewis KA, Lee VM, DeMartino GN, Thomas PJ. A precipitating role for truncated α-synuclein and the proteasome in α-synuclein aggregation: implications for pathogenesis of Parkinson disease. J Biol Chem. 2005;280(24):22670–8.
157. Arai T, Hasegawa M, Akiyama H, Ikeda K, Nonaka T, Mori H, Mann D, Tsuchiya K, Yoshida M, Hashizume Y, Oda T. TDP-43 is a component of ubiquitin-positive tau-negative inclusions in frontotemporal lobar degeneration and amyotrophic lateral sclerosis. Biochem Biophys Res Commun. 2006;351(3):602–11.
158. Cascella R, Capitini C, Fani G, Dobson CM, Cecchi C, Chiti F. Quantification of the relative contributions of loss-of-function and gain-of-function mechanisms in TAR DNA-binding protein 43 (TDP-43) proteinopathies. J Biol Chem. 2016;291(37):19437–48.
159. Lee EB, Lee VM, Trojanowski JQ. Gains or losses: molecular mechanisms of TDP43-mediated neurodegeneration. Nat Rev Neurosci. 2012;13(1):38–50.
160. Capitini C, Conti S, Perni M, Guidi F, Cascella R, De Poli A, Penco A, Relini A, Cecchi C, Chiti F. TDP-43 inclusion bodies formed in bacteria are structurally amorphous, non-amyloid and inherently toxic to neuroblastoma cells. PLoS One. 2014;9(1):e86720.
161. Ayala YM, Zago P, D'Ambrogio A, Xu YF, Petrucelli L, Buratti E, Baralle FE. Structural determinants of the cellular localization and shuttling of TDP-43. J Cell Sci. 2008;121(22):3778–85.
162. Chou CC, Zhang YI, Umoh ME, Vaughan SW, Lorenzini I, Liu F, Sayegh M, Donlin-Asp PG, Chen YH, Duong DM, Seyfried NT. TDP-43 pathology disrupts nuclear pore complexes and nucleocytoplasmic transport in ALS/FTD. Nat Neurosci. 2018;21(2):228–39.
163. Winton MJ, Igaz LM, Wong MM, Kwong LK, Trojanowski JQ, Lee VM. Disturbance of nuclear and cytoplasmic TAR DNA-binding protein (TDP-43) induces disease-like redistribution, sequestration, and aggregate formation. J Biol Chem. 2008;283(19):13302–9.
164. Ling SC, Polymenidou M, Cleveland DW. Converging mechanisms in ALS and FTD: disrupted RNA and protein homeostasis. Neuron. 2013;79(3):416–38.
165. Barmada SJ, Serio A, Arjun A, Bilican B, Daub A, Ando DM, Tsvetkov A, Pleiss M, Li X, Peisach D, Shaw C. Autophagy induction enhances TDP43 turnover and survival in neuronal ALS models. Nat Chem Biol. 2014;10(8):677–85.

166. Scotter EL, Vance C, Nishimura AL, Lee YB, Chen HJ, Urwin H, Sardone V, Mitchell JC, Rogelj B, Rubinsztein DC, Shaw CE. Differential roles of the ubiquitin proteasome system and autophagy in the clearance of soluble and aggregated TDP-43 species. J Cell Sci. 2014;127(6):1263–78.
167. Urushitani M, Sato T, Bamba H, Hisa Y, Tooyama I. Synergistic effect between proteasome and autophagosome in the clearance of polyubiquitinated TDP-43. J Neurosci Res. 2010;88(4):784–97.
168. DeJesus-Hernandez M, Mackenzie IR, Boeve BF, Boxer AL, Baker M, Rutherford NJ, Nicholson AM, Finch NA, Flynn H, Adamson J, Kouri N. Expanded GGGGCC hexanucleotide repeat in noncoding region of C9ORF72 causes chromosome 9p-linked FTD and ALS. Neuron. 2011;72(2):245–56.
169. Fecto F, Yan J, Vemula SP, Liu E, Yang Y, Chen W, Zheng JG, Shi Y, Siddique N, Arrat H, Donkervoort S. SQSTM1 mutations in familial and sporadic amyotrophic lateral sclerosis. Arch Neurol. 2011;68(11):1440–6.
170. Freischmidt A, Wieland T, Richter B, Ruf W, Schaeffer V, Müller K, Marroquin N, Nordin F, Hübers A, Weydt P, Pinto S. Haploinsufficiency of TBK1 causes familial ALS and frontotemporal dementia. Nat Neurosci. 2015;18(5):631–6.
171. Hadano S, Hand CK, Osuga H, Yanagisawa Y, Otomo A, Devon RS, Miyamoto N, Showguchi-Miyata J, Okada Y, Singaraja R, Figlewicz DA. A gene encoding a putative GTPase regulator is mutated in familial amyotrophic lateral sclerosis 2. Nat Genet. 2001;29(2):166–73.
172. Maruyama H, Morino H, Ito H, Izumi Y, Kato H, Watanabe Y, Kinoshita Y, Kamada M, Nodera H, Suzuki H, Komure O. Mutations of optineurin in amyotrophic lateral sclerosis. Nature. 2010;465(7295):223–6.
173. Parkinson N, Ince PG, Smith MO, Highley R, Skibinski G, Andersen PM, Morrison KE, Pall HS, Hardiman O, Collinge J, Shaw PJ. ALS phenotypes with mutations in CHMP2B (charged multivesicular body protein 2B). Neurology. 2006;67(6):1074–7.
174. Pottier C, Bieniek KF, Finch N, van de Vorst M, Baker M, Perkersen R, Brown P, Ravenscroft T, van Blitterswijk M, Nicholson AM, DeTure M. Whole-genome sequencing reveals important role for TBK1 and OPTN mutations in frontotemporal lobar degeneration without motor neuron disease. Acta Neuropathol. 2015;130:77–92.
175. Rea SL, Majcher V, Searle MS, Layfield R. SQSTM1 mutations–bridging Paget disease of bone and ALS/FTLD. Exp Cell Res. 2014;325(1):27–37.
176. Yang YI, Hentati A, Deng HX, Dabbagh O, Sasaki T, Hirano M, Hung WY, Ouahchi K, Yan J, Azim AC, Cole N. The gene encoding alsin, a protein with three guanine-nucleotide exchange factor domains, is mutated in a form of recessive amyotrophic lateral sclerosis. Nat Genet. 2001;29(2):160–5.
177. Cascella R, Fani G, Capitini C, Rusmini P, Poletti A, Cecchi C, Chiti F. Quantitative assessment of the degradation of aggregated TDP-43 mediated by the ubiquitin proteasome system and macroautophagy. FASEB J. 2017;31(12):5609–24.
178. Cascella R, Fani G, Bigi A, Chiti F, Cecchi C. Partial failure of proteostasis systems counteracting TDP-43 aggregates in neurodegenerative diseases. Int J Mol Sci. 2019;20(15):3685.
179. Bang Y, Kang BY, Choi HJ. Preconditioning stimulus of proteasome inhibitor enhances aggresome formation and autophagy in differentiated SH-SY5Y cells. Neurosci Lett. 2014;566:263–8.
180. Wang XJ, Yu J, Wong SH, Cheng AS, Chan FK, Ng SS, Cho CH, Sung JJ, Wu WK. A novel crosstalk between two major protein degradation systems: regulation of proteasomal activity by autophagy. Autophagy. 2013;9(10):1500–8.
181. Seyfried NT, Gozal YM, Dammer EB, Xia Q, Duong DM, Cheng D, Lah JJ, Levey AI, Peng J. Multiplex SILAC analysis of a cellular TDP-43 proteinopathy model reveals protein inclusions associated with SUMOylation and diverse polyubiquitin chains. Mol Cell Proteomics. 2010;9(4):705–18.

182. Hebron ML, Lonskaya I, Sharpe K, Weerasinghe PP, Algarzae NK, Shekoyan AR, Moussa CE. Parkin ubiquitinates Tar-DNA binding protein-43 (TDP-43) and promotes its cytosolic accumulation via interaction with histone deacetylase 6 (HDAC6). J Biol Chem. 2013;288(6):4103–15.
183. Dammer EB, Fallini C, Gozal YM, Duong DM, Rossoll W, Xu P, Lah JJ, Levey AI, Peng J, Bassell GJ, Seyfried NT. Coaggregation of RNA-binding proteins in a model of TDP-43 proteinopathy with selective RGG motif methylation and a role for RRM1 ubiquitination. PLoS One. 2012;7(6):e38658.
184. Liu G, Coyne AN, Pei F, Vaughan S, Chaung M, Zarnescu DC, Buchan JR. Endocytosis regulates TDP-43 toxicity and turnover. Nat Commun. 2017;8(1):2092.
185. Cobley JN, Fiorello ML, Bailey DM. 13 reasons why the brain is susceptible to oxidative stress. Redox Biol. 2018;15:490–503.
186. Reed TT, Sultana R, Butterfield DA. Redox Proteomics of Oxidatively Modified Brain Proteins in Mild Cognitive Impairment. In O. Alzate (Ed.), Neuroproteomics. CRC Press/Taylor & Francis. 2010, Chapter 10.
187. Sohal RS, Weindruch R. Oxidative stress, caloric restriction, and aging. Science. 1996;273(5271):59–63.
188. Davies MJ. Protein oxidation and peroxidation. Biochem J. 2016;473(7):805–25.
189. Schiropoulos H, Beckman JS. Oxidative stress and nitration in neurodegeneration: cause, effect, or association? J Clin Invest. 2003;111(2):163–9.
190. Hans F, Fiesel FC, Strong JC, Jäckel S, Rasse TM, Geisler S, Springer W, Schulz JB, Voigt A, Kahle PJ. UBE2E ubiquitin-conjugating enzymes and ubiquitin isopeptidase Y regulate TDP-43 protein ubiquitination. J Biol Chem. 2014;289(27):19164–79.
191. Schopfer FJ, Baker PR, Freeman BA. NO-dependent protein nitration: a cell signaling event or an oxidative inflammatory response? Trends Biochem Sci. 2003;28(12):646–54.
192. Montine TJ, Neely MD, Quinn JF, Beal MF, Markesbery WR, Roberts LJ II, Morrow JD. Lipid peroxidation in aging brain and Alzheimer's disease. Free Radic Biol Med. 2002;33(5):620–6.
193. Munoz-Sanjuan I, Bates GP. The importance of integrating basic and clinical research toward the development of new therapies for Huntington disease. J Clin Invest. 2011;121(2):476–83.
194. Hannan AJ. Tandem repeats mediating genetic plasticity in health and disease. Nat Rev Genet. 2018;19(5):286–98.
195. Bates GP, Dorsey R, Gusella JF, Hayden MR, Kay C, Leavitt BR, Nance M, Ross CA, Scahill RI, Wetzel R, Wild EJ. Huntington disease. Nat Rev Dis Primers. 2015;1(1):1–21.
196. Walker FO. Huntington's disease. Lancet. 2007;369(9557):218–28.
197. Ciosi M, Maxwell A, Cumming SA, Moss DJ, Alshammari AM, Flower MD, Durr A, Leavitt BR, Roos RA, Holmans P, Jones L. A genetic association study of glutamine-encoding DNA sequence structures, somatic CAG expansion, and DNA repair gene variants, with Huntington disease clinical outcomes. EBioMedicine. 2019;48:568–80.
198. Williams AJ, Paulson HL. Polyglutamine neurodegeneration: protein misfolding revisited. Trends Neurosci. 2008;31(10):521–8.
199. Warby SC, Visscher H, Collins JA, Doty CN, Carter C, Butland SL, Hayden AR, Kanazawa I, Ross CJ, Hayden MR. HTT haplotypes contribute to differences in Huntington disease prevalence between Europe and East Asia. Eur J Hum Genet. 2011;19(5):561–6.
200. Soto C. Unfolding the role of protein misfolding in neurodegenerative diseases. Nat Rev Neurosci. 2003;4(1):49–60.
201. Kar K, Jayaraman M, Sahoo B, Kodali R, Wetzel R. Critical nucleus size for disease-related polyglutamine aggregation is repeat-length dependent. Nat Struct Mol Biol. 2011;18(3):328–36.
202. Qi L, Zhang XD, Wu JC, Lin F, Wang J, DiFiglia M, Qin ZH. The role of chaperone-mediated autophagy in huntingtin degradation. PLoS One. 2012;7(10):e46834.
203. DiFiglia M, Sapp E, Chase KO, Davies SW, Bates GP, Vonsattel JP, Aronin N. Aggregation of huntingtin in neuronal intranuclear inclusions and dystrophic neurites in brain. Science. 1997;277(5334):1990–3.

204. Hipp MS, Patel CN, Bersuker K, Riley BE, Kaiser SE, Shaler TA, Brandeis M, Kopito RR. Indirect inhibition of 26S proteasome activity in a cellular model of Huntington's disease. J Cell Biol. 2012;196(5):573–87.
205. Jeong H, Then F, Melia TJ, Mazzulli JR, Cui L, Savas JN, Voisine C, Paganetti P, Tanese N, Hart AC, Yamamoto A. Acetylation targets mutant huntingtin to autophagosomes for degradation. Cell. 2009;137(1):60–72.
206. Martinez-Vicente M, Talloczy Z, Wong E, Tang G, Koga H, Kaushik S, De Vries R, Arias E, Harris S, Sulzer D, Cuervo AM. Cargo recognition failure is responsible for inefficient autophagy in Huntington's disease. Nat Neurosci. 2010;13(5):567–76.
207. Wong E, Cuervo AM. Autophagy gone awry in neurodegenerative diseases. Nat Neurosci. 2010;13(7):805–11.
208. Hyrskyluoto A, Bruelle C, Lundh SH, Do HT, Kivinen J, Rappou E, Reijonen S, Waltimo T, Petersén Å, Lindholm D, Korhonen L. Ubiquitin-specific protease-14 reduces cellular aggregates and protects against mutant huntingtin-induced cell degeneration: involvement of the proteasome and ER stress-activated kinase IRE1α. Hum Mol Genet. 2014;23(22):5928–39.
209. Sarkar S, Ravikumar B, Floto RA, Rubinsztein DC. Rapamycin and mTOR-independent autophagy inducers ameliorate toxicity of polyglutamine-expanded huntingtin and related proteinopathies. Cell Death Differ. 2009;16(1):46–56.
210. Ramdzan YM, Trubetskov MM, Ormsby AR, Newcombe EA, Sui X, Tobin MJ, Bongiovanni MN, Gras SL, Dewson G, Miller JM, Finkbeiner S. Huntingtin inclusions trigger cellular quiescence, deactivate apoptosis, and lead to delayed necrosis. Cell Rep. 2017;19(5):919–27.
211. Lam HT, Graber MC, Gentry KA, Bieschke J. Stabilization of α-synuclein fibril clusters prevents fragmentation and reduces seeding activity and toxicity. Biochemistry. 2016;55(4):675–85.
212. Guo JL, Covell DJ, Daniels JP, Iba M, Stieber A, Zhang B, Riddle DM, Kwong LK, Xu Y, Trojanowski JQ, Lee VM. Distinct α-synuclein strains differentially promote tau inclusions in neurons. Cell. 2013;154(1):103–17.
213. Sanders DW, Kaufman SK, DeVos SL, Sharma AM, Mirbaha H, Li A, Barker SJ, Foley AC, Thorpe JR, Serpell LC, Miller TM. Distinct tau prion strains propagate in cells and mice and define different tauopathies. Neuron. 2014;82(6):1271–88.
214. Heilbronner G, Eisele YS, Langer F, Kaeser SA, Novotny R, Nagarathinam A, Åslund A, Hammarström P, Nilsson KP, Jucker M. Seeded strain-like transmission of β-amyloid morphotypes in APP transgenic mice. EMBO Rep. 2013;14(11):1017–22.
215. Pujol-Pina R, Vilaprinyo-Pascual S, Mazzucato R, Arcella A, Vilaseca M, Orozco M, Carulla N. SDS-PAGE analysis of Abeta oligomers is disserving research into Alzheimer s disease: appealing for ESI-IM-MS. Sci Rep. 2015;5:14809. https://doi.org/10.1038/srep14809.
216. Woods LA, Platt GW, Hellewell AL, Hewitt EW, Homans SW, Ashcroft AE, Radford SE. Ligand binding to distinct states diverts aggregation of an amyloid-forming protein. Nat Chem Biol. 2011;7(10):730–9.
217. Bleiholder C, Do TD, Wu C, Economou NJ, Bernstein SS, Buratto SK, Shea JE, Bowers MT. Ion mobility spectrometry reveals the mechanism of amyloid formation of Aβ (25–35) and its modulation by inhibitors at the molecular level: epigallocatechin gallate and scyllo-inositol. J Am Chem Soc. 2013;135(45):16926–37.
218. Gandhi J, Antonelli AC, Afridi A, Vatsia S, Joshi G, Romanov V, Murray IV, Khan SA. Protein misfolding and aggregation in neurodegenerative diseases: a review of pathogeneses, novel detection strategies, and potential therapeutics. Rev Neurosci. 2019;30(4):339–58.
219. Kennedy ME, Stamford AW, Chen X, Cox K, Cumming JN, Dockendorf MF, Egan M, Ereshefsky L, Hodgson RA, Hyde LA, Jhee S. The BACE1 inhibitor verubecestat (MK-8931) reduces CNS β-amyloid in animal models and in Alzheimer's disease patients. Sci Transl Med. 2016;8(363):363ra150.
220. Roy SM, Minasov G, Arancio O, Chico LW, Van Eldik LJ, Anderson WF, Pelletier JC, Watterson DM. A selective and brain penetrant p38αMAPK inhibitor candidate for neurologic and neuropsychiatric disorders that attenuates neuroinflammation and cognitive dysfunction. J Med Chem. 2019;62(11):5298–311.

221. DeVos SL, Miller RL, Schoch KM, Holmes BB, Kebodeaux CS, Wegener AJ, Chen G, Shen T, Tran H, Nichols B, Zanardi TA. Tau reduction prevents neuronal loss and reverses pathological tau deposition and seeding in mice with tauopathy. Sci Transl Med. 2017;9(374):eaag0481.
222. Zatsepina OG, Evgen'ev MB, Garbuz DG. Role of a heat shock transcription factor and the major heat shock protein Hsp70 in memory formation and neuroprotection. Cells. 2021;10:1638. [Internet].
223. Vilariño-Güell C, Rajput A, Milnerwood AJ, Shah B, Szu-Tu C, Trinh J, Yu I, Encarnacion M, Munsie LN, Tapia L, Gustavsson EK. DNAJC13 mutations in Parkinson disease. Hum Mol Genet. 2014;23(7):1794–801.
224. Wadhwa R, Ryu J, Ahn HM, Saxena N, Chaudhary A, Yun CO, Kaul SC. Functional significance of point mutations in stress chaperone mortalin and their relevance to Parkinson disease. J Biol Chem. 2015;290(13):8447–56.
225. Koppers M, van Blitterswijk MM, Vlam L, Rowicka PA, van Vught PW, Groen EJ, Spliet WG, Engelen-Lee J, Schelhaas HJ, de Visser M, van der Kooi AJ. VCP mutations in familial and sporadic amyotrophic lateral sclerosis. Neurobiol Aging. 2012;33(4):837–e7.
226. Thirstrup K, Sotty F, Montezinho LC, Badolo L, Thougaard A, Kristjánsson M, Jensen T, Watson S, Nielsen SM. Linking HSP90 target occupancy to HSP70 induction and efficacy in mouse brain. Pharmacol Res. 2016;104:197–205.
227. Brandvold KR, Morimoto RI. The chemical biology of molecular chaperones—implications for modulation of proteostasis. J Mol Biol. 2015;427(18):2931–47.
228. Lindberg I, Shorter J, Wiseman RL, Chiti F, Dickey CA, McLean PJ. Chaperones in neurodegeneration. J Neurosci. 2015;35(41):13853–9.
229. Smith HL, Li W, Cheetham ME. Molecular chaperones and neuronal proteostasis. Semin Cell Dev Biol. 2015;40:142–52. Academic Press.
230. Hervás R, Oroz J. Mechanistic insights into the role of molecular chaperones in protein misfolding diseases: from molecular recognition to amyloid disassembly. Int J Mol Sci. 2020;21(23):9186.
231. Pratt WB, Gestwicki JE, Osawa Y, Lieberman AP. Targeting proteostasis through the protein quality control function of the Hsp90/Hsp70-based chaperone machinery for treatment of adult onset neurodegenerative diseases. Annu Rev Pharmacol Toxicol. 2015;55:353.
232. Prodromou C, Aran-Guiu X, Oberoi J, Perna L, Chapple JP, van der Spuy J. Hsp70-hsp90 chaperone networking in protein-misfolding disease. In: The networking of chaperones by co-chaperones. Cham: Springer; 2022. p. 389–425.
233. Wang B, Liu Y, Huang L, Chen J, Li JJ, Wang R, Kim E, Chen Y, Justicia C, Sakata K, Chen H. A CNS-permeable Hsp90 inhibitor rescues synaptic dysfunction and memory loss in APP-overexpressing Alzheimer's mouse model via an HSF1-mediated mechanism. Mol Psychiatry. 2017;22(7):990–1001.
234. Webster JM, Darling AL, Uversky VN, Blair LJ. Small heat shock proteins, big impact on protein aggregation in neurodegenerative disease. Front Pharmacol. 2019;10:1047.
235. Liu YH, Han YL, Song J, Wang Y, Jing YY, Shi Q, Tian C, Wang ZY, Li CP, Han J, Dong XP. Heat shock protein 104 inhibited the fibrillization of prion peptide 106–126 and disassembled prion peptide 106–126 fibrils in vitro. Int J Biochem Cell Biol. 2011;43(5):768–74.
236. Ciechanover A, Kwon YT. Degradation of misfolded proteins in neurodegenerative diseases: therapeutic targets and strategies. Exp Mol Med. 2015;47(3):e147.
237. Dantuma NP, Bott LC. The ubiquitin-proteasome system in neurodegenerative diseases: precipitating factor, yet part of the solution. Front Mol Neurosci. 2014;7:70.
238. Kulkarni A, Preeti K, Tryphena KP, Srivastava S, Singh SB, Khatri DK. Proteostasis in Parkinson's disease: recent development and possible implication in diagnosis and therapeutics. Ageing Res Rev. 2023;84:101816.
239. Moscovitz O, Ben-Nissan G, Fainer I, Pollack D, Mizrachi L, Sharon M. The Parkinson's-associated protein DJ-1 regulates the 20S proteasome. Nat Commun. 2015;6(1):6609.
240. Zenko D, Marsh J, Castle AR, Lewin R, Fischer R, Tofaris GK. Monitoring α-synuclein ubiquitination dynamics reveals key endosomal effectors mediating its trafficking and degradation. Sci Adv. 2023;9(24):eadd8910.

241. Sweeney P, Park H, Baumann M, Dunlop J, Frydman J, Kopito R, McCampbell A, Leblanc G, Venkateswaran A, Nurmi A, Hodgson R. Protein misfolding in neurodegenerative diseases: implications and strategies. Transl Neurodegener. 2017;6:1–3.
242. De Felice FG, Munoz DP. Opportunities and challenges in developing relevant animal models for Alzheimer's disease. Ageing Res Rev. 2016;26:112–4.
243. Mullard A. Alzheimer amyloid hypothesis lives on. Nat Rev Drug Discov. 2017;16(1):3–6.
244. McDade E, Llibre-Guerra JJ, Holtzman DM, Morris JC, Bateman RJ. The informed road map to prevention of Alzheimer Disease: a call to arms. Mol Neurodegener. 2021;16:1–9.
245. Sukoff Rizzo SJ, Masters A, Onos KD, Quinney S, Sasner M, Oblak A, Lamb BT, Territo PR, MODEL-AD Consortium. Improving preclinical to clinical translation in Alzheimer's disease research. Alzheimers Dement: Transl Res Clin Interv. 2020;6(1):e12038.
246. Rafii MS, Aisen PS. Detection and treatment of Alzheimer's disease in its preclinical stage. Nat Aging. 2023;3(5):520–31.
247. Aisen PS, Jimenez-Maggiora GA, Rafii MS, Walter S, Raman R. Early-stage Alzheimer disease: getting trial-ready. Nat Rev Neurol. 2022;18(7):389–99.
248. Akhtar A, Andleeb A, Waris TS, Bazzar M, Moradi AR, Awan NR, Yar M. Neurodegenerative diseases and effective drug delivery: a review of challenges and novel therapeutics. J Control Release. 2021;330:1152–67.

Future Perspectives and Challenges in the Field of Proteostasis and Neurodegenerative Disorders

18

Suruchi Chaudhary, Vikas Shrivastava, Manish Kumar, Ekansh Agrawal, Niharika Singh, and Rajesh S. Tomar

18.1 Fundamental Principles of Proteostasis

The maintenance of cellular proteostasis, or the regulation of protein biogenesis, maturation, trafficking, and degradation, is a critical process that underlies the proper function of all living organisms [57]. Disruptions to proteostasis have been linked to the aetiology of a wide range of human illnesses, including prion diseases, Parkinson's disease, and Alzheimer's disease. The buildup of misfolded or aggregated proteins is usually indicative of these diseases, which can lead to proteotoxic stress, neuronal dysfunction, and eventual cell death [14, 15].

The interconnected mechanisms of autophagy and the ubiquitin–proteasome system provide the basis of proteostasis, which works in concert to clear damaged or superfluous proteins from the cell. Impairments to these quality control pathways can cause the accumulation of toxic protein species, setting the stage for the neurodegenerative cascade. Additionally, other factors like neuroinflammation, oxidative stress and dysregulation of apoptosis pathways have also been shown to contribute to the pathophysiology of these disorders [43].

The regular operation and vitality of cells depend on the maintenance of protein homeostasis. Damage to the protein quality control system, proteasome, autophagy, and unfolded response can also lead to the deposition of aggregated, misfolded, and oxidised proteins.

Non-coding RNAs have been linked to the pathophysiology of neurodegenerative diseases in recent years, and microRNAs (miRNAs), in particular, have become important regulators of proteostasis. The expression of target genes, such as those involved in protein synthesis, folding, and degradation, can be post-transcriptionally modulated by microRNAs, which are tiny, endogenous, single-stranded RNA

S. Chaudhary (✉) · V. Shrivastava · M. Kumar · E. Agrawal · N. Singh · R. S. Tomar
Amity Institute of Biotechnology, Amity University, Gwalior, Madhya Pradesh, India
e-mail: suruchichaudhary02@gmail.com

N. Puranik (ed.), *Proteostasis*, https://doi.org/10.1007/978-981-96-6202-9_18

molecules. Patients with Huntington's, Parkinson's, and Alzheimer's diseases have shown signs of dysregulation of particular microRNAs in their brains, which may indicate that these microRNAs are involved in the disturbance of proteostasis [38].

For instance, it has been demonstrated that miR-29 targets the mRNA of BACE1, a crucial enzyme that produces amyloid-β, the main building block of the toxic protein aggregates associated with Alzheimer's disease. Likewise, it has been discovered that miR-7 controls the production of α-synuclein, a protein that accumulates and misfolds in Parkinson's disease [30]. Additionally, microRNAs can indirectly affect how cells handle misfolded proteins by modifying the expression of proteostasis regulators such as chaperones, ubiquitin–proteasome system components, and genes linked to autophagy.

Non-coding RNAs have become interesting therapeutic targets for neurodegenerative diseases due to the mounting evidence of their role in proteostasis.

18.2 Emerging Perspectives on Autophagic Dysfunction

There is growing evidence that the onset and course of neuro-degenerative illnesses are mostly influenced by disturbances to the autophagic flux, which is the coordinated process of cargo engulfment, lysosomal breakdown, and recycling [37, 43, 48]. It has been shown that affected neuronal populations exhibit changes in both macroautophagy and chaperone-mediated autophagy, which results in the buildup of harmful protein aggregates and cellular malfunction [37, 43].

The precise mechanisms by which autophagic dysfunction manifests in different neurodegenerative disorders remain an active area of investigation. Some studies have proposed that imbalances in autophagic activity, with either excessive or insufficient autophagy, can be detrimental, underscoring the need for tight regulation of this process. Unravelling the complex interplay between autophagy, proteostasis, and other cellular pathways will be crucial for developing effective therapeutic strategies targeting these pathways.

18.3 Therapeutic Opportunities and Challenges

The recognition of the critical role of proteostasis in neurodegenerative diseases has spurred intense research efforts to develop novel therapeutic interventions that can restore cellular homeostasis and mitigate neurodegeneration. One promising approach is the use of pharmacological modulators of autophagy, which have shown therapeutic potential in preclinical models of Alzheimer's, Parkinson's, and other neurodegenerative conditions [31, 37, 41, 58].

In addition, natural compounds with neuroprotective properties, such as polyphenols and terpenoids, have garnered significant attention due to their ability to ameliorate mitochondrial dysfunction, oxidative stress, and proteostatic imbalances [10, 44]. These natural products may offer a safer and more well-tolerated alternative to synthetic drugs, making them attractive candidates for the treatment of age-related neurodegenerative diseases.

However, the development of effective autophagy-based therapies faces several challenges. First, the complex and dynamic nature of autophagic flux, with its spatiotemporal regulation and crosstalk with other cellular processes, makes it difficult to precisely manipulate this pathway. Second, the heterogeneity of neurodegenerative diseases and their diverse etiologies necessitates a tailored approach to targeting proteostasis, as a one-size-fits-all solution may not be feasible.

Furthermore, the blood–brain barrier and the vulnerability of neuronal cells pose additional obstacles in delivering and maintaining the efficacy of autophagy-modulating therapies in the central nervous system [60]. Overcoming these hurdles will require a multifaceted approach, integrating a deeper understanding of the underlying disease mechanisms, innovative drug delivery strategies, and personalised treatment approaches [43].

18.4 Emerging Roles of Proteostasis Pathways

Recent studies have highlighted the diverse and critical roles that proteostasis pathways, such as the ubiquitin–proteasome system and autophagy, play in the maintenance of neuronal function and the prevention of neurodegenerative disorders [45]. Disruptions to these quality control mechanisms have been frequently observed in the selective neuronal populations affected by conditions like Alzheimer's, Parkinson's, and prion diseases, underscoring their importance in neuronal health and survival [43].

The ubiquitin–proteasome system is responsible for the selective degradation of misfolded, damaged, or short-lived proteins, thereby preventing their accumulation and aggregation. Impairments in this system have been linked to the pathogenesis of many neurodegenerative diseases, as the failure to efficiently clear aberrant proteins can lead to their toxic buildup and the eventual demise of affected neurons.

Autophagy, in particular, has emerged as a key player in the pathogenesis of neurodegenerative diseases. Defects in autophagic flux, the coordinated process of cargo engulfment, lysosomal degradation, and recycling can lead to the accumulation of toxic protein aggregates and cellular dysfunction. Imbalances in autophagic activity, with either excessive or insufficient autophagy, have been observed in different neurodegenerative conditions, suggesting the need for tight regulation of this pathway. The precise mechanisms by which autophagic dysfunction contributes to neurodegeneration are still being elucidated. Some studies have proposed that the buildup of misfolded proteins and damaged organelles, such as mitochondria, can trigger a cascade of cellular stress responses, ultimately leading to neuronal death.

18.5 Proteostasis and Protein Misfolding Diseases

Neurodegenerative diseases cause disruptions in a crucial physiological function called proteostasis, which is the precise balance between protein synthesis, folding, trafficking, and destruction. A common feature of many neurodegenerative diseases is the buildup of misfolded and aggregated proteins, which can overwhelm the body's proteostasis system and cause neuronal death and cellular malfunction.

The link between proteostasis and neurodegenerative diseases is particularly evident in the case of Alzheimer's disease, Parkinson's disease, and prion diseases. In these conditions, the buildup of specific misfolded proteins, such as amyloid-β, tau, α-synuclein, and prion proteins, has been implicated as a key driver of neurodegeneration [19, 43].

Disruptions to the protein quality control mechanisms, including the ubiquitin–proteasome system and autophagy, have been widely observed in these neurodegenerative disorders [19, 31, 43, 49]. The impairment of these pathways can lead to the accumulation of toxic protein aggregates, which can, in turn, trigger a cascade of cellular stress responses, including endoplasmic reticulum stress, oxidative stress, and neuroinflammation [11, 16, 47] (Table 18.1).

18.6 Neurodegenerative Disorders and Proteostasis Dysregulation

Neurodegenerative diseases, such as Alzheimer's, Parkinson's, and prion diseases, are defined by the accumulation of misfolded proteins and the disruption of proteostasis pathways. In these conditions, the buildup of specific misfolded proteins, such as amyloid-β, tau, α-synuclein, and prion proteins, has been identified as a key driver of neurodegeneration [34, 57].

Disruptions to the cellular quality control mechanisms, such as the ubiquitin–proteasome system and autophagy pathways, have been extensively documented in these neurodegenerative conditions. These impairments in the crucial proteostasis regulatory systems can contribute to the buildup of misfolded proteins and the breakdown of cellular homeostasis, ultimately driving the neurodegenerative process [25] (Fig. 18.1).

18.6.1 Role in Neurodegeneration

A steady decrease in UPS activity and the build-up of changed proteins are linked to aging, which is a significant risk factor for neurological disorders. Deposition of ubiquitinated protein aggregates and extensive disruption of cellular proteostasis have been associated with dysfunction of the UPS in the neurodegenerative brain. A number of neurodegenerative illnesses, including Alzheimer's, Parkinson's, and Huntington's diseases, have been linked to UPS impairment [55]. These illnesses

Table 18.1 Misfolded protein diseases: target proteins, folding locations, and mechanisms of toxicity [52]

S. No	Disease	Target protein	Folding location	Mechanism of toxicity
1.	Alzheimer's disease	Amyloid β-protein (Aβ), microtubule-associated protein (tau)	Cytosol	Accumulation of amyloid-beta (Aβ) oligomers disrupts neuronal membranes, leading to cell death and synaptic dysfunction. Aβ also induces oxidative stress and inflammation, contributing to neurodegeneration.
2.	Cystic fibrosis	Cystic fibrosis Transmembrane regulator	Endoplasmic reticulum (ER)	Misfolding of CFTR protein leads to defective chloride channels, resulting in thick mucus accumulation, inflammation, and tissue damage in the lungs and pancreas.
3.	Parkinson's disease	α-Synuclein	Cytosol	Alpha-synuclein aggregation into Lewy bodies impairs mitochondrial and synaptic functions, resulting in neuronal death and movement impairment.
4.	Sickle cell anemia	Hemoglobin	Cytosol	Misfolded hemoglobin (HbS) polymerises under low oxygen conditions, causing red blood cell deformation, vaso-occlusion, and hemolysis, leading to pain and organ damage.
5.	Multiple myeloma	Antibody light chain	Endoplasmic Reticulum (ER)	Accumulation of misfolded immunoglobulin light chains (amyloid light chains) leads to organ damage and impairment of the immune system, contributing to disease progression.
6.	Huntington's disease	Huntington	Cytosol	Mutant huntingtin protein aggregates disrupt cellular functions, including mitochondrial integrity and proteostasis, leading to neuronal cell death.

(continued)

Table 18.1 (continued)

S. No	Disease	Target protein	Folding location	Mechanism of toxicity
7.	Type II diabetes	Amylin	Cytosol	Buildup of misfolded islet amyloid polypeptide (IAPP) in pancreatic beta cells leads to cell dysfunction and apoptosis, contributing to insulin resistance.
8.	Nephrogenic diabetes insipidus (NDI)	Aquaporin-2/ vasopressin	Endoplasmic reticulum (ER)	Mutations in aquaporin-2 (AQP2) lead to misfolding and improper trafficking of the water channel, resulting in impaired water reabsorption and excessive urination.
9.	Retinitis pigmentosa (inherited blindness)	Rhodopsin	Endoplasmic reticulum (ER)	Mutations in rhodopsin or other retinal proteins lead to misfolding and aggregation, causing photoreceptor cell death and progressive vision loss.
10.	Desminopathy	Desmin and α β- crystallin	Cytosol	Misfolding of desmin, a muscle-specific intermediate filament, leads to aggregation and disruption of muscle cell integrity, resulting in muscle weakness and atrophy.

cause cellular malfunction and death as a result of the proteostatic processes being challenged by the buildup of harmful protein aggregates [2, 39, 53].

18.7 Therapeutic Strategies Targeting Proteostasis and Future Perspectives and Challenges

The recognition of the central role of proteostasis dysregulation in neurodegenerative diseases has spurred the development of novel therapeutic strategies aimed at restoring cellular homeostasis and mitigating neurodegeneration. One promising approach is the use of pharmacological modulators of autophagy, which have demonstrated therapeutic potential in preclinical models of Alzheimer's, Parkinson's, and other neurodegenerative conditions [15, 19, 34, 43]. These 20-modulating compounds have shown the ability to enhance the clearance of misfolded proteins and improve neuronal function and survival [29].

However, the development of effective autophagy-based therapies faces several challenges. The complex and dynamic nature of autophagic flux, with its intricate spatiotemporal regulation and crosstalk with other cellular processes, makes it difficult to precisely manipulate this pathway [23].

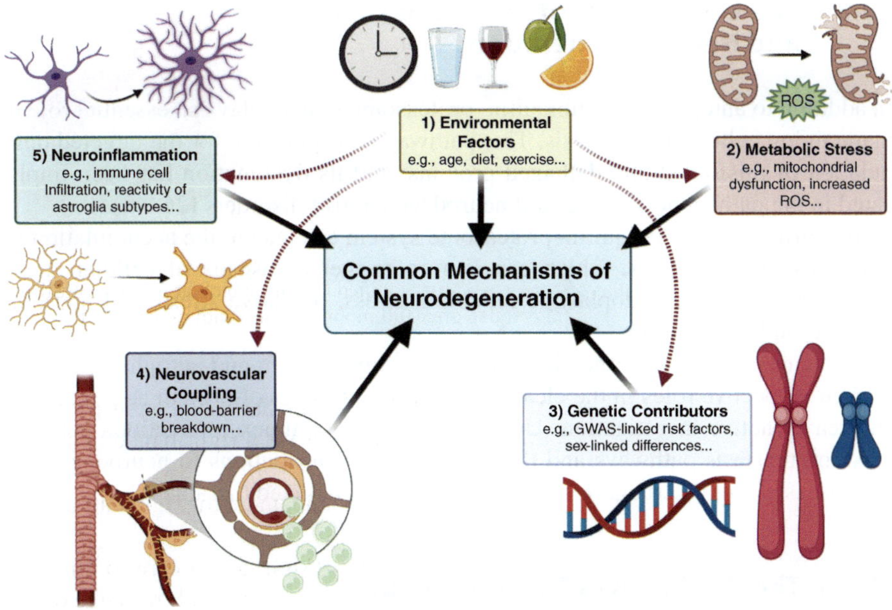

Fig. 18.1 Classic neurodegenerative processes. There are five primary areas of mechanistic overlap in neurodegenerative diseases: (1) environmental factors, such as age, diet, and exercise; (2) metabolic stress, such as mitochondrial dysfunction and increased reactive oxygen species (ROS); (3) genetic contributions, such as sex-linked genetic contributions and genome-wide association study-linked risk alleles (GWAS); (4) neurovascular coupling, such as dysfunctional neurovascular coupling and blood–brain barrier breakdown; and (5) neuroinflammation, such as peripheral immune cell infiltration and increased glial reactivity. Entire mechanistic areas of degeneration are influenced by environmental variables [56]

Additionally, the heterogeneity of neurodegenerative diseases and their diverse etiologies necessitates a tailored approach to targeting proteostasis, as a one-size-fits-all solution may not be feasible.

Furthermore, the unique challenges posed by the blood–brain barrier and the vulnerability of neuronal cells present additional obstacles in delivering and maintaining the efficacy of autophagy-modulating therapies in the central nervous system.

Overcoming these hurdles will require a multifaceted approach, integrating a deeper understanding of the underlying disease mechanisms, innovative drug delivery strategies, and personalised treatment approaches [19, 37, 43, 49].

The use of natural compounds, such as phytochemicals, has emerged as a promising alternative or complementary strategy for addressing proteostasis dysregulation in neurodegenerative diseases. These natural products have been reported to exhibit neuroprotective effects through multiple mechanisms, including the modulation of protein misfolding, the enhancement of antioxidant defences, and the regulation of cellular stress pathways [7, 10, 13, 17].

18.8 The Ubiquitin–Proteasome System and Neurodegeneration

In addition to autophagy, the ubiquitin–proteasome system plays an essential role in maintaining cellular proteostasis. This pathway is responsible for the targeted degradation of misfolded and damaged proteins, and its dysfunction has been implicated in the pathogenesis of several neurodegenerative disorders [26, 48].

Impairment of the ubiquitin–proteasome system can lead to the accumulation of toxic protein aggregates, which can in turn trigger a cascade of cellular stress responses, including endoplasmic reticulum stress, oxidative stress, and neuroinflammation [15, 43, 49].

The interplay between the ubiquitin–proteasome system and autophagy, as well as their respective roles in the clearance of different types of protein aggregates, is an area of active investigation. Understanding the complex interactions between these proteostasis pathways and their differential contributions to neurodegenerative diseases will be crucial for the development of targeted therapies (Fig. 18.2).

18.8.1 The UPS Involves Two Main Steps

18.8.1.1 Ubiquitination

Ubiquitin chains are attached to misfolded or damaged proteins, acting as a signal for proteasomal breakdown. A series of enzymes (E1, E2, E3) that attach ubiquitin to particular lysine residues on target proteins mediate this process.

18.8.1.2 Proteasomal Degradation

The proteins with polyubiquitin chains are recognised by the proteasome and broken into small peptides and different amino acids, which can then be utilised again for the production of new proteins [42].

The ubiquitinated proteins are then transported to the 26S proteasome, a large multi-subunit complex that degrades the proteins into small peptides.

Both the ubiquitination and proteasomal degradation steps are necessary for the correct functioning of the UPS, and disruptions to either process may play a role in the aetiology of neurodegenerative diseases.

18.9 Cellular Chaperone Machinery and Neurodegenerative Diseases

Maintaining protein homeostasis is a critical function of the cellular chaperone machinery, which is made up of a network of molecular chaperones and co-chaperones. The pathophysiology of certain neurodegenerative diseases has been connected to the dysregulation of these chaperone proteins, which are in charge of the appropriate folding, trafficking, and clearance of client proteins.

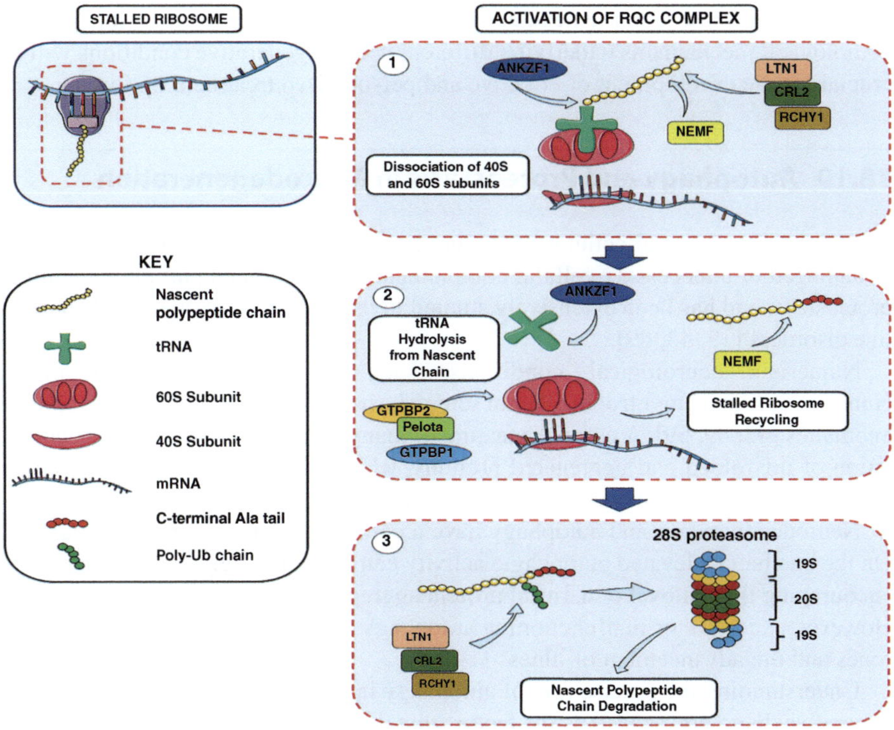

Fig. 18.2 Degradation of abnormal nascent chains and stalled ribosome resolution are the outcomes of activating the RQC machinery. After scanning the arrested ribosomes, the RQC complex causes the small and large ribosomal subunits to separate, the tRNA from the nascent chain to be hydrolysed, the translation machinery to be recycled, and finally the 28S proteasome to degrade the abnormal nascent chains that the stalled ribosomes have produced. The success of these phases is ascribed to the careful control of every molecular player and the redundant role of specific proteins, which ensure exact and thorough monitoring of any possible errors in the translation process [32]

In neurodegenerative diseases, the chaperone machinery can become overwhelmed by the burden of misfolded and aggregated proteins, leading to its impairment and the further accumulation of toxic protein species. This can trigger a vicious cycle of proteostasis collapse and neurodegeneration [46].

Efforts to modulate the chaperone network, either through pharmacological interventions or by targeting specific chaperone proteins, have exhibited promising results in preclinical models of neurodegenerative disorders. By enhancing the capacity of the chaperone machinery to maintain proteostasis, these tactics seek to lessen the harmful consequences of protein aggregation and misfolding.

Future proteostasis-based treatments for neurodegenerative diseases will probably take a multimodal strategy that incorporates the ubiquitin–proteasome system, the cellular chaperone machinery, and targeted autophagy modulation.

Integrating these complementary strategies and tailoring them to the specific pathological mechanisms underlying different neurodegenerative conditions will be crucial for the development of effective and personalised treatment options.

18.10 Autophagy and Proteostasis in Neurodegeneration

Autophagy, an essential cellular procedure liable for the breakdown and recycling of damaged or unnecessary cellular components, plays a crucial part in maintaining proteostasis and has been extensively studied in its connection with neurodegenerative disorders [19, 43, 49].

Numerous neurological conditions, such as Parkinson's, Alzheimer's, Huntington's, and amyotrophic lateral sclerosis, have been found to have autophagy anomalies [43, 49, 59]. A common feature of many of these disorders is an accumulation of misfolded and aggregated proteins, which is frequently accompanied by autophagic cascade imbalance, hindering cellular proteostasis [19, 43, 49].

Neurodegeneration and autophagy have a complicated and nuanced interaction. On the one hand, elevated autophagic activity can work as a defence mechanism by encouraging the removal of harmful protein aggregates and reducing stress on cells. However, excessive or malfunctioning autophagy can also lead to the death of neurones and the advancement of illness [19].

Understanding the precise role of autophagy in the context of different neurodegenerative disorders, as well as the factors that determine whether autophagic activity is beneficial or detrimental, remains an active area of research [19, 37, 43, 49] (Fig. 18.3).

18.11 Therapeutic Potential of Autophagy Modulation

Given the main role of proteostasis disruption in neurodegenerative disorders pathogenesis, targeting autophagy has appeared as a promising curative strategy [8].

Pharmacological modulators of autophagy, such as rapamycin, trehalose, and spermidine, have been investigated for the restoration potential of cellular proteostasis and mitigate neurodegeneration in preclinical models [12, 31, 37, 49].

However, the heterogeneity of neurodegenerative diseases and their diverse etiologies necessitates a tailored approach to targeting proteostasis, as a one-size-fits-all solution may not be feasible [19, 31, 37, 49]. Further studies are needed to clarify the accurate mechanisms by which autophagy dysregulation contributes to the pathogenesis of different neurodegenerative disorders, as well as to develop more selective and potent autophagy-modulating compounds that can be safely and effectively translated to the clinic.

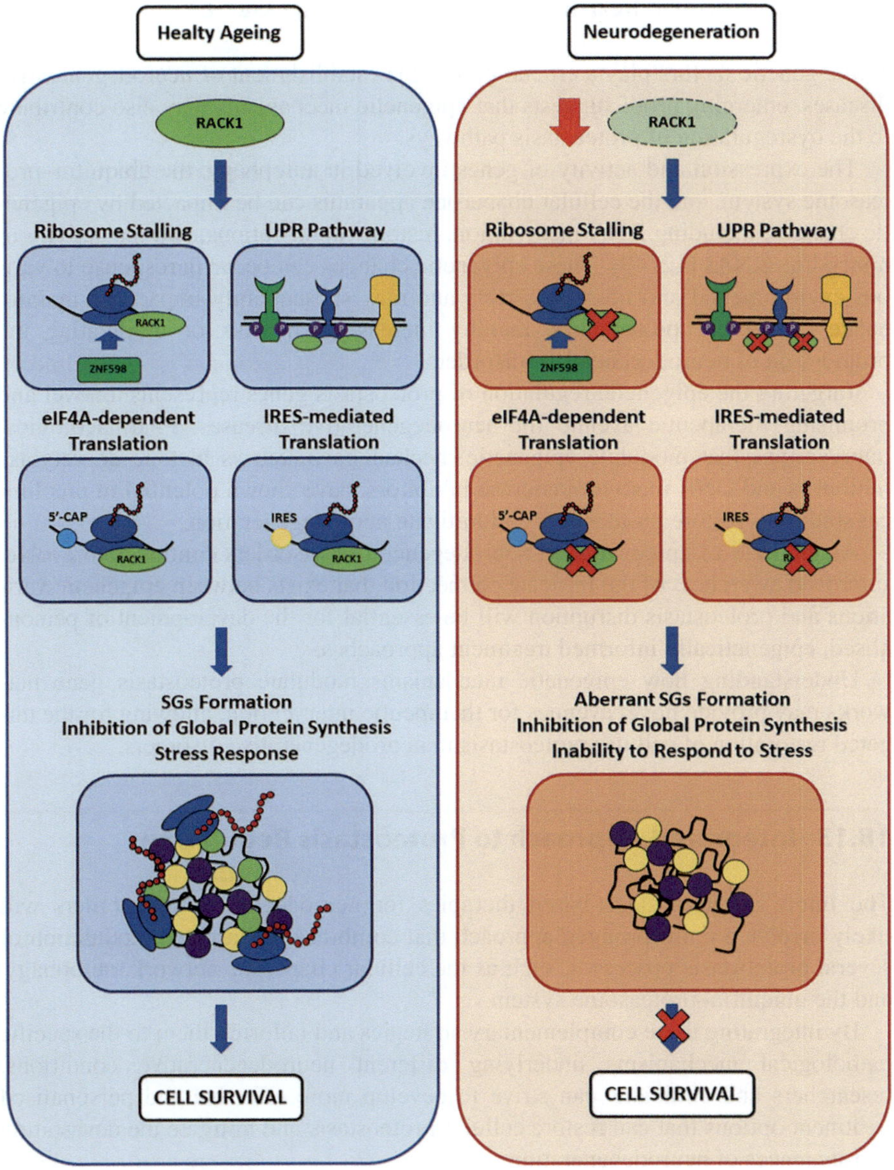

Fig. 18.3 RACK1's ribosome and translation-related roles in healthy ageing and neurodegeneration, as well as how they impact SG-correlated activities. While levels in AD patients are lower than in age-matched healthy controls, RACK1 ribosome residency and stoichiometry remain unchanged during healthy ageing. This article discusses the ribosomal and extra-ribosomal actions of RACK1, which can aid in maintaining proteostasis but whose decreasing levels in a diseased situation can worsen the underlying dysregulation of proteostasis observed in a number of neurodegenerative illnesses [3, 27, 32]

18.12 Epigenetic Regulation of Proteostasis Genes

While genetic factors play a crucial part in the establishment of neurodegenerative diseases, emerging proof suggests that epigenetic mechanisms may also contribute to the dysregulation of proteostasis pathways.

The expression and activity of genes involved in autophagy, the ubiquitin–proteasome system, and the cellular chaperone apparatus can be impacted by epigenetic changes, including DNA methylation, histone modifications, and the activity of noncoding RNAs [12, 49]. These epigenetic changes can occur in response to various environmental and lifestyle factors and may subsequently impact the maintenance of cellular proteostasis, thereby increasing the risk or modulating the progression of neurodegenerative disorders.

Targeting the epigenetic regulation of proteostasis genes represents a novel and promising therapeutic avenue for neurodegenerative diseases. Pharmacological interventions that modulate epigenetic mechanisms, such as histone deacetylase inhibitors and DNA methyltransferase inhibitors, have shown potential in preclinical studies to restore proteostasis and mitigate neurodegeneration.

As the field of epigenetics in neurodegenerative disorders continues to evolve, Improved awareness of the intricate connection that exists between epigenetic variations and proteostasis disruption will be essential for the development of personalised, epigenetically informed treatment approaches.

Understanding how epigenetic mechanisms modulate proteostasis gene networks may provide novel avenues for therapeutic intervention, allowing for the targeted restoration of cellular proteostasis in neurodegenerative disorders.

18.13 Integrated Approach to Proteostasis Regulation

The future of proteostasis-based therapies for neurodegenerative disorders will likely involve a multi-pronged approach that combines the targeted modulation of several proteostasis processes, such as the cellular chaperone network, autophagy, and the ubiquitin–proteasome system.

By integrating these complementary strategies and tailoring them to the specific pathological mechanisms underlying different neurodegenerative conditions, researchers and clinicians can strive to develop more effective and personalised treatment options that can restore cellular proteostasis and mitigate the devastating consequences of neurodegeneration.

18.14 Metabolic Control of Proteostasis Homeostasis

Emerging evidence indicates that the regulation of proteostasis homeostasis may also be significantly influenced by metabolic pathways and cellular energetics.

Metabolic disturbances, such as mitochondrial dysfunction, altered nutrient sensing, and disrupted energy balance, have been associated with the pathogenesis of various neurodegenerative disorders [28, 34]. Understanding how metabolic perturbations influence proteostasis networks, as well as the potential for targeting metabolic pathways to restore proteostasis, represents an exciting frontier in the development of future therapies for neurodegenerative conditions [12, 37, 43, 58].

18.15 Proteostasis Research in Stem Cell Models

The use of stem cell-based models, particularly induced pluripotent stem cells (iPSCs) derived has given scientists so far unprecedented chances to look into the underlying mechanism of proteostasis dysregulation in a patient-specific setting from patients with neurodegenerative disorders [35, 54]. These stem cell models make it possible to investigate proteostasis pathways in a more physiologically relevant context and to examine disease-related cell types, such as neurones and glia [20, 40].

By leveraging stem cell technologies, researchers can gain deeper understanding into the intricate interaction between epigenetic, genetic and environmental factors which contribute to proteostasis disruption in neurodegenerative diseases, ultimately paving the way for the advancement of more efficacious, personalised therapeutic interventions.

Furthermore, stem cell-based platforms can be leveraged for high-throughput screening of pharmacological compounds and genetic modulators, accelerating the recognition of new therapeutic targets and the creation of neurodegenerative disease treatments based on proteostasis [4].

Moreover, stem cell-based platforms can be utilised for high-throughput screening of therapeutic compounds that target proteostasis pathways, accelerating the drug discovery process and paving the way for personalised, proteostasis-based interventions.

18.16 Harnessing Proteostasis for Neuroprotection

The field of proteostasis research has made significant strides in our knowledge of the complex interplay between protein homeostasis pathways and neurodegenerative disorders. As we continue to unravel the molecular mechanisms underlying proteostasis dysregulation in the context of neurodegeneration, the potential for developing targeted, proteostasis-based therapies remains a promising avenue for future research and clinical translation [6].

By harnessing the remarkable plasticity of proteostasis networks and their ability to adapt to various cellular stresses, researchers can explore innovative strategies to restore proteostasis homeostasis and promote neuroprotection in the face of neurodegenerative challenges. These efforts may ultimately result in the creation of more efficient, individualised, and disease-modifying treatments that can improve the quality of life for millions of people impacted by neurodegenerative diseases globally and lessen their terrible effects.

Integration of multi-disciplinary approaches, leveraging emerging technologies, and fostering collaborative efforts, the scientific community can work towards the common goal of harnessing the power of proteostasis to alleviate the burden of neurodegenerative diseases and improve the lives of patients and their families.

18.17 Innovative Imaging Techniques in Proteostasis

The advancement of imaging technologies, such as super-resolution microscopy and in vivo imaging, has provided researchers with unprecedented tools to visualise and monitor the dynamics of proteostasis pathways in both cellular and animal models of neurodegenerative disorders.

These innovative techniques have enabled the real-time observation of protein aggregation, trafficking, and clearance mechanisms, as well as the evaluation of the spatiotemporal regulation of proteostasis networks in the context of neurodegeneration.

The integration of these cutting-edge imaging approaches with other analytical methods, such as proteomics and metabolomics, can further elucidate the complex interplay among proteostasis and other cellular processes that contribute to the pathogenesis of neurodegenerative diseases [5, 12, 22, 43, 58].

18.18 Computational Modelling of Proteostasis Pathways

The complexity of proteostasis networks and their dysregulation in neurodegenerative disorders necessitates the use of computational modelling and systems biology approaches to gain a holistic understanding of these intricate pathways.

Computational models can simulate the dynamic interplay between different proteostasis mechanisms, predict the impact of genetic or pharmacological interventions, and identify key regulatory nodes that could serve as potential therapeutic targets.

Computational modelling can facilitate the creation of more potent, mechanism-based treatments for neurodegenerative illnesses and offer important insights into the complex proteostasis pathways by incorporating experimental data.

Protease networks contain important regulatory hubs and nodes that may make good targets for therapeutic interventions. These can be found by combining multi-omics data genomics, transcriptomics, proteomics, and metabolomics with sophisticated computational methods like machine learning and network analysis.

Moreover, computational models can be used to simulate the dynamic interplay between various proteostasis pathways and predict the potential impact of therapeutic interventions on restoring proteostasis homeostasis in neurodegenerative conditions [58] (Table 18.2).

Table 18.2 Computational approaches in proteostasis research: methods, descriptions, and applications [21]

S. No.	Computational approach	Description	Applications in proteostasis research
1.	Molecular dynamics simulations	Simulates how atoms and molecules move physically over time.	Used to study protein folding, misfolding, and aggregation processes.
2.	Monte Carlo simulations	Uses random sampling to obtain numerical results for physical systems.	Helps in understanding the thermodynamics of protein folding and stability.
3.	Coarse-grained modelling	Simplifies the representation of molecules to study larger systems.	Useful for studying large protein complexes and their interactions.
4.	Bioinformatics tools	Computational tools for analysing biological data, including protein sequences.	Assists in predicting protein structure and function related to proteostasis.
5.	Machine learning algorithms	Algorithms that learn from data to make predictions or decisions.	Employed in predicting protein misfolding and aggregation tendencies.
6.	Network analysis	Studies the interactions and relationships within biological systems.	Analyses proteostasis networks and their regulatory mechanisms.
7.	Statistical mechanics	Applies statistical methods to understand the behaviour of systems with many particles.	Provides insights into the folding landscapes of proteins.
8.	Computational docking	Predicts the preferred orientation of one molecule to another when bound.	Used to identify potential inhibitors that can enhance proteostasis.
9.	Visualisation software	Tools for visualising molecular structures and dynamics.	Aids in the analysis of protein conformations and interactions.
10.	Integrative modelling	Combines data from various sources to create a comprehensive model.	Useful for understanding complex proteostasis networks in cells.

18.19 Collaborative Efforts and Data Sharing

The field of proteostasis research in the context of neurodegenerative disorders requires a collaborative, interdisciplinary approach to effectively translate scientific discoveries into clinical applications.

Fostering strong partnerships between researchers, clinicians, and bioinformaticians, as well as promoting the sharing of data and resources, can accelerate the pace of discovery and the development of novel proteostasis-based therapies.

By embracing a collaborative spirit and a commitment to data sharing, the scientific community can leverage the collective expertise and knowledge to tackle the complex challenges posed by neurodegenerative diseases, ultimately improving patient outcomes and quality of life.

Establishing global networks and data-sharing platforms among researchers, clinicians, and industry partners can facilitate the exchange of knowledge, the integration of diverse datasets, and the development of comprehensive strategies to tackle the challenges in proteostasis-based therapies.

18.20 Translational Challenges in Proteostasis Therapeutics

Although substantial advancements have been made in understanding the role of proteostasis in neurodegenerative diseases, the translation of these insights into effective therapeutic interventions remains a significant challenge.

The complex and multi-factorial nature of proteostasis dysregulation, coupled with the inherent difficulties in drug delivery to the CNS (central nervous system), pose unique obstacles in the development of proteostasis-based therapies.

Overcoming these translational hurdles will require a concerted effort to address issues related to bioavailability, target engagement, and the development of suitable biomarkers to monitor the efficacy of proteostasis-targeted interventions [20].

18.21 Patient-Derived Cellular Models of Neurodegeneration

The study of proteostasis in neurodegenerative diseases has undergone a radical change all thanks to the development of induced pluripotent stem cell (iPSC) technology.

Patient-derived cellular models, such as neurons and glia, offer a unique opportunity to investigate disease-specific pathways and mechanisms of proteostasis dysregulation in a personalised, physiologically relevant context.

The integration of these patient-derived cellular models with high-throughput screening platforms can accelerate the discovery of new substances that modulate proteostasis and the creation of tailored precision-based treatments.

These stem cell-based platforms can be further enhanced by incorporating advanced culture systems, such as 3D organoids and microfluidic devices, to better recapitulate the complex cellular interactions and tissue-level organisation observed in the human brain [40].

18.22 Precision Medicine Approaches in Proteostasis Disorders

The heterogeneous nature of neurodegenerative diseases and the variability in individual patient responses to therapeutic interventions underscore the need for precision medicine approaches in the field of proteostasis research.

By leveraging multi-omics data, advanced computational modelling, and patient-derived cellular models, researchers can identify personalised biomarkers and develop tailored proteostasis-based therapies that take into account the unique behavioural, environmental, and genetic factors that affect how disease manifests and progresses.

Pharmacogenomics and pharmacoepigenomics are two precision medicine techniques that can be used to forecast how a person will react to treatments that target proteostasis, enabling the optimisation of drug dosage, formulation, and delivery to maximise therapeutic efficacy and minimise adverse effects.

By integrating multi-omics data, detailed phenotyping, and patient-derived cellular models, researchers can uncover the unique molecular signatures and underlying biological drivers of proteostasis dysregulation in each patient or patient subgroup.

The integration of advanced analytical methods, computational modelling, and precision medicine approaches has a lot of potential for creating efficient proteostasis-based treatments for neurodegenerative conditions.

This information can then be used to develop tailored therapeutic strategies that target the specific proteostasis pathways and cellular mechanisms impacted in an individual or a defined patient population [18, 24, 33, 36].

The research in the field of proteostasis and its role in neurodegenerative disorders has made significant advancements in recent years, as evidenced by the sources provided [1, 24, 33, 36].

The sources highlight several key aspects of this research area the significance of comprehending the intricate relationship between neurodegeneration and proteostasis pathways [33], the need for innovative technologies and computational modelling to elucidate the dynamics of proteostasis networks [24], the challenges in translating these insights into effective therapies, and the potential of precision medicine approaches to personalise treatment strategies [33, 36, 38, 51]. These discoveries have opened the door for the creation of treatments, but the sector has seen unsatisfactory results, highlighting the necessity of looking into pathways shared by all neurodegenerative illnesses, such as the cellular processes that regulate and remove misfolded proteins [36].

This personalised approach can inform the development of tailored therapeutics, optimise drug selection and dosing, and enhance the possibility that patients will have beneficial clinical results with neurodegenerative disorders.

Harnessing the power of these precision medicine strategies, coupled with continued research into the fundamental mechanisms of proteostasis dysregulation, holds promise for improving the management and treatment of devastating neurodegenerative diseases [24, 33].

This personalised approach can guide the development of tailored, proteostasis-based therapies, improve patient stratification, and enhance the likelihood of successful clinical translation.

18.22.1 Targeting Proteostasis Imbalance for Therapeutic Interventions in Neurodegenerative Diseases

The preservation of cellular protein homeostasis, or proteostasis, is essential for the health of neurones. Proteostasis is frequently disturbed in neurodegenerative conditions, which results in the buildup of misfolded proteins that increase cellular toxicity. Here are several therapeutic strategies aimed at restoring proteostasis and mitigating neurodegenerative diseases based on the provided search results.

Therapeutic Strategies

1. Enhancing Molecular Chaperone Activity
   (a) *Mechanism*: Molecular chaperones assist in the proper folding of proteins and prevent aggregation. Increasing their activity can help refold misfolded proteins [46].
   (b) *Example*: Small molecules that enhance chaperone function are being explored as potential treatments for diseases like Alzheimer's and Parkinson's.
2. Modulating the Ubiquitin–Proteasome System (UPS)
   (a) *Mechanism*: The UPS is responsible for degrading misfolded and damaged proteins. Enhancing UPS activity can promote the clearance of toxic aggregates [42].
   (b) *Example*: Pharmacological agents that stimulate the UPS are being investigated for their capability to decrease protein accumulation in neurodegenerative diseases.
3. Activating Autophagy
   (a) *Mechanism*: A cellular mechanism known as autophagy breaks down and recycles damaged proteins and organelles. Promoting autophagy can help clear aggregated proteins [23].
   (b) *Example*: mTOR inhibitors, which activate autophagy, have shown promise in preclinical models of neurodegeneration.
4. Targeting Protein Aggregation
   (a) *Mechanism*: Toxic aggregates can be avoided by specifically targeting the way that particular proteins assemble.
   (b) *Example*: Small molecules that inhibit the aggregation of amyloid-beta or alpha-synuclein are under investigation.
5. Gene Therapy Approaches
   (a) *Mechanism*: Gene therapy can be used to correct mutations that lead to protein misfolding or to enhance the expression of protective proteins.
   (b) *Example*: Techniques such as CRISPR/Cas9 are being explored to edit genes associated with neurodegenerative diseases.
6. Reducing Oxidative Stress
   (a) *Mechanism*: Oxidative stress can exacerbate proteostasis failure. Antioxidants can mitigate this stress and protect neuronal health [9].

   (b) *Example*: Compounds that reduce oxidative damage are being studied for their neuroprotective effects.
7. Network Pharmacology
   (a) *Mechanism*: This approach considers the complex interactions within the proteostasis network and aims to identify multiple targets for drug development.
   (b) *Example*: Systems biology methods are being used to discover novel therapeutic targets that can restore proteostasis.
8. Biomarker Development
   (a) *Mechanism*: Identifying biomarkers for protein misfolding can help in monitoring disease progression and therapeutic response.
   (b) *Example*: New biomarkers are being developed to facilitate clinical trials and personalised treatment strategies.
9. Interventions in Inflammation
   (a) *Mechanism*: Chronic inflammation can disrupt proteostasis. Targeting inflammatory pathways may help restore balance.
   (b) *Example*: Anti-inflammatory treatments are being explored as adjunct therapies in neurodegenerative diseases.
10. Personalised Medicine Approaches
   (a) *Mechanism*: Tailoring therapies based on individual genetic and proteomic profiles can enhance treatment efficacy.
   (b) *Example*: Precision medicine strategies are being developed to match specific therapies to the unique characteristics of a patient's disease.

18.23 Conclusion

The field of proteostasis and its role in neurodegenerative disorders is a fast-evolving area of research, with critical implications for our understanding of disease pathogenesis and the development of novel theraputic stratergies. The central importance of autophagic dysfunction in these conditions has emerged as a key focus, and ongoing investigations are aimed at elucidating the complex interplay between autophagy, proteostasis, and other cellular processes that contribute to neurodegeneration.

Despite the challenges, the potential for therapeutic breakthroughs in this field is substantial. Continued advancements in our mechanistic understanding, coupled with innovative approaches to modulate proteostasis pathways, hold promise for the development of effective treatments that can restore cellular homeostasis and halt the progression of these debilitating neurological disorders [12, 31, 37, 49].

In conclusion, the field of proteostasis research in the connection of neurodegenerative diseases holds immense potential for the development of novel therapeutic strategies and the improvement of patient outcomes.

By harnessing innovative technologies, computational models, collaborative efforts, and precision medicine approaches, the scientific community can work towards a future where the challenges of neurodegeneration are effectively addressed through the lens of proteostasis [20, 24, 40, 50].

18.24 Summary

An overview of the state of proteostasis research in relation to neurodegenerative illnesses is provided in this book chapter. It draws attention to how crucial computational modelling and systems biology techniques are to comprehending the intricate proteostasis networks and the dysregulation of these networks under certain circumstances. The chapter emphasises the need for collaborative efforts and data-sharing platforms to translate scientific discoveries into clinical applications. It also addresses the significant translational challenges in developing effective proteostasis-based therapies, including issues related to bioavailability, target engagement, and the development of suitable biomarkers. Additionally, the chapter discusses the emerging role of patient-acquired cellular models, such as induced pluripotent stem cell-based platforms, in investigating disease-specific mechanisms and pathways. Finally, the chapter underscores the potential of precision medicine approaches in proteostasis research, leveraging multi-omics data and personalised strategies to improve patient outcomes in neurodegenerative disorders.

References

1. Alshial EE, Abdulghaney MI, Wadan AS, Abdellatif M, Ramadan N, Suleiman AM, et al. Mitochondrial dysfunction and neurological disorders: a narrative review and treatment overview. Life Sci. 2023;334:122257. https://doi.org/10.1016/j.lfs.2023.122257.
2. Ascherio A, Schwarzschild MA. The epidemiology of Parkinson's disease: risk factors and prevention. Lancet Neurol. 2016;15(12):1257–72. https://doi.org/10.1016/S1474-4422(16)30230-7.
3. Battaini F, Pascale A. Protein kinase C signal transduction regulation in physiological and pathological aging. Ann N Y Acad Sci. 2005;1057(1):177–92. https://doi.org/10.1196/annals.1356.011.
4. Boison D, Meier JC, Masino SA. Editorial: metabolic control of brain homeostasis. Front Mol Neurosci. 2017;10:184. https://doi.org/10.3389/fnmol.2017.00184.
5. Boja ES, Fehniger TE, Baker MS, Végvári Á, Rodriguez H. Analytical validation considerations of multiplex mass-spectrometry-based proteomic platforms for measuring protein biomarkers. J Proteome Res. 2014;13(12):5325–32. https://doi.org/10.1021/pr500753r.
6. Braak H, Braak E. Neuropathological stageing of Alzheimer-related changes. Acta Neuropathol. 1991;82(4):239–59. https://doi.org/10.1007/BF00308809.
7. Carrera I, Cacabelos R. Current drugs and potential future neuroprotective compounds for Parkinson's disease. CNS Neurol Disord Drug Targets. 2019;17(3):295–306. https://doi.org/10.2174/1570159x17666181127125704.
8. Chang Y, Kong Q, Shan X, Tian G, Ilieva H, Cleveland DW, et al. Messenger RNA oxidation occurs early in disease pathogenesis and promotes motor neuron degeneration in ALS. PLoS One. 2008;3(8):e2849. https://doi.org/10.1371/journal.pone.0002849.
9. Chen X, Guo C, Kong J. Oxidative stress in neurodegenerative diseases. Neural Regen Res. 2012;7(5):376–85. https://doi.org/10.3969/j.issn.1673-5374.2012.05.009.
10. Cho B, Kim T, Huh Y, Lee J, Lee Y. Amelioration of mitochondrial quality control and proteostasis by natural compounds in Parkinson's disease models. Int J Mol Sci. 2019;20(20):5208. https://doi.org/10.3390/ijms20205208.
11. Choi I, Zhang Y, Seegobin SP, Pruvost M, Wang Q, Purtell K, et al. Microglia clear neuron-released α-synuclein via selective autophagy and prevent neurodegeneration. Nat Commun. 2020;11(1):1386. https://doi.org/10.1038/s41467-020-15195-5.

12. Chun Y, Kim J. Autophagy: an essential degradation program for cellular homeostasis and life. Cells. 2018;7(12):278. https://doi.org/10.3390/cells7120278.
13. Desai S, Doke R, Pansare PA, Sainani SR, Bhalchim V. Natural products: an emerging tool in Parkinson's disease therapeutics. IP Indian J Neurosci. 2019;5(3):95–105. https://doi.org/10.18231/j.ijn.2019.014.
14. Dickson DW, Braak H, Duda JE, Duyckaerts C, Gasser T, Halliday GM, et al. Neuropathological assessment of Parkinson's disease: refining the diagnostic criteria. Lancet Neurol. 2009;8(12):1150–7. https://doi.org/10.1016/S1474-4422(09)70238-8.
15. Dugger BN, Dickson DW. Pathology of neurodegenerative diseases. Cold Spring Harb Perspect Biol. 2017;9(7):a028035. https://doi.org/10.1101/cshperspect.a028035.
16. Dujardin S, Lécolle K, Caillierez R, Bégard S, Zommer N, Lachaud C, et al. Neuron-to-neuron wild-type Tau protein transfer through a trans-synaptic mechanism: relevance to sporadic tauopathies. Acta Neuropathol Commun. 2014;2:14. https://doi.org/10.1186/2051-5960-2-14.
17. Elufioye TO, Berida TI, Habtemariam S. Plants-derived neuroprotective agents: cutting the cycle of cell death through multiple mechanisms. Evid Based Complement Alternat Med. 2017;2017:3574012. https://doi.org/10.1155/2017/3574012.
18. Franco R, Cedazo-Mínguez A. Successful therapies for Alzheimer's disease: why so many in animal models and none in humans? Front Pharmacol. 2014;5:146. https://doi.org/10.3389/fphar.2014.00146.
19. Fujikake N, Shin M, Shimizu S. Association between autophagy and neurodegenerative diseases. Front Neurosci. 2018;12:255. https://doi.org/10.3389/fnins.2018.00255.
20. Gibbs RM, Lipnick S, Bateman J, Chen L, Cousins H, Hubbard EG, et al. Toward precision medicine for neurological and neuropsychiatric disorders. Cell Stem Cell. 2018;23(1):21–4. https://doi.org/10.1016/j.stem.2018.05.019.
21. Gracia KC, Husi H. Computational approaches in proteomics. In: Husi H, editor. Computational biology. Brisbane: Codon Publications; 2019. Available from: https://www.ncbi.nlm.nih.gov/books/NBK550333/, https://doi.org/10.15586/computationalbiology.2019.ch8.
22. Gu S, Cui D, Chen X, Xiong X, Zhao Y. PROTACs: an emerging targeting technique for protein degradation in drug discovery. BioEssays. 2018;40(4):e1700247. https://doi.org/10.1002/bies.201700247.
23. Guo F, Liu X, Cai H, Le W. Autophagy in neurodegenerative diseases: pathogenesis and therapy. Brain Pathol. 2018;28(1):3–13. https://doi.org/10.1111/bpa.12545.
24. Ho D, Quake SR, McCabe ER, Chng WJ, Chow EK, Ding X, et al. Enabling technologies for personalized and precision medicine. Trends Biotechnol. 2020;38(5):497–518. https://doi.org/10.1016/j.tibtech.2019.12.021.
25. Iadecola C, Gottesman RF. Cerebrovascular alterations in Alzheimer disease. Circ Res. 2018;123(4):406–8. https://doi.org/10.1161/CIRCRESAHA.118.313400.
26. Jellinger KA. Interaction between pathogenic proteins in neurodegenerative disorders. J Cell Mol Med. 2012;16(6):1166–83. https://doi.org/10.1111/j.1582-4934.2012.01542.x.
27. Limanaqi F, Biagioni F, Gambardella S, Familiari P, Frati A, Fornai F. Promiscuous roles of autophagy and proteasome in neurodegenerative proteinopathies. Int J Mol Sci. 2020;21(8):3028. https://doi.org/10.3390/ijms21083028.
28. Liu X, Homma H, Sayama N, Seki T, Ueda K, Akagi T, et al. Neurodegenerative disease: models, mechanisms, and a new hope for cure. Neural Regen Res. 2021;16(6):1109–17. https://doi.org/10.4103/1673-5374.299059.
29. Madeo F, Eisenberg T, Kroemer G. Autophagy for the avoidance of neurodegeneration. Genes Dev. 2009;23(19):2253–9.
30. Manecka D, Vanderperre B, Fon EA, Durcan TM. The neuroprotective role of protein quality control in halting the development of alpha-synuclein pathology. Front Mol Neurosci. 2017 Sep;27(10):311. https://doi.org/10.3389/fnmol.2017.00311.
31. Marazziti D, Mao K, Wang K, Zhao M, Klionsky DJ. The role of autophagy in Parkinson's disease. Cold Spring Harb Perspect Biol. 2012;2(4):a009357. https://doi.org/10.1101/cshperspect.a009357.

32. Masi M, Attanzio A, Racchi M, Wolozin B, Borella S, Biundo F, Buoso E. Proteostasis deregulation in neurodegeneration and its link with stress granules: focus on the scaffold and ribosomal protein RACK1. Cells. 2022;11(16):2590. https://doi.org/10.3390/cells11162590.
33. Nair R, Corrochano S, Gasco S, Tibbit C, Thompson D, Maduro C, et al. Uses for humanised mouse models in precision medicine for neurodegenerative disease. Mamm Genome. 2019;30(7–8):173–91. https://doi.org/10.1007/s00335-019-09807-2.
34. Nakamura T, Lipton SA. Cell death: protein misfolding and neurodegenerative diseases. Apoptosis. 2009;14(4):455–68. https://doi.org/10.1007/s10495-008-0301-y.
35. Nguyen HN, Byers B, Cord B, Shcheglovitov A, Byrne J, Gujar P, et al. LRRK2 mutant iPSC-derived DA neurons demonstrate increased susceptibility to oxidative stress. Cell Stem Cell. 2011;8(3):267–80. https://doi.org/10.1016/j.stem.2011.01.013.
36. Niedowicz DM, Nelson PT, Murphy MP. Alzheimer's disease: pathological mechanisms and recent insights. Curr Neuropharmacol. 2011;9(4):674–84. https://doi.org/10.2174/157015911798376181.
37. Ntsapi C, Swart C, Lumkwana D, Loos B. Autophagic flux failure in neurodegeneration: identifying the defect and compensating flux offset. In: Autophagy in current trends in cellular physiology and pathology; 2016. https://doi.org/10.5772/64197.
38. O'Brien DT, Oosten-Hawle PV. Regulation of cell-non-autonomous proteostasis in metazoans. Essays Biochem. 2016;60(2):133–42. https://doi.org/10.1042/ebc20160006.
39. Ottens F, Franz A, Hoppe T. Build-UPS and break-downs: metabolism impacts on proteostasis and aging. Cell Death Differ. 2021;28(3):505–21. https://doi.org/10.1038/s41418-020-00682-y.
40. Pasteuning-Vuhman S, de Jongh RD, Timmers A, Pasterkamp RJ. Towards advanced iPSC-based drug development for neurodegenerative disease. Mol Ther Methods Clin Dev. 2021;27:263–79. https://doi.org/10.1016/j.molmed.2020.09.013.
41. Poewe W, Seppi K, Tanner CM, Halliday GM, Brundin P, Volkmann J, et al. Parkinson disease. Nat Rev Dis Primers. 2017;3:17013. https://doi.org/10.1038/nrdp.2017.13.
42. Rao G, Croft B, Teng C, Awasthi V. Ubiquitin-proteasome system in neurodegenerative disorders. J Drug Metab Toxicol. 2015;6(4):187. https://doi.org/10.4172/2157-7609.1000187.
43. Ribarić S, Ribarić IM. Autophagy and cell death in Alzheimer's, Parkinson's and prion diseases. In: Programmed cell death; 2020. https://doi.org/10.5772/intechopen.86706.
44. Scapagnini G, Vasto S, Abraham GNA, Calogero C, Zella D, Galvano F. Modulation of Nrf2/ARE pathway by food polyphenols: a nutritional neuroprotective strategy for cognitive and neurodegenerative disorders. Mol Neurobiol. 2011;44(2):192–201. https://doi.org/10.1007/s12035-011-8181-5.
45. Schmidt MF, Gan ZY, Komander D, Rape M. Ubiquitin signalling in neurodegeneration: mechanisms and therapeutic opportunities. Cell Death Differ. 2021;28(2):570–90. https://doi.org/10.1038/s41418-020-00706-7.
46. Sharma A, Shah OP, Sharma L, Gulati M, Behl T, Khalid A, et al. Molecular chaperones as therapeutic target: hallmark of neurodegenerative disorders. Mol Neurobiol. 2024;61(7):4750–67. https://doi.org/10.1007/s12035-023-03846-2.
47. Sienes Bailo P, Llorente Martín E, Calmarza P, Montolio Breva S, Bravo Gómez A, Pozo Giráldez A, et al. The role of oxidative stress in neurodegenerative diseases and potential antioxidant therapies. Adv Lab Med. 2022;3(4):342–50. https://doi.org/10.1515/almed-2022-0111.
48. Slanzi A, Iannoto G, Rossi B, Zenaro E, Constantin G. In vitro models of neurodegenerative diseases. Front Cell Dev Biol. 2020;8:328. https://doi.org/10.3389/fcell.2020.00328.
49. Son JH, Shim JH, Kim K, Ha JY, Han J. Neuronal autophagy and neurodegenerative diseases. Exp Mol Med. 2012;44(2):89. https://doi.org/10.3858/emm.2012.44.2.031.
50. Strafella C, Caputo V, Galota MR, Zampatti S, Marella G, Mauriello S, et al. Application of precision medicine in neurodegenerative diseases. Front Neurol. 2018;9:701. https://doi.org/10.3389/fneur.2018.00701.
51. Strianese O, Rizzo F, Ciccarelli M, Galasso G, D'Agostino Y, Salvati A, et al. Precision and personalized medicine: how genomic approach improves the management of cardiovascular and neurodegenerative disease. Genes. 2020;11(7):747. https://doi.org/10.3390/genes11070747.

52. Sweeney P, Park H, Baumann M, Dunlop J, Frydman J, Kopito R, McCampbell A, Leblanc G, Venkateswaran A, Nurmi A, Hodgson R. Protein misfolding in neurodegenerative diseases: implications and strategies. Transl Neurodegener. 2017;6:6. https://doi.org/10.1186/s40035-017-0077-5. PMID: 28293421; PMCID: PMC5348787
53. Thibaudeau TA, Anderson RT, Smith DM. A common mechanism of proteasome impairment by neurodegenerative disease-associated oligomers. Nat Commun. 2018;9:1097. https://doi.org/10.1038/s41467-018-03509-0.
54. Tubsuwan A, Pires C, Rasmussen MA, Schmid B, Nielsen JE, Hjermind LE, et al. Generation of induced pluripotent stem cells (iPSCs) from an Alzheimer's disease patient carrying a L150P mutation in PSEN-1. Stem Cell Res. 2016;16(1):110–2. https://doi.org/10.1016/j.scr.2015.12.015.
55. Wang J, Wang CE, Orr A, Tydlacka S, Li SH, Li XJ. Impaired ubiquitin-proteasome system activity in the synapses of Huntington's disease mice. J Cell Biol. 2008;180(6):1177–89.
56. Wareham LK, Liddelow SA, Temple S, et al. Solving neurodegeneration: common mechanisms and strategies for new treatments. Mol Neurodegener. 2022;17(1):1–29. https://doi.org/10.1186/s13024-022-00524-0.
57. Yang F, Yang YP, Mao CJ, Liu L, Zheng HF, Hu LF, et al. Crosstalk between the proteasome system and autophagy in the clearance of α-synuclein. Acta Pharmacol Sin. 2013;34(5):674–80.
58. Young CL, Robinson AS. Protein folding and secretion: mechanistic insights advancing recombinant protein production in S. cerevisiae. Curr Opin Biotechnol. 2014;30:168–77. https://doi.org/10.1016/j.copbio.2014.06.018.
59. Zare-Shahabadi A, Masliah E, Johnson GV, Rezaei N. Autophagy in Alzheimer's disease. Rev Neurosci. 2015;26(4):385–95. https://doi.org/10.1515/revneuro-2014-0076.
60. Zhao Z, Nelson AR, Betsholtz C, Zlokovic BV. Establishment and dysfunction of the blood-brain barrier. Cell. 2015;163(5):1064–78. https://doi.org/10.1016/j.cell.2015.10.067.

Therefore, a protein quality control ...
maintaining cellular proteostasis through ...
stress responses. Dysfunction in the ...
misfolded proteins contributing to ...

...

... in neuropathic pain ... suggested ...
... Hsp70 ...

... specific Hsp70 isoforms for neuronal ...
... sion of Hsp70 through increased ...
... interventional medications and targeted delivery ...
... for peripheral neurodegenerative ...
... receptors, potentially ...

GPSR Compliance

The European Union's (EU) General Product Safety Regulation (GPSR) is a set of rules that requires consumer products to be safe and our obligations to ensure this.

If you have any concerns about our products, you can contact us on ProductSafety@springernature.com

In case Publisher is established outside the EU, the EU authorized representative is:

Springer Nature Customer Service Center GmbH
Europaplatz 3
69115 Heidelberg, Germany

Batch number: 10703867

Printed by Printforce, the Netherlands